April 17–19, 2013
Budapest, Hungary

I0038158

**Association for
Computing Machinery**

Advancing Computing as a Science & Profession

WiSec'13

Proceedings of the Sixth ACM Conference on
Security and Privacy in Wireless
and Mobile Networks

Sponsored by:
ACM SIGSAC

Supported by:
Symantec, Ericsson, Ukatemi

Association for Computing Machinery

Advancing Computing as a Science & Profession

The Association for Computing Machinery
2 Penn Plaza, Suite 701
New York, New York 10121-0701

Notice to Past Authors of ACM-Published Articles

ISBN: 978-1-4503-2003-0 (Digital)

ISBN: 978-1-4503-2426-7 (Print)

Additional copies may be ordered prepaid from:

ACM Order Department
PO Box 30777
New York, NY 10087-0777, USA

Phone: 1-800-342-6626 (USA and Canada)
+1-212-626-0500 (Global)
Fax: +1-212-944-1318
E-mail: acmhelp@acm.org
Hours of Operation: 8:30 am – 4:30 pm ET

Printed in the USA

ACM WiSec 2013 Welcome Message

The ACM Conference on Security and Privacy in Wireless and Mobile Networks has been a premier venue for researchers in wireless security and privacy to present the latest research in the field. It has also served as a forum for fostering international collaboration to address eminent security threats faced by our society. Over the years, we have seen an evolution in research, from traditional network security to complex, multi-faceted security problems that cannot be effectively addressed using conventional techniques. In particular, we have witnessed the emergence of new wireless systems (e.g., cognitive radios, RFID, vehicular networks, 4G/WiMax, NFC), the widespread deployment of new communication platforms (e.g., smartphones) and of their applications (e.g., social media), as well as an increased awareness of privacy issues associated with these emerging technologies.

In 2013, WiSec broadened its scope, and solicited papers that address issues beyond the traditional WiSec staples of physical, link, and network layer security. In particular, we encouraged the submission of papers focusing on the security and privacy of mobile software platforms and the increasingly diverse range of mobile or wireless applications including the automotive area.

The Call-for-Papers attracted 70 submissions from all around the world. We saw many exciting papers, and after a thorough review process and thanks to our highly competent expert reviewers, we arrived at a collection of 11 full papers and 15 short papers that we felt are mature and ready to be presented to the security community and to be included in the conference proceedings.

In addition to the research papers being presented at the conference, we also have two exciting keynotes, to be delivered by Henning Schulzrinne, US FCC and Columbia University ("The Internet Is Insecure and Will Likely Remain So - Now What?") and Michael John Elster ("How do You Define Security and Privacy for Smart Metering in Europe Today?"), as well as a plenary talk given by Ashkan Soltani ("Mobile Threats to Privacy").

For the first time this year, the program also includes a half-day workshop, HotWiSec 2013, which provides a forum for researchers and practitioners to discuss exciting new directions or out-of-the-box, disruptive or controversial ideas on the exploitation or protection of wireless communications and systems. We hope that the 8 papers accepted for presentation at the workshop will initiate a lively discussion among the participants, which will provide early, useful feedback to the authors, who may decide to mature their work and consider submitting it to future ACM WiSec conferences.

Putting together ACM WiSec'13 was a team effort. We first would like to thank all authors for supporting the conference by submitting their manuscripts and providing the technical content of the program. In the same way, we are thankful to the keynote speakers for accepting our invitation, and sharing their views and ideas with the community. We are grateful to all technical program committee members for their valuable reviews and their efforts in shepherding conditionally accepted papers. Special thanks go to *Wenyuan Xu* and *Panos Papadimitratos* (Publicity Co-Chairs), *Loukas Lazos* (Proceedings Chair), *Matthew Smith* (Poster/Demo Chair), *Mark Felegyhazi* (Web Chair), and *Christian Wachsmann* (support for the submission management system) for their efforts in making WiSec'13 a success. We also thank *April Mosqus*, *Maritza Nichols*, *Lisa M. Tolles* and *Stephanie Sabal* from ACM for their timely assistance and organizational support. We gratefully acknowledge the support of ACM SIGSAC, and our generous corporate supporters, Symantec, Ericsson, and Ukatemi Technologies. Last, but not least, we appreciate the trust and guidance of the WiSec Steering Committee, and in particular *Gene Tsudik* (WiSec Steering Committee Chair).

We hope that you find this year's program interesting and thought-provoking. Enjoy the conference and the spring in Budapest.

Levente Buttyán
General Chair

Ahmad-Reza Sadeghi
Technical Program Chair

Marco Gruteser
Technical Program Chair

Table of Contents

WiSec 2013 Conference Organization

General Chair: Levente Buttyán *(Budapest University of Technology and Economics, Hungary)*

Program Co-Chairs: Ahmad-Reza Sadeghi *(Technische Universitaet Darmstadt, Fraunhofer SIT, Intel-CRISC, Germany)*
Marco Gruteser *(Rutgers University, USA)*

Proceedings Chair: Loukas Lazos *(University of Arizona, USA)*

Publicity Co-Chairs: Wenyuan Xu *(University of South Carolina, USA)*
Panos Papadimitratos *(Kungliga Tekniska Hogskolan, Sweden)*

Poster/Demo Chair: Matthew Smith *(Leibniz Universitat Hannover, Germany)*

Website Chair: Mark Felegyhazi *(Budapest University of Technology and Economics, Hungary)*

Steering Committee Chair: Gene Tsudik *(University of California Irvine, USA)*

Steering Committee: N. Asokan *(University of Helsinki, Finland)*
Levente Buttyan *(BME, Hungary)*
Claude Castelluccia *(INRIA, France)*
Jean–Pierre Hubaux *(EPFL, Switzerland)*
Douglas Maughan *(DHS/HSARPA, USA)*
Adrian Perrig *(Carnegie Mellon University, USA)*
Dirk Westhoff *(HAW Hamburg, Germany)*

Program Committee: N. Asokan *(University of Helsinki, Finland)*
Giuseppe Ateniese *(The Johns Hopkins University, USA)*
Levente Buttyan *(BME, Hungary)*
Srdjan Čapkun *(ETH Zurich, Switzerland)*
Claude Castelluccia *(INRIA, France)*
Yingying Chen *(Stevens Institute of Technology, USA)*
Mauro Conti *(University of Padua, Italy)*
Emiliano De Cristofaro *(Palo Alto Research Center, USA)*
Roberto Di Pietro *(Universita di Roma Tre, Italy)*
William Enck *(North Carolina State University, USA)*
Sebastian Gajek *(NEC)*
Albert Held *(Daimler, Germany)*
Urs Hengartner *(University of Waterloo, Canada)*
Yih-Chun Hu *(University of Illinois at Urbana-Champaign, USA)*
Frank Kargl *(University of Twente, Netherlands)*

WiSec 2013 Sponsor & Supporters

Sponsor:

Supporters:

ERICSSON

Ukatemi
advanced threat
mitigation technologies

A Pilot Study on the Security of Pattern Screen-Lock Methods and Soft Side Channel Attacks

Panagiotis Andriotis[*]
University of Bristol, MVB
Bristol, BS8 1UB, UK
p.andriotis@bristol.ac.uk

Theo Tryfonas
University of Bristol, QB
Bristol, BS8 1TR, UK
theo.tryfonas@bristol.ac.uk

George Oikonomou
University of Bristol, MVB
Bristol, BS8 1UB, UK
g.oikonomou@bristol.ac.uk

Can Yildiz
University of Bristol, UK
canyildiz.2011@my.bristol.ac.uk

ABSTRACT

Graphical passwords that allow a user to unlock a smartphone's screen are one of the Android operating system's features and many users prefer them instead of traditional text-based codes. A variety of attacks has been proposed against this mechanism, of which notable are methods that recover the lock patterns using the oily residues left on screens when people move their fingers to reproduce the unlock code. In this paper we present a pilot study on user habits when setting a pattern lock and on their perceptions regarding what constitutes a secure pattern. We use our survey's results to establish a scheme, which combines a behaviour-based attack and a physical attack on graphical lock screen methods, aiming to reduce the search space of possible combinations forming a pattern, to make it partially or fully retrievable.

Categories and Subject Descriptors

D.4.6 [**Software**]: Operating Systems—*Security and Protection*

General Terms

Security, Human Factors

Keywords

Android, smudge attacks, usability, pattern lock

1. INTRODUCTION

Nowadays, passwords are integrated in people's routines. Humans authenticate themselves using keyboards, fingerprint readers or touchscreens. Smartphones hold an important amount of information about the owner and for this

---

[*]Corresponding Author. Panagiotis Andriotis, Theo Tryfonas and George Oikonomou are with the Bristol Cryptography Group.

reason people tend to lock them using the provided mechanisms. In most cases, phone lock mechanisms are implemented either as a PIN or a password.

Contemporary smartphones using the Android Operating System adopt a type of lock mechanism different to traditional PIN codes. This approach, called 'pattern lock', is based on existing research on graphical passwords [2] and requires the user to form a pattern on the screen by drawing lines in order to unlock the device. Its interface consists of 9 nodes in a 3x3 grid formation. Users start by touching one of the dots to make it the start point and swipe their fingers to add dots and form a pattern. However, there are some constraints while setting a scheme. It takes a minimum of 4 and a maximum of 9 dots to create one, each node can be visited only once and a previously not visited node becomes visited if it is on the way of a horizontal, vertical or diagonal line. Due to these constraints, the total number of possible patterns is 389,112 [1].

There are various types of attacks that can be used against a device to retrieve its pattern lock. Typical security attacks would entail attempts to exploit flaws in the theoretical design of a scheme or brute force a security mechanism. Brute forcing attacks against PINs or pattern locks may be rendered ineffective, if the number of unsuccessful attempts permitted is very limited and the device locks after that. Attacks that do not rely on brute forcing or exploiting a design weakness, but instead, are based on information gained from the physical implementation of a security scheme, are called *side channel attacks*. Some of such physical attacks against pattern locks aim to retrieve a pattern using physical traces left by the user, e.g. fingerprint marks left on the device's screen [11]. Others, such as *psychological attacks*, aim to detect user bias in PIN and pattern setting. Such information could be used to drastically limit the search space of possible combinations, in the same manner that heuristics about the use of meaningful passwords (e.g. familiar words) reduce the search space of a brute-force password attack [5].

In this paper, we attempt to combine physical attacks that relate to traces left from the use of a phone, with heuristics about the way users set lock patterns in order to facilitate attacks on this security mechanism. We use an optical camera and a microscope to analyse oily residues left on the screen, to evaluate the effectiveness of such relatively mid-term lived physical traces. We also exert a thermal camera to analyse heat traces left on the screen after drawing a pattern (shorter-term lived traces). To enhance the effectiveness of the physical examination, we exploit outcomes of trends related to the setting of pattern locks. Therefore, we analyse the average pattern length, the number of direction changes when drawing patterns, the start points, end points and sub-patterns with length one to four nodes. To achieve this we conducted a pilot survey collecting data from 144 participants, indicating that there exist useful detectable trends when people try to form such a password. We therefore demonstrate that the combination of physical and psychological attacks can diminish the security efficiency of the pattern lock, revealing parts or the whole of a pattern.

The rest of this paper is organised as follows. In section 2, we discuss relevant research considering attacks on passwords. In section 3 we present our experiments to attack Android's graphical password scheme. In section 4 we present a preliminary evaluation of the proposed combination of physical and psychological-based attacks. The conclusion is drawn in section 5 and ideas for further work are also discussed.

2. BACKGROUND

2.1 Text-based and Graphical Passwords

Text-based passwords and PIN codes are normally coupled with bank accounts, computational devices etc. Individuals posses several accounts and numerous passwords that need to remember. Thus, the users often have to balance usability with security. As a consequence, they may recall another account's password [8] or even worse use the same across all their accounts [2]. If a word-like password is chosen, it may be possible for an attacker to retrieve it by using dictionary based attacks. On the other hand, if random characters have been set as password, it is highly likely that the user will fail to fully remember the sequence [8]. This renders text-based passwords hard for the legitimate user and easy for the attacker. Another aspect that makes the text-based passwords hard to remember is the way the human brain works. According to Dual Coding Theory, cognition is composed of two separate parts: nonverbal and verbal systems [4]. Having different systems in the brain to process the verbal and nonverbal information, humans perform differently in these two ways when it comes to remembering. Text requires an additional process of associating symbols with a contextual meaning [2].

Graphical passwords may come in much more variety compared to text-based solutions. They can include procedures such as clicking some points on an image, drawing a line or a shape. The most important advantage they provide is the possibility to define a password in a way that is memorisable by the user and yet, still hard to guess by the attacker. However, graphical passwords can also have their weaknesses, if we take into consideration the fact that users may select their graphical passwords with respect to some meaningful process. Thus, human psychology can be associated with the choice of a graphical password. Studies on image-based graphical passwords show that humans tend to choose popular points (called hotspots) on the image [10]. In their experiments, Thorpe and van Oorschot [10] argue that there are some general hotspots and areas on images that people tend to select. Furthermore, they are trying to answer the question if we can successfully build brute-force dictionary attacks on graphical passwords by defining weak password subspaces and applying attacks using complexity properties, such as password length, number of components, and symmetry [7]. Their predictive model leads to password rules and propose a set of precautions to increase security.

Another study provides 9 different face images to users and lets them choose 4 of them in a sequence to form a password [3]. Using this 'face selection' mechanism, they collect passwords from 79 participants. The results are significant and show that a number of passwords set by males can be easily guessed in 2 attempts. The fact that humans have similar preferences on graphical passwords provides reasonable grounds to investigate if there exist sub-patterns preferred by users when forming a pattern lock.

2.2 Methods of Pattern Lock Retrieval

Android's pattern lock mechanism relies on users swiping their finger to unlock the device. This action leaves behind an oily residue or smudges. Relevant research on retrieving lock patterns using standard optics is conducted by Aviv et al. [1]. In their work, they demonstrate how recovering smudges using a light source and a digital camera is possible due to the fact that touchscreen surfaces are reflective rather than diffusive. Experimenting with directional and omnidirectional light sources and testing angles ranging from $0°$ to $180°$, by taking pictures at steps of $15°$, they found out that the smudges were visible in most cases when a directional light source was used. Omnidirectional light sources prove to create a full reflection effect at all angles rendering this type of light source unusable. Apart from the ideal photograph capturing angles to retrieve smudges, the experiments focus on various states of a touchscreen such as: pattern entered using normal or light touches by the user, pattern entered before or after phone usage. Note that the notions of 'normal' and 'light' touches are not quantitative in this study, and thus must be intuitively guessed. For this reason, we assume the light touch stands for intentionally low pressure touches to minimise any smudge left behind, whilst the normal touch is the one made without any concern of leaving a smudge behind.

The smudge persistence of the patterns was tested on two phones. It is indicative that different touchscreen surfaces of Android phones (even from the same manufacturer) may have different properties with respect to capturing and retaining physical traces. When all angle setups are taken into consideration they derive that the best angle to retrieve a pattern is $60°$ with 80% of the lighting scenarios resulting in nearly perfect retrieval [1]. It is also noted that the directionality was discernible which is particularly important because it decreases the number of attempts to unlock the device. Their results highlight that intentionally cleaning with cloth or putting the phone to pocket was not enough to prevent pattern retrieval. It is important to mention that the researchers preferred describing the process as 'simple clothing', which probably means that the results may not hold true when the screen is rubbed thoroughly. Overall the optical method is particularly efficient as all it requires is a directional light source and a digital camera. An attacker can easily and quickly capture a photo of the touchscreen from a useful angle and perform any necessary contrast and brightness adjustments on the photo to retrieve one's pattern lock. As discussed in [1], even if the pattern is only partially retrievable, multiple photos taken in different times may reveal the full pattern.

The use of a thermal camera to retrieve the PIN codes

2

is an already existing attack on other devices such as ATM keypads (Mowery et al. [6]). Although there are two keypads for testing, one being metal and the other plastic, the tests are carried out on the plastic keypad as it is indicated that the metal keypad's conductivity renders it impervious to attack. Data gathered from 21 people and 27 PIN combinations display that the heat transferred to the keypad depends on the amount of pressure exerted to the keys as well as the warmth of the hand. However, the heat of the ATM was not taken into consideration as the keypad is used as an isolated test bed, without being wired to or placed on any electronic device. The thermal image shows a clear distinction between the background and the touched keys and right after the PIN entry, it displays with no hassle which buttons (and in which order) are pressed. An important aspect of thermal images is that they may be useful in situations where it is not possible to retrieve the smudges using standard optics, in a dimly lighted place, for instance.

3. EXPERIMENTS AND RESULTS

3.1 Physical Attacks

We tried to replicate Aviv et al.'s [1] methods using a different camera and smartphone. We used a Samsung Galaxy S featuring a Gorilla Glass screen (which is widely used by different manufacturers), a Panasonic Lumix DMC-TZ5 compact camera to conduct the attacks and a hard light source to achieve edged shadows. The objective in this section is to confirm previous work and for this reason the same person first draws the patterns and then conducts the attack in an open environment, with photos taken from a 60° angle. We performed an optical camera attack on three different surface conditions in terms of cleanliness. The first one was an attempt to retrieve the pattern drawn on a clean screen. It is evident that the pattern can be fully retrieved without any difficulty at this stage. The second test adds a light clean up to the first state. The 'lightness' of the clean up is indeed subjective and, in our case, we aim to mimic a person casually cleaning the device's screen without the specific intent of removing any oily residues. The oily residues turned to be quite resistant against simple cleaning attempts. Therefore, although the pattern can fade slightly, it can still be almost fully retrieved (some nodes might disappear). The final test was conducted on a heavily cleaned up surface. For this test we mimic a person determined to clean all the smudges on the screen at once. In this attempt most of the pattern is lost, except some diagonal lines. Due to increased entropy, it is also not possible to tell the directionality of the pattern. Therefore, we conclude that it is possible to capture patterns using compact optical cameras in most of the cases where the phone is not heavily used or cleaned and where there is efficient lighting.

For the microscope attack experiment, we used a USB microscope with 400x magnification. Our experiments followed the same logic we used for the optical attack. However, in the microscope case, we assume that the attackers have already seized the smartphone and are able to investigate the screen in a laboratory as long as desired. Similar to optical camera results, the microscope performed well during the first two cases. Lines and directionalities were very easily seen (full retrieval). There is, however, loss of some detail after the first clean up. For the heavy clean up case the microscope performed slightly better than the camera providing

more details of smudge residues, but assuming that attackers can make use of a Digital Single-Lens Reflex (DSLR) camera in a controlled environment, it is highly likely to gather similar results without the need of a microscope.

The goal of the thermal image attack was to retrieve the pattern by examining the heat trace left by the finger on the device surface. The camera we used was a FLIR E30 and the experiments have been conducted from a distance of approximately 1 meter. The ambient temperature was 26°C, the light was low and no direct sunlight was coming to or near the device. Since time and heat are the main factors that contribute to form the results, the test cases were different than the previous: drawing a pattern on an idle device and drawing a pattern on a recently used device. The first scenario experiments revealed that it is possible to retrieve parts of a pattern via thermal imaging. We managed to observe the heat trace for 3 seconds after the pattern was drawn. However, we were unable to extract the pattern from a recently used device. When the device runs for a short period of time, its CPU starts to emit a considerable amount of heat. This in turn, heats up the upper and centre parts of the device rendering finger's heat untraceable. Even in the idle state, the CPU part of the device is considerably hotter. Consequently, the top three dots are hard to detect in most circumstances. To conclude, whilst it may not be a preferable attack compared to other options, a thermal attack might be used in the future, as the sizes of manufactured components diminishes and chip voltages are lowered.

3.2 User Tradeoffs between Security and Usability for their Choice of Pattern Locks

In order to study the effect of psychological or behavioural factors on pattern setting we conducted a web-based survey. This method was chosen because the participant does not need to own a specific smartphone. We used JavaScript, PHP, and AJAX to create the web-based survey and on the server side we held a MySQL database to store the given data. The pattern lock simulation utilised RaphaëlJS, which is a vector graphics library for drawing objects. The results presented in our work are calculated after filtering the database from irrelevant entries (144 unique participants). The survey started with basic demographics, continued with questions about participants' smartphone experience and their opinion on the notion of device locking and finalised after two pattern entries. The first was a pattern the user thought would be easy to remember and the second was a pattern that the user thought would be a secure password.

Summarising the findings of the survey we deduce that 65.97% were males and 34.03% females. The majority of the users were aged between 18 and 29 inclusive (81.25%) and the next more frequent age bracket was 30-49 (15.28%). This figure was expected because the survey was promoted through social media and through a university mailing list. 79.86% of the participants have owned a smartphone at least once, 92.17% of which still own a smartphone. Among them, 48.11% currently use iOS and 40.57% use Android. Symbian and Blackberry follow with 5.67% and 3.78% respectively. The people who ever owned an Android smartphone had at least one year of experience with it. 47.22% of the participants with a smartphone use any type of screen lock whereas 52.78% do not. The basic reasons they use a screen lock is to protect personal data and prevent others fiddling with the device. 65.98% of the participants believe that the

Table 1: Average pattern lengths and standard deviations.

Group	Average Length		Standard Deviation	
	Secure	Easy	Secure	Easy
Females	6.16	5.94	1.87	1.75
Males	6.89	6.32	1.91	1.94
Total	6.64	6.19	1.92	1.88

Table 2: Average number of direction changes (all users).

Average Changes		Standard Deviation	
Secure	Easy	Secure	Easy
3.57	2.74	1.65	1.59

highest risk that would compromise a lock is shoulder surfing. Smudges left on the screen and cameras in the room have the same rating of being the highest risk with 15.97%, yet the former has been selected more times as the second highest risk compared to the latter, rendering it the second highest risk after the shoulder surfing. Furthermore, 57.64% of the participants thought the secure pattern they entered is usable in everyday life, while 42.36% did not. Finally, 35.42% of the participants thought that the easy pattern they entered was secure enough, while 64.58% did not.

3.3 Secure Pattern Analysis

3.3.1 Pattern Length and Direction Changes

As part of our analysis, we calculated the average pattern length of the secure and easy patterns (and their standard deviations). The average length calculated by summing the number of dots used and dividing that value to the number of participants. While the average pattern length for a secure pattern drawn by a male participant is 6.88 dots, females averaged 6.16 dots. The same situation can be observed in easy patterns: the average among males is 6.32 dots while among females it is 5.94 dots. The total average lengths for secure and easy patterns are 6.64 and 6.19 respectively (Table 1). Results indicate that there may be a difference in perception of secure pattern length and direction between male and female participants. As part of our future work, we are planning to test the statistical significance of this claim, using a larger number of participants. Another indication of a pattern's security efficiency is the number of direction changes made per pattern (Table 2). We assume that for humans, a direction change is a more difficult move than following a direct line. Consequently, we deduce that when a user makes more direction changes in a pattern, then it gets more complex, hence more secure. The average number of direction changes made in a secure pattern is 3.57 and the average number of direction changes made in an easy pattern is 2.74. This finding demonstrates that secure patterns have more direction changes with respect to their lengths, rendering them more complex.

3.3.2 Entropy

In the following sections of survey data analysis, we calculated the Shannon's entropy while studying sub-patterns, start and end points for the secure patterns. For monograms, start and end points, entropy is calculated based on the probability of point X being selected in the pattern or being the start (or end) point. For N-grams, we calculated

conditional entropy, whereby the probability of point X appearing N^{th} in the pattern is dependent on which N-1 points have been used in the pattern so far.

With that in mind, for conditional entropy calculations (F_N: N-gram entropy), we used Shannon's formula [9]:

$$F_N = -\sum_{i,j} p(b_i, j) \log_2 p(b_i, j) + \sum_i p(b_i) \log_2 p(b_i) \quad (1)$$

in which b_i is an (N-1)-gram (a pattern that consists of N-1 nodes), j is an arbitrary node (following b_1) that has not yet been chosen and $p(b_i, j)$ is the probability of the N-gram b_i, j. Note that in the case of sub-patterns, we consider $p(b_i, j)$ as $p(b_i||j)$, where $||$ stands for concatenation.[1] For instance, if the bigram is $b_i =$ "01" and $j =$ "2", then

$$p(b_i, j) = p(\text{"012"}) = \frac{\# \ of \ occurrences \ of \ trigram \ \text{"012"}}{sum \ of \ occurrences \ of \ all \ trigrams}$$

3.3.3 Start and End Points

An interesting observation from the survey is the way participants preferred to start their secure patterns (Figure 1a). More than half of them (52.08%) started their secure patterns from the top left node. The entropy of the start points is 2.35 bits compared to a maximum of $\log_2 9= 3.17$ bits, for which all the dots must have the same probability. This imposes heavy bias and makes the first dot highly predictable. It is important to note that the survey did not examine whether the user is right-handed, left-handed or ambidextrous (we will examine that in subsequent iterations of the survey). Additionally, the survey could be filled either by using a mobile device or a computer, which means participants might have used a mouse to draw the pattern. Nevertheless, participants consistently chose the top left dot as the starting point. A reason for this clustering can be linked to participants' geographical positions. Most of the entries in our pilot run originate from across Europe and the United States. Most of these countries' native alphabets consist of Latin characters and consequently, their writing starts from the top left corner and ends in the bottom right. In addition, the survey's language is English, which may make the participants think in Latin language style even if they have a non-Latin based native alphabet. As a result, they may be inclined to start from the top left, because this looks like a more natural starting point. The collection of data from participants that have top-to-bottom or right-to-left native languages would provide interesting results in the future.

We then checked the ending dots for secure patterns. Even though there was no single dot on which most participants preferred to end their secure pattern, the bottom right was the most selected node with 20.83% (Figure 1b). The entropy calculated is 3.00 bits for the probabilities of end points. The ending dots were mostly focused on right and bottom. From this observation we deduce that the most frequent paths before the ending node can be found between the top right and the bottom right dot. As expected, these results also conform to the assumption about the Latin alphabet made on the analysis of start points.

3.3.4 Sub-patterns Analysis

One of the main objectives of the current work was to investigate the possibility to find reccurring sub-patterns

---

[1] Android's screen lock pattern nodes are represented with numbers starting with 0 from the top left node.

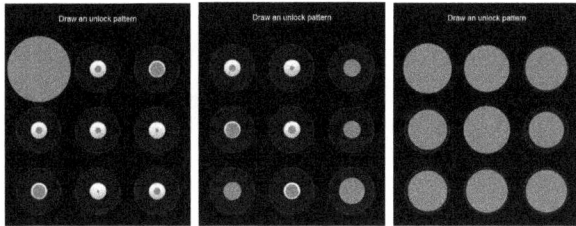

(a) Start points   (b) End points   (c) Monograms

Figure 1: Node usage (radius depicts frequency).

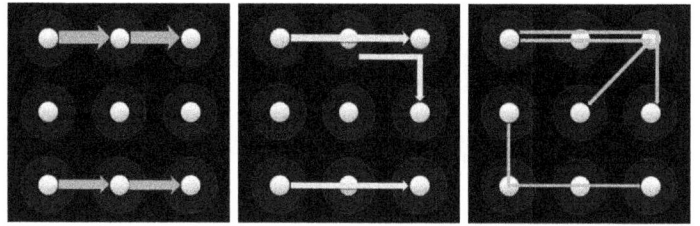

(a) Bigrams   (b) Trigrams   (c) Fourgrams

Figure 2: Most frequently drawn paths ('secure' patterns).

within the responses (focused on secure patterns). Extracting these sub-patterns would allow an attacker to guess a partially retrieved pattern's missing parts easier. In other words, an attacker can incorporate the physical attacks with the behavioural attacks to fully retrieve the pattern. The first step of our sub-pattern analysis involves monograms. We estimated the frequency of appearance of each dot to explore the existence of any particular nodes that are frequently chosen (Figure 1c). The result depicts that there is no significant bias towards any of the dots; they are more or less equally used in patterns, hence monogram entropy is 3.16 bits. We then looked for bigrams, a sub-pattern consisting of two dots. Since bigrams and other longer n-grams create a path, the directionality of the path is taken into account during analysis. There have been some bigrams that occurred especially frequently in patterns collected. In Fig. 2a, path width depicts the frequency of that particular bigram. In this case, the thickest path represents that 32.64% of the participants drew that bigram, while the thinnest represents 23.61%. Using Shannon's entropy, bigram entropy is calculated as 5.47 - 3.16 = 2.31 bits. The maximum entropy for the bigrams is $\log_2 72 = 6.17$bits. Out of 72 possible combinations 64, of them were drawn at least by one participant. Continuing with trigrams, the analysis shows that 18.75% of the participants drew a path from the top left dot to top right dot at one point of their patterns. The thinnest path in Figure 2b represents 14.58% of the participants. The trigram entropy is 6.99 - 5.47 = 1.32 bits. Maximum trigram entropy is $\log_2 504 = 8.98$bits. Out of 504 possible combinations, 203 were drawn at least by one participant. Finally, we conducted a four-gram analysis. Three four-grams stood out of the rest with two of them being drawn by 9.02% of the participants and the other being drawn by 7.64% (Figure 2c). Thus, it is easy to spot a trend towards left to right and top to bottom in these sub-patterns, which contributes to the assumptions made on the psychological behaviour the participants display. The four-gram entropy is 7.75 - 6.99 = 0.76 bits. Maximum fourgram entropy is $\log_2 3024 = 11.56$bits.

4. EVALUATION OF THE RESULTS

Our next step was to integrate our physical experiments and survey findings to propose a scheme, which could increase the success of such a combination of 'soft' and 'hard' side-channel attack on lock patterns. A common physical attack using an efficient optical camera combined with a psychological attack utilising the results of our survey should reduce the number of possible combinations and make pattern retrieval more efficient.

Table 3: Recovery of features.

Optical Attack	Number	Percentage
0 - 49% of pattern	5/22	22.73%
50 - 99% of pattern	5/22	22.73%
100% of pattern	12/22	54.54%
Total Recovery	18/22	81.82%
Phychological	Number	Percentage
Start point	18/22	81.82%
End point	11/22	50.00%
Bigrams	12/22	54.54%
Trigrams	7/22	31.81%
Fourgrams	4/22	18.18%
Direction (C)	14/22	63.63%
Total Retrieval	20/22	90.9%

To evaluate our proposed attack scheme we conducted a new experiment and derived data from a new set of 22 participants, which were not among those who took part on the web-survey. 15 of them (68.2%) were males and 7 (31.8%) females. 86.4% were aged between 21-30 and the age of the rest was 31-40 years old. The participants came from Europe (59.1%), Asia (31.8%) and America (9.1%). The experiment took place at a laboratory. They were asked to think of a secure pattern that they would probably use on their smartphones and then apply it on a real device. We copied and drew their patterns on paper and took photographs of the smartphone screen for further analysis. We marked the drawings with serial numbers before taking the photographs to ensure anonymity. The scenario we investigated assumes light usage of the phone after the pattern was entered, thus, before the photograph was taken we rubbed the screen gently on a cotton surface. We used an HTC Desire smartphone and a Nikon D40x DSLR camera for this experiment.

During the analysis of our data we investigated the correlation between the behavioural trends described in section 3 and smudges left on the screen. We used the following sequence. First, we set the reference standards for this experiment. The presented data in Figure 1 show that the 4 most preferred start points are the corners of the screen. In addition, the 4 most visited end points are those located at the right hand side and also the bottom left node. Finally, we took into account the most preferred N-grams (Figure 2). At the first step of the investigation the nodes and the edges of each pattern were recovered by scholastically analyzing the photograph of the given schema (optical). Then, we compared the findings of the former examination with our reference standards by looking at the drawing we had made

Table 4: Feature recovery of irretrievable patterns.

Physical attack	Number	Percentage
Start point	4/4	100%
End point	3/4	75%
Bigrams	3/4	75%
Trigrams	3/4	75%
Fourgrams	1/4	25%

for the specific pattern. The gathered information answered the question whether the pattern used a common start and end node and whether common N-grams have appeared. If at least one edge of the examined pattern was recovered either by the optical or the behavioral attack, then we can say that we achieved a partial retrieval. Table 3 demonstrates the results of our examinations.

The use of camera revealed, either fully or partially, 18/22 (81.81%) patterns. The psychological attack confirmed that 81.81% of the participants started their passwords using the most common start points and half of them ended their patterns at the expected end points. The average direction changes for males were 3.19, for females 2.83 and average pattern length for males was 7 and for females 6.33. We also saw some of the most common bigrams, trigrams and 4-grams presented in section 3 (popular bigrams were more frequent: 54.5%). Finally, the combination of the two attacks resulted in full or partial retrieval of 20/22 patterns and 100% of the patterns contained at least one of the reference standards.

This statistic shows that combining our web-survey results with well-known physical attacks we can increase the possibility to recover a pattern. We investigated the 4 patterns for which the optical attack did not reveal any information. Table 4 provides the results that justify the assumption that, even without any physical information, the psychological attack can still narrow-down the search space. One can argue about the sample size (4 patterns) but table 4 provides an indication that in most of the cases it is possible to retrieve parts of a pattern. At this table we present the success of the behavioural attack on the patterns that were not recoverable by optical attack. We can see that all of them contain at least one of the reference standards and specifically all contained an expected start point and also, most of them included at least one bigram, trigram and end point.

5. CONCLUSIONS AND FUTURE WORK

We successfully managed to attack an Android pattern lock using various physical attacks. We argue that currently an optical camera or a microscope are the best ways to perform physical attacks and produce quality results. Additionally, we have observed that humans tend to use specific heuristics when they form their lock patterns. We deduce that these heuristics are biased from aspects of peoples' context (e.g. spoken language). Our research demonstrated that it is possible to use the conclusions of our survey to increase the effectiveness of recovering patterns when combined with a successful physical attack. Further work has to be done to create a more global research, which will include other parameters that may be of significance, such as the user's educational level, geographical location and other demographic features. In our evaluation we underlined a trend to clockwise draw a pattern but this is an observation that must be

further examined in the future. It would be also interesting to design a brute force attack model to allow a legitimate user to recover the pattern combining the artifacts found on the screen with the findings of the current research.

6. ACKNOWLEDGMENTS

This work has been supported by the European Union's Prevention of and Fight against Crime Programme "Illegal Use of Internet" - ISEC 2010 Action Grants, grant ref. HOME/2010/ISEC/AG/INT-002.

7. REFERENCES

[1] A. J. Aviv, K. Gibson, E. Mossop, M. Blaze, and J. M. Smith. Smudge attacks on smartphone touch screens. In *Proceedings of the 4th USENIX conference on Offensive technologies*, pages 1–7. USENIX Association, August 2010.

[2] R. Biddle, S. Chiasson, and P. C. Van Oorschot. Graphical passwords: Learning from the first twelve years. *ACM Computing Surveys*, 44(4):1–41, August 2012.

[3] D. Davis, F. Monrose, and M. Reiter. On user choice in graphical password schemes. In *USENIX Assosiation Proceedings of the 13th USENIX Security Symposium*, pages 151–163. USENIX Association, August 2004.

[4] D. J. Delprato. Mind and its evolution: A dual coding theoretical approach. *Psycological Record*, 59(2):295–300, September 2009.

[5] G. Fragkos and T. Tryfonas. A cognitive model for the forensic recovery of end-user passwords. In *Proc. of 2nd Intl. Workshop on Digital Forensics and Incident Analysis*, pages 48–54. IEEE CS Press, August 2007.

[6] K. Mowery, S. Meiklejohn, and S. Savage. Heat of the moment: characterizing the efficacy of thermal camera-based attacks. In *Proceedings of the 5th USENIX conference on Offensive technologies*, pages 6–6. USENIX Association, August 2011.

[7] P. C. v. Oorschot and J. Thorpe. On predictive models and user-drawn graphical passwords. *ACM Trans. Inf. Syst. Secur.*, 10(4):5:1–5:33, January 2008.

[8] M. A. Sasse, S. Brostoff, and D. Weirich. Transforming the 'weakest link' - a human/computer interaction approach to usable and effective security. *BT Technology Journal*, 19(3):122–131, July 2001.

[9] C. Shannon. Prediction and entropy of printed english. *Bell System Technical Journal*, 30(1):50–64, January 1951.

[10] J. Thorpe and P. C. van Oorschot. Human-seeded attacks and exploiting hot-spots in graphical passwords. In *USENIX Assosiation Proceedings of the 16th USENIX Security Symposium*, pages 103–118. USENIX Association, August 2007.

[11] Y. Zhang, P. Xia, J. Luo, Z. Ling, B. Liu, and X. Fu. Fingerprint attack against touch-enabled devices. In *Proceedings of the second ACM workshop on Security and privacy in smartphones and mobile devices*, pages 57–68. ACM, October 2012.

Who do you sync you are?
Smartphone Fingerprinting via Application Behaviour

Tim Stöber
TU Kaiserslautern
t_stoebe@cs.uni-kl.de

Mario Frank
UC Berkley
mfrank@berkeley.edu

Jens Schmitt
TU Kaiserslautern
jens.schmitt@cs.uni-kl.de

Ivan Martinovic
University of Oxford
ivan.martinovic@cs.ox.ac.uk

ABSTRACT

The overall network traffic patterns generated by today's smartphones result from the typically large and diverse set of installed applications. In addition to the traffic generated by the user, most applications generate characteristic traffic from their background activities, such as periodic update requests or server synchronisation. Although the encryption of transmitted data in 3G networks prevents an eavesdropper from analysing the content, periodic traffic patterns leak side-channel information like timing and data volume. In this work, we extract such side-channel features from network traffic generated from the most popular applications, such as Facebook, WhatsApp, Skype, Dropbox, and others, and evaluate whether they can be used to reliably identify a smartphone. By computing fingerprints from ≈ 6 hours of background traffic, we show that 15 minutes of monitored traffic suffice to reliably identify a smartphone based on its behavioural fingerprint with a success probability of 90%.

Categories and Subject Descriptors

C.2.0 [**Computer-Communication Networks**]: General—
Security and protection

Keywords

Smartphone; Authentication; Measurement

1. MOTIVATION

Over the last decade, smartphones became omnipresent. According to [1], 419 million devices have been purchased all over the world during the second quarter of 2012, and half of all US mobile subscribers own a smartphone [2]. The main reasons for such a popularity of smartphone usage are the overall improvement in performance, battery life, and decreasing price of the mobile Internet access over 3G radio networks. In particular, the frequent Internet access is crucial for the success of application markets and a high

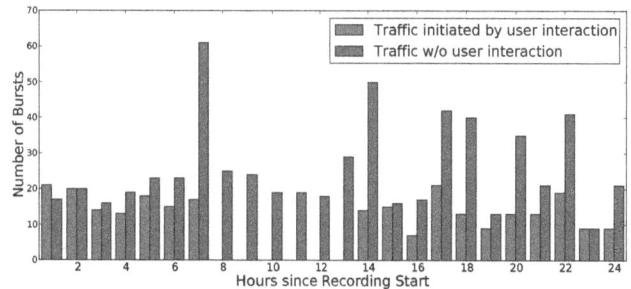

Figure 1: Interactive- and non-interactive smartphone traffic (24 hours). Only about 30% of transmissions are triggered by user interactions.

number of downloaded applications (Apps), such as Email clients, Facebook, WhatsApp, Skype, or Dropbox. Most of these Apps generate traffic, which is not only initiated by the user, but also from the Apps itself to maintain the most current state, receive updates, or synchronise cloud services. For example, Figure 1 shows results of our measurements of network traffic transmitted to and from different smartphones during 24h. We identified that only 30% of the overall smartphone traffic can be attributed to user interactions (we refer to is as *interactive traffic*), and 70% of the traffic belongs to background activities generated by different installed Apps. Many of such background activities result in characteristic traffic patterns, especially in the time domain and in the volume of transmitted data. In addition, the generated traffic highly depends on the multitude of installed applications and their personal configuration.

The 3G/UMTS radio access technology implements encryption at the data link-layer to guarantee the confidentiality of the users' data in the presence of a wireless eavesdropper. Yet, the resulting application-dependent traffic patterns may still pose a privacy risk. A wireless eavesdropper might be able to identify a particular smartphone by analysing only the side-channel information of encrypted traffic. Hence, this motivates the main research question of this work: is it possible to identify a smartphone based on the traffic behaviour of the installed applications, arising from their background activities? Importantly, we assume that an eavesdropper is completely agnostic to the content of the traffic and that security services of the current UMTS radio access network remain unaffected.

In our scenario, the adversary is a passive eavesdropper which is able to capture encrypted wireless 3G/UMTS data from a victim's smartphone and using that traffic to extract smartphone fingerprints. The fingerprints would then allow him to identify the device afterwards and make a point of whether the victim's smartphone is present within a certain UMTS radio cell. While the granularity of such UMTS cell-based tracking does not provide a very fine-grained physical position of the smartphone, it might still reveal important privacy information and help in launching more sophisticated attacks, for example, by detecting if the user is at home or at the workplace.

To assess the threat of identifying smartphones through analysing side-channel information of background traffic, we attempt to answer the following research questions:

- How discriminative are the features of smartphone traffic generated by applications' background activities?
- Are individual configurations of installed Apps sufficient to distinguish between different smartphones?
- How long does it take to identify a smartphone?

2. THREAT MODEL AND ASSUMPTIONS

We assume that the attacker is located within the range of UMTS transmissions. The inherent broadcast characteristic of wireless communications allows him to eavesdrop on UMTS physical signals. Hence, our main assumption is that the attacker is able to demodulate and demultiplex the physical layer and measure side-channel information such as the amount of transmitted data and timing information. In order to acquire such side-channel information, the attacker must extract the users' data streams from the superimposed signal on the corresponding wideband code division multiple access (WCDMA) air interface. In this section, we briefly discuss technical requirements and the complexity to gather side-channel information without the possession of spreading and scrambling codes used in UMTS.

The 3G/UMTS air interface is based on WCDMA, a code multiplexing technique to separate the medium into single channels and enable simultaneous transmissions over the same frequency. The users' data (payload) is transmitted over so-called Dedicated Physical Data Channels (DPDCH), and its correct demodulation requires the knowledge of scrambling and spreading codes as specified in [3]. The spreading codes increase the actual bandwidth of the signal, while the scrambling codes are multiplied chip by chip with the already spreaded signal to achieve orthogonal coding. The scrambling codes separate distinct base stations on the downlink and different user equipments (UEs) on the uplink. However, none of these coding techniques were designed for security reasons and the security services are offered by the higher layers of the UMTS network stack. In particular, both spreading and scrambling codes can be "brute-forced" as their search space is not large (and the assignment of spreading codes on the uplink is almost completely specified in [3]). For example, assuming one base station (i.e., Node-B) using one primary scrambling code for its cell, there are less than 1000 available spreading codes that could be employed after deducting codes for reserved channels [3]. The scrambling code is more expensive to find, as it is generated by 18 bit (downlink) and 24 bit (uplink) seeded shift registers, respectively. Yet, none of these lengths presents a significant computational burden for an adversary. Moreover, the Node-B's scrambling code for downlink is automatically determinable by the smartphone's cell search procedure. In our experiments, we therefore investigate how the availability of only the downlink traffic affects the success probability of correctly identifying a smartphone. Table 1 shows a cost estimate for obtaining the required codes in the downlink and uplink cases.

	Scrambling	Spreading
Up	2^{24} possibilities	max. 7 possibilities
Down	2^{18} possibilities[1]	max. 1000 possibilities

Table 1: Cost estimate for attaining scrambling- and spreading codes. Determination of Node-B's scrambling code is less expensive in contrast to the UEs'.

In summary, we believe it is reasonable to assume that there is a practical way for an adversary to capture the encrypted UMTS traffic and to use it for fingerprint acquisition (i.e., the training phase) and for fingerprint detection (i.e., the attack phase). During fingerprint acquisition, the adversary should know whether the extracted traffic belongs to the victim. One concrete approach to achieve this would be to inject known traffic markers by initiating a call, sending an email, or sending an SMS to the victim. However, the detailed analysis of this approach is out of the scope of this work.

3. DATA ACQUISITION

The first dataset consists of recorded traffic from five distinct users for whom all 3G network communication has been captured in the background for approximately eight hours. During this collection phase, the users interacted with their smartphones without any restrictions.

Due to the low scope of the user dataset, we recorded 8 hours from 20 user devices with different combinations of Apps installed. We call this dataset *non-interactive* because transmissions were captured without any user interactions that could cause traffic.

Our testbed was a Samsung Galaxy Nexus running the Android operating system version 4.0.4. Instead of sniffing traffic on the UMTS link, we captured directly on the devices' 3G interface using `tcpdump`. We randomly picked 20 distinct combinations out of a universe of 14 Apps. Each combination is composed of seven Apps, whereas we assured the marginal cases to be present, meaning two combinations with six common applications and only one differing, as well as two combinations with completely disjoint subsets.

The 14 Apps were selected from the list of top free Android applications from the Google Play Store. We only picked Apps that actually produce background transmissions without user interactions, since otherwise they would have no effect on the traffic behaviour of the device. This comprises cloud services, several Messengers, or Email clients. The complete list of chosen Apps with some additional information from `https://play.google.com/store/apps` (accessed 26/09/2012) can be found in Table 2.

4. FINGERPRINTING

In this section we first introduce the notion of a burst, which plays a central role in our framework, and then we describe the process of creating smartphone fingerprints.

As an input, the process requires traffic extracted from a specific device. The traffic is a chronological sequence of incoming and outgoing packets. Each packet is represented as a vector $p_i = (t_i, s_i, d_i)$. Hence, the only information

---

[1]Node-B's scrambling code determinable on CPICH.

Index	Name	Downloads	Rank
1	Email	-	native
2	Facebook	100 - 500	4
3	WhatsApp	50 - 100	9
4	Skype	50 - 100	12
5	Twitter	50 - 100	14
6	Dropbox	10 - 50	15
7	Instagram	10 - 50	23
8	Flipboard	5 - 10	27
9	Viber	10 - 50	34
10	Evernote	10 - 50	156
11	Spotify	10 - 50	66
12	Wetter.com	5 - 10	not ranked
13	Skydrive	0.1 - 0.5	422
14	ChatON	10 - 50	50
15	Google Account	-	native

Table 2: Universe of applications from which the emulating combinations were chosen.

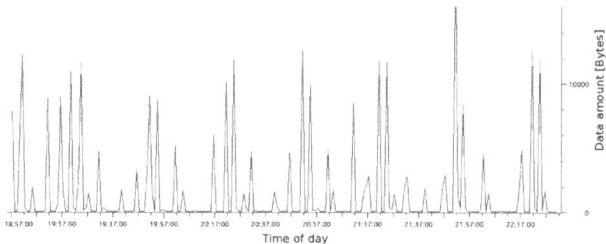

Figure 2: 3.5 hours of smartphone traffic. Idle phases alternating with transmissions illustrate the burstiness. As can be observed similar burst patterns occur in regular intervals.

needed about a packet is its arrival time t_i in the form of an absolute or relative time stamp, its size in bytes s_i and the direction d_i in terms of a Boolean flag distinguishing between incoming and outgoing packets.

Taking a closer look at smartphone traffic, one can observe alternating idle periods followed by short peaks of incoming and outgoing data transfers. This behaviour, which we refer to as burstiness, is illustrated in Figure 2 and has also been observed by Falaki et al. in [10]. A single burst, represented by the peaks of the black curve, consists of a sequence of packets that are mostly semantically connected like, e.g., packets from the same TCP connection.

4.1 Burst Separation

Since we cannot analyse the payload to identify and aggregate packets from the same application (using e.g., TCP flows), the bursts are extracted by only considering the timing information. We define a burst distance to be the minimum length of an idle interval between two packet arrivals. If the arrival time of two subsequent packets is larger, the respective packets are considered to belong to different bursts. Hence, extracting bursts from the captured traffic is equivalent to setting cuts in the packet sequence and aggregating all packets which are within these borders to one burst.

Clearly, selecting the burst distance is an important parameter. On the one hand, we may prefer a small distance because this avoids the aggregation of several unrelated trans-

missions. On the other hand, choosing a too small distance may result in splitting related transmissions.

In agreement to the observations by Falaki et al. [10], we observed that 95% of all packets arrive at most 4.43s after their predecessors (4.5s in [10]). Therefore, we selected a distance threshold of 4.5s for burst separation.

4.2 Burst Characterisation

In the next step, we identify the most discriminative features that distinguish well between bursts generated by different smartphones. In addition to the mean values of the packet inter-arrival times and packet sizes, we also take into account the 20%, 50% (median) and 80% quantiles of the timing and size distributions. In the presence of outliers and non-Gaussian distributions, these measures are more robust in comparison to the simple arithmetic mean. For the same reason, we apply the median absolute deviation (MAD) in terms of packet sizes and inter-arrival times. The MAD is the median of the deviations from the median [14]. Let X be a sample set vector, it is computed as

$$MAD = median(|X - median(X)|). \quad (1)$$

In the following two subsections we examine the individual features in terms of their importance. The purpose of this analysis is not primarily to minimize computational costs in the training phase or to address the curse of dimensionality; we are rather interested in gaining a better understanding of this particular kind of data. The complete list of features is listed in Table 3.

Feature	rMI
Median absolute deviation (MAD) packet size	16.6%
50% quantile packet size	15.7%
Standard deviation packet size	15.6%
80% quantile packet size	14.7%
20% quantile packet size	14.6%
Mean packet size	13.8%
Byte ratio	13.8%
Distance to next burst	8.5%
Number of outgoing bytes	7.2%
80% quantile packet interarrival time	7.0%
Mean packet interarrival time	6.9%
Number outgoing packets	6.9%
Packet ratio	6.5%
50% quantile packet interarrival time	6.2%
Throughput	5.9%
Duration	5.8%
Number of incoming packets	5.7%
Number of incoming bytes	5.6%
Median absolute deviation packet interarrival time	5.5%
Standard deviation packet interarrival time	5.5%
Mean consecutive outgoing packets	4.8%
20% quantile packet interarrival time	4.5%
Mean consecutive incoming packets	4.2%
Random feature	0.9%

Table 3: Feature list with respective relative mutual information (rMI) for the non-interactive dataset.

To determine the importance of individual features, we compute the mutual information $I(F_i; U) = H(U) - H(U|F_i)$ between a feature F_i and the target variable (the user ID) U. To account for the fact that the entropy of target variables can vary, we compute the relative mutual information (rMI) that can be computed as a fraction of entropies:

$$rMI(F_i; U) = I(F_i; U)/H(U) = 1 - H(U|F_i)/H(U) \quad (2)$$

The entropy $H(U)$ quantifies the uncertainty about the ID. The conditional entropy $H(U|F_i)$ quantifies the remaining uncertainty if the value of feature i is known. The difference of $H(U)$ and $H(U|F_i)$ becomes maximal if the feature fully determines the user ID.

Before computing rMI, it is necessary to quantise the continuous feature values. To account for outliers, we divided the 0%- to the 90%-quantile into bins and accounted all outliers to the last quantile. The results for rMI are shown in Table 3. To offer a point of reference, we introduced an artificial feature with random values that should hardly provide any information about the user.

We refrain from choosing only the best ranked features as a result of our analysis. The reason is that even variables with a small independent informativeness can provide rich information when combined [13]. In contrast, pairs of variables with large rMI could be fully redundant.

4.3 Classifiers

As it is common practice in supervised learning, we turn this multi-class classification problem with n user IDs into n individual binary classification problems. In each individual problem, a classifier must decide if the current data comes from the respective phone or not. We consider each phone's classifier as the fingerprint of the phone. For each such problem, the observation matrix, used as an input for training the classifier, consists to 50% of bursts from the legitimate user. The remaining 50% are filled by an equal number of bursts from other users. This way, we only have two class labels, namely *user* and *¬user*. Each observation or burst in the observation matrix is represented by a row whereas each of the 23 columns corresponds to a feature.

In terms of classification, we use the k-nearest neighbors algorithm (kNN) as well as a support vector machine (SVM). The kNN is capable of directly solving the original multi-class classification problems since it simply stores the feature vector of every observation and its corresponding categorial label. A new object is then classified by taking the majority label of the k nearest neighbors in the feature space. To estimate the optimal setting for the parameter k, we perform 5-fold cross-validation on the training data testing all odd numbers from 1 to 13. In contrast to kNN, the SVM does not store all feature vectors but instead computes one or several hyperplanes that divide the set of observations according to their class labels, whereas the distance of the hyperplane to the nearest objects (called support vectors) is maximised.

5. EXPERIMENTATION

5.1 Single Burst Classification

The goal of this experiment is to investigate the feasibility of our identification approach. This requires to test for possible collisions between each user's fingerprint with the other fingerprints.

When referring to a particular user from the non-interactive dataset, we always mean the App combination on the phone of this user.

For each user u out of the dataset, a fingerprint was constructed by training a classifier with 70% of u's bursts and the same amount of bursts from other users, both randomly chosen. After the fingerprints were generated, we started classifying the remaining 30% of u's bursts as well as equally

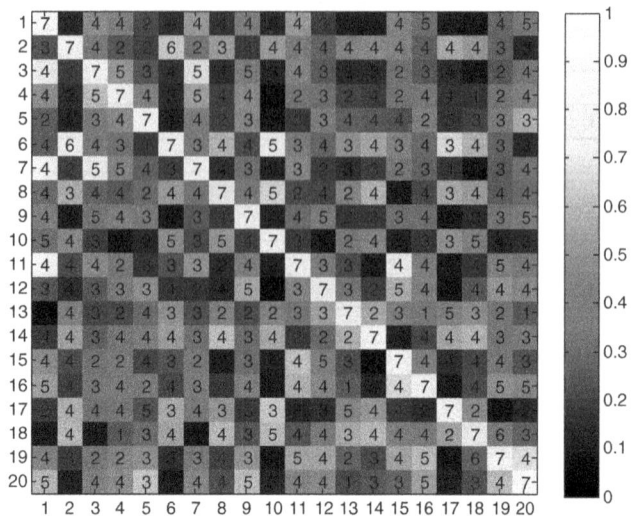

Figure 3: kNN results for matching users against every fingerprint (Non-interactive dataset). The brighter the color, the more bursts were classified to belong to the fingerprint. The numbers inside the cells indicate the count of common Apps of two combinations.

	User dataset		Non-interactive dataset	
FNR	13.60%	16.99%	17.44%	22.85%
FPR	24.02%	31.41%	32.60%	33.46%

Table 4: False negative- and false positive rates for feasibility experiment (SVM left, kNN right).

many random bursts from every single other user, assuring not to employ any burst that has been used for building the classifier. Each classification combination was repeated 20 times. In every round, the training- and test set were populated by newly, randomly selected bursts.

As a result, we obtain the number of false negative- (FN) and false positive (FP) class assignments. FNs for the cases where we match users on their own fingerprints and FPs for the remaining cases. Fig. 3 depicts the results for the kNN classification of the non-interactive dataset. The cells of the colour matrix show how well the traffic of the column-user matched the fingerprint of the row-user. The brighter the colour, the more bursts have been classified to match the fingerprint. The numbers within the cells indicate how many applications the two compared phones have in common.

In general, SVM performs better, especially in terms of the false negative rate (diagonal). The precise false positive- and false negative rates for both classifiers and both datasets are given in Table 4.

The results for SVM on the user dataset are best. This is not only due to the minor scope of the dataset but also because the event of two real users having nearly the same combination of installed Apps and similar configurations is rather unlikely. This makes the UMTS application attractive to an attacker, since he would only have to compare the traffic with a fixed number of users per cell.

Obviously it becomes more and more difficult for the classifier to distinguish users with many shared Apps and, consequently, similar traffic. However, there are also other influences. E.g. the App Spotify tends to generate many bursts. Therefore, matching two phones against each other that both contain Spotify will exacerbate the decision of the

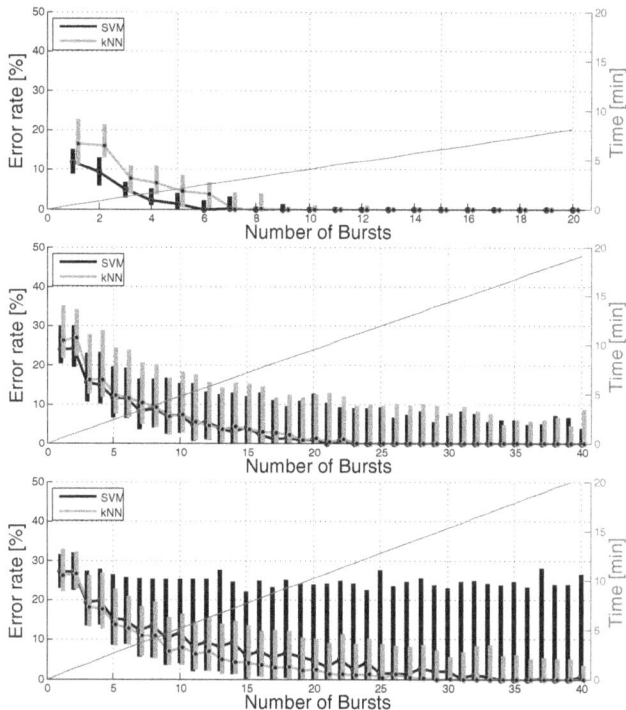

Figure 4: Error rate versus the number of bursts used for classification. To achieve a median classification error rate of 0%, an attacker would have to wait approximately 3 min in case of the user dataset (upper chart), 12 min for the non-interactive dataset (middle chart), and 15.5 min when only non-interactive downlink data is used (lower chart).

classifier. However, we always used Spotify with the same configuration which made the task unrealistically hard. Distinct configurations (Playlists etc.) might affect the traffic and render it more unique. Overall, the investigated setting can be considered rather conservative, since the emulated users have on average 3.5 out of only 7 installed Apps in common. We assume to observe more variety in the wild.

5.2 Effects of Capturing Duration

The objective of this experiment is to find out how much traffic one must capture to make a reliable statement whether it belongs to the victim or not. This means we have to investigate classification results for varying amounts of available traffic. To that end, we must evaluate the burst inter-arrival times to find out how long an attacker should wait to gather a certain amount of traffic.

As in the last experiment, we generate a fingerprint for each user with a training set containing 70% of its bursts and equally many random bursts from other users. Yet, unlike last time, the training set is filled with the remaining 30% of the users bursts as well as the same amount of randomly chosen bursts from other users, assuring to not employ those that have been used for training.

Instead of classifying each single burst separately, we apply a sliding window of size $k \in \{1, 2, \ldots 40\}$ such that we classify k bursts and take the majority vote of the k labels. This procedure is repeated in 20 times for every user to quantify random effects.

The two upper charts in Fig. 4 depict the results for the user dataset and the non-interactive dataset applying the kNN al-

gorithm and the SVM. The bars reach from the 25%-quantile to the 75%-quantile. The dots inside the bars represent the median values. As can be seen, the SVM generally out performs kNN. After 6 bursts for the user dataset and after 23 bursts for the non-interactive data, a median error rate of 0% is reached. In terms of time, depicted by the linear function, an attacker would have to wait merely ~3 minutes and ~12 minutes, respectively.

5.3 Effect of Using Downlink Data Only

In particular for UMTS, where it is easier to capture traffic on the downlink from the Node-B to the user equipments instead of the uplink, it is interesting to restrict the fingerprinting to downlink data. Fig. 4(bottom) illustrates the results of the experiment from Section 5.2 in the downlink-only case. The outcome indicates that the median error only impairs slightly. To achieve an error rate of 0%, the attacker must monitor ~15.5 minutes compared to 12 minutes for the bidirectional case. In this scenario, kNN outperforms SVM by far. This is most likely due to the fact that the SVM performance relies on particular features that are not available anymore in the downlink case like *Packet ratio, #Packets out, Byte ratio, #Bytes out* and *Mean consecutive out*.

6. RELATED WORK

There has been much research in the area of device fingerprinting. In most cases, side-channel information, like hardware and manufacturing inconsistencies, or differences in driver implementations, were exploited as a discriminative component. Probably, one of the first publications in this direction was [15] by Kohno et al. in 2005. They introduced a mechanism to remotely fingerprint and identify devices based on their clock skews, which were in turn based on information from TCP and ICMP timestamps. In [11], the authors succeeded in fingerprinting and identifying wireless device drivers on the basis of a statistical analysis of a device's interarrival rate of IEEE 802.11 probe request frames. Since the standard does not provide a specific value for the scanning intervals of these management frames, distinct drivers tend to differ in their implementations. Based on this work, Loh et al. extended the discriminability of this feature to be even capable of distinguishing between single devices instead of device drivers [7].

In [12, 19, 5, 4], authors exploit manufacturing inconsistencies and hardware imperfections that have effect on the resulting transmission signal in order to differentiate between single entities. In addition, one can find a very detailed review of physical-layer identification systems and state-of-the-art techniques in [6].

All previously mentioned approaches are aiming at remote identification of devices, however the fingerprint does not include any form of application behaviour.

Another related area of research includes various approaches to traffic classification. Some of them use transport layer statistics like packet size, connection duration and ratio of bytes sent in each direction, combined with unsupervised machine learning algorithms [9, 21]. Even though the features are solely from the time and byte dimension, both techniques still rely on the notion of flow or connection, respectively. Clearly, traffic analysis countermeasures like padding can be used to conceal user identities. However, Dyer et al. showed in [8] that bandwidth-efficient, general-purpose traffic analysis countermeasures mostly fail. In [22], the authors use a packet-level classification, only resorting to link layer features which makes their approach applicable even under payload encryption. Similarly to our work, their methodology uses supervised learning algorithms like SVM and neural networks. Yet, their application and network environment is not focused on wireless networks. Related work on detecting network applications discusses more information that can be leaked through traffic analysis, e.g., visited web pages [16], language and spoken phrases [20] or watched videos [18]. Recently, in [17], the authors analysed the security of the TLS Record Protocol and identified attacks against TLS, even if variable length padding (as a countermeasure against SSL/TLS side-channel information leakage) is used. While many smartphone Apps establish an SSL/TLS connection with their servers, in our work, we do not consider SSL/TLS flows, but the overall aggregated traffic from all Apps and without any requirement to identify the SSL/TLS traffic.

7. CONCLUSIONS

In this work, we investigated the question of whether background traffic generated by smartphone applications can be used as a fingerprint to identify and discriminate different smartphones. We based the fingerprint features only on features available as side-channel information such as timing and data volume. These features can be extracted through monitoring of wireless channels used by the UMTS radio technology and without assuming any knowledge of the payload. Our results show that the multitude of installed applications and their background communication generates a unique behaviour that allows an eavesdropper to accurately identify smartphones. To that end, we designed extensive experiments including traffic monitoring from the most popular Apps and demonstrated that even if the smartphones have a large number of the same applications installed, they still can be successfully identified with a very high accuracy. In particular, after the fingerprint is generated, the eavesdropper requires only ≈15 min. of the captured traffic to achieve more than 90% classification accuracy.

These results have direct impact on the user's privacy as they justify that an adversary is able to detect whether a smartphone is associated with a certain UMTS radio cell.

Acknowledgments

This research was partially supported by the Swiss National Science Foundation and by Intel through the ISTC for Secure Computing.

8. REFERENCES

[1] Gartner. http://tinyurl.com/d7ptpqc, 2012. [Online; accessed 11/10/2012].
[2] Nielsen. http://tinyurl.com/8y2e773, 2012. [Online; accessed 11/10/2012].
[3] 3GPP. TS 25.213, Spreading and modulation (FDD). Technical report, 1999.
[4] K. Bonne Rasmussen and S. Capkun. Implications of radio fingerprinting on the security of sensor networks. In *Third International Conference on Security and Privacy in Communications Networks*, SecureComm'07, 2007.
[5] V. Brik, S. Banerjee, M. Gruteser, and S. Oh. Wireless device identification with radiometric signatures. In *Proceedings of the 14th ACM international conference on Mobile Computing and Networking*, MobiCom'08, 2008.
[6] Danev B., Zanetti D., Capkun S. On physical-layer identification of wireless devices.

http://www.syssec.ethz.ch/research/OnPhysId.pdf, 2012. [Online; accessed 22/10/2012].
[7] L. C. C. Desmond, C. C. Yuan, T. C. Pheng, and R. S. Lee. Identifying unique devices through wireless fingerprinting. In *Proceedings of the first ACM conference on Wireless Network Security*, WiSec'08, 2008.
[8] K. P. Dyer, S. E. Coull, T. Ristenpart, and T. Shrimpton. Peek-a-boo, i still see you: Why efficient traffic analysis countermeasures fail. In *IEEE Symposium on Security and Privacy*, SP'12, 2012.
[9] J. Erman, M. Arlitt, and A. Mahanti. Traffic classification using clustering algorithms. In *Proceedings of the 2006 SIGCOMM workshop on Mining Network Data*, MineNet'06, 2006.
[10] H. Falaki, D. Lymberopoulos, R. Mahajan, S. Kandula, and D. Estrin. A first look at traffic on smartphones. In *Proceedings of the 10th annual conference on Internet Measurement*, IMC'10, 2010.
[11] J. Franklin, D. McCoy, P. Tabriz, V. Neagoe, J. Van Randwyk, and D. Sicker. Passive data link layer 802.11 wireless device driver fingerprinting. In *Proceedings of the 15th conference on USENIX Security Symposium*, USENIX-SS'06, 2006.
[12] R. M. Gerdes, T. E. Daniels, M. Mina, and S. F. Russell. Device identification via analog signal fingerprinting: A matched filter approach. In *In Proceedings of the Network and Distributed System Security Symposium*, NDSS'06, 2006.
[13] I. Guyon and A. Elisseeff. An introduction to variable and feature selection. *J. Mach. Learn. Res.*, 2003.
[14] F. R. Hampel. The breakdown points of the mean combined with some rejection rules. *Technometrics*.
[15] T. Kohno, A. Broido, and K. Claffy. Remote physical device fingerprinting. *IEEE Transactions on Dependable and Secure Computing*, 2005.
[16] M. Liberatore and B. N. Levine. Inferring the source of encrypted http connections. In *Proceedings of the 13th ACM conference on Computer and Communications Security*, CCS '06, 2006.
[17] K. G. Paterson, T. Ristenpart, and T. Shrimpton. Tag size does matter: attacks and proofs for the tls record protocol. In *Proceedings of the 17th international conference on The Theory and Application of Cryptology and Information Security*, ASIACRYPT'11, 2011.
[18] T. S. Saponas, J. Lester, C. Hartung, S. Agarwal, and T. Kohno. Devices that tell on you: privacy trends in consumer ubiquitous computing. In *Proceedings of 16th USENIX Security Symposium on USENIX Security Symposium*, SS'07, 2007.
[19] O. Ureten and N. Serinken. Wireless security through rf fingerprinting. *Canadian Journal of Electrical and Computer Engineering*, 2007.
[20] C. Wright, L. Ballard, S. Coull, F. Monrose, and G. Masson. Spot me if you can: Uncovering spoken phrases in encrypted voip conversations. In *IEEE Symposium on Security and Privacy*, SP'08, 2008.
[21] S. Zander, T. Nguyen, and G. Armitage. Automated traffic classification and application identification using machine learning. In *The IEEE Conference on Local Computer Networks*, 2005.
[22] F. Zhang, W. He, X. Liu, and P. G. Bridges. Inferring users' online activities through traffic analysis. In *Proceedings of the fourth ACM conference on Wireless Network Security*, WiSec'11, 2011.

MAST: Triage for Market-scale Mobile Malware Analysis

Saurabh Chakradeo, Bradley Reaves,
Patrick Traynor
School of Computer Science
Georgia Institute of Technology
Atlanta, GA, USA
{schakradeo, brad.reaves}@gatech.edu,
traynor@cc.gatech.edu

William Enck
Department of Computer Science
North Carolina State University
Raleigh, NC, USA
enck@cs.ncsu.edu

ABSTRACT

Malware is a pressing concern for mobile application market operators. While current mitigation techniques are keeping pace with the relatively infrequent presence of malicious code, the rapidly increasing rate of application development makes manual and resource-intensive automated analysis costly at market-scale. To address this resource imbalance, we present the Mobile Application Security Triage (MAST) architecture, a tool that helps to direct scarce malware analysis resources towards the applications with the greatest potential to exhibit malicious behavior. MAST analyzes attributes extracted from just the application package using Multiple Correspondence Analysis (MCA), a statistical method that measures the correlation between multiple categorical (i.e., qualitative) data. We train MAST using over 15,000 applications from Google Play and a dataset of 732 known-malicious applications. We then use MAST to perform triage on three third-party markets of different size and malware composition—36,710 applications in total. Our experiments show that MAST is both effective and performant. Using MAST ordered ranking, malware-analysis tools can find 95% of malware at the cost of analyzing 13% of the non-malicious applications on average across multiple markets, and MAST triage processes markets in less than a quarter of the time required to perform signature detection. More importantly, we show that successful triage can dramatically reduce the costs of removing malicious applications from markets.

Keywords

Mobile application security, Triage, Multiple correspondence analysis

Categories and Subject Descriptors

D.4.6 [**Operating Systems**]: Security and Protection—*Invasive software*

1. INTRODUCTION

Application markets have simplified the process of finding and installing software on smartphones, creating an efficient channel between developers and end users [4, 8, 21, 51]. Unfortunately, they have also provided attackers with an easy point of entry into mobile devices in the form of vulnerable and malicious applications [13, 25–27, 32, 34, 35]. Some markets have responded thus far with a variety of proactive and reactive approaches to this problem (e.g., Apple's manual analysis and Google's Bouncer [31]). Though these approaches have thus far minimized the amount of malware that appears on their respective markets, many alternative Android markets [3,20,49] still remain completely unprotected against malware. Also, the time and monetary cost as well as the need for in-depth inspection of the applications will rise as malware takes greater lengths to avoid detection, and the number of new samples to analyze continues to increase as developers create new and update existing applications. To continue to be effective and work within malware detection budgets, market providers, antivirus companies, and independent researchers must then, more than ever, strategically spend manual effort and computationally expensive program analysis.

Mobile application security is not the first discipline to wrestle with a mismatch of resources and duties. Medical facilities regularly perform triage, or the prioritization of limited resources based on the perceived condition of each individual within the population of patients. Triage is neither diagnosis nor treatment. Rather, it allows medical personnel to immediately deal with patients with the greatest obvious needs while delaying or dismissing treatment for others. Developing such perception and prioritization in the mobile application space, where most applications are in fact benign, would allow scarce resources to be dedicated to the investigation of applications that have the greatest potential to be dangerous.

In this paper, we present the Mobile Application Security Triage (MAST) infrastructure. MAST develops a perception of suspicion for applications through the use of Multiple Correspondence Analysis (MCA) [1], a technique used in the social sciences to determine the statistical correlation between multiple categorical (i.e., qualitative) data. In particular, we develop a "questionnaire" for Android applications that looks for strong relationships between *declared* indicators of application functionality (e.g., permissions, intent filters, the presence of native code, etc.) given the key insight that *these declared indicators are required for malware to perform its malicious functionality.* Effectively, MAST operates under the assumption that the configurations of declared indicators of legitimate app are distinct from malware, which MCA identifies as outliers in a population.

Our methodology is developed using over 15,000 applications from Google Play [21], the Contagio mobile malware repository [43] and malicious applications found by Zhou et al. [53]. We then apply MAST to three third-party application markets: Anzhi (28,760 apps) [3], Ndoo (4,324 apps) [39], and SoftAndroid (3,626 apps) [49].

Our results show that MAST dramatically reduces the effort spent on non-malicious applications. Using MAST, existing detection tools can identify on average 95% of malware in third-party markets at the cost of analyzing 13% of the non-malicious applications in them. As a side-effect of performing this triage, we discovered widespread misuse of the Android default application signing key, appearing in 1,672 applications across the Android Market sample, the malware dataset, and the three third-party markets.

We make the following contributions:

- **Develop the MCA-based MAST Architecture:** As the rate with which mobile apps are added to markets increases, performing deep security analysis of those apps will become a "big data" problem. We develop an infrastructure for directing scarce analysis resources to address this problem. MAST uses MCA to rank applications with a degree of suspicion. Although implemented for Android, this architecture is reusable in any system in which required qualitative declarations of application functionality are available, and is extensible to accommodate new classes of malicious applications.

- **Analyze multiple third-party Android markets:** We train and evaluate the effectiveness of MAST using a corpus of more than 50,000 Android applications from a range of different markets. Even with the diversity in size and malware population of these markets, MAST effectively triages malicious apps—on average 95% of malware present in the market can be detected at the cost of analyzing 13% of the non-malicious applications.

- **Generate rankings faster than any lightweight analysis:** MAST is designed to direct resource intensive operations (e.g., manual analysis), so generating these rankings requires less time than even the most lightweight analysis tools. Specifically, our MAST infrastructure ranks applications more than 4 times faster than signature detection.

MAST is a comprehensive yet lightweight mechanism for identifying malware in Android. Prior work has used rules or simple statistical measures to detect malware [16, 19, 22, 54], while more advanced analyses [44, 47] have reported worse or comparable efficacy, while failing to characterize their efficiency and efficacy on markets with actual threats. The MAST architecture is fundamentally more flexible and robust than a rule-based filtering scheme, and this paper is the first to demonstrate that advanced techniques can be useful in practice.

Finally, MAST is not designed as a replacement for manual analysis (e.g., as in Apple's App Store) or automated malware detection tools (e.g., Google's Bouncer [31]). Rather, *MAST provides a rank-ordered list over which additional, more heavyweight, techniques can be applied.* Alternative Android markets, that have human resources to manually analyze only a small subset of applications, can use MAST to decide which applications deserve the most attention. In the case that markets want to scan all the applications, MAST can aid in deciding which applications require deep, costly analysis and which ones require just a cursory anti-virus scan. The necessity of properly allocating malware analysis resources is going to become evident as smartphone malware authors start adopting techniques such as polymorphic and obfuscated code and malware researchers respond with complex static and dynamic analysis techniques. The goal of MAST is to robustly direct malware analysis resources in a manner that is not dependent on any one characteristic and is thereby more resilient to the "hide-and-seek" games of malware authors. Such an approach is valuable not only to large market operators with extensive resources, but also to smaller markets and even academic and industry researchers as well.

2. RELATED WORK

Application markets currently rely on both proactive and reactive mechanisms for dealing with mobile malware. The proactive approach typically relies on the use of automated tools to detect vulnerabilities. Simple approaches are evadable [38, 42], and complex analyses can become unsustainable as the complexity and number of applications requiring certification grows. Worse still, many proactive analyses require human intervention [37], making them fundamentally unsuitable at market scale. Malicious applications have already exploited this fact and have been seen in multiple markets [25–27, 29]. Reactive approaches recognize this and attempt to recover from such infections. Market controlled "kill-switches" [7, 10] allow malicious apps to be removed post infection; however, significant damage may have already been done.

Various techniques have been proposed to improve security by reducing vulnerabilities in applications [12, 15, 18] and creating stronger detection mechanisms for malware and grayware [13, 15, 16, 23, 55]. Traditional antivirus products [33, 40] use known malware signatures to detect malicious applications; however, they fail to capture new threats. Enck et al. [14] use taint-tracking to detect information flow between sources of private information (e.g. IMEI, phone number) and external data sinks (e.g. internet, SMS). High-level run-time behavior [9] and power [28, 30] have also been explored for malware detection. However, when automating dynamic analysis, it is nontrivial and costly to ensure complete coverage to trigger malicious payloads.

A number of efforts rely on Android permissions to determine the maliciousness of applications. Kirin [16] uses static, conjunctive rule sets to define possible malicious behaviors and warn the user at install time. Barrera et al. [6] use self-organizing maps to analyze permission usage patterns in applications. Felt et al. [19] use permission counts and a comparison of individual permissions to reason about malicious applications. However, all these approaches suffer from high false positive rates and are fundamentally limited by their inability to fully analyze the correlation between high dimensional data associated with applications. DroidRanger [54] and RiskRanker [22] are closer in spirit to MAST. DroidRanger's permission-based filtering had a false positive rate of 40%, after which Zhou et al. [54] had to manually analyze and then add checks for specific Android intents and native code to reduce the false positive rates. Further, DroidRanger does not provide a generalized easily reproducible methodology for selecting such "expert" features. RiskRanker, despite its name, only performs classification into three categories: high, medium, and low risk. Apps are classified as high-risk if they match vulnerability-specific signatures (e.g., root exploits), and medium-risk if they use SMS without a code-path to an *onClick()* callback. MAST improves upon both DroidRanger and RiskRanker by providing a statistical foundation for allocating valuable malware analysis resources based on 208 behavioral features.

The closest research to MAST is concurrent and independent work done by Sarma et al. [47] and Peng et al. [44]. These works test several techniques (but not MCA) over application permissions to provide rankings of relative risks *to users*. Peng et al. show superior results to Sarma et al. Their goals include low user warning rates and developer accessibility; by contrast, MAST uses not only permissions, but application intents and the presence of native code to determine suspicion *for analysts*. It is a strict design goal in [44] that developers understand how to reduce their risk; MAST does not need this property because it is targeted to analysts, not developers. However, *our results are better representative of triage in practice* because we test against third party markets with malware that are unrelated to our training data. Peng et al. train and test

Table 1: A Sample Restaurant Data Set for MCA

Restaurant	Cost	Parking	Attire	Ages	Delivery	Television
R1	High	Valet	Formal	18+	No	No
R2	Med	Valet	Formal	All	No	No
R3	Med	Lot	Casual	18+	No	Yes
R4	Med	Valet	Casual	18+	No	Yes
R5	Med	Valet	Casual	All	No	No
R6	Low	Lot	Casual	All	Yes	No
R7	Low	Lot	Casual	All	No	No

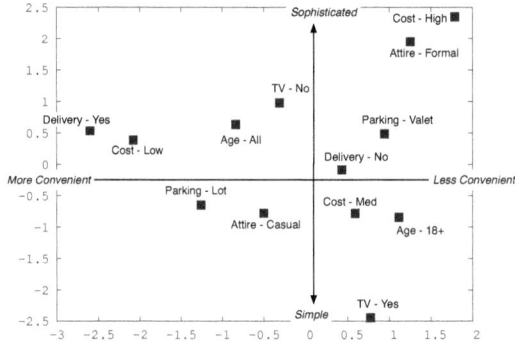

(a) Correlations between restaurant properties (column similarity)

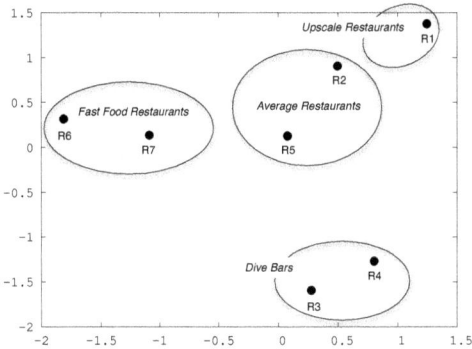

(b) Similar restaurants clustered together (row similarity)

Figure 1: Output of MCA for the example restaurant data

using only the official Google Android market and external malware datasets. We also provide measurement of our run time performance to show that MAST is practical for smaller markets and individual researchers.

Scalability and effective analysis of correlation between multiple attributes are important factors to consider when selecting such "expert" features. Finding these has been a problem in PC malware analysis as well. McBoost [45] uses n-gram analysis and Multi-Layer Perceptron on executable heuristics to detect packed code. ForeCast [41] predicts the information gain of analyzing an executable using a modified linear classifier fed with static executable features and behavioral features from dynamic analysis. BitShred [24] uses feature hashing and co-clustering to enable fast extraction of information from malware samples. However, though these techniques perform well for PC malware, they are not built to handle the nominal categorical data available from mobile applications. On the contrary, Multiple Correspondence Analysis (MCA) [1] is designed to analyze the correlation between multiple categorical attributes. Hence, we design MAST to leverage MCA and provide a lightweight triage technique to narrow down the search space of applications for resource-intensive analyses.

3. MCA BACKGROUND

Multiple Correspondence Analysis (MCA) is an analysis technique used to illuminate the relationships in a dataset with categor-

ical variables. We believe this technique to be more appropriate than generic machine learning techniques (e.g., SVM[1]) as MCA is specifically designed to deal with the categorical data that describes apps. To help the reader understand MCA and gain an intuition for why we use it, we describe an example MCA applied to restaurants and then provide a high-level description of how MCA works. A more rigorous description of MCA is provided in the appendix.

3.1 Example Analysis

Table 1 shows the results of a questionnaire given to seven hypothetical restaurants. In MCA terminology, each restaurant is an "individual", the categorical variables that describe a restaurant are "questions," the values of questions for a given individual are "answers," and the set of questions is termed a "questionnaire." These terms derive from MCA's chief application in the social sciences.

For a given dataset, MCA maps each individual and answer into a set of coordinates in "principal axes." Principal axes condense the information contained in the data sets so that the majority of the information is reflected in only a few axes. The individuals' coordinates in principal axes are scaled in magnitude to reflect how unique their answers are and placed so that the variance of individuals' positions in an axis is maximized. Answers' coordinates in principal axes are scaled and placed based on the individuals that give a particular answer. Related individuals and questions are plotted near each other in the space of principal axes.

Figure 1a and Figure 1b show plots of the answers (characteristics of the restaurants) and the first two principal axes of individuals (restaurants) resulting from the MCA analysis of the data. The left side of Figure 1a shows that low-cost restaurants often offer delivery and are family-friendly; the presence of delivery towards the outside of the plot indicates that it is an unusual feature of a restaurant, while restaurants with casual attire tend to be more common. Likewise, high-cost restaurants are less common, and are strongly correlated with formal attire in this sample (found in the top right corner of Figure 1a). From inspection of Figure 1b, restaurant R1 is a clear outlier as the most unique restaurant in the set.

Principal axes can act as indicators of "hidden variables" that better describe the collection of answers given by an individual. One common interpretation technique in an MCA analysis is to attempt to describe the principal axes of the resulting plot to determine these hidden variables. This can only be done by a human analyst, as the description can be highly subjective. In Figure 1a we have provided an interpretation of the axes. While we display restaurants and their characteristics in separate plots for clarity, both restaurants and characteristics are plotted on the same scales on the same principal axes, so an interpretation of an axis in one plot is valid for the other plot. The horizontal axis in the plots reflects *convenience*; restaurants and characteristics to the left of the vertical axis tend to be cheaper and more accessible, while those to the right of the axis tend to require more effort to enjoy. The vertical axis in the plot reflects *ambiance*; restaurants and characteristics below the horizontal axis tend to be simpler, while those above the horizontal axis tend to more sophisticated. Axis descriptions come from consideration of both restaurants and characteristics and their locations with respect to the principal axes. Thus, we finally are able to say that restaurants R1 is an "upscale restaurant," R2 and R5 are "average restaurants," R3 and R4 are "dive bars," and R6 and R7 would be best classified as "inexpensive fast-food restaurants". This classification is illustrated in Figure 1b.

An alternative to attempting to describe the axes is to compute a distance from the mean point for each individual; this approach is

---

[1]In fact, Sarma et al. [47] use SVM and are less effective at ranking malware than MAST.

used by MAST to avoid the need for semantic description. Had we computed a distance from the center for each restaurant, it would be apparent without examining the plots that R1 is the most unique.

3.2 Informal Description of MCA

This section describes the mathematics of MCA in more general terms, and it follows from and is inspired by Le Roux et al. [46]. First, we consider a set of individuals described by their answers to N questions. If there were only two questions, ($N = 2$), all individuals could be plotted in a two-dimensional plane based on values of these questions. Similarly, individuals described by three questions could be plotted as a "cloud" of points in three-dimensional space. In the case of four or more questions, similar constructions exist for higher dimensions, even if these are hard to visualize.

If one needed to compress the information of the cloud, one could project the cloud into only a few dimensions. A projection can be conceptualized as the shadow that a cloud would cast onto a lower-dimensional space. For example, a three-dimensional cloud of points illuminated from above the cloud would cast a shadow onto a plane. Similarly, that plane can be treated as cloud and projected onto a single line; conceptually, this is like looking at only the x-coordinates of points in a plane.

Given a cloud with large N, certain dimensions frequently provide more insight about the data than others. Often, questions are redundant or strongly correlated, and they only hide more interesting data. A problem in analyzing a large cloud is that the interesting dimensions may not be known *a priori*. In fact, this may be the goal of the analysis. To address this problem, MCA transforms a cloud from a nominal set of N-dimensional coordinates into principal coordinates. Principal coordinates describe a coordinate frame where the first dimension is guaranteed to show more information than the second dimension, the second to show more information than the third, and so on. These new dimensions are termed *principal axes*, and coordinates in these axes are termed *principal coordinates*. Because a majority of the information will be in the first few principal axes, only the first principal axes will need analysis.

MCA uses two insights for transforming a cloud into principal coordinates: a) scaling less-common values to be more distinct than more-common values and b) variance of the data. The process of transforming a cloud from its natural coordinates into principal coordinates (coordinates in terms of the principal axes) consists of two logical steps. In the first step, MCA scales each coordinate in each nominal axis by the probability of the coordinate value being chosen by an individual. This first step skews the shape of the cloud so that uncommon values are further from the origin than common values. The second step starts by projecting the cloud onto a single line. MCA then rotates the cloud in N-dimensions until the variance of the *projection* along the line is maximized. This line becomes the first principal axis. Next, MCA defines a new axis orthogonal to the first, and rotates the cloud in $(N - 1)$-dimensions to maximize the variance on the new axis while keeping the variance on the first axis unchanged. MCA continues this process until N principal axes are defined. Once all axes are defined, MCA describes all points in the cloud in terms of coordinates of each of the principal axes. By construction, the principal axes are in order of decreasing variance.

The result of MCA is a new set of coordinates in N principal axes. These allow for a selection of the most important principal axes to be plotted, analyzed, and interpreted.

4. METHODOLOGY

MAST uses MCA to rank applications. The resulting ranking provides an ordering of relative suspicion. Application security teams can use the MAST ranking to effectively allocate scarce resources (e.g., manual code reviews, automated static program analysis or dynamic analysis).

We develop the MAST architecture based loosely on the principles of boosting [48], a machine learning meta-algorithm. Boosting aims to improve the accuracy of any learning algorithm by creating numerous rough and moderately accurate weak classifiers and combining their results to get an accurate strong learner. Even though the weak classifiers individually are not highly accurate, the merging of their results generates an accurate "boosted" algorithm. MAST creates these rough indicators of suspiciousness by running MCA on multiple subsets of applications. For clarity, we call each subset a "poll." Polls are selected using characteristics of malicious behavior, thereby allowing MCA to identify outliers for that characteristic. Boosting weighs each weak rule depending on its performance. MAST uses a binary weight system — the polls we select have weight one, and all others have weight zero. MAST then merges the results of the individual polls to determine a total ranking that represents a relative degree of suspicion.

Figure 2 shows the MAST architecture. The first step (Section 4.1) identifies application attributes that define interesting security properties. The second step (Section 4.3) combines related sets of attributes to create MCA questions (Section 4.2). The third step runs MCA over multiple polls to generate rough indicators of suspicious behavior. The final merge step (Section 4.4) combines these rough indicators to create an accurate MAST ranking of application suspiciousness.

4.1 Attribute Identification

MAST is designed to be less costly than deep analysis techniques. Therefore, attributes must be easy to obtain from the application package (as opposed to the result of code recovery). We look at Android's package manifest and simple information about files in the package. We chose not to use market-specific metadata such as categories, user ratings, and descriptions, because we do not want to tie MAST to any specific market. That said, deployments of MAST might wish to incorporate market metadata as appropriate. We consider permissions, intent filters, native code and presence of zip files.

Permissions: Android uses permissions to restrict access to security sensitive operations (e.g., sending SMS messages, reading location from GPS, and placing emergency calls). In order to access these resources, the application's developer must declare the corresponding permission in the manifest file. Permissions are only granted at install-time and cannot be changed without upgrading the application. There are several conditions under which permissions are granted (e.g., based on package signatures). However, for the purposes of this paper, the reader may assume all permissions are granted if the user elects to install the application. Interested readers are referred to prior discussions of Android security for additional details [11, 17]. Finally, third-party application developers can define new permissions. MAST currently uses the 114 permissions defined by the Android framework; however, custom permissions can be added if needed.

Intent Filters: Android uses intent messages for interprocess communication (IPC). Intents provide interfaces to core platform functionality as well as interactions between third-party applications. Frequently, intents are addressed to "action strings." Depending on the type of intent, the Android middleware uses action strings to notify applications of the event or resolve the best application for the task. An application indicates in its manifest file the ability to handle an intent by specifying an *intent filter* for a pre-agreed upon

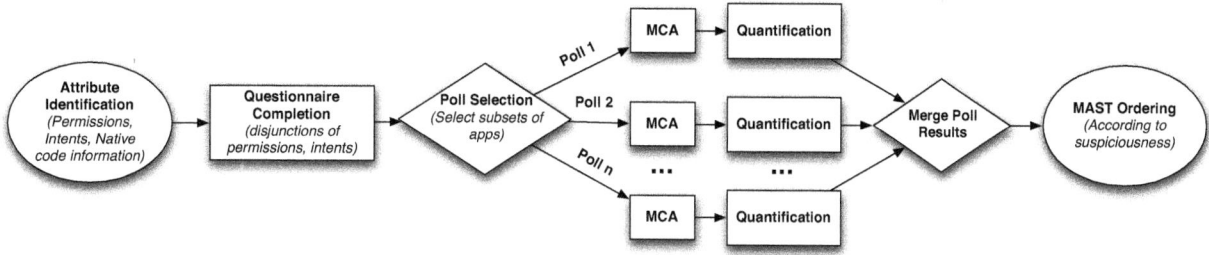

Figure 2: The MAST architecture: Building an analysis architecture using MCA

action string.[2] Intent filters provide a pluggable architecture that allows OEMs and third-party developers to customize the user experience. However, applications can also abuse this extensibility by handling events and tasks in ways that harm the user. MAST currently uses the 92 action strings defined by the Android framework. Similar to permissions, this knowledge base can be trivially extended. Finally, intent filters have an optional *priority* field that influences the order in which Android delivers intents to applications. By defining a high priority intent filter, an application may be able to cancel the intent before other applications receive it. We classify an application's intent filter for each action string as *priority*, *default*, or *none*. For the purposes of our evaluation, we consider an intent filter to be *priority* if it specifies the priority field.

Native Code: Android applications are primarily written in Java; however, the native development toolkit (NDK) allows third-party application developers to include native libraries. While native libraries provide valuable performance benefits for computationally bound applications (e.g., games), they also have been used by malware to exploit root vulnerabilities (e.g., DroidDream). We classify an application based on whether or not it includes native libraries. To identify native libraries, we search the .apk archive for files with the native library magic number using libmagic. Note that we make no attempt to identify downloaded libraries. Doing so statically would require program analysis, which is computationally too expensive for triage.

Zip Files: Android applications are distributed as .apk archive files. Archiving reduces the amount of data that needs to be downloaded when installing an application. However, as no restrictions are put on the type of data these application archives can contain, they have been used to carry malicious payloads as zip files (e.g., BaseBridge carries an entire malicious application as its payload). As a final attribute, we classify an application based on the presence or absence of zip files inside the main application archive. We do not recursively analyze the contents of the zip file.

4.2 MCA Questionnaire

MAST carefully defines an MCA questionnaire to aid security triage. We identified many different attributes: 114 permissions, 92 intent types (with and without priority), the existence of native code, and the presence of zip files. Creating a question for each attribute produces a very high dimensional categorical data set with relatively limited interrelationships. In practice, this results in the "being unique is common" phenomenon.

We considered two methods of combining attributes: conjunctive questions and disjunctive questions. A conjunctive question is true if an app has *all* attributes in a set. Similarly, a disjunctive question is true if an app has *at least one* of the attributes in a set.

Enck et al. [16] define nine conjunctive questions for their Kirin system. These questions (called rules) are the results of a mal-

---

[2]Dynamic broadcast receivers that define intent filters at runtime are excluded as they require costly program analysis to retrieve.

ware oriented security requirements engineering of the Android platform. As we show in Section 5, these rules do not perform as well as the disjunctive questions used in MAST.

MAST uses disjunctive questions to collapse attributes into more general descriptions of functionality. For example, when identifying malware, it is often sufficient to know that an application has permission to perform an SMS related operation as opposed to the specific types of SMS operations. From the MCA perspective, a disjunctive question increases the likelihood that two applications have a property in common, and therefore clusters applications based on their functionality.

The questionnaire consists of permission questions, intent filter questions and native code and zip file questions. The questions group attributes by functionality. Permission, native code, and zip file questions are true or false questions. For example, the answer to the "SMS" question is true if an application requests at least one of the SMS-related permissions. In contrast, the intent filter questions have possible answers of "priority," "default," and "none." An answer of "priority" indicates that the application has at least one intent filter for a listed action string that defines the priority field. If the application does not have any matching priority intent filters, but it does have an intent filter for one of the listed action strings, the answer is "default." Otherwise, the answer is "none."

Table 5 in the Appendix shows our complete MCA Questionnaire. Four questions are not listed in Table 5: a generic permission question, a generic intent filter question, the "contains native code" question, and the "contains zip files" question. A generic question simply groups unrelated attributes that we empirically found to be less important. To avoid the additional "noise" resulting from creating an additional question for each attribute, we grouped them into a single disjunctive question. Therefore, applications having one of these attributes will have something in common. The generic permission and intent filter questions simply contain all of the remaining permissions and action strings, respectively.

4.3 Poll Selection and Quantification

MAST runs MCA on multiple polls to create rough indicators of malicious behavior. Each poll asks the MCA questionnaire to a specific subset of applications. Polls fill two purposes: 1) they group related applications together so that uniqueness within the group is more specific, and 2) they allow MAST to select which characteristics are most effective in identifying current malware trends.

The polls used by MAST should reflect current malware trends, and therefore could change as malware evolves. Furthermore, poll selection characteristics are not necessarily the same as the characteristics used for MCA questions. That said, all of our MCA questions are potential poll characteristics. However, we select only those polls that reflect a certain inclination to malicious behavior. The MAST polls selected by our training process described in Section 5.3 are shown in Table 2.

Note that the application subsets defined by polls are intentionally not disjoint—an application that has both SMS and PACKAGE

Table 2: Characteristics used to define polls. Each poll defines a subset of applications directed to MCA.

Characteristic	Possible Malicious Behavior
SMS perms	SMS trojans, SMS spam
PACKAGE perms	Installing malicious apps, Uninstalling antivirus apps
BOOT intent-filter (default)	Spyware apps that want to autorun on startup
BOOT intent-filter (priority)	Spyware apps that want to autorun before anti-malware apps are started
Native code	Native exploits (rage-against-the-cage)
Zip files	Zipped payload containing malicious apps
COMM. intent-filter (default)	Applications that use incoming calls or SMS as activation triggers
COMM. intent-filter (priority)	Applications that hide incoming calls or SMS from users

permissions will be analyzed in two polls. When MAST merges the MCA results for the selected polls, an application that exhibits multiple malware-requisite characteristics will thus stand out further in the MAST ranking.

Finally, MAST quantifies the MCA results for each poll. Section 3 demonstrated how MCA can be used as a visualization tool. However, MCA can also be used to quantify the deviation of an application from the norm. We quantify a ranking for each MCA by calculating the χ^2 distance of each application from the barycenter (point of commonality) of all apps in the analysis. This quantification provides a *poll score* for each analyzed application.

4.4 Merging Poll Results

MAST merges the individual poll scores to create an accurate MAST ranking. The merge must ensure: 1) results for one poll do not overshadow another; 2) the number of apps in a poll is considered; and 3) the number of polls an application participates in influences the total ordering. We now describe this merge process.

Normalization: Once each poll is ranked, MAST normalizes the poll scores by scaling each poll i from domain $[min_i, max_i]$ to domain $[0, 1]$, where $min_i \geq 0$.

Poll Size Scaling: To account for the number of applications in a poll, MAST scales the normalized poll scores such that the smaller the set of apps in a poll, the larger the contribution of the outlier applications. For each poll, MAST scales the normalized poll score using the subset of applications $A_i \subseteq A$ present in the poll:

$$[0, 1] \rightarrow \left[\left(1 - \frac{|A_i|}{|A|} \right), 1 \right]$$

Combining and Sorting Results: Finally, to determine the overall ranking, MAST calculates a *MAST score* for each application by summing its normalized and scaled poll scores. The greater the number of subsets an application appears in, the larger its MAST score, which reflects a higher degree of relative suspicion. Applications are then sorted by MAST score, producing the *MAST ranking*.

4.5 MAST Ranking

The final output of MAST is a list of applications ranked in order of their dissimilarity to the population of other applications. Due to poll selection, this list indicates a relative degree of suspicion. Applications ranked higher in the list are more likely to be malicious than those appearing at the end. Thus, when performing triage, investigation should start with the highest MAST ranking.

Note that it may then make sense to classify apps as being in a "percentile of suspicion;" that is, if an app is within the top 1% of apps in the MAST ranking, it could be said that that application is within the top 1% of suspicious apps. An app can be defined as suspicious if it exhibits behaviors common to malicious apps. We caution that MAST is not Bayesian — MAST does not specify or even imply the actual likelihood of maliciousness of any application, even given a percentile of suspicion. It is certainly *not* the case that an application in the first percentile of suspicion (1%) is 99% likely to be malicious.

4.6 MAST Implementation

Our implementation of MAST consists of two main steps: obtaining metainformation from applications and processing that metainformation. A Python program unzips every application package, checks all files for zip archives and native code, decompresses the `AndroidManifest.xml` file, computes the questionnaire results, and writes that information to a "MAST table". Table computation is the major bottleneck, but is fortunately trivially parallelizable (though our implementation processes apps serially). Once the questionnaire is complete, the table is read by a program written in R, which is a popular language for statistical analysis that also has an MCA library. The R program parses the MAST table, selects apps based on polls, runs an MCA of every poll, quantifies the poll results, then merges them to produce a final MAST ranking.

5. TRAINING MAST

MAST merges the MCA poll results that roughly indicate maliciousness to magnify the malicious characteristics of malware. These polls are directed at applications that exhibit specific characteristics found in malware (but also benign applications). Choosing which polls MAST should use requires careful consideration. From a high level, we want to choose polls that cover the characteristics of known malware, but we do not want to over-train MAST for any specific malware type. At the same time, we want to select polls for characteristics that are more common in malware than benign applications. In this section, we describe our process of selecting polls for MAST (previously discussed in Table 2).

5.1 Training Data

In order to train MAST, we created a simulated market. This market consists of 14,888 popular free applications from Google Play, 141 samples of known malware from the Contagio mobile malware repository [43] as of October 2011 and 591 malware samples found by Zhou et al. [53]. Here, we assume the Android Market applications are mostly benign, but keep in mind the potential for malicious and questionable applications when training MAST.

Google Android Market: Properly training MAST requires a very large set of benign applications. Manually downloading these applications on a phone is not an option, so we developed a standalone snapshot tool using the unofficial Android Market API [2]. Our tool downloaded the majority of the top 500 free applications within each of the 34 market categories. By selecting most popular applications, we maximize our chances of selecting benign applications. Our dataset, taken on January 20, 2012, included applications for the "T-Mobile" carrier, the "passion" device (Nexus One), and Android API level 8 (Froyo) and below. Due to failed downloads and categories that had fewer applications, our snapshot contains a total of 14,888 applications. This dataset represents approximately 2.1% of the estimated over 700,000 applications in Google Play [52]. Results of real-world triage in Section 6 show that even 2.1% of the market is sufficient to accurately train MAST.

Combined malware training set: Our combined malware set of 732 applications includes malware samples that steal personal data and receive commands from an attacker's command and control server, send SMS to premium numbers, place calls to premium numbers, and/or otherwise spy on the user (including tracking the user's location). Some malware samples display more than one of these malicious behaviors. Android malware authors often embed the same exploit in multiple applications, which leads to the existence of malware families. In addition to malware, the malware set contains several examples of grayware. Two grayware examples include applications designed to spy on the SMS or GPS activity of one's spouse without his/her consent, and applications designed to gain root access for the user. Because there are no guarantees that these samples do not abuse their abilities for malicious purposes, and because MAST is focused on triage for potentially malicious behavior, we include the grayware in our training data set. Table 6 in the appendix presents the composition of the 732 samples in our combined malware set.

5.2 Evaluation Metrics

MAST ranks a given set of applications according to their relative suspiciousness. This ordering indicates where malware researchers should first allocate their resources. To measure the effectiveness of MAST, we use receiver operating characteristic (ROC) curves, which provide a plot of true positive (TP) and false positive (FP) rates with regard to a threshold parameter. In the case of MAST, the threshold parameter is the top percentage of apps in the MAST ranking that are scanned. The positives in the ROC curve are the apps that are chosen to be scanned, with the true positive being the malicious ones and false positives being the non-malicious ones.

When evaluating the performance of MAST, we compare three triage techniques.

- *No Triage*: When no intelligent approach can be applied, triage is equivalent to random guessing. We model the "no triage" scenario by assuming the malicious applications are uniformly distributed within a random ordering of the applications. The corresponding ROC curve is the line of no discrimination and has a slope of one.

- *Kirin-based Triage*: Kirin [16] defines possible malicious behaviors using nine permission-based conjunctive rules. Two rules are concerned with detecting SMS malware. Used as a triage technique, malware researchers apply Kirin to create two sets of applications: high priority and low priority. The high priority set consists of applications that fail any of the Kirin rules. We model Kirin-based triage by assuming the malware binned in each set is uniformly distributed in a random order. The corresponding ROC curve consists of two slopes, indicating the division between the high and low priority sets.

- *MAST-based Triage*: MAST provides an ordering for triage. We use the MCA questionnaire defined in Section 4.

5.3 Selecting Polls

As discussed in Section 4.3, a poll is essentially a subset of applications that have similar functionality. The purpose of MAST training is to select a combination of polls that individually act as rough indicators of suspicious behavior and that improve the overall accuracy of the MAST ranking. As MAST questions are directed towards functionality, each question and answer pair defines a MAST poll. However, we narrowed down the set of possible polls to those that contained at least 10% of the malware in our training set. On the 15 polls that showed a weak correspondence to malware, we ran a brute-force analysis to determine the optimal

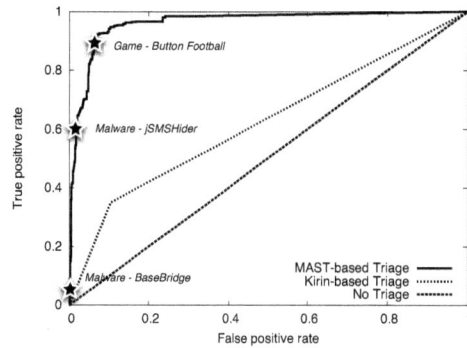

Figure 7: Training data ROC curve (732 malapps in 15,620 apps)

combination of polls. Note that we did not need to run MAST completely for each experiment, but only the merging of the individual poll results. Table 2 presents our final poll selection.

An interesting observation from our poll selection process is that all the polls selected in our optimal combination show good individual poll performance. Figure 3a shows that the SMS permission poll alone ranks nearly 70% of the malware samples from our combined malware training dataset with a false positive rate of 5%. The package permission poll only includes approximately 30% of the malware; however, it ranks that malware with a false positive rate of 0.8%. However, as all polls with good individual performance do not improve overall MAST performance, we can only use this observation to further reduce the training search space of future experiments by discarding polls with poor individual performance (Figure 3c, 3d). Discarding polls with poor performance does not affect the ability of MAST to detect new kinds of malware, as long as we use a sufficient number of polls that are indicative of malicious functionality. We validate the optimality of our poll count by observing that MAST results tend to peak at poll sets having sizes between 8 and 10 and drop as the number of polls is decreased or increased.

5.4 Combining Polls

We previously claimed that: a) MCA classifies malware as outliers and, b) combining multiple polls reduces analysis effort. We now prove these claims by tracking three applications as they are processed by MAST: two malware samples (BaseBridge and jSMS-Hider) and one randomly selected app, Button Football (a soccer game app), that appeared in two polls. In these plots, the tracked applications are highlighted with a black star.

Figure 4 tracks the BaseBridge malware sample in four polls: SMS permissions, BOOT intent-filter (priority) poll, zip file poll, and COMMUNICATION intent-filter (priority) poll. The crosshair in the plot denotes the barycenter (i.e., point of commonality). The BaseBridge sample is a clear outlier in each of the SMS, BOOT-priority, zip file, and COMMUNICATION-priority polls. Being a distinct outlier, the BaseBridge application is ranked 0.25%.

Figure 5 tracks the jSMSHider malware sample in two polls: PACKAGE permissions and zip file filter. The sample is outside the main cluster in the PACKAGE permission poll, but not far from the barycenter in the zip file poll. In this case, combining the two polls has a clear advantage for ranking the malware sample.

Figure 6 tracks Button Football, a soccer game application in the two individual polls matching its characteristics: BOOT intent poll and Native code poll. Here, Button Football is grouped in the main clusters, contributing to its relatively lower MAST rank.

Finally, Figure 7 shows the ROC curve after combining our selected polls. MAST ranks BaseBridge and jSMSHider malware apps very high (0.25% and 4.4%). These results demonstrate how

19

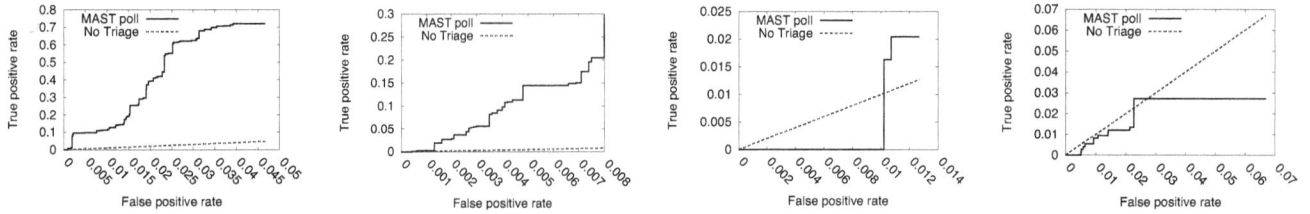

(a) SMS Permission Poll   (b) PACKAGE Permission Poll   (c) CALENDAR Perm. Poll   (d) ACCOUNT Permission Poll

Figure 3: Isolated Poll Performance. CALENDAR and ACCOUNT were not selected due to poor performance.

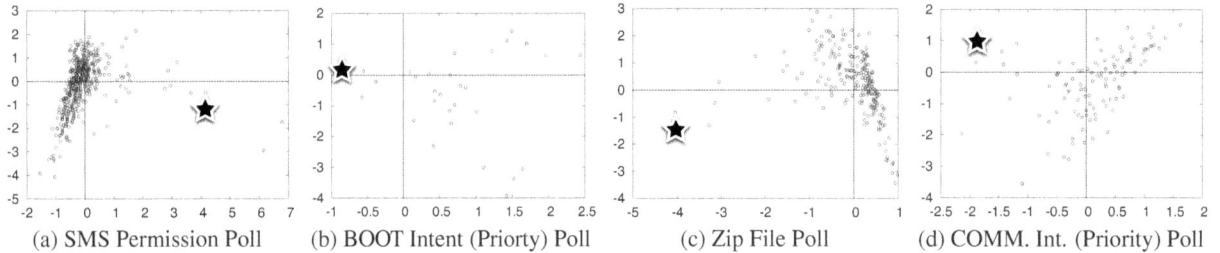

(a) SMS Permission Poll   (b) BOOT Intent (Priorty) Poll   (c) Zip File Poll   (d) COMM. Int. (Priority) Poll

Figure 4: Tracking BaseBridge malware (MAST rank: 0.25%)

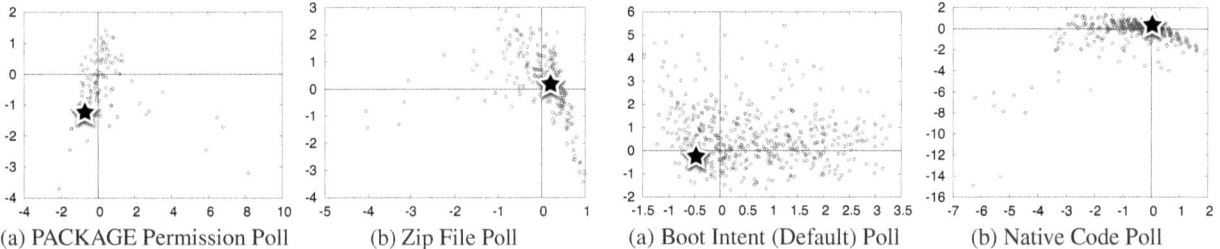

(a) PACKAGE Permission Poll   (b) Zip File Poll

Figure 5: Tracking jSMSHider malapp (MAST rank: 4.4%)

(a) Boot Intent (Default) Poll   (b) Native Code Poll

Figure 6: Tracking Button Football (MAST rank: 9.7%)

the multiple polls aid in giving malware higher rankings. Button Football has a lower ranking (9.7%), but is still within the top 10% of applications to receive analysis. However, we note that Button Football's inclusion in the BOOT intent-filter default poll is interesting for a game application. In the spirit of diagnosis after triage, we analyzed the Button Football app to find that it uses AirPush, an aggressive notification ad library. AirPush goes beyond in-app ads by generating notifications that persuade users to download apps. Further, it starts itself at boot, so the user can see ads even when he is not using any application. The higher MAST ranking of Button Soccer highlights that, from the point of view of MAST, the distinction between undesirable and malicious behavior is thin.

Comparing the ROC curves for MAST-based triage, Kirin-based triage, and no triage, the advantage of MAST is clear. MAST gathers 90% of the malware samples while suffering a false positive rate of just 6.5%. Furthermore, it collects 95% of the malware samples at a false positive rate of just 11.2%. This result merely verifies that our training process is sound. Section 6 provides a real-world evaluation of MAST.

6. REAL MARKET TRIAGE

We now measure the effectiveness of MAST on real third-party markets, *which were not used for training.*

6.1 Experimental Setup

We selected three application markets to evaluate MAST: Ndoo, Anzhi, and Softandroid. All three make their apps available free of charge online, therefore we were able to download all apps in each market. *All markets were found to host some amount of malware.*

To validate MAST's ability to highly rank malware, we need

a perception of which apps in a market are malicious. We stress that this knowledge is not required for MAST to rank the applications in the market. We use two tools to identify malicious apps: Androguard and Virus Total. Androguard is a well-known, open-source project with volunteer-submitted definitions and provides a lightweight signature-based malware detection tool. Our analysis is based on the signature definitions from February 5, 2012. Virus Total is an online service that scans submitted files with tens (often more than 40) of up-to-date commercial antivirus products and provides the results from each AV product. Our markets were last scanned by Virus Total on February 11, 2012.

Androguard has a low true positive rate as its malware signatures are updated very slowly. On the other hand, some AV products used in Virus Total have a high false positive rate. To overcome these limitations, we use a hybrid approach that tags applications as malware if either Androguard or at least three out of the "Top 5" AV products in Virus Total tag the application as malware. The "Top 5" AV products (GData, Avast, Kaspersky, BitDefender, and FSecure) were chosen based on their accuracy in the detection of known malware samples. The distribution of the malware we discovered in these markets is presented in Table 6 in the appendix.

Ndoo: The Ndoo market is a Chinese app market. On October 25, 2011, the market contained 4,324 apps. Of these apps, 26 were considered malicious by Androguard or Virus Total. At the time of data collection, Ndoo claimed that its "comprehensive software testing & certification procedures contribute to the reputation of [the] developer." No further details were provided on its English website about its security evaluation procedures.

Anzhi: Like Ndoo, Anzhi is a market catering to Chinese Android users. Anzhi contains far more apps though: on January 31, 2012,

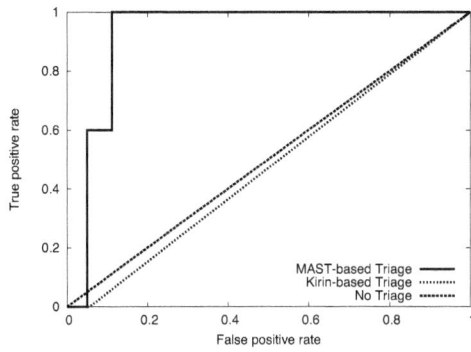

Figure 8: Softandroid ROC curve (5 malapps in 3,626 apps)

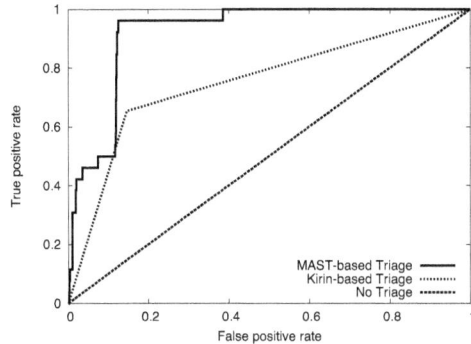

Figure 9: Ndoo ROC curve (26 malapps in 4,324 apps)

the market contained 28,760 apps. 166 of these were considered malicious by Androguard or Virus Total (six of these were later found to be false positives).

Softandroid: Softandroid is a Russian application market hosting 3,626 apps on February 7, 2012. VirusTotal or Androguard marked six of these as malicious (one was later found to be a false positive).

6.2 MAST Results

Figures 8, 9, and 10 show ROC curves for the Softandroid, Ndoo, and Anzhi markets, respectively.

As the SoftAndroid market contained only five instances of malware, its ROC curve is the easiest to analyze. Three of the apps marked as malicious are within the top 5% of the MAST rankings. The remaining two apps are within the top 11.2% of the MAST rankings. For this market, MAST is able to highly rank all malicious applications, even when no malicious applications violate a Kirin rule. This validates the choice of disjunctive polls.

The Ndoo market contained 26 malicious apps. MAST ranks 25 of these – 96.1% – in the top 12.9% of the MAST rankings. Half of these malicious apps are ranked in the top 7.5%, and the highest-ranked malicious app is the 8th app in the rankings. While Kirin ranks a majority of the malicious apps in this market, MAST catches more and orders them earlier than Kirin.

The Anzhi market contained 160 malicious apps. 60% of these were ranked in the first 2.1% of the market. 85% of the malicious apps are located in the top 10% of the MAST rankings, and 95.5% of the malicious apps are located in the top 25% of the rankings. As with Softandroid, Kirin rules would have failed to rank the majority of malicious apps in the Anzhi market.

We compare MAST's ROC curves to those of Peng et al.'s best technique, HMNB [44], using their area under the curve (AUC) figures. The AUC of a ROC curve gives an overall indication of quality of the curve. Peng et al. [44] cite an AUC of 0.9281 when testing against market data that does not overlap with their testing data, while the AUC for Anzhi, SoftAndroid, and Ndoo are 0.9362,

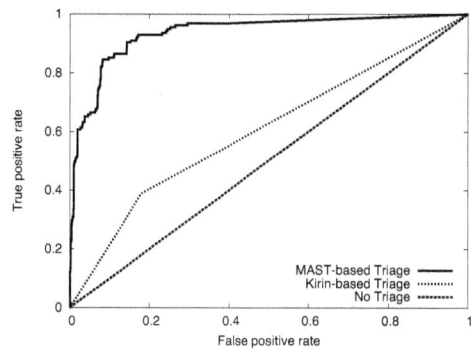

Figure 10: Anzhi ROC curve (160 malapps in 28,760 apps)

0.9252, and 0.9207 respectively. (None of these markets have overlapping apps, according to the SHA-256 of their packages). These results show that MAST provides comparable or better malware triage than HMNB would. We emphasize that Peng et al. [44] test on two snapshots of the Android market with malware from an external dataset added, while we test on real markets with malware actually present in the market.

6.3 Performance Analysis

Because the purpose of a triage is to act as a fast first step to optimize a later heavy analysis, we compare MAST and other analysis tools. All tests were run in a VirtualBox VM configured with a single CPU and 2GB of memory with a 10GB virtual hard disk running Ubuntu Linux 10.04 (Windows XP SP3 in the case of the TrendMicro test). Market files were stored on the host's 16TB 12-drive RAID 6 array and accessed using VirtualBox shared folders[3]. The host machine has 2 Intel E5620 Xeon quad-core processors and 48 GB RAM and runs Debian Squeeze.

Table 3 provides a comparison of the time to run MAST compared to heavier, more comprehensive methods. We note that MAST is meant to complement these methods, not replace them. The comparison is made only to prove that MAST is light-weight enough that the cost of using MAST is amortized by more efficient analysis of apps. First, MAST performance numbers are provided for two markets: Ndoo and Android. The runtime of MAST is overwhelmingly bound by collecting attributes from each app; performing the MCA for all polls requires less than 1% of the total time presented in the table. MAST is comparable in time to Baksmali [5], a popular open-source disassembler for Dalvik bytecode. Disassembly is a first minimum step for manual analysis, so it represents a reasonable lower bound on manual analysis. Actual manual analysis would of course take minutes or hours per application. MAST is 4.3 times faster than the command line version of TrendMicro Antivirus [50] used to perform signature detection over the Ndoo market. On a selection of 23 randomly selected apps from the Google market, ded [15] (a decompiler for Android applications) runs four orders of magnitude slower than MAST.

For compute-intensive analyses of markets, MAST increases the likelihood of identifying malicious apps in a reasonable amount of time. Moreover *the cost of MAST is so low compared to techniques such as ded that the overhead of running MAST is negligible.*

6.4 Post-triage Analysis

After having successfully performed triage, we made two interesting observations by performing deeper analysis across suspicious applications from each of the markets.

---

[3]Virtual folders were found by microbenchmarks to be faster than accessing from the virtual hard disk.

Table 3: Analysis Technique Timing Comparison

Tool	Input	Total Time	Time/App
Baksmali	Ndoo Market	24.63m	0.34s
MAST	Ndoo Market	32.18m	0.45s
MAST	Android Market	111.5m	0.45s
TrendMicro AV	Ndoo Market	140.7m	1.95s
ded	23 Random Apps	487.7m	1272.29s

Table 4: Rampant use of the default key from the Android codebase to sign applications across markets

Market	Number of applications signed using the default key
Google Play	11
Contagio + Malware	156
Softandroid	245
Ndoo	510
Anzhi	750

False positives in commercial antivirus tools: After the initial evaluation of MAST, we found five applications in our malware training set, six in Anzhi, and one in SoftAndroid that were classified as SMS trojans (HippoSMS and RogueSPPush) by antivirus tools, yet were ranked poorly by MAST. Upon inspection, none actually had SMS permissions, meaning that it could not exercise any malicious code that might be present. In effect, these are false positives.[4]. To verify this, we selected one of these apps (classified as RogueSPPush) for further manual analysis. We found that this app did have the malicious RogueSPPush code, but was effectively "dead" as it did not declare the SMS intent filter that executes that code. We did not find evidence of a root exploit or any other method to violate Android's permission system. This finding highlights that just as MAST highly ranks apps that could exhibit malicious behavior, it ranks apps without this explicit ability lower.

Default Android key used to sign applications: Android requires all applications to be signed to ensure that an application can be upgraded only by the developer who created it. Thus, there is no need for a certificate authority or PKI — self-signed certificates suffice, as long as the security of the private key is maintained by the developer. However, we found that 1,672 applications across the markets (distribution in Table 4) used the default key present in the Android codebase to sign their applications.

In terms of security, using the default private key is equivalent to posting your private key in a public forum. Any malicious author can update the applications signed with the default key and replace them with malicious code. The fact that we found these applications in the official Google market indicates the need to check for use of the default key when applications are submitted to the market. Given the use of this key in applications in the alternative markets, we highly recommend that the package installer in production Android phones blacklists the default key.

We also found a significant number of applications across markets signed with the same private key as malicious applications. However, we leave further analysis of those results as future work.

6.5 MAST ranking limitations

Triage in principle is a "best effort approach". With MAST, we aim to increase the likelihood of finding malware in the lowest percentiles of suspicion, but our results can only be as good as the polls we conduct. Our current efforts attempt to select classes of attributes (e.g., permissions, intents, etc.) that are absolutely necessary for malware attempting to perform a specific task to declare (e.g., SMS trojans must ask for SMS permissions, or their malicious code simply will not run). However, if new classes of malware attempt to abuse other protected or unregulated [36] inter-

---
[4]We exclude these apps from malware counts throughout the paper.

faces, MAST is unlikely to rank them highly. This is analogous to a medical triage case where a patient has no symptoms of illness, but is infected with an unknown disease. These scenarios can be easily addressed in MAST as soon as new "zero-day classes" are discovered. Specifically, new polls can be added to the infrastructure, and new malicious applications can just as quickly be flagged.

7. CONCLUSION

Application markets simplify the distribution of consumer software. The benefits have not been lost on malware authors: application markets are the primary means of distributing smartphone malware. Preventing malware in markets is extremely difficult. Market maintainers simply do not have the computational or personnel resources to thoroughly or deeply inspect the large number of applications submitted each day. To address this challenge, we propose MAST. The goal of MAST is to direct available analysis resources to the most suspicious applications. To do this, MAST uses Multiple Correspondence Analysis (MCA) to measure the correlation between declared indicators of functionality required to be present in application packages. We described how to parameterize MAST using current malware trends and then demonstrated its value by using it to successfully perform triage on three third-party markets.

The concepts underlying MAST transcend malware discovery in two ways. First, we show that MCA, a tool primarily used in the social sciences, has tremendous potential for security, specifically when the adversary must declare (i.e., commit to) functional specifications. The key contribution of MCA is its ability to establish relationships between otherwise incomparable information. Second, we believe that security triage tools that quantify perception are essential to protect systems from intelligent adversaries. Triage is neither diagnosis nor treatment. Rather, such tools make security analysis more methodological, and less reliant on a "gut feeling."

8. ACKNOWLEDGEMENTS

The authors gratefully acknowledge the support of VirusTotal for this work. This work was supported in part by the US National Science Foundation under grant numbers DGE-1148903, CNS-0916047, CNS-0952959, and TWC-1222699. Any opinions, findings, and conclusions or recommendations expressed in this material are those of the authors and do not necessarily reflect the views of the National Science Foundation.

9. REFERENCES

[1] H. Abdi and D. Valentin. Multiple correspondence analysis. In *Encyclopedia of Measurement and Statistics*, page 13. Sage, California, 2007.
[2] Android market API. http://code.google.com/p/android-market-api/.
[3] Anzhi Market. http://www.anzhi.com.
[4] Apple app store, 2012. http://www.apple.com/iphone/from-the-app-store/.
[5] Baksmali, 2012. http://code.google.com/p/smali/.
[6] D. Barrera, H. G. Kayacik, P. C. van Oorschot, and A. Somayaji. A methodology for empirical analysis of permission-based security models and its application to android. In *Proceedings of the 17th ACM conference on Computer and communications security*, page 73. ACM Press, 2010.
[7] C. Beaumont. Apple iPhone 'kill switch' discovered, August 2008. http://www.telegraph.co.uk/technology/3358115/Apple-iPhone-kill-switch-discovered.html.
[8] Blackberry app world, 2012. http://appworld.blackberry.com/webstore/.
[9] A. Bose, X. Hu, K. G. Shin, and T. Park. Behavioral detection of malware on mobile handsets. In *Proceeding of the 6th international conference on Mobile systems, applications, and services*, page 225. ACM Press, 2008.
[10] T. Bray. Exercising Our Remote Application Removal Feature, June 2010. http://android-developers.blogspot.com/2010/06/exercising-our-remote-application.html.
[11] J. Burns. Developing Secure Mobile Applications for Android. iSEC Partners, Oct. 2008. http://www.isecpartners.com/files/iSEC_Securing_Android_Apps.pdf.

[12] E. Chin, A. P. Felt, K. Greenwood, and D. Wagner. Analyzing inter-application communication in android. In *Proceedings of the 9th international conference on Mobile systems, applications, and services*, MobiSys '11, pages 239–252, New York, NY, USA, 2011. ACM.

[13] M. Egele, C. Kruegel, E. Kirda, and G. Vigna. PiOS: Detecting Privacy Leaks in iOS Applications. In *Proceedings of the ISOC Network & Distributed System Security Symposium (NDSS)*, 2011.

[14] W. Enck, P. Gilbert, B. Chun, L. P. Cox, J. Jung, P. McDaniel, and A. Sheth. TaintDroid: An information-flow tracking system for realtime privacy monitoring on smartphones. In *OSDI*, pages 393–407, 2010.

[15] W. Enck, D. Octeau, P. McDaniel, and S. Chaudhuri. A study of Android application security. In *Proceedings of the 20th USENIX Security Symposium*, San Francisco, CA, USA, 2011.

[16] W. Enck, M. Ongtang, and P. McDaniel. On lightweight mobile phone application certification. In *Proceedings of the 16th ACM conference on Computer and communications security*, page 235. ACM Press, 2009.

[17] W. Enck, M. Ongtang, and P. McDaniel. Understanding Android Security. *IEEE Security & Privacy Magazine*, 7(1):50–57, January/February 2009.

[18] A. P. Felt, E. Chin, S. Hanna, D. Song, and D. Wagner. Android permissions demystified. In *Proceedings of the 18th ACM conference on Computer and communications security*, CCS '11, Chicago, Illinois, USA, Oct. 2011.

[19] A. P. Felt, M. Finifter, E. Chin, S. Hanna, and D. Wagner. A survey of mobile malware in the wild. In *ACM Workshop on Security and Privacy in Mobile Devices*, Chicago, Illinois, USA, Oct. 2011.

[20] GFan Market. http://www.gfan.com/.

[21] Google play, 2012. https://play.google.com/store/apps.

[22] M. Grace, Y. Zhou, Q. Zhang, S. Zou, and X. Jiang. RiskRanker: Scalable and Accurate Zero-day Android Malware Detection. In *Proceedings of the International Conference on Mobile Systems, Applications, and Services (MobiSys)*, June 2012.

[23] P. Hornyack, S. Han, J. Jung, S. Schechter, and D. Wetherall. These Aren't the Droids You're Looking For: Retrofitting Android to Protect Data from Imperious Applications. In *Proceedings of the ACM Conference on Computer and Communications Security (CCS)*, 2011.

[24] J. Jang, D. Brumley, and S. Venkataraman. Bitshred: feature hashing malware for scalable triage and semantic analysis. In *Proceedings of the 18th ACM conference on Computer and communications security*, CCS '11, pages 309–320, New York, NY, USA, 2011. ACM.

[25] X. Jiang. Questionable Android Apps – SndApps – Found and Removed from Official Android Market, July 2011. http://www.csc.ncsu.edu/faculty/jiang/SndApps/.

[26] X. Jiang. Security Alert: New Android SMS Trojan – YZHCSMS – Found in Official Android Market and Alternative Markets, June 2011. http://www.csc.ncsu.edu/faculty/jiang/YZHCSMS/.

[27] X. Jiang. Security Alert: New Stealthy Android Spyware – Plankton – Found in Official Android Market, June 2011. http://www.csc.ncsu.edu/faculty/jiang/Plankton/.

[28] H. Kim, J. Smith, and K. G. Shin. Detecting energy-greedy anomalies and mobile malware variants. In *Proceeding of the 6th international conference on Mobile systems, applications, and services*, page 239. ACM Press, 2008.

[29] A. Kingsley-Hughes. So that's what happens when you highlight an iOS security hole, November 2011. http://www.zdnet.com/blog/hardware/so-thats-what-happens-when-you-highlight-an-ios-security-hole/16078.

[30] L. Liu, G. Yan, X. Zhang, and S. Chen. VirusMeter: preventing your cellphone from spies. In *Recent Advances in Intrusion Detection*, volume 5758, pages 244–264, Berlin, Heidelberg, 2009. Springer Berlin Heidelberg.

[31] H. Lockheimer. Android and Security. Google Mobile Blog, Feb. 2012. http://googlemobile.blogspot.com/2012/02/android-and-security.html.

[32] Lookout Mobile Security. Mobile threat report. Technical report, Lookout Mobile Security, Aug. 2011.

[33] Lookout mobile security, 2012. https://www.mylookout.com/.

[34] J. Lowensohn. iPhone lock-screen password app pulled, June 2011. http://news.cnet.com/8301-27076_3-20071405-248/iphone-lock-screen-password-app-pulled/.

[35] K. Mahaffey. Security Alert: DroidDream Malware Found in Official Android Market, March 2011. http://blog.mylookout.com/2011/03/security-alert-malware-found-in-official-android-market-droiddream/.

[36] P. Marquardt, A. Verma, H. Carter, and P. Traynor. (sp)iPhone: Decoding Vibrations From Nearby Keyboards Using Mobile Phone Accelerometers. In *Proceedings of the ACM Conference on Computer and Communications Security (CCS)*, 2011.

[37] P. McDaniel and W. Enck. Not so great expectations: Why application markets haven't failed security. *IEEE Security & Privacy*, 8(5):76–78, Oct. 2010.

[38] Min Zheng, Patrick P.C. Lee, and John C.S. Lui. ADAM: an automatic and extensible platform to stress test android anti-virus systems. In *Proceedings of the 9th Conference on Detection of Intrusions and Malware & Vulnerability Assessment (DIMVA'12)*, Heraklion, Crete, Greece, July 2012.

[39] Ndoo market. http://www.nduoa.com/.

[40] NetQin Mobile Security, 2012. http://www.netqin.com/en/.

[41] M. Neugschwandtner, P. M. Comparetti, G. Jacob, and C. Kruegel. Forecast: skimming off the malware cream. In *Proceedings of the 27th Annual Computer Security Applications Conference*, ACSAC '11, pages 11–20, New York, NY, USA, 2011. ACM.

[42] Nicholas J. Percoco and Sean Schulte. Adventures in BouncerLand. In *Blackhat USA*, Las Vegas, NV, 2012.

[43] M. Parkour. Contagio mobile malware MiniDump. http://contagiominidump.blogspot.com/.

[44] H. Peng, C. Gates, B. Sarma, N. Li, Y. Qi, R. Potharaju, C. Nita-Rotaru, and I. Molloy. Using probabilistic generative models for ranking risks of android apps. In *Proceedings of the 2012 ACM conference on Computer and communications security*, CCS '12, page 241–252, New York, NY, USA, 2012. ACM.

[45] R. Perdisci, A. Lanzi, and W. Lee. Mcboost: Boosting scalability in malware collection and analysis using statistical classification of executables. In *Proceedings of the 2008 Annual Computer Security Applications Conference*, ACSAC '08, pages 301–310, Washington, DC, USA, 2008. IEEE Computer Society.

[46] B. L. Roux and H. Rouanet. *Multiple Correspondence Analysis*. Number 163 in Quantitative Applications in the Social Sciences. SAGE Publications, Los Angeles, California, USA, 2010.

[47] B. P. Sarma, N. Li, C. Gates, R. Potharaju, C. Nita-Rotaru, and I. Molloy. Android permissions: a perspective combining risks and benefits. In *Proceedings of the 17th ACM symposium on Access Control Models and Technologies*, SACMAT '12, page 13–22, New York, NY, USA, 2012. ACM.

[48] R. E. Schapire. The Boosting Approach to Machine Learning: An Overview. In *Nonlinear Estimation and Classification*. Springer, 2003.

[49] SoftAndroid Market. http://softandroid.ru.

[50] Trend Micro Command Line Antivirus Scanner, 2012. http://esupport.trendmicro.com/solution/en-us/0117058.aspx.

[51] Windows Phone: Marketplace. 2011. http://www.windowsphone.com/en-US/marketplace.

[52] B. Womack. Google says 700,000 applications available for android, Oct. 2012.

[53] Y. Zhou and X. Jiang. Dissecting Android Malware: Characterization and Evolution. In *Proceedings of the IEEE Symposium on Security and Privacy (OAKLAND)*, 2012.

[54] Y. Zhou, Z. Wang, W. Zhou, and X. Jiang. Hey, You, Get off of My Market: Detecting Malicious Apps in Official and Alternative Android Markets. In *Proceedings of the Network and Distributed System Security Symposium*, Feb. 2012.

[55] Y. Zhou, X. Zhang, X. Jiang, and V. W. Freeh. Taming information-stealing smartphone applications (on android). In *TRUST*, pages 93–107, 2011.

APPENDIX

A. FORMAL DESCRIPTION OF MCA

MCA has been independently discovered numerous times [1]; accordingly, descriptions of the methods and the terms used differ from author to author. We follow the description of Abdi et al. [1], with terminology and insights borrowed from La Roux et al. [46].

MCA constructs a cloud of points representing individuals by encoding the data as a matrix with a column for each possible *answer*, not question. Each of these columns is termed a "category". Let K be the set of categories. The contents of an individual element of this matrix δ_{ik} for $i \in I$, $k \in K$ will be a "1" for an individual choosing a category, and "0" otherwise. This matrix is called an *indicator matrix* and is represented as $\mathbf{X}$. This indicator matrix will be of dimensions $I \times K$, and is defined such that each row sum is constant:

$$\sum_{k=1}^{K} \delta_{ik} = N$$

for all $i \in I$. The restaurant example would have two categories describing "Attire" ("Formal" and "Casual") and three categories for "Cost" ("High", "Med", and "Low").

Once an indicator matrix of the data is constructed, MCA computes a *probability matrix* $\mathbf{Z} = |I|^{-1}\mathbf{X}$ and a supplemental probability matrix $\mathbf{Z}_0 = \mathbf{Z} - \mathbf{rc}^{\mathbf{T}}$, where $\mathbf{r}$ and $\mathbf{c}$ are column vectors with $\mathbf{r_i} = \sum_{k=1}^{K} \delta_{ik}$ for all i in I and $\mathbf{c_k} = \sum_{i=1}^{I} \delta_{ik}$ for all k in K. Essentially $\mathbf{r}$ and $\mathbf{c}$ are the respective row and column sum vectors of $\mathbf{Z}$. Then MCA constructs the matrix $\mathbf{H} = \mathbf{D_r}^{-\frac{1}{2}}\mathbf{Z}_0\mathbf{D_c}^{-\frac{1}{2}}$ where $\mathbf{D_r} = diag(\mathbf{r})$ and $\mathbf{D_c} = diag(\mathbf{c})$.

To arrive at the new coordinates for the individuals represented by the rows of the indicator matrix, MCA computes the singular

Table 5: MCA Questionnaire* composed of disjunctive questions.

Permission Question	Included permissions (All permissions have the "android.permission" prefix.)
ACCOUNT	ACCOUNT_MANAGER, AUTHENTICATE_ACCOUNTS, GET_ACCOUNTS, MANAGE_ACCOUNTS, USE_CREDENTIALS
AUDIO	RECORD_AUDIO, MODIFY_AUDIO_SETTINGS
BOOKMARKS	WRITE_HISTORY_BOOKMARKS, READ_HISTORY_BOOKMARKS
CALENDAR	WRITE_CALENDAR, READ_CALENDAR
CONTACTS	WRITE_CONTACTS, READ_CONTACTS
FILESYSTEM	MOUNT_FORMAT_FILESYSTEMS, MOUNT_UNMOUNT_FILESYSTEMS
GENERIC_SETTINGS	WRITE_SECURE_SETTINGS, WRITE_SETTINGS
INTERNET	INTERNET
LOCATION	ACCESS_FINE_LOCATION, ACCESS_COARSE_LOCATION, ACCESS_LOCATION_EXTRA_COMMANDS, AC-CESS_MOCK_LOCATION, CONTROL_LOCATION_UPDATES, INSTALL_LOCATION_PROVIDER
NETWORK	ACCESS_NETWORK_STATE, ACCESS_WIFI_STATE, BLUETOOTH, BLUETOOTH_ADMIN, BROADCAST_WAP_PUSH, CHANGE_NETWORK_STATE, CHANGE_WIFI_MULTICAST_STATE, CHANGE_WIFI_STATE, NFC, RECEIVE_WAP_PUSH, WRITE_APN_SETTINGS
PACKAGE	DELETE_PACKAGES, INSTALL_PACKAGES, BROADCAST_PACKAGE_REMOVED, GET_PACKAGE_SIZE
PHONE_STATE	MODIFY_PHONE_STATE, READ_PHONE_STATE
CALLS	CALL_PHONE, CALL_PRIVILEGED, PROCESS_OUTGOING_CALLS, USE_SIP
SMS	READ_SMS, SEND_SMS,WRITE_SMS, RECEIVE_SMS, BROADCAST_SMS, RECEIVE_MMS
Intent Filter Question	**Included action strings from intent filters** (Unless otherwise indicated, all actions strings have the "android.intent.action" prefix.)
BOOT	BOOT_COMPLETED, REBOOT
COMMUNICATIONS	ANSWER, CALL, CALL_BUTTON, DIAL, android.provider.Telephony.SMS_RECEIVED, PHONE_STATE
MEDIA	MEDIA_BAD_REMOVAL, MEDIA_BUTTON, MEDIA_CHECKING, MEDIA_EJECT, MEDIA_MOUNTED, MEDIA_NOFS, MEDIA_REMOVED, MEDIA_SCANNER_FINISHED, MEDIA_SCANNER_SCAN_FILE, MEDIA_SCANNER_STARTED, MEDIA_SHARED, MEDIA_UNMOUNTABLE, MEDIA_UNMOUNTED
PACKAGE	MANAGE_PACKAGE_STORAGE, MY_PACKAGE_REPLACED, NEW_OUTGOING_CALL, PACKAGE_ADDED, PACKAGE_CHANGED, PACKAGE_DATA_CLEARED, PACKAGE_FIRST_LAUNCH, PACKAGE_INSTALL, PACKAGE_REMOVED, PACKAGE_REPLACED, PACKAGE_RESTARTED, MANAGE_PACKAGE_STORAGE, MY_PACKAGE_REPLACED, NEW_OUTGOING_CALL, PACKAGE_ADDED, PACKAGE_CHANGED, PACKAGE_DATA_CLEARED, PACKAGE_FIRST_LAUNCH, PACKAGE_INSTALL, PACKAGE_REMOVED, PACKAGE_REPLACED, PACKAGE_RESTARTED
POWER	BATTERY_CHANGED, BATTERY_LOW, BATTERY_OKAY, ACTION_POWER_CONNECTED, ACTION_POWER_DISCONNECTED, POWER_USAGE_SUMMARY, ACTION_SHUTDOWN
WALLPAPER	WALLPAPER_CHANGED, SET_WALLPAPER

* Discussed in Section 4.2, we additionally include a generic permission question, a generic intent filter question, a native code question, and a zip file question

Table 6: Malware distribution across markets

Malware Family	Malware Description	Training Set	Anzhi	Ndoo	Softandroid
ADRD	Information Stealer	28		1	
Anserver Bot	Downloads malicious payloads, Information Stealer	5			
Asroot	Native root exploit	8			
BaseBridge	Root exploit, SMS, CALL Trojan	116	19		
Bgserv	Information Stealer, SMS Trojan	7		1	
Boxer	SMS Trojan				2
DroidDream	Root Exploit, Information Stealer	20			
DroidDreamLight	Information Stealer	23	18		
DroidKungFu	Root exploit, Downloads malicious payloads	179	53	19	
Geinimi	SMS Trojan, Information Stealer	72			3
GoldDream	SMS, CALL spy, Bot capabilities	27	12		
GPSSMSSpy	Location and SMS spy	6			
HippoSMS	SMS Trojan	5			
jSMSHider	SMS Trojan targetting custom ROMs	18			
KMIN	Information Stealer	40			
NickySpy	SMS, GPS, CALL spy	3			
Pjapps	Information Stealer, SMS Trojan	52			
Plankton	Downloads malicious payloads, Information Stealer	10	3		
RogueSPPush	Automatically subscribes to premium SMS services	3			
SndApps	Information Stealer	10			
YZHC	SMS Trojan	34	1		
zHash	Native root exploit	12			
Zsone	SMS Trojan	12	1	1	
Other	-	42	53	4	
Total	-	732	160	26	5

value decomposition of $\mathbf{H}$: $\mathbf{H} = \mathbf{P}\mathbf{\Delta}\mathbf{Q^T}$. The principal coordinates of each individual and category are described by the matrices

$$\mathbf{Y_I} = \mathbf{D}_r^{-\frac{1}{2}}\mathbf{P}\mathbf{\Delta} \; and \; \mathbf{Y_K} = \mathbf{D}_c^{-\frac{1}{2}}\mathbf{Q}\mathbf{\Delta}$$

From this, the χ^2 distance of an individual from the barycenter of all individuals is computed as $\mathbf{d} = diag(\mathbf{Y_I}\mathbf{Y_I^T})$

Eigenvalues are also computed from the singular value decomposition of $\mathbf{H}$. The eigenvalues, termed *inertia* in MCA, are the variances of the principal axes, and are defined as $\mathbf{\Lambda} = \mathbf{\Delta}^2$. These variances sum to the total variance of the cloud formed by the indicator matrix: $Var(\mathbf{X}) = \mathbf{e_l}diag(\mathbf{\Lambda})$ where $\mathbf{e_l}$ is the column vector of ones of appropriate size. The concept of inertia is important to the interpretation of an MCA analysis because it describes which principal axes are relevant. In most cases, the relevant principal axes are determined as those with inertia greater than $\frac{Var(\mathbf{X})}{l}$ where l is the total number of eigenvalues. The effectiveness of MCA is predicated on the fact that the number of principal axes with sufficient inertia is quite small (2 or 3) even with large N.

Tap-Wave-Rub: Lightweight Malware Prevention for Smartphones using Intuitive Human Gestures

Haoyu Li[1], Di Ma[1], Nitesh Saxena[2], Babins Shrestha[2], and Yan Zhu[1]

[1]University of Michigan-Dearborn
{haoyul,dmadma,yanzhu}@umd.umich.edu
[2]University of Alabama at Birmingham
{saxena,babins}@cis.uab.edu

ABSTRACT

We introduce a lightweight permission enforcement approach –
Tap-Wave-Rub (TWR) – for smartphone malware prevention. TWR
is based on simple human gestures (implicit or explicit) that are
very quick and intuitive but less likely to be exhibited in users'
daily activities. Presence or absence of such gestures, prior to
accessing an application, can effectively inform the OS whether
the access request is benign or malicious. In this paper, we focus
on the design of an *accelerometer-based phone tapping detection*
mechanism. This implicit tapping detection mechanism is geared
to prevent malicious access to NFC services, where a user is usually required to tap her phone with another device. We present a
variety of novel experiments to evaluate the proposed mechanism.
Our results suggest that our approach could be very effective for
malware prevention, with quite low false positives and false negatives, while imposing no additional burden on the users. As part of
the TWR framework, we also briefly explore explicit gestures (finger tapping, rubbing or hand waving based on proximity sensor),
which could be used to protect services which do not have a unique
implicit gesture associated with them.

Categories and Subject Descriptors

H.4 [**Information Systems Applications**]: Miscellaneous; C.2.0
[**Computer Systems Organization**]: Computer-Communication
Networks—*General, Security and Protection*

Keywords

malware; mobile devices; NFC; context recognition; sensors

1. INTRODUCTION

Smartphones are undoubtedly becoming ubiquitous. They are
not only used as (traditional) mobile phones for phone calling and
SMS messaging, but also for many of the same purposes as desktop computers, such as web browsing, social networking, online
shopping and banking. Also, smartphones are incorporating more
and more sensors and communication interfaces. Such new capabilities enable smartphones with many new unique functionali-

ties that desktop computers lack. For example, many smartphones
are beginning to incorporate Near Field Communication (NFC)
chips [15], which allows short, paired transactions with other NFC-
enabled devices in close proximity. The use of NFC-equipped smartphones as payment tokens (such as Google Wallet) is considered to
be the next generation payment system and the latest buzz in the
financial industry [4].

Due to their popularity, smartphones are becoming a burgeoning
target for malicious activities. There has been a rapid increase in
mobile phone malware targeting different smartphone platforms [9,
10, 21, 14, 19]. Newer functionalities of smartphones only make
them more attractive to malware writers. For example, the incorporation of NFC chips on smartphones provides malware authors
another (possibly much easier) way to deploy their attacks through
the NFC interface [20]. Especially, due to the ease with which financial transactions can take place using NFC, it is predicted that
NFC will become a popular target for malware aiming at credential and credit card theft [11]. Indeed, a proof-of-concept Trojan
Horse electronic pickpocket program under the cover of a tic-tac-
toe game has already been developed by Identity Stronghold [2].
In this attack, the game containing the malware is downloaded and
installed on a NFC-enabled smartphone. Once activated (when the
game is played), the malware accesses the NFC chip and enables
the RFID (Radio Frequency ID) reading functionality. This reader
then surreptitiously scans tags (e.g. RF tagged credit card) around it
and reports the acquired information to the malware owner through
e-mail once a victim tag is found in proximity.

While the security community has been battling with PC malware for many years, malware detection on smartphones turns out
to be an even more challenging problem [3]. This is partially due to
the resource constraints of smartphones (especially limited battery
power). Thus, existing malware defenses for desktop computers
cannot be applied directly on the smartphone platform. Much of
the existing research focuses on optimizing desktop based defenses
for mobile phones [25, 23, 24, 6, 3].

In practice, to protect mobile phones from malware attacks, major mobile phone manufacturers, such as Google, Apple, and Nokia,
employ permission models to prevent malware from being installed
at the first place. However, this approach relies upon user diligence
and awareness, while most computer users lack these traits in practice. Instead of relying on user permissions, smartphone manufacturers also rely upon application review before releasing to people
for download. However, application review process can be cumbersome and prone to human error [3].

1.1 Motivation and Rationale

We argue that existing malware defenses, without considering
the special characteristics of smartphone malware and that of smart-

phones themselves, might not be sufficient to detect sophisticated malware, such as the pickpocket malware targeting NFC mentioned previously.[1] First, the pickpocket malware [2], under the cover of tic-tac-toe, is quite stealthy. Its surreptitious scanning may not cause substantial changes (such as sharp increase in the number of emails sent or in power consumption) to the normal behavioral profile and therefore behavioral detection schemes will not be effective. Moreover, most existing malware detection schemes employ a *posteriori approach*. That is, malicious attacks are detected after they took place as traces need to be collected and trained before they can be compared with profiles to find abnormalities. Because of the sensitive (financial) nature of the NFC service, it is quite risky to adopt such a *posteriori* detection approach. Instead, it is desired to develop a preventive approach which can constantly monitor, identify, and then stop such potentially malicious activity *before* it is launched so as to minimize damage or loss.

This motivates us to design a novel approach for malware prevention through contextual awareness. Our rationale is as follows. Smartphones are personal devices. That is, the end user is a human being. Thus, (legitimate) access to sensitive/valuable services such as premium calls, SMS or NFC usually involves different types of human activities such as dialing a phone number, typing a message, or clicking an application icon on the screen (to execute the application). In contrast, one common pattern followed by malware found on mobile phones is that it attempts to access sensitive services without the user's awareness and authorization (thus user activity is very unlikely to be involved). Therefore, one way to detect such unauthorized, thus potentially malicious behavior, is to validate *whether an action is initiated by pure software or purposefully by a human user*.

Since legitimate access to sensitive services usually involves different types of hand movements, we explore the use of gestures to differentiate between pure software and human-initiated activities. In particular, in this paper, we propose *Tap-Wave-Rub (TWR)*, a lightweight malware detection mechanism for smartphones based on intuitive human gesture recognition, using sensors already available on current smartphones with little or no additional user involvement.

The proposed gesture-based detection mechanism serves as an extension to the currently adopted permission model used by major smartphone OSs. That is, whenever a sensitive service is requested, a particular gesture needs to be detected (to make sure it is a human generated activity) before the request can be granted. Otherwise, the activity is very likely generated by malware. As gesture detection is enforced every time a sensitive request is received, the proposed mechanism provides continuous monitoring of sensitive resources and services from unauthorized access attempts by malware. We note, the latest Android Jelly Bean 4.2 has an added security feature, Premium SMS Confirmation, that includes a giant list of premium shortcodes for each country and alerts a user anytime an app tries to send a message to a shortcode [1]. Our TWR permission model follows a similar approach but the security decision is based on presence or absence of gestures.

1.2 Our Contributions

The main contributions of this paper are summarized as follows.

1. We propose TWR, a novel approach for malware prevention with an exclusive focus on the smartphone platform based on

intuitive gesture recognition. As part of this system, we propose a implicit light-weight *phone tapping detection mechanism based on accelerometer data*, which is geared for NFC applications where a user is usually required to tap her phone with another device.

2. We outline how Tap-Wave-Rub can reside within the *kernel-level middle layer* between sensitive services and applications trying to access these services, and be integrated specifically with the existing *Android permission model*. This TWR-enhanced permission model provides continuous enforcement of access control to sensitive resources and services even after an application is installed on the platform.

3. To evaluate our approach, we conduct experiments to simulate the behavior of malware and normal user usage activity. Our experiment results show that the proposed mechanism can successfully detect malicious attempts to access sensitive services with high detection rates, while imposing minimal usability burden.

4. As part of the TWR framework, we also explore lightweight explicit gestures based on proximity sensor data, that could broadly appeal to many applications (e.g., SMS). These include tapping or rubbing a finger near the top of phone's screen or waving a hand close.

2. RELATED WORK

The most closely related work to ours is the one proposed in [5]. It shares similar philosophy as ours. It utilizes whether there are hardware interruptions to differentiate pure software initiated action and human initiated action [5]. It aims at detecting malware specifically targeting SMS and audio services. These services usually start with user's pressing or touching the keypad or touchscreen which generate hardware interruptions for each key/screen press event. A purely software generated activity (or malware generated activity), on the other hand, will not explicitly generate a hardware interrupt. Although this approach is believed to be effective for malware detection, it cannot help detect a more sophisticated malware such as the pickpocket malware. This is because the pickpocket malware gets activated by user's playing the tic-tac-toe game, which already involves touch screen activity that can generate hardware interrupts. The difference between [5] and our work can be summarized as follows. [5] attempts to check whether there is (any) user activity whereas our goal is to check whether there is a **special user-aware** activity. So our approach provides more fine-grained access control to sensitive services and thus can detect even sophisticated malware.

Another work that parallels to ours was recently presented in [22]. It proposes an approach of user-driven access control by granting permission to the application when user's permission granting intent is captured. It introduces access control gadgets (ACGs) which are UI elements exposed by each user-owned resource for applications to embed. The user's authentic UI interaction with corresponding ACG grant the permission to an application to access the corresponding resource. A fundamental difference between [22] and our work is that the design proposed by [22] grants the permission to an application when user's authentic **UI interaction** with corresponding ACG is captured whereas our design grants permission to an application when a specific user **gesture** (tapping/waving) is captured. The design proposed by [22] not only requires *kernel level changes* but also necessitates *application level modifications*. It also requires Resource Monitor (RM) to be incorporated for each resource such as the device drivers. Moreover, it

---

[1]Throughout the paper, we will center our malware mitigation design based on properties observed from the pickpocket malware [2]; however, our approaches, being fundamental in nature, will be applicable to a broad range of future malware.

requires additional composition ACG (C-ACG) along with composition RM if an application requires different resources to be accessed/used. Our work, in contrast to [22], has an advantage in that it neither requires application level changes nor requires resource monitor to be added for each resource. Note that if there are many resources that can be used by an application, then the number of C-ACG and C-RM will become extremely large. Another advantage of our work over [22] is that our design supports "services" (such as NFC) which do not have any specific UI elements or ACGs associated with them. For the approach of [22] to work with services like NFC, additional ACG for UI interaction will need to be added, which will significantly hamper the usability of such services. In contrast, in our case, implicit permission granting intent is acquired by capturing the tapping gesture.

Gesture recognition has been extensively studied to support spontaneous interactions with consumer electronics and mobile devices in the context of pervasive computing. Due to the uniqueness of gestures to different users, personalized gestures have been used for various security purposes. Gesture recognition has been used for user authentication to address the problem of illegal use of stolen devices [12, 7]. [18] reports a series of user studies that evaluate the feasibility and usability of light-weight user authentication based on gesture recognition using a single tri-axis accelerometer. Gesture recognition is also used to defend against unauthorized reading and Ghost-and-Leech relay attacks in RFID systems [8, 13].

3. BACKGROUND

3.1 Threat Model

In our model, we assume that the mobile phone is already infected with malware. As in the pickpocketing attack of [2], the malware could reside within a benign looking application (e.g., a game) which the user may have downloaded from an untrusted source. Our model covers a broad range of malware and does not impose any restriction on malware behavior except that an action from the malware is not human-triggered. For example, the malware may want to access a service or resource (such as NFC, SMS or GPS) available on the phone itself, or to communicate with an external entity, such as an attacker-controlled remote server (botmaster).

We assume that the OS kernel is healthy and immune to malware infection. In particular, the malware is not able to maliciously alter the kernel control flow. Also, the phone hardware is assumed to be malware-free. Specifically, we assume that the malware can not manipulate the input to, and output from, the phone's on-board sensors.

We do not impose any restriction as to how frequently the malware attempts to access a given service. However, in order to remain stealthy, constantly attempting access would not be feasible for the malware, and rather random or periodic sampling is expected.

In addition to the user space level control of the phone, the malware may collude and synchronize with an entity in close physical proximity of the phone (and its user). This external entity may attempt to manipulate the physical environment in which the phone is present or interfere with the user per se. We do not, however, allow this attacker to have physical access to the phone. That is, if the attacker has physical access to the phone, then he can lock/unlock a resource just like the phone's user. In other words, our mechanisms are not meant for user authentication and do not provide protection in the face of loss or theft of phone.

3.2 Design Goals

For our malware prevention approach to be useful in practice, it must satisfy the following properties:

- **Lightweight-ness:** The approach should be lightweight in terms of the various required resources available on the phone, such as memory, computation and battery power.

- **Efficiency:** The approach should incur little delay. Otherwise, it can affect the overall usability of the system. We believe that no more than a few seconds should be spent executing the approach.

- **Robustness:** The approach should be tolerant to errors. Both the False Negative Rate (FNR) and False Positive Rate (FPR) should be quite low. A low FNR means that a user would, with high probability, be able to execute an application (which accesses some sensitive services) without being rejected. Low FNR also implies a better usability. On the other hand, Low FPR means that there should be little probability to grant access to a sensitive service when a user does not intend to do so. Low FPR clearly implies a little chance for malware to evade detection.

- **Usage Model Consistency:** The solution should require little, or no change, to the usage model of existing smartphone applications. Ideally, if the use of a particular phone service can be commonly associated with a particular (unique) gesture (e.g., phone tapping for NFC), this gesture may be used to specially protect the said service. In this case, no changes to the adopted usage model will be necessary. It is also possible that there is no unique gesture pattern that can be found to use a certain service (e.g., Bluetooth). In such a situation, an intuitive gesture template can be associated with that service and a user will be required to explicitly perform the hand movements defined by that gesture. In this case, only minor changes to the adopted usage model will be imposed.

4. TWR-ENHANCED PERMISSION MODEL

Permission models have become very common on smartphone operating systems to provide access control to sensitive services for installed third party application. The Android platform has the most extensive permission system and poses to become a market leader. Thus, we base the design of our TWR system on the Android platform.

The idea of the TWR system is to add another layer of permission check before the original Android permission check. As stated in Section 3.1, we assume the adversary is not able to maliciously alter the kernel control flow. So gesture detection forms a trusted path with the OS. Intercepted permission requests are handled by the five components in the TWR's architecture: *TWR PermissionChecker*, *TWR GestureManager*, *TWR GestureExtractor*, *TWR TemplateCreator*, and *TWR GestureDatabase*. The architecture of TWR is depicted in Figure 1. In the following paragraphs, we present the role of each TWR component by describing the possible interaction between them and the outside world.

The *TWR PermissionChecker* stands in front of the original Android Permission check. When an application initiates a request to access a sensitive service, the request is intercepted by *TWR PermissionChecker*. This component interacts with *TWR GestureManager* to check whether the requested service is protected by a certain gesture. If not, the request is forwarded to the Android Permission Check as usual. Otherwise, *TWR GestureManager* interacts with the *TWR GestureExtractor* to begin collecting gesture

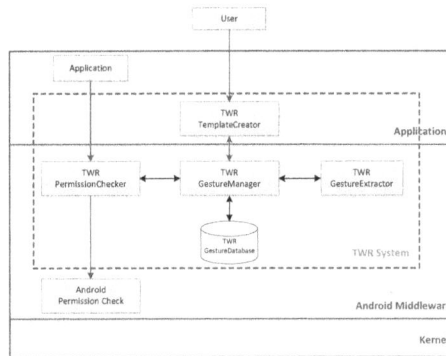

Figure 1: The TWR Architecture

data (tapping, rubbing, or waving in this paper). The captured data is then sent to the *TWR GestureManager* for further process.

Here we distinguish between two types of gesture recognition: user-dependent and user-independent. As their names suggest, a gesture is user-dependent if there is significant variation among gesture data from multiple participants for the same predefined gesture; while a gesture is user-independent if either there is no apparent difference or the recognition process does not differentiate among user data from multiple participants. Our phone tap gesture recognition is user-dependent as it captures the user's own features of the tapping movement. Given that users can hold a phone in different ways and use different forces to tap, the phone tap gesture is thus user-dependent. Our hand wave or finger tap/rubbing recognition scheme is user-independent as it infers user activity by checking whether a special location (in our context, the place where the proximity sensor is located) is touched or not (instead of the potentially biometric feature of human movement).

For user-dependent gesture recognition, we usually need to create a gesture template which is used as a reference in the actual recognition stage. The user can interact with the *TWR Template-Creator* to register a new gesture template, to update and delete exiting gesture templates. *TWR TemplateCreator* is an Android application which allows interaction between TWR and the user. When the user creates, deletes, or modifies the gesture information, it needs to retrieve and store the information to *TWR Gesture-Database* via *TWR GestureManager*. *TWR GestureManager* is the only component that has access to *TWR GestureDatabase*.

So depending on the type of gesture recognition scheme (user-dependent or user-independent), the *TWR GestureManager* processes the gesture data from *TWR GestureExtractor* differently. If the gesture is user-dependent, it compares the similarity between the newly captured data with the corresponding gesture template stored in the *TWR GestureDatabase*. If the gesture is user-independent, the *TWR GestureManager* determines directly whether a gesture is performed or not by utilizing information in the captured data without the help of a template. In either case, if a required gesture is detected, the request is forwarded to Android Permission Check for further check. Otherwise, the request for service access is rejected.

5. TWR GESTURE DETECTION

As we mentioned in Section 2, the use of hardware interruption to differentiate between pure software initiated activity and human initiated activity is not effective to prevent malware hidden under the cover of a victim app, since activating the victim app already involves keypad click or screen touch which can generate hardware interrupts. This motivates us to use app-specific user events to distinguish between hidden malware and an app initiated by a human being. That is, instead of simply using general key/screen press

events to infer human activity, we try to recognize whether it involves the *right* activity a user needs to do to access a sensitive service, such as the access to NFC.

A smartphone is a personal hand-held device installed with a lot of apps. Most of the time, these apps are activated by specific phone/hand movements. For example, when a user wants to place a call, she needs to unlock the screen, activates the phone app, inputs the number (or clicks on a name in the contact list), and then puts the phone near the ear to start the call. Also, to use the NFC to scan a smart poster, a user needs to unlock the screen, activates the NFC reader app, and taps the phone on the smart poster to read information. Since "tapping" (touching the phone against an RFID tag, or another NFC device) is a gesture which users commonly need to perform to use the NFC functionality, as an illustrative example, we can use tapping to determine whether an NFC access is human-initiated or not. Intuitively, tapping on a smart-poster should be different from other user phone activities (such as keyboard click or screen touch) and user physical activities (such as walking or running).

One advantage of this tapping approach is that it does not require any additional user activity besides what is being used commonly, and thus transparently recognizes user activity when a user taps a smart poster to obtain information. However, it may exhibit false positive rates and not fully prevent the pickpocket malware activation since normal user activities (such as playing the game) may generate motions similar to tapping. To achieve higher prevention rate, we can try other intuitive user-aware gestures similar to tapping, such as "tapping twice" or "tapping thrice" in succession.

To recognize tapping, we utilize the on-board accelerometer data. An accelerometer sensor measures the forces applied to the phone (minus the force of gravity) on the three axes: x, y, and z. Let (a_x, a_y, a_z) denote the values corresponding to the 3 axes from the accelerometer.

Our detection algorithm consists of two phases: training phase and recognition phase. In the training phase, a user performs the target action (tapping) multiple times, and accelerometer data of the action is recorded and processed to generate a tapping template. The template serves as a reference to be compared later with real-time user movement data: a match indicates the recognition of user tapping activity; otherwise, it is inferred that either there is no user activity or the activity is not the "valid" user activity to grant NFC access.

After the training phase, the system compares a newly observed movement with the template. The system records the accelerometer data, from the moment the user activates the NFC reader app, until she taps on a smart poster. To recognize tapping, the system computes the cross-correlation C of the acceleration data A against the template T, both of size n data points as shown in Equation 1.

The cross-correlation C computed from Equation 1 when comparing two time series is a real value, representing a similarity measure. The higher the value of C, the higher the similarity between the two series. The maximum value is obtained when the two series under comparison are identical. A movement is considered a valid tapping activity when the computed cross-correlation C exceeds a certain cross-correlation threshold which is usually obtained through empirical study.

$$C(A,T) = \frac{\sum_{i=1}^{n}(a_{x_i} - \bar{a}_x)(T_{x_i} - \bar{T}_x)}{\sqrt{\sum_{i=1}^{n}(a_{x_i} - \bar{a}_x)^2}\sqrt{\sum_{i=1}^{n}(T_{x_i} - \bar{T}_x)^2}}$$
$$= \frac{\sum_{i=1}^{n}(a_{y_i} - \bar{a}_y)(T_{y_i} - \bar{T}_y)}{\sqrt{\sum_{i=1}^{n}(a_{y_i} - \bar{a}_y)^2}\sqrt{\sum_{i=1}^{n}(T_{y_i} - \bar{T}_y)^2}}$$
$$= \frac{\sum_{i=1}^{n}(a_{z_i} - \bar{a}_z)(T_{z_i} - \bar{T}_z)}{\sqrt{\sum_{i=1}^{n}(a_{z_i} - \bar{a}_z)^2}\sqrt{\sum_{i=1}^{n}(T_{z_i} - \bar{T}_z)^2}} \quad (1)$$

where $\bar{a}_x$ denotes the means of time series a_{x_i} for $i \in [1,n]$ (others follow the same notation).

Here we describe one way to determine the cross-correlation threshold. Suppose we have m traces of tapping movements T_1, ..., T_m. The threshold C_T can be estimated as the minimum cross-correlation between any two series T_i and T_j ($i, j \in [1,m]$ and $i \neq j$). That is:

$$C_T = min_{i,j=1}^{m}(C(T_i, T_j)) \quad (2)$$

The aforementioned phone tapping detection mechanism is geared for NFC applications. Unlike NFC, most other services/resources on the phone may not be associated with a unique implicit gesture. In such situations, an explicit human involvement would be necessary. As part of the TWR framework, we have also designed and implemented lightweight explicit gestures based on proximity sensor data, that could broadly appeal to many applications (e.g., SMS). These include tapping or rubbing a finger near the top of phone's screen or waving a hand close. We note the design of an explicit gesture detection scheme is not trivial. The challenge is to keep the gestures very simple for the users to perform, and lightweight for the system to identify. We report the details of our explicit gestures in the extended version of this paper [16].

6. IMPLEMENTATION AND EVALUATION

To evaluate the feasibility of the TWR approach for malware prevention, we developed a prototype application on the Android platform. The phone tapping detection scheme was implemented and installed on a Google Nexus S Android phone. This Nexus phone comes equipped with NFC chip, therefore a good target device for an eventual deployment of our approach.

In this section, we report on evaluation of tapping based user activity recognition scheme outlined in Section 5. This scheme is specially designed to protect against malware targeting NFC reading services, since tapping is a natural hand movement which a user needs to perform to use the reader function of NFC.

Since "tapping" (touching the phone against an RFID tag, or another NFC device) is a very simple hand movement, we hypothesize it might be confused with other user movements such as those users perform when they play games, and thus have higher false positive rate, FPR (or lower prevention rate). In a hope to achieve higher prevention rate, we also experiment with two other intuitive user-aware gestures similar to tapping: tapping twice and tapping thrice in succession. We call these three tapping gestures as "tapping once", "tapping twice", and "tapping thrice", respectively.

First, to determine the cross-correlation detection threshold, we collected 30 traces of accelerometer data for each tapping gesture. Each of our trace contains 100 data points and is recorded over a 2-second time period (we wanted our schemes to be efficient). Each trace is then used as a template, which is compared with all the other 29 traces to calculate a serial of C values. The smallest C value is chosen as the threshold value. This threshold value is then stored with the corresponding tapping template and a matched posture needs to yield a C value larger than this threshold. These traces were collected by the experimenter while performing NFC tapping gesture 30 times. Such data collection and testing methodology is in-line with related prior security work, e.g., Secret Handshakes [8]. Our methodology captures a realistic usage scenario whereby each user can be trained "once by their phone" and can create their template, e.g., when they purchase the phone.

We first test the performance of the three tapping gestures to identify which one can have higher recognition rate (thus lower false negative rate, FNR). To do this, we collect a total of 150 traces for each tapping gesture, 30 traces every day for 5 days. We then use the template and the threshold calculated above to determine the recognition rate. The successful recognition rate is listed, in the form of a confusion matrix, in Table 1. It shows that "tapping once" achieves high recognition rate 94.67% (or a low FNR 5.33%) compared to the other two tapping gestures.

We next test the performance of the three tapping gestures to identify which one is the least to be confused with other user or phone movements and thus has low FPR. It is important to evaluate the FPR. If a tapping gesture can be very similar to a certain other movement (accidental or manipulated by an attacker), the malware may circumvent the gesture detection process.

To determine the FPR, we compare tapping postures with many phone/user movements. These movements might be just normal user activities, or activities coerced by a nearby attacker. They include user movements such as: walking, walking stairs, screen-touch activities (text messaging and surfing Internet), phone-moving activity (motion gaming and picking up calls), as well as, the scenarios where phone is left still. For each movement, we also collect a total of 150 traces, 30 traces every day for 5 days. The error rates are all listed in Table 1.

Our experiment result shows "tapping once" is very unlikely to be confused with walking, walking stairs, still, and screen-touch activities such as text messaging or Internet browsing. However, it might occasionally be confused with phone motion caused when the user plays game or picks up a phone call with a false positive rate of 2%. "tapping twice" and "tapping thrice", on the other hand, are very resilient to phone motions but they resemble motions when a user walks on stairs. Nevertheless, all achieve satisfying low false positive rate.

One potential reason why the false positive rate is low might be that tapping is a type of user-aware movement. When performing such a gesture, the user is believed to be aware of her hand movement. So gestures are performed in a more-or-less controlled way, e.g., the phone is always held in the similar way when a user performs tapping. In non-user-aware movements, on the other hand, the phone can be tilted in any position. The reference template is usually collected in a reference coordinate system. However, once the phone is tilted, movement data collected from the device is no longer in the reference coordinate system and the corresponding movement will not be detected correctly. In this way, user-aware gesture is very unlikely to be similar with user-unaware movements, and thus has low false positive rate. Previous studies on gesture recognition also suggest certain gestures can be quite unique and different from other gestures [8, 17]. So tapping can be

Table 1: Tapping Detection Results (rates at which a gesture shown on each row matches with gesture/activity shown on each column

	Tapping Once	Tapping Twice	Tapping Thrice	Walking	Walking Stairs	Still	Screen-touch Activities	Phone Movement
Tapping Once	94.67%	NA	NA	0%	0%	0%	0%	2%
Tapping Twice	NA	92.67%	NA	0%	1.33%	0%	0%	0%
Tapping Thrice	NA	NA	96.67%	0%	5.33%	0%	0%	0%

distinguished from other user-aware movements such as "picking up the call".

Our experiment result, contrary to our hypothesis that "tapping once" may have high false positive rate, shows that "tapping once" actually achieves both high recognition rate and low false positive rate, and has a performance comparable to "tapping twice" and "tapping thrice". However, "tapping once" outperforms the other two tapping gestures in term of efficiency, and has better usability since it does not require any change to the usage model of NFC.

7. CONCLUSIONS AND FUTURE WORK

In this paper, we introduced a lightweight permission enforcement approach – Tap-Wave-Rub (TWR) – for smartphone malware detection and prevention. TWR is based on simple human gestures that are very intuitive but less likely to be exhibited in users' daily activities. Specifically, we presented the design of a phone tapping detection based on accelerometer data. This mechanism is geared for NFC applications, which usually require the user to tap her phone with another device. In addition, we present the TWR Android permission model, the prototypes implementing the underlying gesture recognition mechanisms, and a variety of novel experiments to evaluate them. Our results suggest the proposed approach could be very effective for malware prevention, with quite low false positives and false negatives, while imposing little to no additional burden on the users. The false negatives are expected to further reduce significantly as users become more familiar with the underlying gestures, especially since they are quite intuitive. In addition, the false positives can also be carefully avoided in most cases, for example, by detecting the orientation of the device. Our future effort will be focused on realizing the TWR approach in practice and further evaluate it with a wide range of smartphones and smartphone users.

Acknowledgements: We thank William Enck and WiSec'13 anonymous reviewers for their thoughtful feedback.

8. REFERENCES

[1] R. Amadeo. Exclusive: Android 4.2 alpha teardown, part 2: SELinux, VPN lockdown, and premium SMS confirmation. Available online at http://www.androidpolice.com/2012/10/17/exclusive-android-4-2-alpha-teardown-part-2-selinux-vpn-lockdown-and-premium-sms-confirmation/, Oct. 2012.

[2] W. Augustinowicz. Trojan horse electronic pickpocket demo by identity stronghold. Available online at http://www.youtube.com/watch?v=eEcz0XszEic, June 2011.

[3] I. Burguera, U. Zurutuza, and S. Nadjm-Tehrani. Crowdroid: Behavior-based malware detection systems for Android. In *ACM CCSW Workshop*, 2011.

[4] M. Calamia. Mobile payments to surge by $670 billion by 2015. Available online at http://www.mobiledia.com/news/96900.html, Jul. 2011.

[5] A. Chaugule, Z. Xu, and S. Zhu. A specification based intrusion detection framework for mobile phones. In *ACNS'11*, 2011.

[6] J. Cheng, S. Wong, H. Yang, and S. Lu. Smartsiren: virus detection and alert for smartphones. In *5th International Conference on Mobile Systems, Applications and Services (MobiSys'07)*, 2007.

[7] M. Conti, I. Zachia-Zlatea, and B. Crispo. Mind how you answer me!: transparently authenticating the user of a smartphone when answering or placing a call. In *ASIACCS'11*.

[8] A. Czeskis, K. Koscher, J. Smith, and T. Kohno. RFIDs and secret handshakes: Defending against Ghost-and-Leech attacks and unauthorized reads with context-aware communications. In *ACM Conference on Computer and Communications Security*, 2008.

[9] F-Secure. Bluetooth-worm:symbos/cabir. Available online at http://www.f-secure.com/v-descs/cabir.shtml.

[10] F-Secure. Worm:symbos/commwarrior. Available online at http://www.f-secure.com/v-descs/commwarrior.shtml.

[11] A. P. Felt, M. Finifter, E. Chin, S. Hanna, and D. Wagner. A survey of mobile malware in the wild. In *ACM CCSW Workshop*, 2011.

[12] D. Gafurov, K. Helkala, and T. Sndrol. Biometric gait authentication using accelerometer sensor. *Journal of Computers*, 1(7):51–59, 2006.

[13] T. Halevi, S. Lin, D. Ma, A. Prasad, N. Saxena, J. Voris, and T. Xiang. Sensing-enabled defenses to rfid unauthorized reading and relay attacks without changing the usage model. In *PerCom'12*, 2012.

[14] J. Han, E. Owusu, T.-L. Nguyen, A. Perrig, and J. Zhang. ACComplice: Location Inference using Accelerometers on Smartphones. In *Proc. of COMSNETS*, Jan. 2012.

[15] ISO. Near field communication interface and protocol (nfcip-1)——iso/iec 18092:2004. Available online at http://www.iso.org/iso/catalogue_detail.htm?csnumber=38578, 2004.

[16] H. Li, D. Ma, N. Saxena, B. Shrestha, and Y. Zhu. Tap-wave-rub: Lightweight malware prevention for smartphones using intuitive human gestures. Extended Technical Report, Available online at http://arxiv.org/abs/1302.4010, Feb. 2013.

[17] J. Liu, Z. Wang, L. Zhong, J. Wickramasuriya, and V. Vasudevan. uWave: Accelerometer-based personalized gesture recognition and its applications. *Pervasive and Mobile Computing*, 5(6):657–575, December 2009.

[18] J. Liu, L. Zhong, J. Wickramasuriya, and V. Vasudevan. User evaluation of lightweight user authentication with a single tri-axis accelerometer. In *MobileHCI'09*, 2009.

[19] P. Marquardt, A. Verma, H. Carter, and P. Traynor. (sp)iPhone: decoding vibrations from nearby keyboards using mobile phone accelerometers. In *Proc. of ACM CCS*, 2011.

[20] C. Mulliner. Vulnerability analysis and attacks on NFC-enabled mobile phones. In *1st International Workshop on Sensor Security (IWSS) at ARES*, 2009.

[21] E. Owusu, J. Han, S. Das, A. Perrig, and J. Zhang. ACCessory: Keystroke Inference using Accelerometers on Smartphones. In *Proc. of HotMobile*), Feb. 2012.

[22] F. Roesner, T. Kohno, A. Moshchuk, B. Parno, H. J. Wang, and C. Cowan. User-driven access control: Rethinking permission granting in modern operating systems. In *33rd IEEE Symposium on Security and Privacy (Oakland 2012)*, 2012.

[23] A.-D. Schmidt, R. Bye, H.-G. Schmidt, J. Clausen, O. Kiraz, K. Yksel, S. Camtepe, and A. Sahin. Static analysis of executables for collaborative malware detection on Android. In *ICC 2009 Communication and Information Systems Security Symposium*, 2009.

[24] A. S. Shamili, C. Bauckhage, and T. Alpcan. Malware detection on mobile devices using distributed machine learning. In *20th International Conference on Pattern Recognition (ICPR'10)*, 2010.

[25] D. Venugopal. An efficient signature representation and matching method for mobile devices. In *WICON'06*, 2006.

Counter-Jamming Using Mixed Mechanical and Software Interference Cancellation

Triet D. Vo-Huu   Erik-Oliver Blass   Guevara Noubir

College of Computer and Information Science
Northeastern University, Boston, MA
{vohuudtr|blass|noubir}@ccs.neu.edu

ABSTRACT

Wireless networks are an integral part of today's cyber-physical infrastructure. Their resiliency to jamming is critical not only for military applications, but also for civilian and commercial applications. In this paper, we design, prototype, and evaluate a system for cancelling jammers that are significantly more powerful than the transmitting node. Our system combines a novel mechanical beam-forming design with a fast auto-configuration algorithm and a software radio digital interference cancellation algorithm. Our mechanical beam-forming uses a custom-designed two-elements architecture and an iterative algorithm for jammer signal identification and cancellation. We have built a fully functional prototype (using 3D printers, servos, USRP-SDR) and demonstrate a robust communication in the presence of jammers operating at five orders of magnitude stronger power than the transmitting node. Similar performance in traditional phased arrays and radar systems requires tens to hundreds of elements, high cost and size.

Categories and Subject Descriptors

C.2.1 [**Network Architecture and Design**]: Wireless communication

General Terms

Design, Experimentation, Security

Keywords

anti-jamming; beam forming; software radio

1. INTRODUCTION

Over the last decades, wireless communication proved to be an enabling technology for an increasingly large number of applications. The convenience of wireless and its support of mobility has revolutionized the way we access data, information services, and interact with the physical world. Beyond enabling mobile devices to access information and data services ubiquitously, it is today widely used in cyber-physical systems such as air-traffic control [42], power plants synchronization, transportation systems, and human body implantable devices [13]. For example, the United States Congress recently passed an FAA bill that speeds up the switching to GPS-based air traffic control [24]. The trend of wireless communication utilization in the electricity grid is already visible with over 20 millions smart meters already installed in the US and over 70 million worldwide [26]. Wireless Remote Terminal Units (W-RTU) with long-range wireless communication capabilities have been used for many years and several companies are increasingly switching to Wireless RTUs, e.g., vMonitor [38], Industrial Control Links [14], Synetcom [34], and Semaphore [29].

This pervasive elevated wireless communication systems to the level of critical infrastructure. Jamming is a prominent security threat as it cannot only lead to denial of service attacks, but can also be the prelude to sophisticated spoofing attacks against cellular, WiFi, and GPS system [6]. While basic jamming hardware against GPS, Cellular Systems, and WiFi are already a commodity that can be found on Internet online websites for few tens of dollars, more powerful jammers can also easily be made given that they do not necessitate to generate precise, clean RF signals. A \$7 magnetron generates a 1KWatt interfering signal (covering hundreds of meters) and can be tuned to a wide range of frequencies by slightly modifying its resonant cavity [4]. Various websites have online and YouTube tutorials to re-purpose the magnetron of a \$50 microwave oven and build High Energy RF guns (HERF). In addition to its use in war zones, jamming recently caused sufficient concerns to trigger an FCC campaign to enforce anti-jamming laws as stated by the chief of the FCC's Enforcement Bureau on February 2011 [10, 23].

We consider a setup of jamming where the spread spectrum and coding gain are not sufficient to counter the jammer. This paper focusses on the case of a single jammer/antenna adversary. We assume a fairly narrowband signal (few MHz) and that mechanical steering components are possible as is the case on many military vehicles or as widely used around the world in motorized dish antennas. Our system operates in two stages (Figure 1):

- **First stage – Antenna Auto-Configuration:** We introduce a novel two-element antenna that dynamically reconfigures to track the jammer and to weaken its signal by up to 28 dB (Fig. 1b). Our design with two moving elements is *simple*, *low-cost*, and has unique characteristics unexplored in mechanically steerable an-

(a) Without anti-jamming

(b) Our system with 2-element antenna (1st stage)

(c) Our system, digital cancellation (2nd stage)

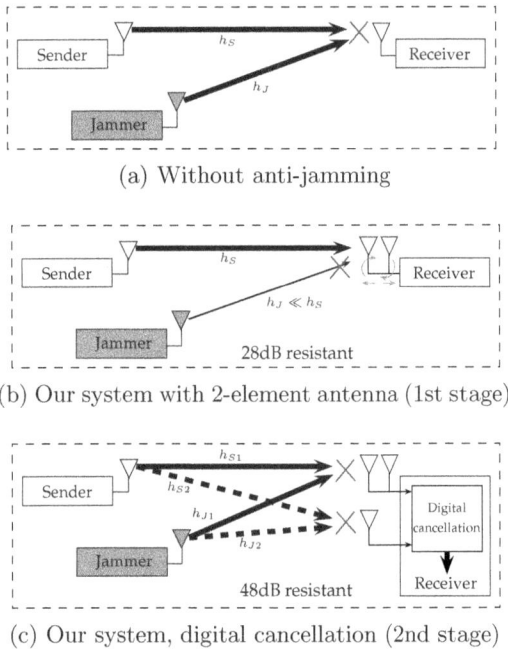

Figure 1: Jamming and its effect in a traditional system and in our system with two stages of anti-jamming.

Figure 2: System prototype.

tennas. Our configuration algorithm allows to converge on elements separation/rotation that maximizes the signal-to-jamming ratio (SJR) within 20 seconds.

- **Second stage – Digital Jamming Cancellation:** To further eliminate the jammer's signal, we also use a single-element antenna to get an additional copy of the jamming signal and develop a MIMO-like interference cancellation techniques tailored for anti-jamming. Unlike traditional MIMO and beam-forming techniques we do not rely on training sequences. We demonstrate a reliable communication equivalent to reducing the impact of a jammer by 48 dB.

Our contributions are:

- *Anti-jamming adversaries with significantly more power than transmitting nodes:* We are able to efficiently remove unknown jamming signals up to almost five orders of power higher than legitimate user's signals and recover the user data with an acceptable bit error rate.
- *Zero-knowledge anti-jamming:* We neither require knowledge about the legitimate signals (no additional preamble, no training sequence), nor knowledge about the jammer (unknown location, variable jamming power).
- *Environment adaptiveness:* The system works efficiently in both outdoor as well as indoor environments and can handle multipath jamming.

While the techniques used in this system are rooted in techniques developed for MIMO communication [36] and phased array antenna [22, 35], fields that have been extensively studied over several decades, the characteristics of our setup and design require new algorithms and techniques. Our digital jamming cancellation algorithms target *powerful unknown jammers*, unlike traditional MIMO techniques that operate over user-designed transmission signals of similar powers, allowing adequate channel estimation through training sequences.

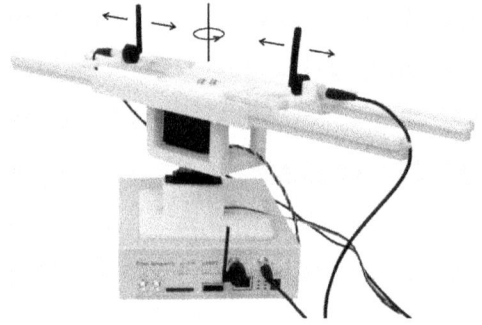

Previous work on *phased array antennas* uses fixed elements and primarily aims at producing a directed beam that can be repositioned electronically/digitally. Adaptive beam-forming with algorithms such as MVDR and MMSE beam-formers aims at minimizing the impact of the sidelobe and considered to be more adequate for radar systems. Phased array systems used in radar systems achieving our jamming cancellation gain use a large number of elements and size (sometimes hundreds [3, 4]). In contrast, our design allows the formation of a large number of beam patterns that are impossible for an electronically steerable fixed two-element antenna [18, 35] and is combined with a MIMO-like digital interference cancellation.

One starting point for our approach is that increasing the distance between the antenna elements increases the number of beams and reduces their width. A traditional two-element electronically steerable antenna can only *rotate* the beam patterns [18]. With our setup, the large number of controlled "nulls" allows to potentially cancel even several simultaneous jammers and their multipaths. In addition, the large number of beams allows the creation of connected networks.

2. MODELS AND APPROACH

2.1 Communication and adversarial models

We consider a communication setup (Figure 1a) with two legitimate communicating nodes and one jamming node.

2.1.1 Communication nodes

The two communicating nodes operate over an open, fairly narrowband channel (few MHz), using a pre-agreed modulation scheme. The sender transmits data at a constant power via a single-element antenna. The receiver uses a two-element antenna and a single-element antenna for signal reception. Both nodes are unaware of each other's location and their own location, also the presence of jammer. During the communication, the users remain in fixed locations.

2.1.2 Adversary

The jammer is equipped with a single transmit antenna that can emit a powerful jamming signal to interfere with the communication between the users. The jamming signal can be either noise or a modulated signal. The jammer can purposely start or stop jamming at any time, or adjust the transmit power to variable levels during the jamming period. The jammer does not change its location while jamming.

2.1.3 Communication channel

The users' communication channel is assumed to be narrowband with slow-fading. Typically the channel is few MHz wide at the 2.4GHz band and the modulation is BPSK or QPSK. We consider both outdoor and indoor environments in this setup.

2.2 Approach

In a traditional system, where the receiver R has only a single antenna, the simultaneous transmission of both sender and jammer causes interference at the receiver: $R = h_S S + h_J J$, where h_S and h_J are the channel gains from sender S and jammer J to the receiver, respectively. If the jammer interference is strong, the signal-to-jamming ratio (SJR) at the receiver is low (equivalently h_S/h_J is small), the signal S is undecodable.

In our system, the receiver has an additional two-element antenna. To decode the data, the receiver operates in two stages to increase the SJR. In the *first stage* (Figure 1b), the two-element antenna is used for signal reception. The configuration algorithm adjusts the two-element antenna (Figure 2) such that the distance between the two elements (*element separation*) and the rotational direction (*angle*) of the antenna increases the received legitimate's signal power, while, at the same time, reducing the received jammer's signal power. As a result, the SJR is increased, allowing successful data decoding.

As the antenna angle and element separation are adjustable, the receive pattern is dynamically configurable. In fact, we can construct a *large* number of different receive patterns (shown in Section 3), in comparison with fixed-position electronically steerable arrays. Our experiments show that our system can cope with a jammer with up to 28dB stronger power than legitimate users. At the heart of the first stage, we introduce an algorithm that dynamically configures the angle and the element separation of the antenna to maximize the received SJR. The flexibility of our custom-designed antenna allows the auto-configurability of the system to work effectively in both outdoor and indoor environments, where the latter often incurs problems to electronically steerable antenna arrays and results in poor performance. We also show that our configuration algorithm significantly outperforms a brute-force configuration in speed and converges within 20 seconds.

However, our purpose is to allow communication in the presence of jammers *beyond* the 28dB limit. In the *second stage*, we extend our model by using digital interference cancellation techniques (Figure 1c) to eliminate the jamming signal. Equation (1) illustrates the idea of the jamming cancellation techniques applied to the received signals at the two-element antenna and the single-element antenna. We obtain two different copies of the transmitted signal at the receiver: R_1 from the two-element antenna and R_2 from the single-element antenna.

$$R_1 = h_{S1}S + h_{J1}J$$
$$R_2 = h_{S2}S + h_{J2}J \quad (1)$$

One major difference between our setup and MIMO systems is that MIMO systems use training sequences to estimate the channel gains. This is not possible in our setup since we do not have control over the jamming signal. Instead, we propose a technique specific to this model to estimate the channel gain ratio $a = h_{J2}/h_{J1}$ in order to recover

the legitimate signal, as shown in the following equation:

$$bS = aR_1 - R_2, \quad (2)$$

where $b = ah_{S1} - h_{S2}$. Knowing a, we can decode S. The ratio a depends on the channel characteristics such as attenuation, multipath and the power of the jamming signal. The factor b is considered as a new channel gain of the residual signal after eliminating the jamming signal, and does not introduce any difficulty for the decoder, thus requires no estimation.

In summary, the high-level idea of our approach is to build a hybrid system consisting of two levels of anti-jamming techniques: analog signal cancellation by mechanical means of our custom-designed antenna and digital interference cancellation by software-based signal processing techniques. The robustness of our system highly depends on the performance of the configuration algorithm and digital interference cancellation algorithm. In the next sections, we will discuss the following problems:

- What is the optimal antenna configuration (separation, angle) that maximizes the SJR?

- How to estimate the channel characteristics to optimize the performance of the digital jamming cancellation technique against unknown jamming signals?

3. ANTENNA CONFIGURATION

Increasing the SJR at the receiver is a key goal of our system. We will now present a new, efficient algorithm for reconfiguring the two-element antenna, such that the receiver is able to reduce a significant portion of the jammer's power. For ease of understanding, we first introduce our notation:

DEFINITION 1. *A* configuration *of the two-element antenna, denoted as* (L, ϕ), *consists of element* separation L *and a rotational* angle ϕ.

A configuration specifies the angular position of the two-element antenna and the physical separation of its elements. The antenna is able to rotate within a range $[\phi_{\min}, \phi_{\max}]$, and adjust the separation between the limits $[L_{\min}, L_{\max}]$. In practice, depending on the mechanical devices' capabilities, the number of possible configurations, L and ϕ for given ranges, is finite. We denote $P(L, \phi)$ as the received signal's power at the two-element antenna with configuration (L, ϕ).

DEFINITION 2. (L, ϕ) *is a* minimizing *(or maximizing)* configuration, *if* $P(L, \phi) \leq P(L', \phi')$ *(or* $P(L, \phi) \geq P(L', \phi')$*) for all other configurations* (L', ϕ').

DEFINITION 3. L_ϕ *is called a* minimizing separation *for an angle* ϕ, *if* $P(L_\phi, \phi) \leq P(L, \phi)$ *for all* $L \in [L_{\min}, L_{\max}]$. *Similarly,* L_ϕ *is a* maximizing separation *for an angle* ϕ, *if* $P(L_\phi, \phi) \geq P(L, \phi)$ *for all* $L \in [L_{\min}, L_{\max}]$.

DEFINITION 4. ϕ_L *is called a* minimizing angle *for a separation* L, *if* $P(L, \phi_L) \leq P(L, \phi)$ *for all* $\phi \in [\phi_L - \theta, \phi_L + \theta]$. *The parameter* θ *denotes the desired search range for the antenna control algorithm. Similarly,* ϕ_L *is a* maximizing angle *for a separation* L, *if* $P(L, \phi_L) \geq P(L, \phi)$ *for all* $\phi \in [\phi_L - \theta, \phi_L + \theta]$.

Intuitively, minimizing angles are directions inside the nulls where the received power is minimized, and maximizing angles are directions inside the lobes where the received power is maximized.

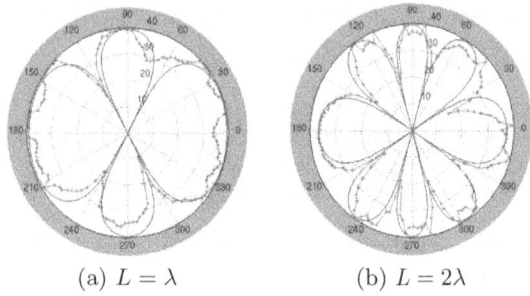

(a) $L = \lambda$ (b) $L = 2\lambda$

Figure 3: Outdoor receive patterns of the two-element antenna. Experimental gains (lines with plus signs) are compared to theoretical values.

3.1 Pattern analysis

We first study the basic characteristics of the two-element antenna. Signals received at two elements can be added constructively or destructively depending on their phase difference when arriving at the elements. When the signals add up together, we have lobes in the receive pattern. When the signals eliminate each other, we have nulls. In free-space communications, the phase difference between arriving signals can be determined based only on the element separation L. The following two theorems give the locations and number of lobes and nulls in the receive pattern of the two-element antenna.

THEOREM 1. *The receive pattern of the two-element antenna in a free-space communication has maximizing angle at ϕ_k and minimizing angle at ϕ_m, which satisfies*

$$\begin{aligned} \cos \phi_k &= \tfrac{k}{K} & k \in \mathbb{Z} \\ \cos \phi_m &= \tfrac{2m+1}{2K} & m \in \mathbb{Z}, \end{aligned}$$

where $K = L/\lambda$ is the ratio between the separation and the carrier wavelength, k and m are integers. Besides, if $\{K\} \geq \tfrac{1}{2}$, where $\{K\}$ denotes the fractional part of K, 0 and π are 2 additional maximizing angles; otherwise, they are 2 additional minimizing angles.

THEOREM 2. *The number of maximizing angles of the two-element antenna in a free-space communication is equal to the number of minimizing angles and equal to*

$$\begin{aligned} &4K, & \text{if } K \in \mathbb{Z} \\ &2\lfloor 2K \rfloor + 2, & \text{if } K \notin \mathbb{Z} \end{aligned}$$

where $K = L/\lambda$, and $\lfloor K \rfloor$ is the largest integer not greater than K.

3.1.1 Outdoor Experiment

We conducted an experiment to measure the received power at the two-element antenna. The transmitter is placed at distance 10m to the receiver. Figure 3 shows the measured receive patterns for separation $L = \lambda = 12.5$cm and $L = 2\lambda = 25$cm ($f = 2.4$GHz). The results show that the outdoor environments have very similar characteristics to the theoretical patterns in free-space communications. Our antenna design is featured with the capability of adjusting the element separation, by which the two-element antenna can change the *locations* and the *number* of lobes and nulls in the receive pattern (according to Theorem 1 and Theorem 2).

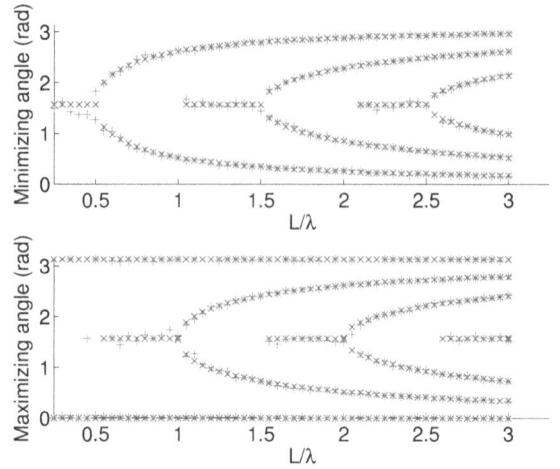

Figure 4: Locations of minimizing and maximizing angles for separations between $\lambda/4$ and 3λ in an outdoor environment. Plus ($+$) indicates experimental results, and cross (X) indicates theoretical predictions.

To study the *change* of locations of lobes and nulls when adjusting the separation, we conducted another experiment, in which the separation is adjusted from minimum value $L_{\min} = \lambda/4 = 3.1$cm to maximum value $L_{\max} = 3\lambda = 37.5$cm. Figure 4 shows the locations of maximizing and minimizing angles for each separation value found in both experimental and theoretical cases. Note that, since the pattern is almost symmetric, only the maximizing and minimizing angles found in one half $[0, \pi]$ of the pattern are shown. As an example, the receive pattern for separation $L = \lambda$ has 4 minimizing angles at $\pm\pi/3$, $\pm 2\pi/3$ and 4 maximizing angles at 0, $\pi/2$, π, and $3\pi/2$, which imply 4 nulls and 4 lobes in Figure 3a. When the separation is increased by a small value to $L' = L + \Delta L$ with $\Delta L \approx 0.6$cm, 2 more minimizing angles and 2 more maximizing angles appear in the pattern (in Figure 4 we see 1 more minimizing angle and 1 more maximizing angle in $[0, \pi]$), which comply with the results of Theorem 2. In addition, Theorem 1 implies that if $K' \approx K$, $\cos \phi' \approx \cos \phi$, then $\phi' \approx \phi$, i.e., the locations of the maximizing and minimizing angles deviate *slightly* from the previous locations. We call this the *continuity* property of the receive pattern. This property is important for the antenna configuration algorithm described later.

3.1.2 Indoor Experiment

In an indoor environment, the receive patterns become more unpredictable due to reflecting and blocking objects. Figure 5 shows that the indoor receive patterns (at different separation values) highly depend on the indoor environment. The locations and the number of lobes and nulls do not always comply with the results of Theorem 1 and Theorem 2. However, similarly to the outdoor scenario, the indoor receive patterns also have the *continuity* property of maximizing angles and minimizing angles, which can be observed from Figure 6: a small adjustment of separation results in a small change of maximizing angles and minimizing angles; in other words, the maximizing angles (or minimiz-

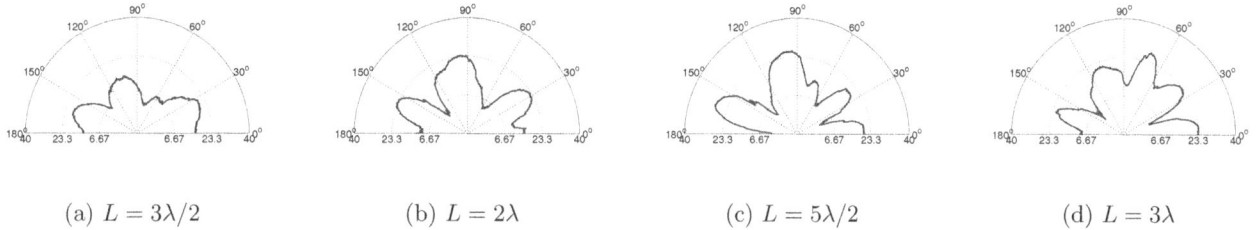

| (a) $L = 3\lambda/2$ | (b) $L = 2\lambda$ | (c) $L = 5\lambda/2$ | (d) $L = 3\lambda$ |

Figure 5: Experimental indoor receive patterns.

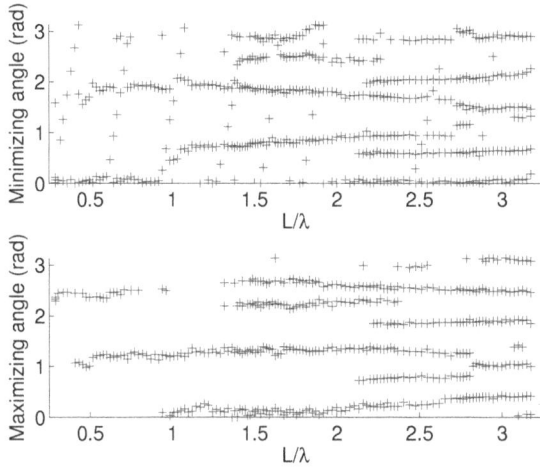

Figure 6: Locations of minimizing and maximizing angles for separation values between 0.25λ and 3.25λ measured in an indoor experiment.

ing angles) of two close values of separation are likely not to deviate much from each other.

3.2 Antenna configuration algorithm

In this section, we derive the algorithm for controlling the two-element antenna to maximize the SJR at the receiver. We note that if both jammer and sender are in the same (or tiny range of) angles relatively to the receiver, Theorem 1 implies that there is no configuration resulting in significantly changing the portion of jamming power in the received signal, as the gains to the transmitters are always (almost) the same. We consider a situation in which the jammer is located in a different direction with respect to the sender.

3.2.1 Outdoor and known locations

In an outdoor environment, if the locations of the communicating and jamming nodes are known, we can maximize the SJR by determining the maximizing angles according to the relative locations between sender and receiver in order to maximize the received power from the sender, and at the same time, determining the minimizing angles according to the relative locations between jammer and receiver in order to minimize the received power from the jammer. Based on the results of Theorem 1, the maximizing and minimizing angles can be precomputed, cf. Algorithm 1.

Algorithm 1 Precomputable configuration for outdoor and known locations

for $L \in [L_{\min}, L_{\max}]$ **do**
  $A_J \leftarrow$ minimizing_angles_to_jammer(L)
  $A_S \leftarrow$ maximizing_angles_to_sender(L)
  for $\phi \in A_J \cap A_S$ **do**
    if $SJR(L, \phi) > SJR(L_{\mathrm{opt}}, \phi_{\mathrm{opt}})$ **then**
      $(L_{\mathrm{opt}}, \phi_{\mathrm{opt}}) \leftarrow (L, \phi)$
    end if
  end for
end for
return $(L_{\mathrm{opt}}, \phi_{\mathrm{opt}})$

In Algorithm 1, minimizing angles and maximizing angles are computed based on the element separation L and relative locations of the nodes and returned as two sets: $A_J = [\phi_{m_1} - \theta, \phi_{m_1} + \theta] \cup \ldots \cup [\phi_{m_k} - \theta, \phi_{m_k} + \theta]$ for minimizing jammer's power and $A_S = [\phi_{k_1} - \theta, \phi_{k_1} + \theta] \cup \ldots \cup [\phi_{k_n} - \theta, \phi_{k_n} + \theta]$ for maximizing sender's power, where k and n are the number of minimizing and maximizing angles found by above theorems, respectively. As for each separation L, there are multiple positions that maximize the SJR, the SJR corresponding to each angle in the intersection of A_J and A_S are compared to find the best configuration. The advantage of Algorithm 1 is that the computations can be done offline, therefore requiring minimal setup time in a real-world deployment.

3.2.2 Unknown locations

For outdoor environments and unknown locations of nodes, Algorithm 1 is not applicable. For indoor environments, even if the locations of nodes are known, the channel highly depends on the specific environment and results in unpredictable patterns. In this section, we present the antenna configuration algorithms that work for both outdoor and indoor environments.

Our goal is to maximize the SJR at the receiver. According to Theorem 1, changing separation results in new locations of maximizing and minimizing angles, therefore yielding different gains for the jammer and the sender (as they are not in the same direction). Consider a powerful jammer whose power dominates the received signal. Changing the antenna configuration to null the jammer, we would reduce the received signal's power. Thus, *maximizing* the SJR implies *minimizing* the total received power at the receiver. For low-power jammer, this implication is not applied, however the algorithms described below are still useful when combining with the digital cancellation technique to recover the user data.

35

Brute-force algorithm.

To minimize the total received power, a "brute-force" search would yield the best configuration: this search would measure the received power at the two-element antenna for all possible configurations and select the one corresponding to the minimum power.

Algorithm 2 Brute-force for unknown node locations

---

function bruteforce($L_{\min}, L_{\max}, \phi_{\min}, \phi_{\max}$)
    for $\phi_{\min} \leq \phi \leq \phi_{\max}$ **do**
        for $L_{\min} \leq L \leq L_{\max}$ **do**
            if $P(L, \phi) < P(L_{\mathrm{opt}}, \phi_{\mathrm{opt}})$ **then**
                $(L_{\mathrm{opt}}, \phi_{\mathrm{opt}}) \leftarrow (L, \phi)$
            end if
        end for
    end for
    return $(L_{\mathrm{opt}}, \phi_{\mathrm{opt}})$

Without knowledge of node locations, we cannot rely on Theorem 1 to compute the optimal configuration. Instead, the brute-force approach tries each configuration by varying the rotational angle and the element separation within the physical limits and measuring the received power. Given the large number of separation values and angle values, such approach would take a significant amount of time to find the best configuration.

Fast algorithm.

Recall the continuity property of the receive pattern: continuously changing the separation results in new locations of maximizing angles and minimizing angles in the small vicinity of the previous ones. Based on this property, we propose the *fast algorithm*, cf. Algorithm 3.

Algorithm 3 Fast algorithm for unknown node locations

---

function fast($L_0, L_1, L_2, \phi_0, \phi_1, \phi_2$)
    $L_{\mathrm{opt}} \leftarrow L_0, \phi_{\mathrm{opt}} \leftarrow \phi_0$
    repeat
        for $\phi_1 \leq \phi \leq \phi_2$ **do**
            if $P(L_{\mathrm{opt}}, \phi) < P(L_{\mathrm{opt}}, \phi_{\mathrm{opt}})$ **then**
                $\phi_{\mathrm{opt}} \leftarrow \phi$
            end if
        end for
        for $L_1 \leq L \leq L_2$ **do**
            if $P(L, \phi_{\mathrm{opt}}) < P(L_{\mathrm{opt}}, \phi_{\mathrm{opt}})$ **then**
                $L_{\mathrm{opt}} \leftarrow L$
            end if
        end for
        $L_1 \leftarrow L_{\mathrm{opt}} - \Delta L; L_2 \leftarrow L_{\mathrm{opt}} + \Delta L$
        $\phi_1 \leftarrow \phi_{\mathrm{opt}} - \theta; \phi_2 \leftarrow \phi_{\mathrm{opt}} + \theta$
    until $(L_{\mathrm{opt}}, \phi_{\mathrm{opt}})$ unchanged
    return $(L_{\mathrm{opt}}, \phi_{\mathrm{opt}})$

To find the optimal configuration, Algorithm 3 is run with $L_1 = L_{\min}$, $L_2 = L_{\max}$, $\phi_1 = \phi_{\min}$, $\phi_2 = \phi_{\max}$, and the current configuration (L_0, ϕ_0). The configuration search is, first, started by rotating the antenna between the given range while fixing the separation at the given separation value L_0. By measuring the received power at each angular position, we locate the angle ϕ_{opt} that gives the minimum received power for the current separation value L_0. We know

that ϕ_{opt} found in this step is not necessarily the best one for other separation values. Therefore, in the next step, different separations within the given range $[L_1, L_2]$ are tried to improve the configuration. The configuration search in these two steps relies on the continuity property: if there is a better configuration, it is likely to be found in small vicinity of the most recently optimal configuration. We repeat these steps until no better configuration is found. We note that, before repeating the search, we reduce the search range by setting new values for L_1, L_2, ϕ_1, ϕ_2.

Algorithm 3 is much faster than brute-force, as it probes the optimal angle and separation values separately. We emphasize that the configuration returned by the fast algorithm is not essentially the best configuration, however as shown in Section 6, is comparable to brute-force.

To have a hybrid anti-jamming system, we use the fast algorithm to control the two-element antenna in parallel with digital processing. The fast algorithm is performed to reduce the received to such power levels that the signal received at the two-element antenna can be directly decoded.

4. DIGITAL JAMMING CANCELLATION

The digital jamming cancellation improves the system, in the case that the *first stage* cannot completely remove the jamming signal when the jammer power is extremely high (over 28 dB). The *second stage* comprises digital processing components as shown in Figure 7. The main idea of digital jamming cancellation is to eliminate the jamming signal from equation (1) to obtain the decodable user signal by equation (2). Therefore, the most important component in this stage is the *gain ratio estimation* component, which estimates the gain ratio a. In MIMO systems [36], the channel characteristics are usually estimated by training sequences. This technique is not applicable in our scenario, as the jamming signal is unknown to the receiver. Consequently we derive our own estimate technique described as follows.

4.1 Gain ratio estimation

In general, the channel gains affected by the communication medium are represented as complex numbers which introduce magnitude and phase change in the received signals. Our digital processing techniques are applied to sequences of samples taken from the analog input at discrete time $t = t_0, t_0 + \tau, t_0 + 2\tau, \ldots$ where τ is the sampling period and t_0 is the time when the signals arrive at the receiver input. Equation (1) can be rewritten in the time domain:

$$R_1(t) = h_{S1}(t)S(t) + h_{J1}(t)J(t)$$
$$R_2(t) = h_{S2}(t)S(t) + h_{J2}(t)J(t)$$

Removing the jamming signal involves the estimation of $a(t) = \frac{h_{J2}(t)}{h_{J1}(t)}$.

4.1.1 Magnitude estimation

The received power at the two-element antenna in the sampling time range $[t_0 - (n-1)\tau, t_0]$ of the past n samples before t_0 is

$$P_1(t_0) = \frac{1}{n} \sum_{t=t_0-(n-1)\tau}^{t_0} |h_{S1}(t)S(t) + h_{J1}(t)J(t)|^2$$

Since the jammer's signal and user's signal are highly uncorrelated, i.e. $\sum h_{S1}(t)h_{J1}(t)S(t)J(t) = 0$, the received

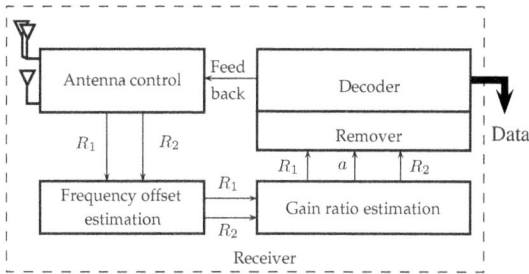

Figure 7: Receiver components.

power at the two-element antenna is rewritten as:

$$\begin{aligned}
P_1(t_0) &= \frac{1}{n}\left(\sum_{t_0}|h_{S1}(t)S(t)|^2 + \sum_{t_0}|h_{J1}(t)J(t)|^2\right) \\
&= \frac{1}{n}\left(|h_{S1}|^2\sum_{t_0}|S(t)|^2 + |h_{J1}|^2\sum_{t_0}|J(t)|^2\right)
\end{aligned}$$

where the second equality comes from the fact that the channel gains are considered constant during the period $[t_0 - (n-1)\tau, t_0]$, due to slow-fading characteristics in a narrowband communication [36], i.e., $h_{S1}(t) = h_{S1}$, $h_{S2}(t) = h_{S2}$, $h_{J1}(t) = h_{J1}$, $h_{J2}(t) = h_{J2}$. Similarly, the power received at the single-element antenna can be represented as $P_2(t_0) = \frac{1}{n}\left(|h_{S2}|^2\sum_{t_0}|S(t)|^2 + |h_{J2}|^2\sum_{t_0}|J(t)|^2\right)$.

If the portion of jamming power in $P_1(t_0)$ and $P_2(t_0)$ were significantly greater than that of the sender, one could estimate $|a| = \left|\frac{h_{J2}}{h_{J1}}\right| = \frac{P_2(t_0)}{P_1(t_0)}$. In order to estimate $|a|$ in more general cases, we apply another approach, in which the receiver is assumed to be able to determine at a specific time instant whether there is a data transmission or whether there is an interference, and therefore, can record the signals level in those periods. We note that in simple scenarios, where the jammer only emits noise, the jammer can be identified when the received signal is undecodable. In complex scenarios, if the jammer is capable of transmitting "user-like" data (e.g., the jammer is a compromised user), the system needs more sophisticated methods to identify whether the received signal is the jammer's signal. By rate adaptation algorithm, the sender can transmit the signal at different levels of power at different time, which also affects the accuracy of the jammer identification process. We leave those complex scenarios for future work. In this work, we consider a sender with basic constant-power modulation scheme (BPSK, QPSK) and a "dump" with unknown signal jammer in the following two cases:

Sender transmitted before collision.

If the sender transmitted before the jammer causes interference, the receiver estimates $|a|$ by the following steps:

- Measures $P_i(t_0)$ in the period t_0, which contains only the power of the sender's signal received at both (two-element and single-element) antennas, $P_{Si}(t_0) = P_i(t_0) = \frac{1}{n}|h_{Si}|^2\sum_{t_0}|S(t)|^2$ ($i = 1$ denotes the two-element antenna, and $i = 2$ denotes the single-element antenna). As the sender's power is constant, we obtain $P_{Si}(t) = P_{Si}(t_0) = P_{Si}$ for any other period $t > t_0$.

- Measures $P_i(t_1)$ in the interference period t_1, $|a|$ can be computed: $|a| = \left|\frac{h_{J2}}{h_{J1}}\right| = \sqrt{\frac{P_2(t_1)-P_{S2}}{P_1(t_1)-P_{S1}}}$.

Jammer transmitted before collision.

If the jammer is known to jam before the collision time t_0, the receiver measures the power at both (two-element and single-element) antennas before the collision, $P_i(t_0) = \frac{1}{n}|h_{Ji}|^2\sum_{t_0}|J(t)|^2$. In the collision period t_1, the receiver measures $P_i(t_1)$. Since the time period immediately before and after the collision is short, the jammer's power remains almost constant, i.e., $P_{Ji}(t_1) \approx P_{Ji}(t_0)$. This allows the sender's power at each antenna to be found by $P_{Si} = P_i(t_1) - P_i(t_0)$. Knowing P_{Si}, $|a|$ can be estimated by the last step described in the first case.

4.1.2 Phase estimation

The phase difference ϕ between $R_1(t)$ and $R_2(t)$ is determined by $\phi = \tan^{-1}\left(-\frac{\sum_t[I_1(t)Q_2(t) - I_2(t)Q_1(t)]}{\sum_t[I_1(t)I_2(t) + Q_1(t)Q_2(t)]}\right)$, where $I_1(t) = \text{Re}[R_1(t)]$, $Q_1(t) = \text{Im}[R_1(t)]$, $I_2(t) = \text{Re}[R_2(t)]$, $Q_2(t) = \text{Im}[R_2(t)]$ represent the real and imaginary parts of the received signals. Similarly to the approach used in estimating the magnitude, we derive ϕ based on the phase difference ϕ in the periods before and after the collision. In software-defined radio, for both magnitude and phase estimation, the signal processing operations are done for chunks of n samples taken from the analog input.

4.2 Removing and Decoding

When the gain ratio a is estimated correctly, the jamming signal can be removed completely from the received signals by solving equation (2). The residual signal $b \cdot S$ is sent to the decoder to decode the data. The gain b of the residual signal is considered as a new channel gain of the signal after removing the jamming signal. Therefore, the data can be decoded by the decoder with well-known decoding techniques [12, 27] in software-defined radio. Consequently, estimation of b is not required.

4.3 Practical issues

In practice, we need to address the issue of frequency offset between the received signals which are unavoidable in real devices. Moreover, the multipath problem is always an interesting part of systems working indoor.

Frequency offset estimation.

With the goal of providing a zero-knowledge anti-jamming system, manual calibration for compensating the frequency offset is not desired in our system. The frequency offset between the received signals is estimated in real-time by $\Delta f^* = \text{argmax}_{\Delta f}|\mathscr{F}\{R_1(t)R_2^*(t)\}|$, for which the performance is optimized when the chunk size is a power of 2.

Dealing with multipath.

Our system also works efficiently in indoor environments (see Section 6). Due to space limits, we will now only provide a short, intuitive justification. In an indoor environment, due to reflection, multiple copies of the transmitted signals arrive at the receive antennas

$$\begin{aligned}
R_1 &= \left(\sum_k h_{S1}^{(k)}\right)S + \left(\sum_k h_{J1}^{(k)}\right)J \\
R_2 &= \left(\sum_k h_{S2}^{(k)}\right)S + \left(\sum_k h_{J2}^{(k)}\right)J
\end{aligned} \tag{3}$$

where $h_{Si}^{(k)}$, $h_{Ji}^{(k)}$ denote the channel gain of the k-th path from the sender and the jammer to the receiver, respectively.

By letting $h_{Si} = \sum_k h_{Si}^{(k)}$ and $h_{Ji} = \sum_k h_{Ji}^{(k)}$, equation (3) becomes equivalent to equation (1). Thus, sums R_1 and R_2 are now considered as line-of-sight signals transmitted from a different location. Recall that Algorithm 3 is designed to deal with unknown location transmitters, so it is applicable in this scenario in an attempt to reduce the multipath jamming signal.

Low-power jammer

As mentioned in Section 3, the antenna control algorithms rely on the implication of minimum received power. In case of low-power jammer, minimizing total received power does not necessarily maximizes the SJR at the two-element antenna. However, the antenna algorithms result in the change in portion of jammer power in the total received power at the two-element antenna compared to that at the single-element antenna, i.e., $h_{J1}/h_{S1} \neq h_{J2}/h_{S2}$, which allows obtaining the residual signal in equation (2). Therefore, when combining with the digital stage, the antenna algorithms help eliminating the jamming signal. Although the first stage does not necessarily reduce the jamming power, it helps the second stage to derive the residual signal for successful decoding, thus is useful even for low-power jammers.

Variable-power jammer

Recall the estimation of the gain ratio; as soon as the sender's power portion is determined, it can be used to derive the jammer's power portion (and hence their ratio a). Therefore, as long as the antenna remains in the same configuration, the power of the signal received from the sender is constant during the collision period, allowing the system to remove the variable-power jamming signal.

5. PROTOTYPE AND IMPLEMENTATION

Our system consists of one receiver node and two transmitter nodes. We use a software-defined radio [12] to deploy our testbed. The digital signal processing is done by a host computer connected to the receiver.

Nodes: Each node is deployed on an Ettus USRP device [28] with RFX2400 daughterboards. The jammer and sender use a single-element antenna for transmission. The receiver has a single-element and a two-element antenna for signal reception. All antenna elements are Titanis 2.4 GHz dipole Swivel SMA antennas. The receiver transfers digital samples to the host computer through an Ethernet link.

Two-element antenna: Our two-element antenna (Figure 2) comprises two Titanis antennas. Signals received from two elements are added together through a HyperLink Technologies SCW02 combiner, which is then connected to one input of the receiver (the other input is connected to the single-element antenna). To build the antenna frame, we used Autodesk Inventor 2012 to design it and built it using a uPrint Plus 3D printer [37]. The mechanical movement of the two-element antenna is controlled by two servos.

- **Rotation:** To rotate the antenna frame, we use a Hitec HS-485HB servo and attach the antenna frame to its rotating shaft. The HS-485HB servo is capable of rotating up to 200 degrees. However, we only need 180 degrees for half-circle rotation of the antenna, as two elements of the antenna are attached into the frame symmetrically with respect to the shaft. We set $\phi_{\min} = 0$ and $\phi_{\max} = \pi$ for the configuration algorithms.

Table 1: Comparison of brute-force and fast algorithm

	Brute-force	Fast
Reduction of power	15-30dB	15-28dB
Reduction compared to brute-force in each experiment	–	<6dB
Running time	> 5mins	5-18s

- **Separation:** We use a Hitec HS-785HB servo (capable of rotating up to 3.5 circles) to transform the rotation movement to element separation by using a combination of gears and racks adjustable on the antenna frame. The frame allows the separation adjusted from $L_{\min} = 3.1$cm to $L_{\max} = 37.5$cm.

The servos operate based on pulse-width modulation signals given to their input. To generate those signals from the host computer, we use a Crossbow TelosB mote for receiving commands from the host and generating signals with appropriate pulse-width.

6. EVALUATION

In this section, we evaluate our system for indoor environments using three nodes: jammer, sender, and receiver. In our testbed environment, there are usual blocking objects and reflectors, such as walls, desks, metallic cabinets, and office space separators. We run the testbed at a fixed frequency of 2.4GHz ($\lambda \approx 12.5$cm).

6.1 Antenna configuration

Basic operations

Two basic operations of the two-element antenna are the rotation and the separation adjustment. We measure the performance of those operations in terms of running time. The half-circle rotation takes roughly 1 second to rotate the antenna frame from 0 to π. The rotation servo is capable of rotating in sub-degree step. The separation adjustment takes about 2 seconds to increase the separation from 3.1cm to 37.5cm. The minimal separation step is ≈ 3.5 mm.

Brute-force algorithm

The brute-force algorithm is evaluated in terms of running time and capability of reducing jamming power. We deploy three nodes in a typical indoor environment. The jammer is set to transmit at 30dB higher power than the sender. Figure 8 shows the running time versus the power received at the two-element antenna relatively to the minimum value during the brute-force search. In this specific setup, using brute-force can eliminate up to almost 30dB of the jammer's power. Depending on the environment, the optimal configuration can be found at different time and the capability of reducing jamming power may vary. The total time to complete the brute-force search is more than 5 minutes as it tries all possible configurations.

Fast algorithm

In order to evaluate the performance of the fast algorithm, we run the fast algorithm with the same setup (same node locations and same settings of transmit power). The capability of reducing jamming power is shown explicitly in two steps in Figure 9. While the first step (rotation only)

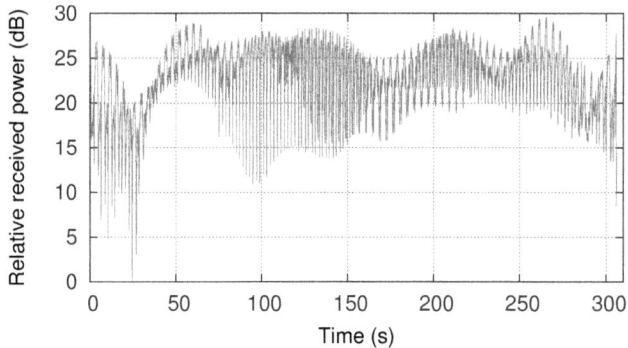

Figure 8: Brute-force: total received power relative to total received power's minimum value during search.

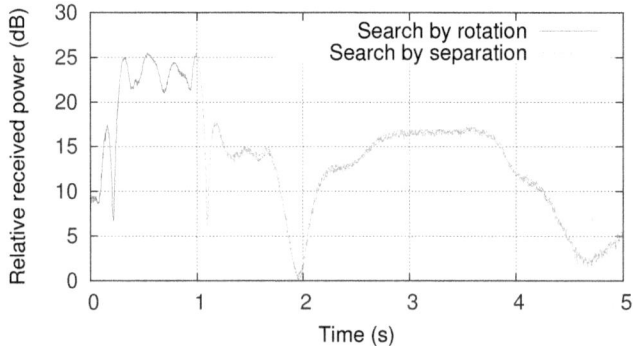

Figure 9: Fast: total received power relative to total received power's minimum value during search.

can find a configuration that reduces the received power to more than 15dB, the second step (separation adjustment only) helps improving the power reduction of the jammer to roughly 25dB, which is not far compared to the performance of the brute-force algorithm. The running time of the fast algorithm in this experiment takes only 5 seconds to complete. We note that the running time of the fast algorithms depend on the environment. Table 1 summarizes the performance of the brute-force and fast algorithms in various experiments with different setups.

6.2 Anti-jamming performance

We investigate the performance of our system by examining the probability of bit error of the decoded data after removing the jamming signal. In this experiment, we use basic DBPSK modulation for data transmission between sender and receiver and for generating the jamming signal of the jammer. The bit rate used by sender is 500kbps. The receiver runs continuously during the experiment. In order to investigate the probability of bit error, sent and received signals are recorded at each node and later transferred to the host computer to compare and count the error bits. In the experiment, we keep the power of the sender constant and increase the power of the jammer gradually after each run to a threshold that the data becomes undecodable.

To evaluate our system's performance, we compare three cases: (a) decode the received signal directly from the receiver's single-element antenna, i.e. without any anti-jamming technique, (b) decode the received signal from the receiver's two-element antenna, and (c) decode the residual signal after applying the digital jamming cancellation. The average

probability of bit error is presented in Figure 10a. We visualize the BER in absolute (not log-scale) to make it easier to show the relative gain between combinations of techniques. Without the antenna auto-configuration capability (AA) and digital jamming cancellation (DC), the probability of bit error at the single-element antenna increases quickly when the jamming-to-signal ratio is greater than 3dB. Using the antenna auto-configuration with fast algorithm, the receiver can resist the jammer up to 28dB. The overall anti-jamming performance of the system is around 48dB when we combine two stages. The results demonstrate that our system is able to work efficiently in indoor environments.

DQPSK modulation

To study the effects of a higher-rate modulation on the performance of our system, we repeat the above experiments with DQPSK modulation at a doubled bit rate of 1Mbps.

Figure 10a compares the probability of bit error between DBPSK and DQPSK modulation. The performance of the system, when using DQPSK modulation, is around 42dB. Compared to the case of DBPSK modulation, the efficiency of the anti-jamming capability drops around 4 to 5dB. This is not surprising, since the constellation of the DQPSK modulation has a smaller minimum distance which results in higher probability of bit error [27]. Considering only the performance of the digital jamming cancellation, there is no significant difference in the capability of jamming cancellation between the two cases. This shows the efficiency of the estimation techniques applied in the second stage.

Variable power jammer

In the above experiments, the jammer transmitted at constant transmission power. To evaluate our system against a variable-power jammer, we modify the jammer such that after every 40 bytes it changes the transmit power to a random level within the range of 10 dB compared to the specified average power in each run. For this experiment, we use DBPSK modulation. We note that during the experiment, the antenna configuration does not change and is capable of removing a portion of about 28 dB in jamming power. Figure 10b shows the comparison between variable and constant jamming power cases in probability of bit error versus the average power in each run. The results show a performance degradation of 5-6 dB, demonstrating that the gain estimation is adaptive to the change of jamming power as long as the sender's power and the antenna remain unchanged.

7. DISCUSSION AND RELATED WORK

Based on our hybrid system, one can envision nodes with the two-element antenna that are capable of simultaneous signal transmission and reception. For example, in previous work [15], a design based on a balanced/unbalanced transformer could completely eliminate the self-signal. Similar components can augment our system, therewith enabling *full duplex* wireless communication.

In the context of *multi-hop wireless networks*, the two-element antennas open new opportunities for communicating nodes to enable selective transmission to desired destinations while at the same time avoiding jamming from concurrent or malicious nodes. This allows multiple simultaneous senders and increases the network's total throughput.

Anti-jamming has been an active area of research for decades. Techniques developed at the physical layer [30] in-

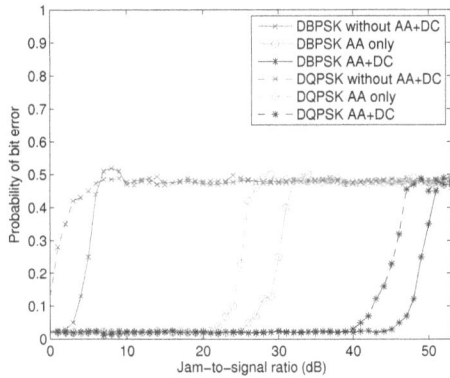

(a) DBPSK and DQPSK modulation

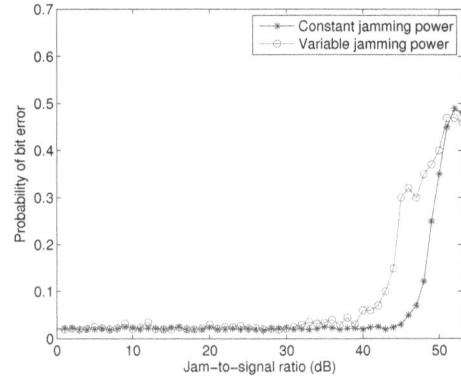

(b) Variable vs. constant power jamming

Figure 10: Bit error measurements.

clude directional antennas [18], spread spectrum communication, and power, modulation, and coding control. More recently, research has also addressed higher layers [1, 2, 7, 9, 11, 16, 17, 20, 21, 25, 31–33, 39–41]. However, given the ease of building *high power* jamming devices, there is still a strong need for efficient and flexible techniques operating at the physical layer. There is a demand for low-cost solutions mitigating the effects of jammers that are orders of power stronger than legitimate communication.

While spread spectrum has been a solution of choice for anti-jamming, it suffers from a need for pre-shared secrets between the communicating nodes. Several solutions were recently proposed for alleviating the need for pre-shared secrets [5, 11, 16, 19, 21, 32, 33]. However, they are not designed to tackle powerful jammers (meaning jammer with power 4-5 orders of magnitude higher than the transmitting node).

Other recent work has demonstrated mechanisms for cancelling interference. This work has found applications in protecting the confidentiality of communication [8, 13]. However cancelling *powerful, unknown* jammers results in several challenging problems such as jammer signal identification and channel estimation.

The closest related work to our system consists of phased array antennas and MIMO systems. Phased array antennas were very well studied since the 1950s [3, 4, 18, 35]. Likewise, multiple input multiple output systems (MIMO) were also very well studied since the 90s [36]. Our design and approach have unique characteristics that distinguishes them from prior work. Similar performance phased array antennas consist of a fairly large number of *fixed*-position elements aiming at creating a directed beam that can be electronically and digitally repositioned. A major goal is to minimize the impact of side lobes. Adaptive beamforming with algorithms such as MVDR and MMSE beamformers aims at minimizing the impact of the sidelobe, using a fairly large number of antennas and are considered to be more adequate for radar systems. In contrast, our system's goal is to create one or multiple nulls to minimize the jammer's impact while maximizing the legitimate user signal power and preparing the signal for a digital MIMO-like second stage of interference cancellation. To the best of our knowledge, our two-elements mechanical beam-forming design is new and is supplemented with an automatic configu-

ration algorithm that achieves 28dB jammer cancellation in less than 20 seconds – both in indoor and outdoor environments. The system reaches 48dB cancellation when combined with our second-stage digital jamming cancellation. Existing phased array antennas achieving a gain of 48dB require hundreds of elements even with high-end, expensive 7-bit phase shifters [3, 22]. Our two-elements mechanical steering can be controlled with low-cost micro controllers instead of requiring expensive DSP boards. Our second-stage digital jamming cancellation is in principle similar to MIMO. However, existing algorithms assume that the incoming signals are of similar power, transmitted by a cooperating node, with the possibility to embed training sequences for the channel estimation.

8. CONCLUSION AND FUTURE WORK

The availability of software radios and commodity jammers are making jamming of wireless communication a problem of increasing importance for many cyber-physical applications. To mitigate the problem of jammers that are significantly more powerful than the transmitting nodes, we have designed, physically built, and evaluated a hybrid system of mechanical beam/null-forming and MIMO-like digital interference cancellation. Our novel antenna design and algorithms have several important characteristics and advantages compared to phased array antennas and MIMO techniques e.g., simplicity, low-cost, convergence speed. It allows a flexible creation of multiple nulls to cancel the effects of multi-path jamming. We have developed several techniques to effectively cancel the remaining interference digitally and verified their effectiveness in practice. To the best of our knowledge, this is the first academically published low-cost system that reduces the effects of powerful unknown jammers by almost five orders of power. As future work, we plan to extend our techniques from BPSK/QPSK modulation to multi-carrier BPSK/QPSK OFDM and higher order modulation. We plan to extend our antenna configuration algorithm and analytically quantify the worst-case gain loss as a function of speed in comparison with the brute-force configuration. We also believe that the unique beam-forming characteristics of our system results in new research problems in the context of multi-hop wireless network topology control in the presence of malicious interference.

References

[1] B. Awerbuch, A. Richa, and C. Scheideler. A jamming-resistant mac protocol for single-hop wireless networks. In *PODC*, pages 45–54, 2008.

[2] M. Bender, M. Farach-Colton, S. He, B. Kuszmaul, and C. Leiserson. Adversarial contention resolution for simple channels. In *SPAA*, pages 325–332, 2005.

[3] E. Brookner. Phased arrays and radars – past, present and future. *Microwave Journal*, 49(1):24–46, 2006. ISSN 01926225.

[4] E. Brookner. Phased-array radar: Past, astounding breakthroughs, and future trends. *Microwave Journal*, 51(1):30–50, 2008. ISSN 01926225.

[5] A. Cassola, T. Jin, G. Noubir, and B. Thapa. Efficient spread spectrum communication without pre-shared secrets. *IEEE Transactions on Mobile Computing*, to appear.

[6] A. Cassola, W. Robertson, E. Kirda, and G. Noubir. A practical, targeted, and stealthy attack against wpa enterprise authentication. In *Proceedings of NDSS*, 2013.

[7] J. Chiang and Y.-C. Hu. Dynamic jamming mitigation for wireless broadcast networks. In *INFOCOM*, pages 1211–1219, 2008.

[8] J. Choi, M. Jain, K. Srinivasan, P. Levis, and S. Katti. Achieving single channel, full duplex wireless communication. In *MOBICOM*, pages 1–12, 2010.

[9] J. Dong, R. Curtmola, and C. Nita-Rotaru. Practical defenses against pollution attacks in intra-flow network coding for wireless mesh networks. In *In Proceedings of WiSec*, pages 111–122, 2009.

[10] FCC. Jammer enforcement, 2012. http://www.fcc.gov/encyclopedia/jammer-enforcement, http://transition.fcc.gov/eb/News_Releases/DOC-304575A1.html.

[11] S. Gilbert, R. Guerraoui, and C. Newport. Of malicious motes and suspicious sensors: On the efficiency of malicious interference in wireless networks. In *OPODIS*, 2006.

[12] GNU. Gnu radio. http://www.gnuradio.org.

[13] S. Gollakota, H. Hassanieh, B. Ransford, D. Katabi, and K. Fu. They can hear your heartbeats: Non-invasive security for implantable medical devices. In *SIGCOMM*, pages 2–13, 2011.

[14] iClinks. Scada and industrial automation, ethernet scada and ethernet i/o. http://www.iclinks.com/.

[15] M. Jain, J. Choi, T. Kim, D. Bharadia, S. Seth, K. Srinivasan, P. Levis, S. Katti, and P. Sinha. Practical, real-time, full duplex wireless. In *Proceedings of international conference on Mobile computing and networking*, pages 301–312, Las Vegas, USA, 2011.

[16] T. Jin, G. Noubir, and B. Thapa. Zero pre-shared secret key establishment in the presence of jammers. In *In Proceedings of ACM MobiHoc*, pages 219–228, 2009.

[17] C. Koo, V. Bhandari, J. Katz, and N. Vaidya. Reliable broadcast in radio networks: The bounded collision case. In *PODC*, pages 258–264, 2006.

[18] J.-D. Kraus. *Antennas*. Mcgraw Hill Higher Education; 3rd edition, 2001.

[19] A. Liu, P. Ning, H. Dai, Y. Liu, and C. Wang. Defending dsss-based broadcast communication against insider jammers via delayed seed-disclosure. In *Proceedings of ACSAC'2010*, 2010.

[20] S. Liu, L. Lazos, and M. Krunz. Thwarting inside jamming attacks on wireless broadcast communications. In *In Proceedings of ACM WiSec*, pages 29–40, 2011.

[21] Y. Liu, P. Ning, H. Dai, and A. Liu. Randomized differential dsss: jamming-resistant wireless broadcast communication. In *INFOCOM*, pages 695–703, 2010.

[22] R. Mailloux. *Phased Array Antenna Handbook*. Artech Print on Demand, 2005.

[23] R. Miller. FCC steps up crackdown on cell jammers, 2012. http://www.securitysystemsnews.com/article/fcc-steps-crackdown-cell-jammers.

[24] NPR. Congress passes FAA bill that speeds switch to GPS, 2012. http://www.npr.org/.

[25] K. Pelechrinis, S. V. Krishnamurthy, C. Gkantsidis, and I. Broustis. Ares: An anti-jamming reinforcement system for 802.11 networks. *CoNEXT*, pages 181–192, 2009.

[26] PG & E. Smart meters by the numbers, 2011. http://www.pge.com/myhome/customerservice/smartmeter/deployment/.

[27] J. G. Proakis and M. Salehi. *Digital Communications*. McGraw-Hill, 5 edition, 2007.

[28] E. Research. Universal software radio peripheral. http://www.ettus.com/.

[29] SEMAPHORE. Integrated scada, control, and communication solutions. http://www.cse-semaphore.com/, 2011.

[30] M. K. Simon, J. K. Omura, R. A. Scholtz, and B. K. Levitt. *Spread Spectrum Communications Handbook*. McGraw-Hill, 2001.

[31] D. Slater, P. Tague, R. Poovendran, and B. J. Matt. A coding-theoretic approach for efficient message verification over insecure channels. In *Proceeding of ACM WiSec*, 2009.

[32] M. Strasser, C. Popper, and S. Capkun. Efficient uncoordinated FHSS anti-jamming communication. In *Proceedings of ACM MobiHoc*, 2009.

[33] M. Strasser, C. Popper, S. Capkun, and M. Cagalj. Jamming-resistant key establishment using uncoordinated frequency hopping. In *Proceedings of IEEE Symposium on Security and Privacy*, 2008.

[34] Synetcom. Synetcom industrial wireless systems. http://www.synetcom.com/.

[35] H. V. Trees. *Detection, Estimation, and Modulation Theory, Part I*. Wiley & Sons, 2001.

[36] D. Tse and P. Viswanath. *Fundamentals of wireless communication*. Cambridge University Press, New York, NY, USA, 2005.

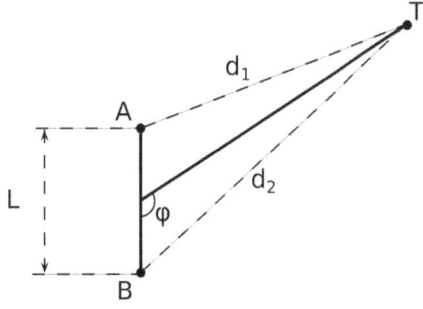

Figure 11: The two-element antenna in a free-space communication with one transmitter.

[37] uPrint. Uprint plus.
http://www.uprint3dprinting.com/.

[38] vMonitor. Scada wireless systems.
http://www.vmonitor.com/, 2011.

[39] M. Wilhelm, I. Martinovic, J. B. Schmitt, and
V. Lenders. Short paper: reactive jamming in wireless
networks: how realistic is the threat? In *Proceedings
of ACM WiSec*, 2011.

[40] W. Xu, K. Ma, W. Trappe, and Y. Zhang. Jamming
sensor networks: attack and defense strategies. *IEEE
Network*, 20(3):41–47, 2006.

[41] W. Xu, W. Trappe, and Y. Zhang. Channel surfing:
defending wireless sensor networks from interference.
In *Proceedings of IPSN*, 2007.

[42] W. Zhang, M. Kamgarpour, D. Sun, and C. Tomlin.
A hierarchical flight planning framework for air traffic
management. *Proceedings of the IEEE*, 100(1), 2012.

APPENDIX

A. PROOFS

PROOF. [Theorem 1] In Figure 11, we consider the antenna configuration (L, ϕ) and the signals transmitted from T and received at antenna elements A and B. We assume a narrowband slow fading communication channel, therefore the signal received at A and B does not significantly differ in frequency offset, channel attenuation, fading, etc. The received signals at the two elements are represented by $r_A(t) = g(x) \cos(2\pi f t + 2\pi \frac{d_1}{\lambda})$, $r_B(t) = g(x) \cos(2\pi f t + 2\pi \frac{d_2}{\lambda})$, where $g(x)$ contains the transmitted data, f is the carrier frequency, λ is the carrier wavelength, and t denotes the receiving time. The sum of two signals at the output of the combiner, $r(t) = r_A(t) + r_B(t) = 2g(x) \cos(\pi \frac{d_1-d_2}{\lambda}) \cos(2\pi f t + \pi \frac{d_1+d_2}{\lambda})$, is a signal of amplitude $|2g(x) \cos(\pi \frac{d_1-d_2}{\lambda})|$. Regardless of the transmitted data, the amplitude of $r(t)$ depends on the value of $|\cos(\pi \frac{d_1-d_2}{\lambda})|$. Since the distances between the transmitter and the receiver elements are much larger than the element separation, i.e., $d_1 \gg L$, $d_2 \gg L$, we have $d_1 - d_2 \approx L \cos \phi$. Let $h(\phi) = \cos^2(\pi K \cos \phi)$, $K = L/\lambda$. We investigate the amplitude of $r(t)$ indirectly by investigating $h(\phi)$. Note that by definition of maximizing angles and minimizing angles, the maximum and minimum values of $|r(t)|$ are not necessarily equal to 0 or 1. In fact, they are the roots of $h'(\phi) = 0$, where $h'(\phi) =$

$2\pi K \sin \phi \sin(2\pi K \cos \phi)$ is the derivative of $h(\phi)$. Roots of $h'(\phi) = 0$ satisfy the following conditions:

$$\sin \phi = 0 \tag{4}$$
$$\text{or} \quad \sin(2\pi K \cos \phi) = 0 \tag{5}$$

Letting $h_1(\phi) = 4\pi^2 K^2 \sin^2 \phi \cos(\pi K \cos \phi)$ and $h_2(\phi) = 2\pi K \cos \phi \sin(\pi K \cos \phi)$, we have $h''(\phi) = h_2(\phi) - h_1(\phi)$.

First, we consider the equation (4). Let ϕ_1 be a root of (4), i.e., $\sin(\phi_1) = 0$, then $\cos(\phi_1) = \pm 1$, and $\phi_1 = 0$ or $\phi_1 = \pi$. As a result, $h_1(\phi_1) = 0$, and $h_2(\phi_1) = \pm 2\pi K \sin(\pm 2\pi K) = 2\pi K \sin(2\pi K)$ (the last equality is due to x having same sign as $\sin x$). Now that $h''(\phi_1) = h_2(\phi_1) = 2\pi K \sin(2\pi K)$.

- If $\{K\} \leq \frac{1}{2}$, $h''(\phi_1) \geq 0$, then ϕ_1 is a minimizing angle.
- If $\{K\} \geq \frac{1}{2}$, $h''(\phi_1) \leq 0$, ϕ_1 is a maximizing angle.

Now consider the equation (5). Let ϕ_2 be a root of (5), i.e., $\sin(2\pi K \cos \phi_2) = 0$, then we have $h_2(\phi_2) = 0$, and $h''(\phi_2) = -h_1(\phi_2) = -4\pi^2 K^2 \sin^2(\phi) \cos(\pi K \cos \phi)$. Note that $\cos(2\pi K \cos \phi_2) = \pm 1$.

- If $\cos(2\pi K \cos \phi_2) = 1$, or $\cos \phi_2 = k/K$, then $h''(\phi_2) < 0$, and ϕ_2 is a maximizing angle.
- If $\cos(2\pi K \cos \phi_2) = -1$, or $\cos \phi_2 = (k + \frac{1}{2})/K$, then $h''(\phi_2) > 0$, and ϕ_2 is a minimizing angle.

In conclusion, ϕ is a maximizing angle, if $\cos \phi = k/K$, or a minimizing angle, if $\cos \phi = (k + \frac{1}{2})/K$, $k \in \mathbb{Z}$. In addition, if $\{K\} \geq \frac{1}{2}$, we have two more maximizing angles at 0 and π; otherwise, they are two additional minimizing angles. □

PROOF. [Theorem 2] First, we observe that there is always one null between two lobes, and one lobe between two nulls, that is the number of minimizing angles equals the number maximizing angles. Therefore, it is enough to only determine the number of maximizing angles of the receive pattern given ratio K between the separation and the carrier wavelength. We prove the theorem by counting the number of maximizing angles.

If K is integer, according to Theorem 1, we have maximizing angles at ϕ, $\cos \phi = \frac{k}{K}$, $k = -K, \ldots, 0, \ldots, K$, $k \in \mathbb{Z}$.

- For $k = \pm K$, we have maximizing angles at 0 and π.
- For each $k \in S_1 = \{-K+1, \ldots, 0, \ldots, K-1\}$, $|S_1| = 2K-1$, there are two maximizing angles at $\phi = \arccos \frac{k}{K}$ and $\phi = \pi - \arccos \frac{k}{K}$.

In total, we have $2 + 2|S_1| = 4K$ maximizing angles.

If K is a non-integer, for each $k \in S_2 = \{-\lfloor K \rfloor, \ldots, 0, \ldots, \lfloor K \rfloor\}$, $|S_2| = 2\lfloor K \rfloor + 1$, we have 2 maximizing angles at $\phi = \arccos \frac{k}{K}$ and $\phi = \pi - \arccos \frac{k}{K}$. The number of those maximizing angles is $2|S_2|$.

- If $\{K\} \leq \frac{1}{2}$, we have no more maximizing angles (Theorem 1), so the total number of maximizing angles is $2|S_2| = 2 \cdot (2\lfloor K \rfloor + 1) = 2\lfloor 2K \rfloor + 2$.
- If $\{K\} \geq \frac{1}{2}$, we have two additional maximizing angles at 0 and π (Theorem 1), which increase the total number of maximizing angles to $2|S_2|+2 = 2 \cdot (2\lfloor K \rfloor + 1) + 2 = 4\lfloor K \rfloor + 4 = 2\lfloor 2K \rfloor + 2$.

Therefore, the total of maximizing angles for the case of non-integer K is $2\lfloor 2K \rfloor + 2$. Note that the above formulas are established based on the following claim: *"for any number x, if $\{x\} < \frac{1}{2}$, then $\lfloor 2x \rfloor = 2\lfloor x \rfloor$; otherwise $\lfloor 2x \rfloor = 2\lfloor x \rfloor + 1$".* □

Detection of Reactive Jamming
in DSSS-based Wireless Networks

Domenico Giustiniano,* Vincent Lenders,† Jens B. Schmitt,‡
Michael Spuhler,* and Matthias Wilhelm‡
*ETH Zürich, Switzerland
†armasuisse, Switzerland
‡TU Kaiserslautern, Germany
domenico.giustiniano@tik.ee.ethz.ch, vincent.lenders@armasuisse.ch,
{jschmitt,wilhelm}@cs.uni-kl.de, spuhlemi@student.ethz.ch

ABSTRACT

We propose a novel approach to detect reactive jammers in
direct sequence spread spectrum (DSSS) wireless networks.
The key idea is to use the chip error rate of the first few
jamming-free symbols at the DSSS demodulator during the
signal synchronization phase of regular packet reception to
estimate the probability of successful packet delivery. If the
estimated probability is significantly higher than the actual
packet delivery ratio, we declare jamming. As a proof of concept, we implement a prototype in a network of three USRP
software-defined radios (transmitter, receiver, and jammer)
and evaluate the feasibility, responsiveness, and accuracy of
our approach in a controlled lab environment. Our experiments with IEEE 802.15.4 DSSS-based communication show
that for links with a jamming-free packet delivery probability above 0.5, the false positive and negative detection rates
remain below 5 %.

Categories and Subject Descriptors

C.2.0 [**Computer Communication Networks**]: General—
Security and protection (e.g., firewalls)

Keywords

Jamming detection; reactive jamming; 802.15.4; DSSS

1. INTRODUCTION

Wireless networks are built upon a shared medium, which
makes them vulnerable to jamming attacks. Jamming attacks are accomplished by emitting interfering RF signals
that do not adhere to the rules of an underlying MAC protocol [17]. When such jamming signals interfere with the
transmissions of legitimate transmitters at the receiver, the
signals collide and render the originally transmitted data
signals uninterpretable at the receiver.

In contrast to traditional security primitives such as authentication, confidentiality, or integrity that can be addressed with the application of cryptographic techniques,
jamming attacks cannot be entirely fended off by conventional security mechanisms. While spread spectrum communication techniques are able to mitigate the effect of narrowband sources of interference, a jammer can always disturb the communication by emitting broadband signals that
exceed the power of legitimate signals at the receiver.

Jammers may employ a wide range of strategies to disturb wireless communication [3, 8, 9, 16, 17]. Among these
existing strategies, *reactive* jammers have been shown to be
not only the hardest to detect, but also the most energy-
efficient approach, making them a serious threat in wireless
networks. In addition, [15] demonstrated that reactive jammers can be implemented on inexpensive COTS platforms
such as the USRP2 from Ettus Research, and that reactive jamming can be triggered selectively on any field of the
packet header, making them a *realistic threat* for wireless
communication.

Since jamming cannot be prevented by design, it is important to understand how it works and, in turn, how to
detect its presence. This paper proposes a novel method to
detect reactive jamming in direct sequence spread spectrum
(DSSS) systems. In DSSS systems, bits or symbols at the
transmitter are spread to higher-order chip sequences. To
detect the presence of jamming, our approach accounts for
chip errors in the preamble at the output of the demodulator
to model the probability of packet losses. If the experienced
packet loss rate exceeds the one estimated from chip errors
in the preamble, a reactive jammer is likely jamming parts of
the packet, and we thus declare jamming. Since the preamble of a packet represents the very first chips being sent for
synchronization purposes, it significantly reduces the probability that a reactive jammer will jam these chips because it
requires very fast reactivity, low signal propagation delays,
and prevents a jammer from making jamming decisions according to physical, MAC, or payload based rules [15].

At the core of our detection scheme is an accurate packet
delivery estimation model based on chip errors in the preamble, which is independent of the received signal strength
(RSS) that is being used by existing detection schemes [14,
17]. Our approach does not require any modification to the
communication system or standard and works even when the
reactive jammer targets the synchronization phase of packet
transmissions. We implement a reactive jamming detector

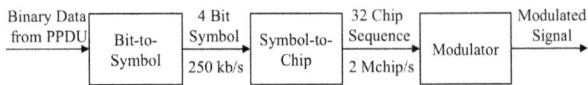

Figure 1: DSSS modulation in IEEE 802.15.4.

for IEEE 802.15.4 on the USRP software-defined radio platform from Ettus Research and we evaluate its performance in a controlled lab environment with the reactive jammer from [15]. Our results show that our detection scheme is able to accurately detect reactive jammers on fading wireless links with a jamming-free packet delivery probability above 0.5. The false positive and negative detection rates remain below 5 % for these links.

The rest of this paper is organized as follows. In the next section, we briefly review important aspects of the IEEE 802.15.4 standard, introduce the attacker model, and describe the experimental setup used in the evaluation. Section 3 explores the feasibility to model the packet delivery with limited information from chip errors in the preamble. In Section 4, we introduce our jamming detection scheme. Section 5 covers the evaluation of the detection accuracy. Related work is discussed in Section 6, and Section 7 concludes the paper.

2. BACKGROUND AND ATTACKER MODEL

In this section, we briefly review important aspects of the IEEE 802.15.4 standard, introduce the attacker model and describe the experimental setup that we use for evaluation.

2.1 Background on IEEE 802.15.4

Our work on jamming detection focuses on direct sequence spread spectrum (DSSS) communication systems, and is practically demonstrated for the IEEE 802.15.4 standard [1]. IEEE 802.15.4 defines a 16-ary quasi-orthogonal DSSS modulation technique. This modulation spreads a low rate sequence of bits to a higher rate sequence of so-called chips in the following way: binary source data is divided into groups of 4 bits (referred to as *symbols*) and mapped to a quasi-orthogonal 32-chip pseudo-noise sequence $(b_0, b_1, b_2, b_3) \mapsto (c_0, c_1, \ldots, c_{31})$, resulting in a chip rate of $2\,\mathrm{MChips/s}$ as shown in Figure 1. The effect of this spreading is an increased robustness against fading and in-band interference: DSSS systems can tolerate a certain number of chip errors and still receive symbols correctly.

Our proposed detection scheme relies on the fact that the packet error probability can be predicted accurately using the number of chip errors in the first few symbols in a packet. An IEEE 802.15.4 packet consists (as shown in Figure 2) of a physical layer header with a preamble sequence for symbol synchronization (eight 0 symbols), a start of frame delimiter (SFD; symbols 7 and 10) and a frame length field indicating the duration of the frame, followed by a MAC protocol data unit (MPDU). The MPDU contains a MAC header, data payload, and ends with a frame check sequence (FCS) that is used to detect transmission errors. IEEE 802.15.4 does not mandate the use of error correction mechanisms, and any received packet with an incorrect FCS is hence discarded.

To receive a packet, the receiver first synchronizes with the preamble sequence to detect the symbol boundaries, i.e., the time instants when chip sequences start. This timing information is subsequently used to detect the SFD and frame length field. The rest of the signal is decoded using a cor-

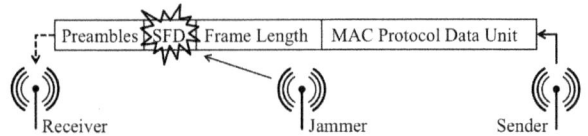

Figure 2: Reactive jamming: an attacker jams the start-of-frame delimiter (SFD) to disturb the synchronization of the packet at the receiver.

relator to map the received 32 chips back to symbols. The received chip sequence R may contain errors caused by fading or interference. It is compared to the 16 predefined chip sequences $C_i, i = 0, 1, \ldots, 15$. The receiver chooses the best match, i.e., the C_i for which $h(R, C_i)$ is minimal, where $h(\cdot, \cdot)$ is the Hamming distance (number of positions containing different chips) between the two arguments. However, if too many chips are flipped, the expression $h(R, C_i)$ may be minimal for the wrong chip sequence C_i such that the receiver interprets the chip sequence as a wrong symbol.

2.2 Attacker Model

We consider reactive jammers that aim to minimize their jamming duration to only a few symbols in order to remain undetected and to save energy. We assume a jammer that is able to sniff any symbol of the packet over the air in real-time and react with a jamming signal that flips selected bits at the receiver with high probability. An attacker may therefore pursue different reactive jamming strategies [15]. It may jam *(i)* the MPDU, *(ii)* the packet length field, *(iii)* the frame synchronization field (SFD), or *(iv)* the preamble of the packet. The first two strategies cause packet losses because of resulting FCS errors, while the last two strategies introduce synchronization failures, causing the entire packet to be missed by the receiver. Figure 2 illustrates jamming strategy *(iii)* that targets the SFD.

The jamming reaction time τ denotes the time difference between the arrival of the original signal and the jammer signal at the receiver. The minimal reaction time $\tau_{\min}$ is bounded by the sum of the signal propagation delay between sender and jammer, the reaction delay of the jammer to process the incoming signal and to make a jamming decision, and the signal propagation delay between jammer and receiver. It is therefore safe to assume that the minimum reaction time $\tau_{\min}$ is greater than the duration of one symbol (e.g., $16\,\mu s$ in IEEE 802.15.4). Otherwise it would not be possible to assess the channel state prior to jamming. In fact, [15] showed that the reaction time of a realistic jammer is significantly larger than this minimum reaction delay because of the inherent hard- and software delays to detect, demodulate, process, and trigger jamming signals according to particular jamming rules. While it might be technically feasible to implement reactive devices with lower reaction delays than the duration of one symbol duration (for example by using simple power detectors with analog parts [7,11]), reactive jammers of that kind are not able to use the semantics of the signals to perform smart jamming decisions like jamming only selected packets according to specific rules (e.g., matching packet modulation or header properties).

2.3 Experimental Setup

We rely on measurements to study the performance of packet delivery models and to evaluate the proposed jamming detection. Our experimental setting considers point-to-point data transmissions in a network consisting of three nodes: sender, receiver, and jammer. Our experiments are based on a software-based implementation of IEEE 802.15.4. As hardware platform, we use the USRP software-defined radio from Ettus Research. For the software, we use a slightly optimized version of the UCLA IEEE 802.15.4 implementation [12] that runs on the GNU Radio framework. We have performed multiple tests in indoor lab environments, which are referred to as *cable, static line-of-sight, static non-line-of-sight*, and *mobile*. In the *cable* experiments, sender and receiver are connected by a shielded 60 cm coaxial cable with a 30 dB attenuator. In the *static* experiments, a stationary sender and receiver communicate using omni-directional antennas. The *mobile* experiments are similar to the static scenario except that the sender is kept stationary while the receiver is moving. The receiver is placed on a cart and moved at a constant speed of maximum $v = 1$ cm/s away from, and back towards, the sender.

In each experiment run, 40,000 packets of 26 bytes length are sent during 40 seconds from the transmitter to the receiver at constant rate. Various link conditions in the cable and static experiment runs are obtained by adjusting the transmit power and by changing the position of nodes. The true packet delivery ratio (PDR) at time t is calculated by averaging the number of received packets over a window of 100 packets centered around t. A window size of 100 packets assures that the true PDR is calculated over a time window that is smaller than the channel coherence time when moving the receiver at maximum $v = 1$ cm/s and at a frequency of 2.4 GHz.[1] Note that the mobility experiments have a relatively low node speed of maximum 1 cm/s for the sake of determining the true PDR. We intentionally kept the node mobility low such that the channel coherence time is larger than the window size of 100 packets that are used to calculate the true PDR. Our results are thus relatively conservative with respect to mobility.

As a jammer, we use the reactive jammer from [15], which also runs on the USRP2 platform. It can be configured to jam according to strategies *(i)* to *(iv)*. The detection and decision logic are implemented on the FPGA of the USRP2, resulting in a minimal reaction delay of $\tau_{min} = 19\,\mu s$.

3. CHIP ERROR BASED PDR MODEL

Our jamming detection technique relies on a statistical model of packet delivery from chip errors in the first few symbols of the preamble [13]. This section provides experimental results that show that the packet delivery ratio in DSSS-based wireless networks can be modeled accurately using such limited information. We further show that our model significantly outperforms RSS-based PDR estimators, which constitute the basis of current jamming detection schemes.

Our statistical model exploits the strong correlation between DSSS chip errors in the preamble, observed at the output of the demodulator of the receiver, and the experienced packet delivery ratio. Figure 3 shows this correlation for four experiments in different environments (cable, static line-of-sight, static non-line-of-sight, and mobile). As we can

---

[1]The coherence time is the time duration for which the channel impulse response is considered to be stationary and is approximately $\frac{1}{4D}$, where D is the Doppler spread.

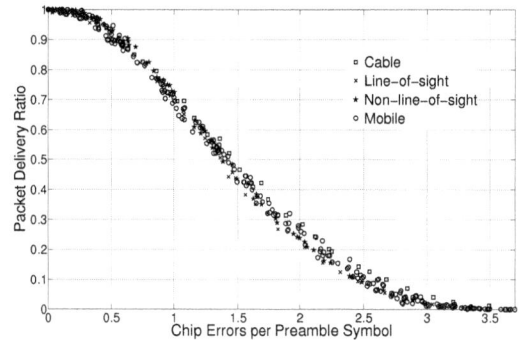

Figure 3: Correlation between average chip errors in the preamble and packet delivery ratio.

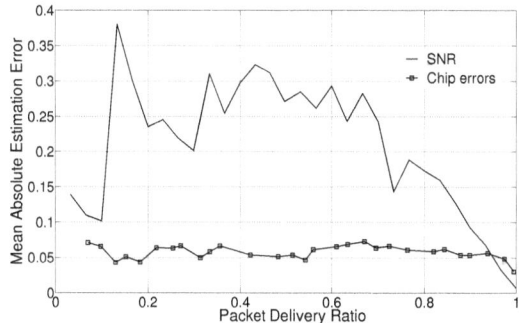

Figure 4: Comparison of mean absolute estimation error of the packet delivery ratio for a model that relies on the average number of chip errors in the preamble versus a model that relies on the RSS.

see, the average number of chip errors per preamble symbol is highly correlated in the entire range of PDRs as indicated by a Pearson correlation coefficient of -0.965.[2] Note that the average number of chip errors does not exceed 4 because the receiver we used makes hard decoding on preamble symbols with this threshold. We varied the hard decoding threshold for the preambles to values ranging from 1 to 6 in order to evaluate the effect on the distribution: while the distribution gets shifted when changing this threshold, the strong correlation still remains.

This correlation is well suited to predict the PDR, as shown in Figure 4 for the case of *mobile* scenarios. The figure compares the mean absolute packet delivery estimation error of our model that relies on the chip errors in the preamble to a model based on the signal-to-noise ratio (SNR) [5]. Our chip error based model estimates the PDR using a regression with a polynomial function $g_{CER}(p)$ that has a root mean square (RMS) error below 3 % across all considered environments. This regression function $g_{CER}(p)$ maps the average number of chip errors per preamble symbol p to the respective PDR. The polynomial function with the smallest degree is of the form

$$g_{CER}(p) = a_0\,p^5 + a_1\,p^4 + a_2\,p^3 + a_3\,p^2 + a_4\,p + a_5,$$

where $a_0 = 0.016$, $a_1 = -0.33$, $a_2 = 2.41$, $a_3 = -7.26$, $a_4 = 8.83$, $a_5 = -3.24$. Similarly, the SNR-based model estimates the PDR also using a polynomial regression function, but fitted to the empirical SNR–PDR distribution. Selecting

---

[2]Values close to 0 indicate a low correlation and values close to ± 1 represent a high linear dependence of two variables.

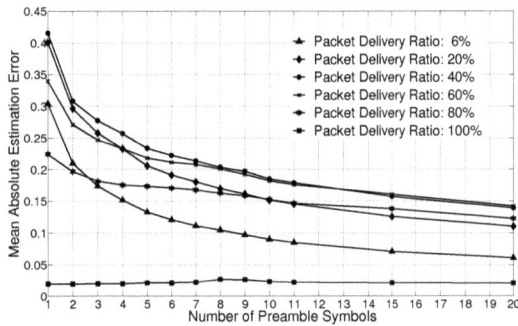

Figure 5: Mean absolute estimation error of the packet delivery for various PDRs and different number of preamble symbols in the model.

the polynomial function $g_{SNR}(SNR)$ as the one with RMS error below 3% across all considered environments, we have:

$$g_{SNR}(SNR) = b_0\ SNR - b_1,$$

where SNR is the signal-to-noise-ratio expressed in dB and the coefficients are $b_0 = 0.12$ and $b_1 = -1.7$.

Using the above models, Figure 4 shows that the mean absolute PDR estimation error is significantly lower for the chip error based model across almost the entire range of delivery ratios (for $PDR > 0.95$, the absolute error is slightly lower for the SNR-based model). Existing jamming detection schemes that rely on the RSS thus suffer inherently from this estimation error. The fact that RSS-based models of packet delivery are generally not very accurate in real-world wireless networks has also been reported previously in the literature [2, 13].

As we cannot control the reaction time τ of the adversary, it is crucial that the proposed model of packet delivery manages to estimate with as few preamble symbols as possible. Figure 5 evaluates the mean absolute estimation error of the packet delivery versus a varying number of preamble symbols used in the estimation. Preamble symbols can be accumulated over multiple transmissions, i.e., they do no have to be from the same packet, hence enabling a number of preamble symbols larger than 8. As we can see, the error quickly converges, hence providing a useful estimator even for a model that accounts for just a few symbols.

4. JAMMING DETECTION

In this section, we describe our jamming detection scheme using the packet delivery model of the previous section. The basic idea is to continuously monitor the traffic over a link and determine two metrics. The first metric is the *observed* packet delivery ratio $PDR_o(t)$ at time t, which is calculated by counting the ratio of correctly received packets over the total number of transmitted packets in a sliding observation window:

$$PDR_o(t) = \frac{\#\ \text{of correct packets in } [t - W, t]}{\#\ \text{of transmitted packets in } [t - W, t]}$$

To determine the number of correctly received packets the receiver checks the FCS of all received packets and, if correct, increments a counter. Determining the total number of transmitted packets at the receiver must take into account that a reactive jammer might successfully jam all SFDs of the transmitted packets, thus preventing any successful packet synchronization at the receiver. The only reliable in-

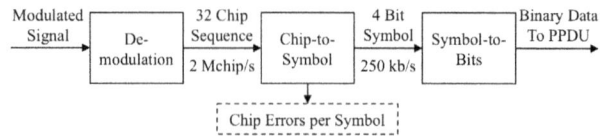

Figure 6: Chip errors in the preamble symbols are determined during the chip-to-symbol mapping of the receiver.

formation source is therefore the preamble when the reactive jammer has not yet started. The receiver counts the received preamble symbols and increments its counter of transmitted packets when at least one symbol 0 is detected within a sliding time window of the size of the preamble. The observed PDR_o should be calculated over a time window that is shorter than the channel coherence time but sufficiently long to capture enough packets to derive a statistically relevant average. In this work, we fix this window size to $W = 100$ ms, corresponding to roughly 100 data packets at the actual transmit rate of the sender.

The second metric is an *estimated* PDR based on the preamble chip errors. As shown in Figure 6, the IEEE 802.15.4 receiver demodulates the incoming signal and attempts to map the demodulated 32-chip sequence to a known symbol. When the receiver is not synchronized yet, it attempts to map the incoming sequences to symbol 0. This is done with hard-decision decoding, that is, the receiver checks if the Hamming distance of the received chip sequence is smaller than a threshold value. This threshold value (4 for our receiver) is usually significantly below the mean Hamming distance of the symbols to prevent the receiver to synchronize on noise. To calculate a statistically relevant chip error rate, the receiver averages the Hamming distances of multiple preamble symbols. We point out again that the calculated average is not constrained to include only preamble symbols from a single packet. For example, when a jammer is reacting very quickly and jams symbols at positions 2 to 8 in the preamble, the received chip sequences 2 to 8 are not accounted for the statistics because, due to chip flipping, their Hamming distance becomes greater than the hard decoding threshold and these symbols are hence not interpreted as 0. Similarly, when the link conditions are poor, a receiver might miss multiple symbols per preamble. After receiving enough 0 symbols, the estimated PDR is calculated as

$$PDR_e = g_{CER}\left(\frac{\sum_{j=1}^{|\mathcal{S}|} h(R_j, C_0)}{|\mathcal{S}|}\right),$$

where R_j is the jth received 32-bit chip sequence that has been interpreted as a 0 with hard decoding, C_0 is the chip sequence of symbol 0, $h(\cdot, \cdot)$ is the Hamming distance, $\mathcal{S}$ is the set of received preamble symbols within a sliding window, and $g_{CER}(\cdot)$ is a function that models the empirical distribution of the PDR versus chip errors per preamble symbol as defined in Section 3. To assure that the set $\mathcal{S}$ is large enough irrespectively of the channel quality and the jammer reaction time, we do not determine PDR_e based on a fixed sliding time window but rather on a fixed set size. We have set this size to $|\mathcal{S}| = 10$ (i.e., 10 symbols 0) in our work as it has proven to provide a reasonable tradeoff between accuracy and reactivity of jamming detection.

We define a hypothesis test based on the relative difference

Δ between the expected and observed PDR:

$$\Delta = \frac{\text{PDR}_e - \text{PDR}_o}{\text{PDR}_e}.$$

Let us define the null hypothesis H_0 and the alternative hypothesis H_1 as

$$H_0 : \text{"Normal transmission,"}$$
$$H_1 : \text{"Jammed transmission."}$$

Then the test is as follows:

$$\text{accept } H_1, \text{ if } \Delta > \epsilon,$$
$$\text{stay with } H_0, \text{ if } \Delta \leq \epsilon,$$

where ϵ represents a tolerance level which directly affects the false positive and false negative detection rates. For small tolerance level values ϵ, the jamming detection is more sensitive, but at the price of higher false negative rates. For higher values of ϵ, the false negative rates may be reduced, but, in turn, at the price of higher false positive rates. To determine a good value for ϵ, we perform a maximum likelihood estimation using our measurements as follows. Let $\Lambda(\epsilon)$ be the sum of the false positive and false negative detection rates for a given PDR:

$$\Lambda(\epsilon) = P(H_0 \mid \text{jammer on}) + P(H_1 \mid \text{jammer off}).$$

Through exhaustive search using our measurements, we perform a maximum likelihood estimation that minimizes $\Lambda(\epsilon)$ for any value of $\epsilon > 0$ and PDR $\in [0, 1]$. The result is that $\Lambda(\epsilon)$ is minimized when $\epsilon = 0.5$ for all PDR $\in [0, 1]$. This agrees with the theoretical expectation that the error threshold lies in the geometric center of the decision region. ·

5. EVALUATION

Our evaluation focuses on quantifying the detection performance in terms of false positives and false negatives under realistic wireless fading channel conditions. For this purpose, we test our detection algorithm on software-defined radios with real traffic over the air.

5.1 Evaluated Jammer

For the performance evaluation, we consider a reactive jammer that jams all packets. We further study the robustness of our approach under the condition that the jammer does not succeed to jam all packets, but is still able to destroy 90 % of the packets. Figure 7 shows the impact of these two forms of reactive jamming on the correlation between the PDR and the chip errors in preamble symbols for $|\mathcal{S}| = 10$. The dark curve in the middle of the figure is the regression curve $g_{\text{CER}}(\cdot)$ derived previously. As expected, if the transmission is not affected by the jammer, the points are spread around this curve. If the jammer is active, the position of these points changes and the strong correlation between the observed PDR and chip error distribution fades away. The points then coincide with the horizontal axis (for the 100% reactive jammer) or are spread around this axis (for the 90% reactive jammer). Another finding is that the detection of reactive jammers that successfully jam 90 % of the packets is more challenging as the PDR gets poorer, because the Euclidean distance between the PDR in presence and absence of jamming is reduced. In the region with higher number of chip errors per preamble symbol, this may be erroneously interpreted as links with poor quality (e.g., where losses are caused by a low SNR).

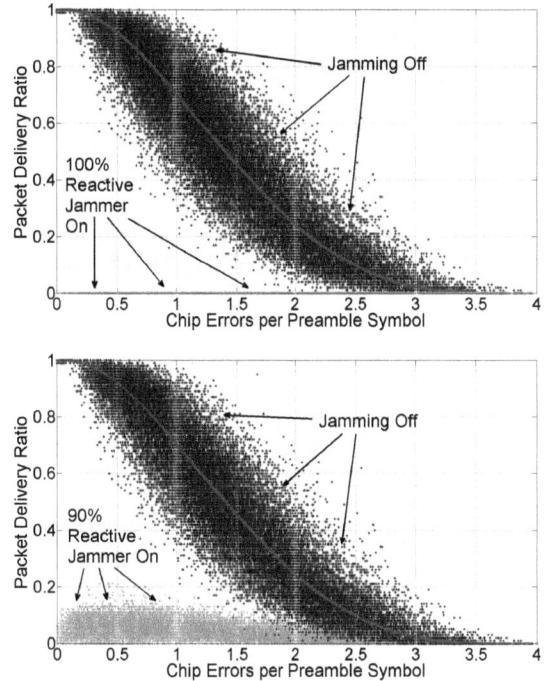

Figure 7: Impact of jamming on the correlation between the PDR and the preamble chip errors. Above we have the case of jamming all packets, below the one of jamming 90 % of packets.

5.2 Detection Performance

The false positive and false negative rates are evaluated in Figure 8. The jammer is configured in these experiments to react and hit the SFD of transmitted packets. This jamming strategy is of particular interest because packet synchronization fails and existing detection mechanisms are not able to cope with this type of reactive jamming. Both the false negative and positive error rates have probabilities below 5 % for links ranging from perfect to a PDR_e of 0.5. Below a PDR_e of 0.35, the reactive jammer causes false negatives over 10 %, constantly increasing for worse links. The false positives rate stays very small as well for good links and exceeds the error threshold of 10 % for PDR_e below 0.35 and then increases similarly for worse link qualities.

This general observation of increasing false positive and false negative rates in poor link environments for the jamming scenario is because PDR_o and PDR_e tend to overlap. A PDR_o obtained in poor link environments is more difficult to assign to either a jammed poor link quality situation or an ordinary poor link quality state. However, it has to be considered that the benefit in detecting jammers in poor link qualities conditions is not that crucial because low quality links are generally not used by higher layer network and application protocols. For good links with $\text{PDR}_e > 0.5$, an accurate jamming detection is more valuable. In this region, we measure that the reactive jammer has a false negative error rate below 5 %.

6. RELATED WORK

To the best of our knowledge, this work is the first to provide a jamming detection scheme that can cope with so-

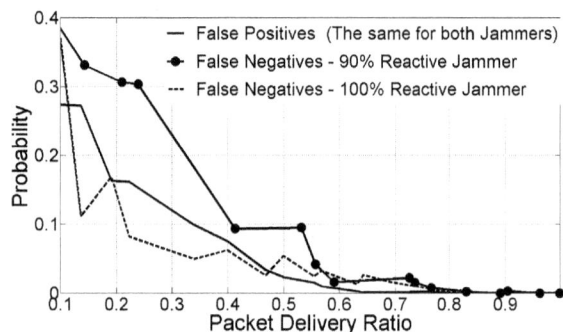

Figure 8: Performance evaluation of reactive jammer detection with respect to the false positive and false negative rates.

phisticated reactive jamming attacks targeting packet synchronization. Strasser et al. [14] propose a jamming detection scheme for sensor networks that enables a per-packet detection of reactive (single-bit) jamming. The main idea is to identify the cause of individual bit errors within a packet by analyzing the RSS of each received bit in the packet. A limitation of this approach is that it relies on a successful packet synchronization. Thus it is not able to detect SFD jamming attacks because decoded MPDU symbols are unavailable at the receiver due to the synchronization prevention. A further challenge is to localize bit errors in a packet. The authors propose to either use *a priori* knowledge of the bit stream sent, the use of error detecting/correcting codes, with drawbacks such as additional overhead and transmission costs, or to acquire the error position based on limited, short-range sensor node wiring in the form of wired node chains. Because our approach is not relying on error positions in a packet, it does not suffer from these restrictions.

Xu et al. [17] propose the usage of the PDR along with either RSS or device location information as a consistency check for proactive and reactive jamming detection. In the first case, jamming is detected if the PDR is low although the RSS is high. In the second case, the PDR is low although the sender–receiver distance is small. Unlike our work, these techniques are not able to detect reactive jamming that targets the physical layer header, or jammers that affect only a few bits per packet.

Xuan et al. [18] describe a method to identify so-called *trigger nodes* that are in the vicinity of reactive jammers and thus trigger jamming. This information is subsequently used to exclude such nodes and route around jammed areas. The authors assume that the detection of jamming on a per-packet level is feasible without error, such that the challenges treated in this work are avoided.

Chiang and Hu [4] leverage the properties of orthogonal spreading codes to achieve jamming detection and mitigation. In contrast to our work, their mode of operation is CDMA and the codes are long and confidential such that the attacker cannot interfere with all transmissions. We assume DSSS systems with public (or compromised) codes.

Finally, Qin et al. [10] suggest that the chip error rate might be a better channel quality indicator than signal power based metrics, particularly in the presence of interference. However they do not propose any estimator nor do they evaluate the feasibility to estimate the PDR from chip error measurements as we do in this work. CEPS [6] models the

PDR from chip errors in the payload of successfully received packets. In contrast, we model the PDR from chip error measurements in the synchronization phase at the preamble and show that this information is already sufficient for detecting reactive jamming.

7. CONCLUSION

We have proposed a novel approach to detect sophisticated reactive jamming attacks that may target any part of a packet transmission. Our approach is based on chip errors of a few initial symbols during the synchronization phase of a packet transmission in order to predict the link packet delivery, which makes it suitable to even detect jammers that target the physical layer header of packets. Our experiments under real-world channel conditions showed that it is possible to predict the packet delivery accurately using the chip error rate derived from just a few preamble symbols. We further showed that we can detect reactive jammers with a false negative rate below 5 % for PDRs over 0.5.

8. REFERENCES

[1] IEEE Standard 802 Part 15.4: Wireless medium access control and physical layer specifications for low-rate WPANs.
[2] D. Aguayo, J. Bicket, S. Biswas, G. Judd, and R. Morris. Link-level measurements from an 802.11b mesh network. In *Proc. of ACM SIGCOMM '04*, pages 121–132, Aug. 2004.
[3] M. Çakiroglu and A. T. Özcerit. Jamming detection mechanisms for wireless sensor networks. In *Proc. of ICST InfoScale '08*, pages 1–8, June 2008.
[4] J. T. Chiang and Y.-C. Hu. Cross-layer jamming detection and mitigation in wireless broadcast networks. *IEEE/ACM Trans. Netw.*, 19(1):286–298, Jan. 2011.
[5] D. Halperin, W. Hu, A. Sheth, and D. Wetherall. Predictable 802.11 packet delivery from wireless channel measurements. *Proc. of ACM SIGCOMM '10*, pages 159–170, Aug. 2010.
[6] P. Heinzer, V. Lenders, and F. Legendre. Fast and accurate packet delivery estimation based on DSSS chip errors. In *Proc. of IEEE INFOCOM '12*, pages 2916–2920, Mar. 2012.
[7] M. Kuhn, H. Luecken, and N. O. Tippenhauer. UWB impulse radio based distance bounding. In *Proc. of WPNC '10*, pages 28–37, Mar. 2010.
[8] Y. W. Law, M. Palaniswami, L. V. Hoesel, J. Doumen, P. Hartel, and P. Havinga. Energy-efficient link-layer jamming attacks against wireless sensor network MAC protocols. *ACM Trans. Sensor Netw.*, 5(1):6:1–6:38, Feb. 2009.
[9] A. Proaño and L. Lazos. Packet-hiding methods for preventing selective jamming attacks. *IEEE Trans. Dependable Secure Comput.*, 9(1):101–114, Jan. 2012.
[10] Y. Qin, Z. He, and T. Voigt. Towards accurate and agile link quality estimation in wireless sensor networks. In *Proc. of IFIP Med-Hoc-Net '11*, pages 179–185, June 2011.
[11] K. B. Rasmussen and S. Čapkun. Realization of RF distance bounding. In *Proc. of USENIX Security '10*, pages 389–402, Aug. 2010.
[12] T. Schmid. GNU Radio 802.15.4 en- and decoding. Technical Report TR-UCLA-NESL-200609-06, UCLA NESL, Sept. 2006.
[13] M. Spuhler, V. Lenders, and D. Giustiniano. BLITZ: Wireless link quality estimation in the dark. In *Proc. of EWSN '13*, pages 99–114, Feb. 2013.
[14] M. Strasser, B. Danev, and S. Čapkun. Detection of reactive jamming in sensor networks. *ACM Trans. Sensor Netw.*, 7(2):16:1–16:29, Aug. 2010.
[15] M. Wilhelm, I. Martinovic, J. B. Schmitt, and V. Lenders. Reactive jamming in wireless networks: How realistic is the threat? In *Proc. of ACM WiSec '11*, pages 47–52, June 2011.
[16] A. D. Wood, J. A. Stankovic, and G. Zhou. DEEJAM: Defeating energy-efficient jamming in IEEE 802.15.4-based wireless networks. In *Proc. of IEEE SECON '07*, pages 60–69, June 2007.
[17] W. Xu, W. Trappe, Y. Zhang, and T. Wood. The feasibility of launching and detecting jamming attacks in wireless networks. In *Proc. of ACM MobiHoc '05*, pages 46–57, May 2005.
[18] Y. Xuan, Y. Shen, N. P. Nguyen, and M. T. Thai. A trigger identification service for defending reactive jammers in WSN. *IEEE Trans. Mob. Comput.*, 11(5):793–806, May 2012.

Selfish Manipulation of Cooperative Cellular Communications via Channel Fabrication

Shrikant Adhikarla, Min Suk Kang, and Patrick Tague
Carnegie Mellon University, Pittsburgh, PA
{sadhikar, minsukk, tague}@andrew.cmu.edu

ABSTRACT

In today's cellular networks, user equipment (UE) have suffered from low spectral efficiency at cell-edge region due to high interference from adjacent base stations (BSs), which share the same spectral radio resources. In the recently proposed cooperative cellular networks, geographically separated multiple BSs cooperate on transmission in order to improve the UE's signal-to-interference-plus-noise-ratio (SINR) at cell-edge region. The service provider of the system dynamically assigns the cluster of BSs to achieve higher SINR for the UE while optimizing the use of system radio resources. Although it is the service provider that makes the the clustering decision for the UE, the service provider relies on the UE's input to the decision; i.e., the channel states from the adjacent BSs to the UE. In essence, the operation of the cooperative cellular netwokrs heavily relies on the trust in the UEs. In this paper, we propose a new selfish attack against the cooperative cellular networks; an adversary reprograms her UE to report fabricated channel information to cause the service provider to make a decision that benefits the adversary while wasting its system resources. We evaluate the proposed attack in a cooperative cellular network having various performance goals on the simulation-based experiments and show that the adversary can trick the service provider into expending 3.7 times more radio resources for the adversary and, accordingly, the adversary achieves up to 16 dB SINR gain. Finally, we propose a threshold-based countermeasure for the service provider to detect the attack with approximately 90% of accuracy.

Categories and Subject Descriptors

C.2.0 [**Computer-Communication Networks**]: General— *security*; C.2.1 [**Computer-Communication Networks**]: Network Architecture and Design—*wireless communication*

General Terms

Security, Algorithms, Reliability, Verification

Keywords

Cooperative cellular networks; channel fabrication; heuristic attack strategies; anomaly detection

1. INTRODUCTION

In traditional cellular networks, each cell uses a different set of radio frequencies from neighboring cells to avoid inter-cell interference [11] [4]. In today's systems, such as LTE, for better usage of scarce radio resources, all the cells in the system share the single frequency band [2] and thus the inter-cell interference management is important to guarantee the quality-of-service (QoS) to the user equipment (UE) in cell-edge regions [6,8,9,12].

Recently proposed *cooperative communication* technique in cellular networks has gained significant interests for it can implement tighter interference coordination among adjacent multiple base stations (BSs) [5,16][1]. In cooperative cellular networks, a *cluster* of multiple geographically separated BSs *cooperate* on transmission in order to improve spectral efficiency of the UEs. The major advantage of the coordination is to increase the received signal-to-interference-plus-noise-ratio (SINR) by reducing the amount of the interference and increasing the amount of the signals. The more BSs join the cluster for a UE, the higher SINR the UE achieves. However, in the persepctive of system operations, utilizing more coordinated BSs for the specific UE implies that the system uses additional radio resources that could have been used for other UEs. Therefore, the service provider of the cellular network should carefully determine the cluster in order to efficiently use the system resources.

In order for the BS to make clustering decisions for UE, it needs to know the current channel state at the UE. Due to the information asymmetry between UE and BS, the BS has to rely on the channel reported by the UE.[2] In essence, the operation of the cooperative cellular network heavily relies on the trust in the UE's.

In this paper, we propose a selfish attack on the cooperative cellular networks; the goal of the adversary is to maximize the received SINR at the adversary's UE by increasing the system radio resources allocated to the UE and the re-

---

[1]In the literature, it is possible to find cooperative communication systems labeled as "Network MIMO", "Multicellular MIMO", "Multicellular cooperation", "CoMP", or "Distributed Antenna System" [23] [13] [1] [10].

[2]This channel information asymmetry holds only in frequency-division duplex (FDD) systems, which is the more popular system configuration for today's cellular networks. Time-division duplex (TDD) systems do not pose this problem.

quired capability to launch the attack is to reprogram (or execute the malicious codes on her UE to reprogram) her UE's network adapters. We propose three heuristic attack strategies for fabricating the channel information and performe simulation-based experiments to evaluate the proposed attack strategies in a cooperative cellular network model. We design two dynamic clustering models for our cooperative cellular network model; guaranteed minimum SINR model and maximum normalized throughput model. Our evaluation shows that the adversary can obtain 3.7 times more radio resources and, accordingly, achieve up to 16 dB SINR gain. To the best of our knowledge, the proposed selfish attack is first discovered attack against the cooperative cellular networks.

In order to mitigate the proposed attack, we present a threshold-based countermeasure for the service provider to detect the attack with approximately 90% of accuracy. The countermeasure requires only a simple modification to the cluster decision model to detect of abnormal channel reports.

2. SYSTEM MODEL: COOPERATIVE CELLULAR NETWORKS

We assume a multi-cell cellular network, where multiple geographically separated BSs manage their own cells and the UEs that belong to them. In this paper, we assume that BSs and UEs are equipped with single antenna. In our system model, a UE is associated with a *primary* base station BS_0 and a set of adjacent base stations, identified by the index set $\mathcal{A} = \{1, \ldots, A\}$. The signal y received by the UE is a combination of the data signals x_i transmitted by these base stations and is given by $y = \sqrt{P_0} h_0 x_0 + \sum_{i \in \mathcal{A}} \sqrt{P_i} h_i x_i + n$, where x_i ($i = 0, \cdots, A$) is the data signal transmitted from the BS_i, h_i ($i = 0, \cdots, A$) is the channel coefficient from the BS_i to the UE, P_i ($i = 0, \cdots, A$) is the transmitted power at the BS_i, and n is the noise signal at the UE. In non-cooperative conventional multi-cell cellular networks, the signals x_i ($i = 0, \ldots, A$) are distinct, so those x_i for $i \geq 1$ act as interference from the adjacent cells. In this case, the SINR at the UE is given as $\gamma = \frac{P_0 |h_0|^2}{\sum_{i \in \mathcal{A}} P_i |h_i|^2 + N}$, where N is the noise spectral density. In this paper, we assume that all the BS's have the same fixed power transmission level. Thus, the SINR becomes $\gamma = \frac{|h_0|^2}{\sum_{i \in \mathcal{A}} |h_i|^2 + N_0}$, where N_0 is given by N/P_0.

For cooperative communication we assume that among the set of all adjacent BSs $\mathcal{A}$, the cluster of cooperative BSs $\mathcal{H}$ is selected and the BS's in $\mathcal{H}$ transmit the same x_0 to the UE. The cooperation decision is made at the central entity, which is assumed to be located at the primary BS, by using the channel information vector $\mathbf{h} = \{h_0, h_1, \cdots, h_A\}$. Upon the channel vector $\mathbf{h}$ reported by the UE, the BS decides the cluster $\mathcal{H}$ based on a decision algorithm, denoted here by $\mathcal{D}$ such that $\mathcal{H} = \mathcal{D}(\mathbf{h})$.

Therefore, the SINR at the UE with the BS cooperation is given as

$$\gamma(\mathcal{H}, \mathbf{h}) = \frac{|h_0|^2 + \sum_{c \in \mathcal{H}} |h_c|^2}{\sum_{i \in (\mathcal{A} \setminus \mathcal{H})} |h_i|^2 + N_0}. \quad (1)$$

In this paper, we assume the perfect synchronization among multiple BSs to the UE.

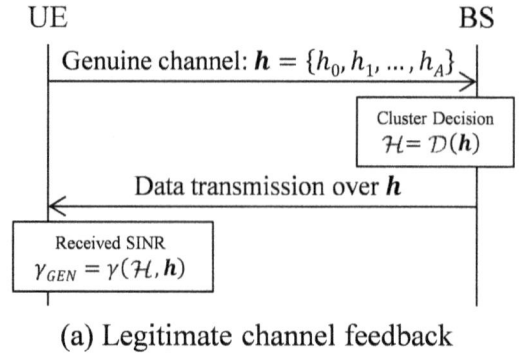

(a) Legitimate channel feedback

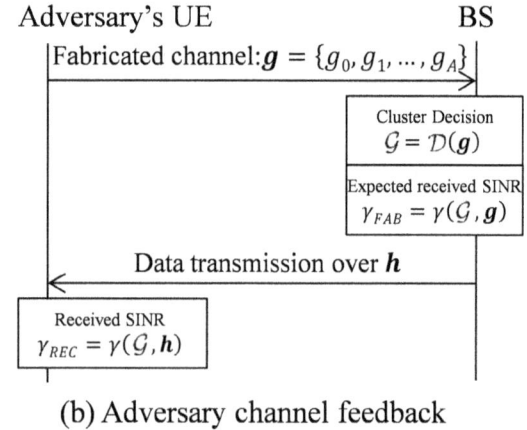

(b) Adversary channel feedback

Figure 1: An adversary can manipulate the BS cooperation model using an SCF attack, fabricating channel information to achieve an increased SINR.

2.1 Clustering Decision Models

The clustering decision process, which determines the set of BSs to cooperate for a UE, is based on the implementation or management of the service providers. As there currently doesn't exists any concrete decision model, as a part of our system model, we design the two following clustering decision models.

Guaranteed Minimum SINR Model: This model ($\mathcal{D}_{th}$) aims to guarantee at least a minimum SINR (γ_{th}) for each UE when deciding the BS cooperation. This model can be used in aiming to guarantee quality of service to each user. If the SINR provided in the absence of cooperation already satisfies the threshold, i.e. $\gamma(\emptyset, \mathbf{h}) \geq \gamma_{th}$, then no cooperation is required. Otherwise, the system chooses the smallest set $\mathcal{H}$ that satisfies the following equation:

$$\mathcal{H} = \underset{\mathcal{H} \neq \emptyset}{\arg \min} |\mathcal{H}| \quad \text{s.t.} \quad \gamma(\mathcal{H}, \mathbf{h}) \geq \gamma_{th}. \quad (2)$$

γ_{th} is a pre-determined threshold value and we note that its choice is critical in determining the number of UE devices that can be supported.

Maximum Normalized Throughput Model: This model ($\mathcal{D}_{max}$) aims to maximize the normalized throughput when deciding the BS cooperation. This model can be used in aiming to guarantee highly efficient resource utilization by the system. In this formulation, the system aims to maximize the Shannon-Hartley capacity [20] per cooperating base station. Independent of the channel bandwidth, this model is specified formally as

$$\mathcal{H} = \arg\max_{\mathcal{H} \neq \emptyset} \log_2 \left(1 + \gamma(\mathcal{H}, \mathbf{h})\right) / |\mathcal{H}| \qquad (3)$$

$$\text{s.t.} \qquad \gamma(\mathcal{H}, \mathbf{h}) > 0. \qquad (4)$$

In the above formulation, the system can break ties according to a preference for smaller $|\mathcal{H}|$, or an additional penalty could be imposed to artificially force the cooperating set to be small.

3. SELFISH ATTACK AGAINST COOPERATIVE CELLULAR NETWORKS

We present *Selfish Channel Fabrication*, or SCF, attack. Figure 1 depicts the SCF attack process, described as follows.

The foundation of the SCF attack relies on the fact that the BS determines the cooperation set $\mathcal{H}$ using the decision model $\mathcal{D}$ as a function of the channel information $\mathbf{h}$ provided by the UE. Instead of reporting the truly measured channel quality indicators $\mathbf{h}$, however, a selfish UE can report *fabricated* channel quality indicators $\mathbf{g}$ given by

$$\mathbf{g} = \{g_0,\ g_1, \cdots,\ g_A\}. \qquad (5)$$

The adversary fabricates the value of $\mathbf{g}$ such that use of cooperation set $\mathcal{G} = \mathcal{D}(\mathbf{g})$ instead of $\mathcal{H} = \mathcal{D}(\mathbf{h})$. Because of the deception of the adversary, the system is forced into provisioning for the SINR, such that,

$$\gamma(\mathcal{G}, \mathbf{g}) = \frac{|g_0|^2 + \sum_{c \in \mathcal{G}} |g_c|^2}{\sum_{i \in (\mathcal{A} \setminus \mathcal{G})} |g_i|^2 + N_0} > \gamma(\mathcal{H}, \mathbf{h}). \qquad (6)$$

We note, however that the SINR $\gamma(\mathcal{G}, \mathbf{g})$, is only what the BS *believes* it is providing to the UE, while the *true channel* from the BS to the UE behaves according to the actual channel indicators $\mathbf{h}$. Therefore, the *actual* SINR achieved by the selfish UE is $\gamma(\mathcal{G}, \mathbf{h})$, so the three SINR values involved in the attack formulation are as follows

- $\gamma(\mathcal{H}, \mathbf{h})$: SINR that the UE would get when it reports the *genuine* channel indicators $\mathbf{h}$.

- $\gamma(\mathcal{G}, \mathbf{g})$: SINR that the BS *believes* the UE would get when the *fabricated* channel indicators $\mathbf{g}$ are reported.

- $\gamma(\mathcal{G}, \mathbf{h})$: SINR that the UE would *actually* get when it reports the *fabricated* channel indicators $\mathbf{g}$.

The goal of the selfish UE is thus to choose $\mathbf{g}$ as a function of $\mathbf{h}$ such that

$$\gamma(\mathcal{H}, \mathbf{h}) < \gamma(\mathcal{G}, \mathbf{g}) < \gamma(\mathcal{G}, \mathbf{h}). \qquad (7)$$

The first inequality in (7) implies that the SINR with the fabricated channel indicators should be greater than that with the genuine channel indicators. This must hold because otherwise the selfish UE does not have any motivation in sending the fabricated channel indicators. The second inequality in (7) implies that the SINR that is used for adaptive modulation and coding (AMC) at the BS should be lower than the actual SINR that the UE measures when receiving the packet. This also must hold otherwise the selfish

UE cannot decode the packet with low packet error probability[3].

In order to optimize the attack the adversary needs to find the optimum value of $\mathbf{g}$, which satisfies the inequalities in in (7). Furthermore, the attack is only useful if it can be done in a timely manner; specifically, it should be done faster than the cellular system frame time (generally of the order of 5ms). Because of the complex form of the inequality constraints in in (7), the adversary has to rely on the use of heuristic approaches for choosing $\mathbf{g}$, which we address in the next sub-section.

3.1 Heuristic SCF Attack Strategies

In this section, we propose three heuristic strategies for timely computation of $\mathbf{g}$ in the SCF attack. Each strategy provide a different method for mapping the true $\mathbf{h}$ to the fabricated $\mathbf{g}$.

Over-Projecting Interference Channel Indicators (OPICI). The adversary generates $\mathbf{g}$ by increasing the magnitude of interference channel indicators according to $\mathcal{H} = \mathcal{D}(\mathbf{h})$. Thus, when the BS calculates the SINR for the selfish UE using the fabricated and reported $\mathbf{g}$, the BS is tricked into believing that the user needs additional cooperation using the stronger signals that are currently acting as interference. For a given over-projection factor $\alpha > 1$, the OPICI strategy is given by

$$\boxed{\begin{aligned} &\textbf{Given: } \alpha > 1 \text{ ; } \textbf{Choose: } g_i \text{ as} \\ &g_i = \begin{cases} h_i & \text{if } i \in \mathcal{H} \\ \alpha h_i & \text{if } i \in \mathcal{A} \setminus \mathcal{H}. \end{cases} \end{aligned}} \qquad (8)$$

Under-Projecting Signal Channel Indicators (UP-SCI). The adversary generates $\mathbf{g}$ by decreasing the magnitude of signal channel indicators according to $\mathcal{H} = \mathcal{D}(\mathbf{h})$. Thus, when the BS calculates the SINR for the selfish UE using $\mathbf{g}$, it will be similarly tricked into believing that an even great extent of cooperation is needed. For a given under-projection factor $\beta > 1$, the UPSCI strategy is given by

$$\boxed{\begin{aligned} &\textbf{Given: } \beta > 1 \text{ ; } \textbf{Choose: } g_i \text{ as} \\ &g_i = \begin{cases} \beta h_i & \text{if } i \in \mathcal{H} \\ h_i & \text{if } i \in \mathcal{A} \setminus \mathcal{H}. \end{cases} \end{aligned}} \qquad (9)$$

Controlling Minimum Channel Indicators (CMCI). The adversary generates $\mathbf{g}$ by increasing any signal or interference channel indicator that falls below a threshold C_{min}, a parameter that can be fine-tuned in such a way that the resulting SINR triggers extended cooperation. For a given threshold C_{min}, the CMCI strategy is given by

$$\boxed{\begin{aligned} &\textbf{Given: } C_{min} \text{ ; } \textbf{Choose: } g_i \text{ as} \\ &g_i = \begin{cases} h_i & \text{if } |g_i| \geq C_{min} \\ C_{min} & \text{if } |g_i| < C_{min}. \end{cases} \end{aligned}} \qquad (10)$$

---

[3]In typical cellular systems, the BS uses the expected received SINR to adaptively encode and modulate its packet so that the packet can be decoded and demodulated at a UE with low packet error rate ($< 10\%$). However, when the actual received SINR is smaller than the expected received SINR, the packet error probability becomes high.

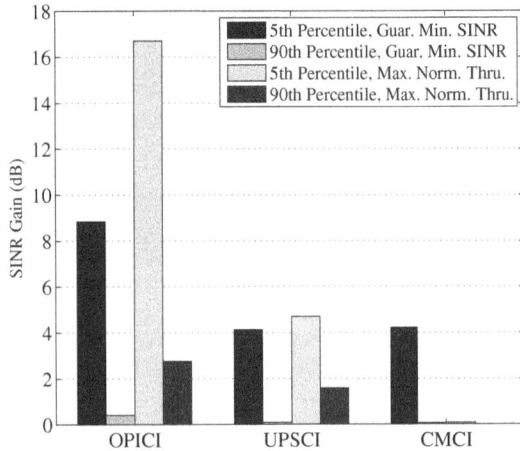

(a) SINR Gain for Adversary

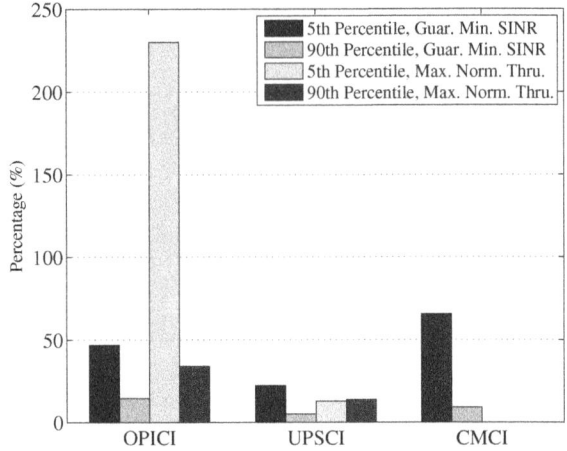

(b) Percentage of Additional Resources Used by Adversary

Figure 2: The three proposed strategies are evaluated under the Guaranteed Minimum SINR and Maximum Normalized throughput model in terms of (a) SINR gain and (b) Additional resource allocation.

4. SIMULATION RESULTS

In our simulation study, we implement seven BSs for a multicell cellular network and multiple UEs. The UEs are randomly distributed and form wireless channels to the seven BSs. The wireless channel model is composed of path loss model, large-scale fading model, and small-scale fading model. We employed the detailed simulation parameters from the IEEE evaluation methodology document [22]. Figure 2 shows the effectiveness of the three heuristic attack strategies over the two cooperation models, for 5th and 90th percentile users.

4.1 Results for 5th Percentile Users

When considering the guaranteed minimum SINR model for cell edge users, we can see that OPICI strategy gives adversary a gain of nearly 9 dB using about 50% of additional resources, which is far better as compared to CMCI and OPICI, both of which just have a gain of 4 dB using 65% and 25% of additional resources respectively. Now considering the maximum normalized throughput model for the cell edge users, we can see that the OPICI strategy gives adversary a huge gain of 17 dB as compared to UPSCI and CMCI strategy, which have a gain of about 6 dB and 0 dB respectively. Although, the resource usage is on a higher side for OPICI strategy in this model, which increases the risk of being detected by anomaly detection but still the enormous gain achieved would be worth the risk. Hence, for an adversary at the cell edge, OPICI clearly is a better attack strategy.

4.2 Results for 90th Percentile Users

The 90th percentile users are representative of users who already have a stronger signal from the BS, so there would be a lesser margin for SINR gain for an adversary as a 90th percentile user. Thus, considering the guaranteed minimum SINR model for 90th percentile users, we can see that the

OPICI strategy gives the adversary a gain lying between 0.45 to 0.5 dB using about 15% of additional resources, which is far better as compared to UPSCI and CMCI which have a gain of 0.07 and 0.05 dB respectively. Again when we consider the maximum normalized throughput model, we can see that OPICI strategy gives the adversary a gain of 2.6 dB using just 33% additional resources, which is better as compared to UPSCI and CMCI which have a gain of 1.4 and 0 dB respectively. Hence, for an adversary with already high SINR, again OPICI turns out to be a better attack strategy.

4.3 Discussion of Results

We first discuss the comparison across heuristic attack strategies. From the figures given, we see that OPICI outperforms both UPSCI and CMCI in terms of the SINR gain that is achieved, under all cases. This is basically because most of the interference channel indicator values that are over projected by the adversary, while reporting the fabricated channel, become a part of the signal channel indicators after the new cooperation cluster is released. Therefore, OPICI turns out to be better than others.

We next discuss the comparison across cooperation decision models. From the figures given, we see that the guaranteed minimum SINR cooperation model is less vulnerable to the heuristic attacks, as compared to the maximum normalized throughput model. This is because the guaranteed minimum SINR cooperation model maintains a definitive system parameter, making it more difficult to manipulate.

5. SELFISH BEHAVIOR DETECTION

One possible deterrent to selfish behavior is through the use of a trusted platform module (TPM) in each UE to allow the BS to verify the UE's operation [15] by mechanisms like secure boot. Equipping each UE with a TPM allows system developers to build trusted software and guarantee the

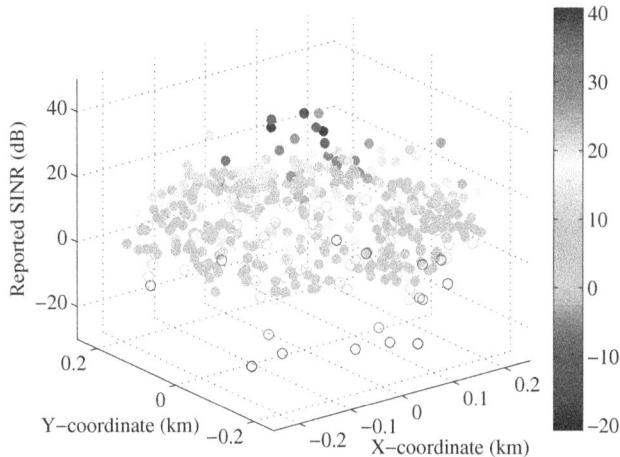

Figure 3: We illustrate the typical distribution of SINR values reported by benign and selfish users in a single cell. Filled dots represent reports from benign users, and hollow dots represent reports from selfish users using our proposed OPICI attack algorithm.

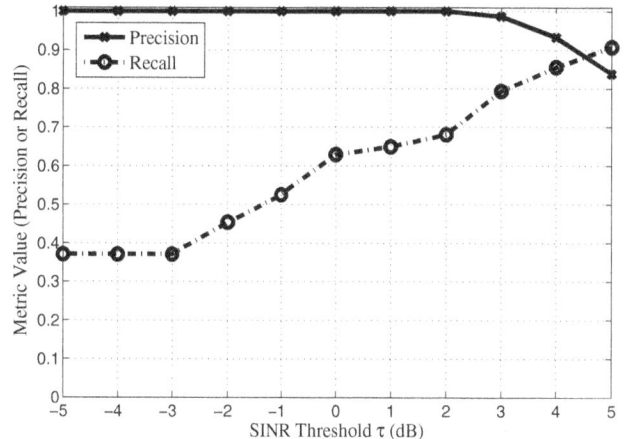

Figure 4: The precision and recall are evaluated as a function of the SINR threshold τ.

software is operating as intended. However, TPMs are not yet common in mobile phones, so an alternative approach is needed. After doing a detailed analysis of the impact of the proposed selfish misbehavior techniques, we have identified a potential detection strategy that BSs can either incorporate into the decision model $\mathcal{D}$ or use to evict users from the system.

5.1 Detection of Inconsistent SINR Reports

We propose the use of spatial consistency checking and a distance-dependent threshold to provide a course detection capability at each BS. If the network keeps track of a relatively accurate location for each UE, it can check to see whether UEs in similar locations or similar distances from the BSs are experiencing similar channel conditions. Any UE with vastly dissimilar SINR reports can be flagged as potentially misbehaving, and appropriate action can be taken. As the network may need to support a large number of UEs that may move relatively quickly, continual correlation analysis across UEs is likely an overly complex task. Instead, we propose to characterize the average behavior of UEs in the cell and compare each UE's SINR report against this model. Figure 3 illustrates the typical distribution of reported SINR values over the cell region for both benign and selfish behavior. We thus propose the use of a distance-dependent threshold $\tau(d)$ such that an SINR report less than $\tau(d)$ from a UE at a distance d from the BS will be flagged as inconsistent with the model.

We note that several factors should be included in the design of the threshold function $\tau(d)$. If the threshold is too high, UEs that are genuinely experiencing poor channel conditions may be flagged as selfish, yielding false positives, essentially defeating the purpose of cooperative communication. If the threshold is too low, selfish UEs will defeat the detection mechanism, yielding false negatives. In a practical scenario, it seems that systems employing cooperative com-

munication may err on the side of setting the threshold on the low side, accepting a certain amount of selfish behavior in order to provide better service for UEs with unfavorable conditions.

5.2 Evaluation of Threshold Detection

In order to evaluate the value of our proposed distance-dependent threshold detection mechanism, we simulate the threshold decision-making in the context of our earlier simulation study using parameters according to the IEEE EMD [22]. We randomly select 20% of the UEs as selfish users, fabricating channel vectors according to the OPICI attack algorithm described in Section 3.1. We consider the case that BSs employ the Maximum Normalized Throughput decision model, as described in Section 2.1. To measure the effectiveness of our detection mechanisms, we compute the resulting precision (fraction of detection results that are correct) and recall (fraction of misbehavior events that are detected) as a function of the constant threshold τ which is shown in Figure 4, noting that a distance-dependent threshold $\tau(d)$ will likely improve performance.

6. RELATED WORK

In recent years, there have been a number of studies on the security of cellular networks, especially focused on 3G networks [7,14,19]. Considering the up-link channel model, Sridharan et al. showed that malicious users can cause interference for normal users, by varying their own power transmission levels [21]. In contrast, the work by Racic et al. focused on down-link bandwidth in 3G networks, in order to exploit the vulnerabilities in scheduling algorithms such as proportional fairness (PF) to gain majority of the time slots in the 3G networks [18]. Bali et al. demonstrated the need for a robust scheduling algorithm by showing that TCP throughput can be reduced by as much as 25 to 30% by a single malicious user [3]. Unlike all the above discussed work which mostly operate on 3G networks, our work focuses on demonstrating selfish behavior by a user equipment in base station cooperation models [17]. Our SCF attack presented in Section 3 is built to attack base station cooperation. The core idea behind the attack is to take advantage of primary

BS's trust over UE. And through the attack, the selfish UE aims to obtain the maximum SINR gain.

7. CONCLUSIONS

In this paper, we identified a fundamental vulnerability of cellular networks using BS cooperations and backed our description with simulation results. As there exists no standardized models for base station to date, hence in our work we first propose two possible cooperation models that base station could potentially use and then further propose attack strategies over those cooperation models. We showed how the selfish UE, fabricated the channel information which degrades the cellular network's performance while benefiting the user's own quality of service. Our proposed heuristic attack strategies demonstrates that the gain for the adversary at cell-edge could go up to 40 times more than the actual received SINR. The results of our research also show that the guaranteed minimum SINR cooperation decision model is less vulnerable to the proposed attacks than the maximizing resource utilization cooperation decision model. Finally, we presented a threshold-based mechanism for BSs to detect SINR fabrication, thereby providing effective mitigation against the fabrication attacks.

8. REFERENCES

[1] 3GPP. TR 36814-900 Further advancements for E-UTRA - Physical Layer Aspects, 2010.

[2] D. Astély, E. Dahlman, A. Furuskar, Y. Jading, M. Lindstrom, and S. Parkvall. Lte: the evolution of mobile broadband. *Communications Magazine, IEEE*, 47(4):44–51, 2009.

[3] S. Bali, S. Machiraju, H. Zang, and V. Frost. A measurement study of scheduler-based attacks in 3G wireless networks. In *Proceedings of the 8th international conference on Passive and active network measurement*, PAM'07, pages 105–114, Berlin, Heidelberg, 2007. Springer-Verlag.

[4] R. Bernhardt. Macroscopic diversity in frequency reuse radio systems. *Selected Areas in Communications, IEEE Journal on*, 5(5):862–870, 1987.

[5] M. Boldi, C. Botella, F. Boccardi, V. D'Amico, E. Hardouin, M. Olsson, H. Pennanen, P. Rost, V. Savin, T. Svensson, and A. Tolli. Intermediate report on CoMP and relaying in the framework of CoMP, June 2011.

[6] C. Botella, L. Cottatellucci, V. D'Amico, M. Doll, R. Fritzsche, D. Gesbert, J. Giese, N. Gresset, H. Halbauer, E. Hardouin, H. Khanfir, M. L. Pablo, S. Saur, T. Svensson, and W. Zirwas. Definitions and architecture requirements for supporting interference avoidance techniques, August 2010.

[7] A. Bovosa. Attacks and counter measures in 2.5G and 3G cellular IP networks. In Juniper White Paper, 2004.

[8] M. Bublin, E. Hardouin, O. Hrdlicka, I. Kambourov, R. Legouable, M. Olsson, S. Plass, P. Skillermark, and P. Svac. Interference averaging concepts, June 2011.

[9] C. Carneheim, S. O. Jonsson, M. Ljungberg, M. Madfors, and J. Naslund. *FH-GSM Frequency Hopping GSM*, pages 1155–1159. 1994.

[10] W. Choi and J. G. Andrews. The capacity gain from intercell scheduling in multi-antenna systems. *IEEE Transactions on Wireless Communications*, 7(2):714–725, 2008.

[11] D. Cox. Cochannel interference considerations in frequency reuse small-coverage-area radio systems. *Communications, IEEE Transactions on*, 30(1):135–142, 1982.

[12] E. Dahlman, S. Parkvall, J. Skold, and P. Beming. *3G Evolution, Second Edition: HSPA and LTE for Mobile Broadband*. Academic Press, 2 edition, 2008.

[13] D. Gesbert, S. Hanly, H. Huang, S. S. Shitz, O. Simeone, and W. Yu. Multi-cell MIMO cooperative networks: a new look at interference. *IEEE J.Sel. A. Commun.*, 28(9):1380–1408, Dec. 2010.

[14] K. Kotapati, P. Liu, Y. Sun, and T. F. La Porta. A Taxonomy of Cyber Attacks on 3G Networks. Technical Report NAS-TR-0021-2005, Network and Security Research Center, Department of Computer Science and Engineering, Pennsylvania State University, University Park, PA, USA, January 2005.

[15] J. M. McCune, Y. Li, N. Qu, Z. Zhou, A. Datta, V. Gligor, and A. Perrig. Trustvisor: Efficient TCB reduction and attestation. In *Proceedings of the 2010 IEEE Symposium on Security and Privacy*, SP '10, pages 143–158, Washington, DC, USA, 2010. IEEE Computer Society.

[16] Y.-H. Nam, L. Liu, Y. Wang, J. C. Zhang, J. Cho, and J.-K. Han. Cooperative communication technologies for LTE-advanced. In *ICASSP*, pages 5610–5613. IEEE, 2010.

[17] A. Osseiran, J. Monserrat, and W. Mohr. *Mobile and Wireless Communications for IMT-Advanced and Beyond*. John Wiley & Sons, 2011.

[18] R. Racic, D. Ma, H. Chen, and X. Liu. Exploiting opportunistic scheduling in cellular data networks.

[19] F. Ricciato. Unwanted traffic in 3G networks. *SIGCOMM Comput. Commun. Rev.*, 36(2):53–56, Apr. 2006.

[20] C. E. Shannon. A mathematical theory of communication. *SIGMOBILE Mob. Comput. Commun. Rev.*, 5(1):3–55, Jan. 2001.

[21] A. Sridharan, R. Subbaraman, and R. Guerin. Uplink scheduling in the ev-do rev. a system: An initial investigation. In Sprint ATL Research Report Nr. RR06-ATL080139, 2006.

[22] R. Srinivasan, J. Zhuang, L. Jalloul, R. Novak, and J. Park. IEEE 802.16m evaluation methodology document (EMD), July 2008.

[23] S. Venkatesan, A. Lozano, and R. Valenzuela. Network mimo: Overcoming intercell interference in indoor wireless systems. In *Signals, Systems and Computers, 2007. ACSSC 2007. Conference Record of the Forty-First Asilomar Conference on*, pages 83–87. IEEE, 2007.

6LoWPAN Fragmentation Attacks
and Mitigation Mechanisms

René Hummen, Jens Hiller, Hanno Wirtz, Martin Henze,
Hossein Shafagh, Klaus Wehrle
Communication and Distributed Systems, RWTH Aachen University, Germany
{hummen, hiller, wirtz, henze, shafagh, wehrle}@comsys.rwth-aachen.de

ABSTRACT

6LoWPAN is an IPv6 adaptation layer that defines mechanisms to make IP connectivity viable for tightly resource-constrained devices that communicate over low power, lossy links such as IEEE 802.15.4. It is expected to be used in a variety of scenarios ranging from home automation to industrial control systems. To support the transmission of IPv6 packets exceeding the maximum frame size of the link layer, 6LoWPAN defines a packet fragmentation mechanism. However, the best effort semantics for fragment transmissions, the lack of authentication at the 6LoWPAN layer, and the scarce memory resources of the networked devices render the design of the fragmentation mechanism vulnerable.

In this paper, we provide a detailed security analysis of the 6LoWPAN fragmentation mechanism. We identify two attacks at the 6LoWPAN design-level that enable an attacker to (selectively) prevent correct packet reassembly on a target node at considerably low cost. Specifically, an attacker can mount our identified attacks by only sending a single protocol-compliant 6LoWPAN fragment. To counter these attacks, we propose two complementary, lightweight defense mechanisms, the *content chaining scheme* and the *split buffer approach*. Our evaluation shows the practicality of the identified attacks as well as the effectiveness of our proposed defense mechanisms at modest trade-offs.

Categories and Subject Descriptors

C.2.0 [**Computer-Communication Networks**]: General—
Security and protection (e.g., firewalls)

Keywords

Internet of Things; 6LoWPAN; Fragmentation; Hash chains

1. INTRODUCTION

6LoWPAN [23, 32] is an IETF-standardized IPv6 adaptation layer that enables IP connectivity over low power, lossy network links. It is envisioned as the main building block

for a number of network scenarios in the Internet of Things including home automation, industrial control systems, and smart cities [27]. Accordingly, a wide range of applications in these scenarios employ 6LoWPAN for IP-based communication via standard or special-purpose upper layer protocols.

As its main task, 6LoWPAN adjusts IPv6 packets to the unique characteristics and requirements of wireless multi-hop communication between low-power devices. The variety of applications thereby requires 6LoWPAN to support both small-sized transmissions, e.g., for sensor data or control commands, and large transmissions, e.g., for firmware updates or security protocol handshakes [38, 24, 19].

To enable the transmission of large IPv6 packets over size-constrained link layer technologies such as IEEE 802.15.4 [26], 6LoWPAN provides fragmentation support at the adaptation layer. However, the design of the 6LoWPAN fragmentation mechanism renders buffering, forwarding and processing of fragmented packets challenging on resource-constrained devices. Specifically, malicious or misconfigured nodes may send duplicate or overlapping fragments. Due to the lack of authentication at the 6LoWPAN layer, recipients are unable to distinguish these undesired fragments from legitimate ones for packet reassembly. Moreover, reassembling nodes have to optimistically store fragments of a packet and rely on a timeout mechanism to discard incomplete packets. This, however, may cause the scarce memory of a node to be occupied with incomplete packets due to missing fragments. Thus, lossy links as well as malicious or misconfigured nodes can block the processing of newly received fragmented packets by spuriously occupying buffer resources.

Our contribution in this paper is the detailed security analysis of the 6LoWPAN fragmentation mechanism for networks that consist of resource-constrained devices. We identify two attacks that a malicious node can mount against the 6LoWPAN layer. First, an eavesdropping attacker can reactively prevent the successful processing of fragmented packets by duplicating an overheard fragment with the *fragment duplication attack*. Second, an attacker without overhearing capabilities can pro-actively block processing of any fragmented packet at the target node by sending a single 6LoWPAN fragment with the *buffer reservation attack*.

To protect resource-constrained devices against these attacks, we propose two complementing, lightweight mechanisms. The *content-chaining scheme* mitigates the fragment duplication attack by offering efficient *per-fragment* sender authentication. Moreover, the *split buffer approach* fosters competition for the scarce buffer resources between legitimate nodes and an attacker on a *per-packet* basis. Our

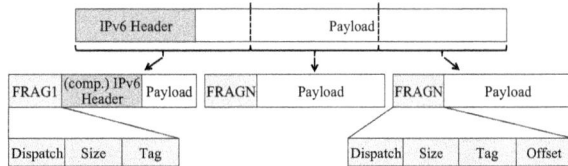

Figure 1: 6LoWPAN packet structure of a first fragment FRAG1 and subsequent fragments FRAGN.

packet discard strategy for the split buffer purges packets with suspicious sending behavior from the buffer in case of a buffer overload situation. Our evaluation shows that these mechanisms mitigate the identified attacks at low cost.

The structure of this paper is as follows. We first give a brief overview of the 6LoWPAN fragmentation mechanism in Section 2. We then describe the network scenario and the attacker model, and provide a detailed security analysis of the 6LoWPAN fragmentation mechanism in Section 3. Based on these findings, we introduce our complementary, lightweight countermeasures in Sections 4 and discuss their security considerations in Section 5. In Section 6, we show that our proposed approaches protect against the identified attacks and discuss their trade-offs. Finally, Section 7 discusses related work and Section 8 concludes our paper.

2. 6LoWPAN PACKET FRAGMENTATION

As a basis for our security analysis, we now give a brief overview of the 6LoWPAN fragmentation mechanism. We also discuss the packet routing mechanisms supported by 6LoWPAN and their implications on fragment forwarding.

2.1 Fragmentation Mechanism

The 6LoWPAN adaptation layer is located between the network and the link layer. It provides header compression and packet fragmentation functionality for IPv6 packets. In case of packet fragmentation, each 6LoWPAN fragment carries information that allows for in-place reassembly, even for out-of-order fragments. In contrast to regular IP fragments, 6LoWPAN fragments only include IP header information in the initial fragment of a packet.

When an IPv6 packet at a sending node exceeds the available link layer payload size, the 6LoWPAN fragmentation mechanism treats the (compressed) IPv6 packet as a single data field and iteratively segments this field into fragments according to the maximum frame size at the data link layer. Each fragment includes a fixed-size *fragment header*. The remaining space of the link-layer frame is iteratively filled with the IPv6 packet content (see Figure 1).

This process implies that only the first fragment (*FRAG1*) contains end-to-end routing information. Hence, a receiving node needs to correlate the remaining fragments (*FRAGN*) to the FRAG1 in order to derive IP-based routing or processing decisions for these fragments. To this end, 6LoWPAN fragments contain a *datagram tag* that is included in each fragment header and is unique per sender and fragmented packet. Thus, the datagram tag enables a receiving node to look up routing information for all fragments belonging to a fragmented packet after the FRAG1 has been received.

Each fragment also carries information that allows for in-place reassembly of fragmented packets at a receiving node. The *datagram size* of the unfragmented (and uncompressed) IPv6 packet enables a receiving node to reserve buffer space

for reassembly of the whole packet. The *datagram offset* indicates the position of the current payload within the original IPv6 packet and thus the reassembly buffer.

2.2 Fragment Forwarding Mechanisms

6LoWPAN supports three routing mechanisms [32, 2]. *Mesh-under* routing offers packet forwarding based on link-layer routing schemes. In contrast, *route-over* delegates routing to the network layer on a per-packet basis, which *enhanced route-over* optimizes by applying forwarding decisions on a per-fragment basis after the FRAG1 is received.

With *mesh-under* routing, the 6LoWPAN layer prepends each fragment with a mesh routing header. This header contains the end-to-end source and destination link layer addresses. As the link layer routing scheme at a forwarding node can immediately use this information to derive a routing decision on a per-fragment basis, mesh-under routing is oblivious to packet fragmentation. As a result, individual fragments may take different paths towards the destination.

In contrast to mesh-under routing, *route-over* routing does not require additional header information and derives forwarding decisions at the network layer. To this end, a receiving node first reassembles the entire packet before passing the packet to the upper layers for processing. If the packet is destined for another node, the receiving node looks up the next hop in its IPv6 routing table and passes the packet to the 6LoWPAN layer for re-fragmentation. As forwarding nodes apply the routing decision on a per-packet basis, all fragments of a packet are sent along the same path.

To afford mesh-under-like forwarding efficiency, *enhanced route-over* [2] proposes an optimization of the route-over approach. Enhanced route-over derives forwarding decisions directly based on the IP header information in the FRAG1. It then stores the forwarding decision, forwards the FRAG1, and applies the same forwarding decision on reception of a FRAGN that belongs to the same IPv6 packet. Hence, while FRAGNs can be forwarded individually, they are transmitted along the same path, similar to route over.

We now proceed with the security analysis of the presented 6LoWPAN fragmentation and routing mechanisms.

3. SECURITY ANALYSIS

Our security analysis focuses on how an attacker can misuse the 6LoWPAN fragmentation and routing mechanisms in order to deny the correct processing of legitimate fragmented packets. Specifically, we focus on the challenging case of an *in-network, standard-compliant* attacker that directly exploits vulnerabilities of the 6LoWPAN protocol design and characteristics of the respective network scenario.

We make no assumptions regarding the attacker's hardware resources. Thus, our identified attacks are feasible even for resource-constrained devices that are similar or equal to the actual devices in the network. Furthermore, the apparently benign behavior at the 6LoWPAN layer makes the attacker and the attacks themselves hard to detect. As such, these attacks are complementary to research on network-external attacks such as jamming-based attacks [33, 39].

We now describe the network scenario and the attacker model as the basis of our security analysis. We then proceed with our discussion of the identified attacks and analyze the topological position of a node that can be targeted depending on the routing scheme and the location of the attacker.

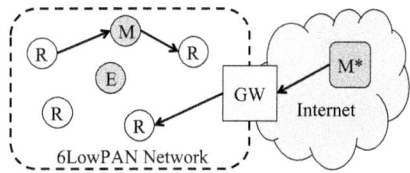

Figure 2: Network scenario with resource-constrained nodes (R) that connect to the Internet via a gateway (GW). The attackers Eve (E), Mallory (M), and Malice (M*) are marked in dark grey. Arrows indicate specific forwarding paths.

3.1 Network Scenario

In our security analysis, we abstract from specific device types, link layer technologies, and network topologies and only regard generic network characteristics. Still, our focus lies on resource-constrained devices. Hence, our network scenario consists of devices with only a few MHz of computational power and tens of kilobytes of RAM. These devices communicate over low-power wireless links and may optionally use security mechanisms provided at the link layer, e.g., based on network-wide keys. We assume the 6LoWPAN network to be connected by a gateway to a backbone infrastructure such as the Internet (see Figure 2).

Most notably, the ability of the resource-constrained devices in our network scenario to process and store fragmented packets for reassembly is very limited. For example, the default configuration of the Contiki operating system [14] for the Tmote Sky platform is restricted to the reassembly of only a *single fragmented packet at a time* with a maximum IPv6 packet size of 240 bytes.

3.2 Attacker Model

We distinguish between three different types of attackers: *Eve*, *Mallory*, and *Malice*. Eve and Mallory are both network-internal attackers who participate in the 6LoWPAN network. To join a network without link layer security, Eve and Mallory can simply be placed within radio range of the target network. In case of a protected network, an attacker must first gain admission to the network, e.g., by extracting the security keys from a legitimate node [20, 1].

Both, Eve and Mallory, participate in the routing structure and thus can send messages to any node in the network. However, with respect to the forwarding path of specific fragmented packets, they are situated in different network locations (see Figure 2). Eve is located *besides* the forwarding path of the fragmented packets. Thus, she can overhear the communication channel and send packets in reaction to overheard messages. Mallory is located *on* the forwarding path. Hence, she has Eve's capabilities and can also delay, reorder, alter or simply drop legitimate packets. The capabilities of Mallory allow her to mount at least the attacks that are viable for Eve. Thus, we do not mention her explicitly in our security considerations when discussing Eve.

In contrast to Eve and Mallory, Malice is located outside the 6LoWPAN network and has significantly more resources than the resource-constrained nodes. This enables her to simply flood a resource-constrained node with numerous large packets [21]. The fragmentation of these packets at the gateway further amplifies this attack by increasing the number of packets that the target node has to process.

To protect against such flooding-based attacks from Malice, the gateway may employ authenticated tunnels to ex-

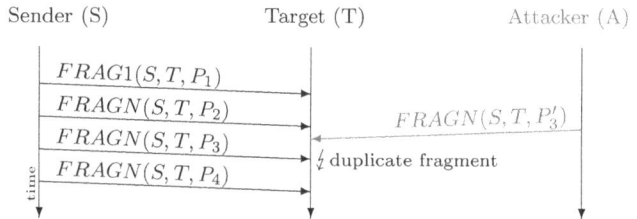

Figure 3: Packet diagram depicting the fragment duplication attack. A target must decide which fragment payload (P_3 or P_3') to use during reassembly.

ternal hosts as well as rate limitation for large packets from authenticated sources. Authenticated tunnels enable the gateway to exclude external hosts from communication if they do not behave correctly. Furthermore, rate limitation at the gateway prevents the 6LoWPAN network or a single 6LoWPAN node from being overloaded due to the vast difference in network resources. However, these mechanisms at the gateway still leave the 6LoWPAN network vulnerable to network-internal attackers. We therefore focus our discussion on attacks that Eve and Mallory can mount.

3.3 Fragment Duplication Attack

The fragment duplication attack leverages the fact that a recipient cannot verify at the 6LoWPAN layer if a fragment originates from the same source as previously received fragments of the same IPv6 packet. Thus, the recipient cannot distinguish legitimate fragments from spoofed duplicates at the time of reception. Instead, it has to process all fragments that appear to belong to the same IPv6 packet according to the sender's MAC address and the 6LoWPAN datagram tag.

Eve can exploit this fact to *selectively* block the reassembly of specific fragmented packets at a target node. For example, she may aim at preventing secure communication by blocking handshake packets of the DTLS protocol. To do so, she inspects the wireless medium for fragments that contain a DTLS message type. She then injects spoofed FRAGNs with random payload and a fragment header that links her fragments to the legitimate 6LoWPAN packet, as illustrated in Figure 3. As the target node cannot distinguish spoofed and legitimate FRAGNs, it cannot decide which fragments to use during packet reassembly at the 6LoWPAN layer.

In general, Eve can block the delivery of *any* fragmented IPv6 packet in her vicinity by injecting FRAGNs for each observed packet. In addition, higher layer protocols may retransmit lost application data in order to ensure reliable delivery, e.g., for confirmable messages in CoAP [37]. Eve's blocking of retransmitted packets then further depletes the energy resources of the forwarding and the target nodes.

Due to the complexity of dealing with duplicate fragments, the 6LoWPAN standard suggests to drop corrupt IPv6 packets. This, however, allows Eve to force her target to drop fragmented packets by sending a single duplicate FRAGN.

Upper layer information of the *reassembled packet* (e.g., message authentication codes) could be used to identify the correct fragment combination for packets with duplicate fragments. However, such an approach shows significant shortcomings. Most importantly, it requires the recipient to store all received fragments and to reassemble them after each fragment has been received at least once. Spoofed duplicates thus cannot be detected early during fragment reception and may overload the scarce buffer space at the receiver.

Sender (S)　　　　　Target (T)　　　　Attacker (A)

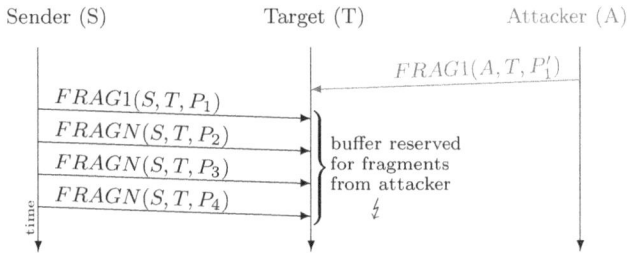

$FRAG1(A, T, P_1')$

$FRAG1(S, T, P_1)$
$FRAGN(S, T, P_2)$
$FRAGN(S, T, P_3)$
$FRAGN(S, T, P_4)$

buffer reserved
for fragments
from attacker

time

Figure 4: Packet diagram illustrating the buffer reservation attack. After the attack, the reassembly buffer of the target node is occupied by attacker fragments until the reassembly timeout expires.

3.4 Buffer Reservation Attack

The buffer reservation attack targets the scarce memory of resource-constrained nodes and leverages the fact that the recipient of a fragmented packet cannot determine a-priori if all fragments will be received correctly. Hence, a receiving node must optimistically reserve buffer space for the reassembly of the complete packet as indicated in the 6LoWPAN header. Other fragmented packets are dropped by the recipient if the reassembly buffer is already occupied. As the buffer reservation attack affects an individual reassembly buffer, the effort for an attacker to mount this attack grows linearly with the number of buffers at the target node. For our discussion of the buffer reservation attack, we assume that the target node operates with a single reassembly buffer as is the case for the Contiki operating system.

6LoWPAN defines a reassembly timeout of up to 60 seconds in order to handle fragment loss on the communication path. This timeout aims to prevent the reassembly buffer from being occupied by an incomplete packet indefinitely. Hence, when this timeout expires on a reassembling node, it must drop an incomplete packet from its reassembly buffer in order to free memory for new fragmented packets. Eve can exploit this mechanism to mount a DoS attack against memory-constrained nodes by maliciously reserving the reassembly buffer with incomplete packets.

To mount a buffer reservation attack, Eve generates a *single FRAG1* with arbitrary payload and sends it towards her target as shown in Figure 4. If the buffer of the target is not yet occupied by another fragmented packet, the received FRAG1 reserves the buffer for the reassembly of Eve's fragmented packet. Eve now either does not send the remaining FRAGNs or releases them sporadically in order to occupy the buffer resources until the reassembly timeout expires. During this time, no additional fragmented packets can be processed by the target node. Furthermore, the target node cannot distinguish Eve's attack fragments from fragments of benign senders with intermittent delays induced by the low-power characteristics of 6LoWPAN networks.

To *continuously* mount the buffer reservation attack, Eve either needs to constantly send FRAG1 fragments or she has to time her attack according to the reassembly timeout value of the target node. In case of a timing attack, the next FRAG1 must be received by the target immediately after the timeout of the previous attack has expired. To learn the exact reassembly timeout value, Eve first has to probe the target's timeout in preparation of her attack. When probing the target, Eve sends an attack FRAG1 followed by a sequence of complete fragmented packets that trigger a response from the target. For example, she may

send ICMP echo requests with a large data field that causes packet fragmentation. If the reservation attack succeeds, Eve only receives replies from the target node for fragmented packets that arrived after the reassembly timeout expired. Hence, Eve can estimate the reassembly timeout of the target node by measuring the time between the FRAG1 and the first reply. Eve is then able to mount a continuous reservation attack by sending single FRAG1s and possibly sporadic FRAGNs according to the probed timeout value.

3.5 Susceptibility of the Routing Schemes

The prerequisite for the identified fragment duplication and the buffer reservation attacks is that the target node reassembles fragmented packets for forwarding purposes or as the 6LoWPAN destination. Hence, the routing mechanism used in the 6LoWPAN network determines which resource-constrained nodes Eve can target with her attacks.

For *route-over*, Eve can interfere with the fragment processing of her one-hop neighborhood. This is because each node reassembles fragmented packets to derive routing decisions at the network layer. Thus, both attacks enable Eve to block all fragmentation-based communication that traverses the target nodes. However, Eve cannot target the reassembly of nodes that are located topologically farther away as her immediate neighbors do not forward attack packets.

In case of *enhanced route-over* or *mesh-under* routing, forwarding nodes do not reassemble fragmented packets, but directly forward attack fragments towards the destination. This enables Eve to mount the identified attacks against arbitrary 6LoWPAN destinations without topological restrictions. Hence, she can target *every* node in the 6LoWPAN network instead of only her one-hop neighborhood.

4. SECURE 6LoWPAN FRAGMENTATION

In this section, we propose lightweight security mechanisms that protect resource-constrained nodes against the fragmentation-based attacks we identified in Section 3. We note that nodes could protect themselves against the fragment duplication attack if the 6LoWPAN layer allowed them to distinguish legitimate and spoofed attack fragments on a *per-fragment* basis. We achieve this property with our *content-chaining scheme* that cryptographically binds the content of a fragmented packet to its FRAG1.

We also introduce a *split buffer approach* with fragment-sized buffer slots in order to enable processing of legitimate fragmented packets in spite of malicious nodes that pretend to require reassembly buffer resources during the buffer reservation attack. We combine this split buffer approach with a packet discard strategy that disposes of fragmented packets with suspicious sending behavior in case of a buffer overload situation. We now present a detailed description of our proposed mechanisms and refer the reader to Section 6 for the discussion of the specific trade-offs.

4.1 Content Chaining Scheme

Resource-constrained nodes could defend against the fragment duplication attack if they were able to *identify the sender* on a per-fragment basis. However, even networks with link layer security based on network-wide keys only offer group authentication. Hence, a network-internal attacker can still spoof fragments. In contrast, pairwise keys at the link layer would prevent spoofing of link layer addresses. As route-over-based routing mechanisms use the

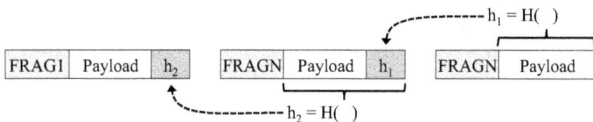

Figure 5: Example of a content chain for a packet consisting of three fragments.

link layer source address in combination with the datagram tag to identify the original packet of a fragment, this would help mitigating the fragment duplication attack. However, the necessary key management is non-trivial and would require central coordination [4], public key cryptography, e.g., a Diffie-Hellman key exchange, or alternative approaches such as probabilistic key pre-distribution [16, 6]. Furthermore, pairwise keys would be required on an end-to-end basis for mesh-under routing as here the mesh header source address is used to identify the original packet of a fragment.

To avoid the overhead of a pairwise-key management, we propose a *content-chaining scheme* that binds the content of a fragmented packet to its FRAG1 instead of binding fragments to cryptographic sender identities. To this end, the legitimate sender adds an authentication token to each fragment during the 6LoWPAN fragmentation procedure. This allows the recipient to cryptographically verify the link between fragments at the time of reception and to discard maliciously duplicated fragments early on the forwarding path.

4.1.1 Content Chain Construction

While the 6LoWPAN standard does not define the sending order of a fragmented packet, *in-order transmission* starting from the FRAG1 has recently been recognized as highly advisable, especially in networks employing (enhanced) route-over [2]. The design of our *content chaining scheme* takes advantage of this new development. Specifically, the legitimate sender cryptographically commits to the content of a fragmented packet in the corresponding FRAG1 by means of a *content chain*. Content chains are based on the concept of hash chains [17, 5], a lightweight, efficient mechanism to authenticate the sender of a data stream of finite length. Thus, they naturally fit the properties of fragmented packets when treating each packet as a fragment stream.

The elements h_i of a hash chain are generated by iteratively applying a cryptographic hash function $H(\cdot)$ to the output of the previous iteration: $h_i = H(h_{i-1})$. For the first iteration, a (random) seed value h_0 is used as input. The last token $h_n = H^n(h_0) = H(H(\dots H(h_0)))$ is referred to as the anchor element of the hash chain. Hash chains are used in reverse order of their generation, i.e. starting with h_n, as the one-way property of the hash function then prevents others from computing undisclosed tokens that are closer to the seed value. Thus, a token represents a cryptographic commitment to all previous tokens of the hash chain.

However, simply appending the hash chain elements h_i to 6LoWPAN fragments as one-time tokens does not suffice to bind these fragments to the legitimate sender as FRAGNs may be received out-of-order on the forwarding path. Specifically, if an attacker received an out-of-order fragment containing token h_{j-k} ($k > 0$), she could compute the token h_j of a previous fragment that has not yet been received on the forwarding path. The attacker could then use this token to create a valid duplicate fragment. Hence, we extend the general structure of a hash chain and include the actual

fragment content in the hash chain generation in order to enable the secure verification of out-of-order fragments.

When generating a content chain, the legitimate sender uses the payload of the last *FRAGN* as the seed value for the hash chain. He then appends the resulting content token to the previous *FRAGN* and computes the hash digest over the fragment content including the appended token (see Figure 5). After iteratively performing this procedure over all packet fragments, the FRAG1 contains a token that commits to the overall packet content. Once the content chain construction has finished, the legitimate sender transmits each fragment with its respective content token.

4.1.2 Content Chain Verification

When a node receives a FRAG1 with a content token, it processes the packet normally and stores the contained token for verification purposes of subsequent FRAGNs. For each received FRAGN, it then validates the current fragment by computing the hash over the received packet content. If the computed hash digest matches the stored token, the verification has been successful and the stored token is replaced by the currently verified one. Otherwise, the fragment is regarded as spoofed and can be dropped immediately.

As the FRAG1 transitively commits to the subsequent fragments, target nodes are able to detect spoofed fragments after receiving the FRAG1. This prevents an attacker from interfering with the packet reassembly by sending duplicates.

4.1.3 Processing Out-of-order Fragments

Content chaining enables a verifying node to cryptographically determine a single valid 6LoWPAN fragment combination for an IPv6 packet despite the reception of malicious duplicate fragments. Notably, our content chaining scheme is not required to be robust against token loss because a lost fragment invalidates the entire packet according to the 6LoWPAN standard. However, a verifying node may not be able to verify out-of-order FRAGNs directly on reception because the previous fragment, and thus the commitment to the content of the current fragment, may still be missing.

Hence, when processing out-of-order FRAGNs, a verifying node follows a simple policy. First, it only forwards fragments that have been verified successfully. Second, it stores out-of-order FRAGNs without prior verification until all previous FRAGNs have been received and verified.

This simple policy enables a malicious node to fill the reassembly buffer of a verifying node with fragments that appear to be out-of-order fragments of a legitimate packet. To counter such an attack, the verifying node discards the fragment with the largest datagram offset when reaching a buffer overload situation. This fragment has the largest distance from the last correctly verified fragment. Since a node only forwards already verified fragments, this fragment is least likely to be the result of one-hop reordering.

4.1.4 Implementation Considerations

The hash function used in our content chaining scheme requires memory resources for its implementation as well as computational resources per token generation and verification. To decrease these overheads, we propose to construct the hash function based on a block cipher-based one-way compression function [3] with a length-padded Merkle-Damgård construction [31, 10]. This allows us to use the hardware acceleration support for the AES block cipher that

59

many embedded platforms provide due to built-in security functionality of the IEEE 802.15.4 radio interface.

Content tokens generate storage and transmission overheads. We propose to decrease these overheads by truncating tokens to 8 bytes. Due to the short validity period of a token, i.e., the reassembly timeout, the remaining cryptographic strength prevents an attacker from calculating a valid pre-image for a given token. Moreover, each hash operation is salted with the actual fragment content. This and the relative shortness of content chains (an IPv6 packet of 1280 bytes requires less than 20 tokens) makes our construction more robust against cycles, i.e., re-occurring elements, than the simple hash chain structure described above [5].

4.2 Split Buffer Approach

The buffer reservation attack allows to block the reassembly buffer of a target node for the timespan of the reassembly timeout at exceptionally low cost, i.e., with a single well-timed fragment. In this section, we propose mechanisms to increase these costs for an attacker such that she has to continuously send complete fragmented packets in short bursts in order to prevent legitimate packets from being processed at the target node. In this case, the buffer reservation attack resembles a flooding attack. Hence, the attacker does not benefit significantly from sending fragmented packets over unfragmented packets and must have sufficient resources to mount a flooding-based DoS attack against the target node.

4.2.1 Fragment-sized Buffer Slots

We observe that a single reassembly buffer forces a node to make a decision whether to replace fragments of a partially received packet with a new fragmented packet as soon as the first new fragment arrives. This prevents the node from optimistically storing fragments from multiple senders and deferring the decision which packet to discard to a point when an *actual* buffer overload situation is reached. Instead, it suffices for an attacker to *pretend* that she will use the available buffer resources by sending a single fragment.

Even with multiple reassembly buffers, the reassembling node would allocate memory resources as indicated in the 6LoWPAN header for each buffer. Thus, an attacker could occupy all buffer resources by sending multiple incomplete packets with an indicated high packet size. However, if a node stored *individual fragments of multiple packets* in its reassembly buffer, legitimate and malicious packets would compete for the available buffer resources based on the actually used buffer space. An attacker would then have to follow up on her pretense by transmitting further fragments.

To enable direct competition for the buffer resources between legitimate nodes and an attacker, we propose to split the reassembly buffer into *fragment-sized buffer slots*. Each slot has the maximum size of a 6LoWPAN fragment for a given link layer. Buffer slots are filled until either a packet has been fully received or an overload situation is reached. In case of a complete packet, the reassembling node assembles the packet in-order in the buffer and processes the packet normally. In case of a buffer overload situation, the node has to decide which packet to discard. To this end, the reassembling node can base its decision on the *observed sending behavior* for packets located in the split buffer.

4.2.2 Packet Discard Strategy

We propose a discard strategy for packets in the split buffer that is based on *per-packet scores*, capturing the extent to which a packet is completed along with the continuity in the sending behavior. In case of a buffer overload situation, the node then discards the packet with the lowest score. If two or more packets share the lowest score, the selection is performed randomly between these packets.

We identify three fundamentally different sending behaviors that an attacker may show during the buffer reservation attack. First, the attacker may send only the first fragment and skip the remaining ones. This sending behavior requires the least commitment of energy resources from an attacker. However, a long gap after the first fragment would indicate a loss of the remaining fragments in case of normal sending behavior. Thus, a reassembling node should preferably process other fragmented packets in an overload situation even before the reassembly timeout expires.

An attacker may also choose between two more sophisticated sending patterns. On the one hand, an attacker may immediately send all but a few fragments in a short burst in order to occupy as many buffer resources at the target node as possible. She may then transmit the remaining fragments shortly before the reassembly timeout expires. However, the long gap after the first fragment burst again indicates a loss of the remaining fragments. On the other hand, an attacker may stretch fragment transmission across the timespan of the reassembly timeout. With this pattern, an attacker appears most legitimate compared to the other behaviors, but only fills the buffer resources of the target node slowly.

Our *discard strategy* for the split buffer takes advantage of these observations. It forces an attacker to send complete packets in short bursts during the buffer reservation attack.

Percentage of Completion. We first show how to prioritize packets, that are sent in a short burst, during a buffer overload situation. To this end, a node scores each packet in the split buffer based on the *percentage of completion*. Short packet bursts increase this score quickly. Likewise, small packets that only require few buffer resources promptly get a high score as each received fragment contributes significantly to the packet completion. In contrast, attack fragments that are received at a low rate increase the score slowly and only add little to the score as an attacker must send large packets to cover (a substantial portion of) the reassembly timeout. Hence, in an overload situation, short packets as well as large packets that are sent in bursts are likely to have a higher score than slowly arriving attack packets. This renders low sending rates unattractive for an attacker.

Sending Behavior. An attacker may also change her sending behavior after she occupies buffer resources at the target with a high score. She may, e.g., stop transmitting after an incomplete burst of fragments. To penalize such a change in sending behavior, we additionally incorporate the time domain in the discard strategy. Specifically, we consider the *average elapsed time between two consecutive fragments* (a) of a packet and the *elapsed time since the last fragment* (l). Reassembling nodes penalize senders by reducing the score of a packet if the currently elapsed time l differs significantly from the expected time a. Nodes therefore store and update the time values l and a after each fragment reception.

To reflect a change in sending behavior in the packet score computation, we introduce a window w around the expected fragment reception time a. As long as the sending behavior does not change considerably ($a - w < l < a + w$), the score is calculated as before. However, if the reassembling node receives a fragment earlier than expected ($l <= a - w$), it decreases the packet score by half. Likewise, if the fragment arrives later than expected ($l >= a + w$), the score is halved for each presumably missing fragment ($\lfloor l/a \rfloor$):

$$score_{i+1} = \begin{cases} \frac{\text{fragment bytes}}{\text{total bytes}} & \text{if } i = 0 \\ score_i + \frac{\text{fragment bytes}}{\text{total bytes}} & \text{if } a - w < l < a + w \\ \frac{score_i}{2^{\max\{1; \lfloor l/a \rfloor\}}} & \text{else} \end{cases}$$

The score of a packet in the split buffer increases proportionally to the contribution of each fragment to the overall packet completion. If the sending behavior for a packet changes significantly, the score decreases depending on the intensity of this change. The window parameter w thereby allows to calibrate the maximum tolerated change in sending behavior to the specific network characteristics.

When a buffer overload situation occurs, the reassembling node has to compare the score of the packets in the split buffer *at the time of the discard decision*. To consider the elapsed time since the last fragment reception, the node computes the current score similarly as described above:

$$score_{compare} = \begin{cases} score & \text{if } a - w < l < a + w \\ \frac{score}{2^{\max\{1; \lfloor l/a \rfloor\}}} & \text{else} \end{cases}$$

The node then discards the packet with the lowest score or randomly chooses between packets with the lowest score.

With our proposed split buffer approach, a reassembling node prioritizes competing fragmented packets that are sent in short bursts. Furthermore, significant changes in sending behavior are severely penalized. This forces an attacker to effectively flood the target node with short fragment bursts and considerably limits the practicality of the attack.

At the same time, our split buffer approach allows for a more efficient use of the reassembly buffer resources for legitimate communication than a single or multiple reassembly buffers of the same size. As buffer resources are assigned on a per-fragment basis, this even allows a node to process interleaved packets that, combined, would otherwise exceed the overall buffer resources. We note that legitimate nodes that slowly send large fragmented packets may be at a disadvantage in network scenarios that rely heavily on the transmission of large fragmented packets and where interleaved packet reception occurs often. This is a trade-off of our split buffer approach for the gained robustness against the buffer reservation attack. Still, our split buffer approach effects a considerable, network-wide throughput increase in such network scenarios as a slowly sending node would block nodes with high sending rates for a notable amount of time.

5. SECURITY CONSIDERATIONS

We now identify and briefly discuss attacks that adversaries Eve and Mallory can mount against our proposed approaches. As noted earlier, the capabilities of Mallory allow her to mount at least the attacks that Eve can mount.

Impact of an on-path attacker. Mallory may buffer legitimate fragments before forwarding them towards the 6LoW-PAN destination. This would allow her to replace the legitimate fragment payload and to compute a valid content chain for the altered packet content. However, the token included in the FRAG1 only commits to a single fragment combination. Hence, her attack does not result in duplicate fragments. Instead, it resembles dropping of the entire original packet and creating a new one. These, however, are inherent capabilities of an on-path attacker. Furthermore, we consider the integrity protection of the overall packet a task for upper layer protocols.

Likewise, Mallory may forward legitimate fragmented packets with varying artificial delays between the individual fragments during a buffer reservation attack. As a result, the forwarded packets would have a lower score than her attack packets. However, she would achieve the same result if she dropped the legitimate packets instead of forwarding them.

Content chaining and FRAG1 spoofing. Eve may overhear a legitimate FRAG1 with a valid token and generate spoofed FRAG1s that include this token. Eve may then inject these spoofed FRAG1s on the forwarding path. To enforce a decision for a single FRAG1 in case of duplicates, a node only considers the first received FRAG1. Other FRAG1s with a matching source address and datagram tag are dropped immediately. As the direct transmission between two (legitimate) nodes is typically faster than through the off-path attacker Eve, this prevents her from attacking the recipient of the overheard FRAG1 with spoofed FRAG1s. However, Eve may be able to attack the 6LoWPAN destination in a mesh-under-based network if she knows a faster forwarding path to the destination node. Notably, in case of (enhanced) route-over, Eve must be in direct communication range of a next hop on the forwarding path and transmit her FRAG1s prior to the legitimate forwarder. Especially for enhanced route-over routing, this considerably limits Eve's capability to mount an attack as nodes immediately forward the legitimate FRAG1s after reception.

Content chaining and a reordering attacker. Eve may take advantage of the fact that out-of-order fragments cannot be validated immediately at a verifying node. To this end, she may send fragments with a large offset in order to occupy buffer resources until all previous fragments have been received and the fragment is discarded due to an invalid content chaining token. However, the discard policy of the content chaining scheme would cause such fragments to be dropped first in case of a buffer overload situation. Hence, Eve may send fragments that are close to the current fragment offset of the legitimate packet. Still, these fragments are discarded upon reception of the legitimate fragments.

Split buffer and unfair competition. Eve may try to maintain a high packet score at the target node by sending large fragments at a low rate. As soon as she detects that another node sends a fragmented packet, she may change her sending behavior to a high sending rate for the remaining fragments in order to block the majority of the target's buffer slots. We account for such behavior in our packet discard strategy by penalizing senders that suddenly change the sending rate. As the packet score is halved for each fragment that is received early, Eve's score would decrease significantly. This makes newly received legitimate packets competitive despite the lack of an initial score.

6. EVALUATION

For our evaluation, we implemented our proposed defense mechanisms for the Contiki operating system version 2.5. We used Tmote Sky motes that are equipped with an 8 MHz MSP430 microcontroller, 10 kB of RAM, 48 kB of ROM, and an IEEE 802.15.4 radio interface as our evaluation platform. As the network setup depends on the evaluated property, we describe the network topology in each section individually.

Regarding Contiki, we used the standard configuration where possible. We only decreased the number of neighbors in the RPL neighborhood table from 20 to 6 in order to have sufficient memory resources to evaluate the behavior and overheads of our proposed mechanisms for IPv6 packets of up to 1280 bytes[1]. While this change allowed us to increase the reassembly buffer from 240 to 1280 bytes, it limits the maximum connectivity of a node. As we expect deployment scenarios to require tailored trade-offs, we highlight the results for the default size in the discussion of our results.

As the hash function for the content chaining scheme, we implemented a Davies-Meyer one-way compression function [3] with a length-padded Merkle-Damgård construction. Hence, we could leverage the AES hardware support of the CC2420 radio interface when computing hash digests.

For our evaluation, we did not consider explicit overheads due to link layer security operations such as encryption and decryption. However, we considered implicit overheads resulting from the *maximum length* of the security header at the link layer by decreasing the available 6LoWPAN payload size by 21 bytes. As a result, FRAG1s contained up to 88 bytes of IPv6 header information and payload, whereas FRAGNs contained between 1 and 72 bytes of payload. Hence, our results indicate worst case overheads for our proposed mechanisms due to an increased number of fragments.

6.1 Defense Against the Identified Attacks

We now show the practical existence of our identified attacks and the effectiveness of our defense mechanisms.

Fragment Duplication Attack. As a proof of concept for the fragment duplication attack, we implemented a simple sender that transmits a constant stream of fragmented UDP packets. To simulate the behavior of a spoofing attacker Eve, this node *additionally* sends one fragment of each legitimate packet with an altered 6LoWPAN payload. A second node receives packets and counts the correctly received packets.

During our evaluation, we ran two different configurations on both nodes: one with an unmodified Contiki implementation and the other one additionally using our content chaining scheme. The sender periodically transmitted 100 fragmented packets of 240 bytes, each consisting of 4 legitimate fragments and 1 duplicate attack fragment.

With an unmodified Contiki, the receiver experienced complete packet loss with a packet delivery rate (PDR) of 0 % at the UDP layer. For each received packet, the receiver detected an invalid UDP checksum as legitimate packet content was overwritten during the attack. In contrast, our content chaining scheme achieves a PDR of 100 %. Each received fragment was verified correctly. The number and the order of spoofed fragments do not influence these results as our content chaining scheme discards unverified fragments first if a buffer overload situation arises. Hence, we conclude that the content chaining scheme effectively mitigates the otherwise viable fragment duplication attack.

Buffer Reservation Attack. To confirm the existence of the buffer reservation attack and the effectiveness of our split buffer approach, we consider a network setup consisting of three nodes: a sender, an attacker Eve, and a target node that also denotes the destination of the legitimate packets.

The buffer reservation attack only succeeds if the attacker

---

[1] The IPv6 standard [11] requires every link to have a maximum transmission unit of 1280 bytes or greater.

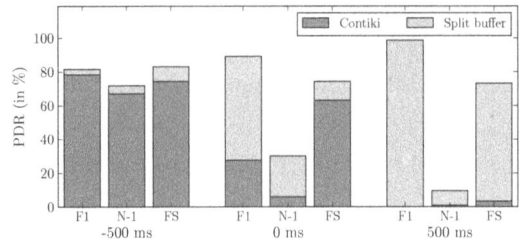

Figure 6: PDR for legitimate packets at the target node during the buffer reservation attack depending on the sending behavior and the sending offset.

reserves buffer resources at the target node *before or while* the legitimate packet is received. Hence, we evaluated the attack for different relative reception times of legitimate and attack packets at the target node. We analyzed the PDR at the UDP layer for a sender who transmits packets 500 ms before, simultaneous to, or 500 ms after the first attack fragment is sent. While the -500 ms offset allows for legitimate transmissions without malicious buffer reservation, the 500 ms offset represents a situation where the attacker successfully occupies the desired resources of the reassembly buffer. For simultaneous transmissions, the order of reception for legitimate and attack packets is not pre-determined.

We also analyzed the following three sending behaviors of an attacker: i) exclusive transmissions of FRAG1s (F1), ii) fragment bursts excluding the last FRAGN (N-1), and iii) fragment spreading across the reassembly timeout (FS). Moreover, we considered packet sizes of 240 and 1280 bytes for the sender. However, as the differences in the results are negligible in case of no protection and further improve for 240 byte packets when using our split buffer approach, we only discuss the results for packets of 1280 bytes.

To compare an unmodified Contiki and the split buffer approach, the target node was configured with these two functionalities respectively. For the split buffer approach, the window value w was set to 250 ms. We measured the PDR at the target node for 10 runs with 25 legitimate packets for each combination of the above configurations.

With an unmodified Contiki, the attack succeeds for all attack behaviors if the first attack fragment is received before the legitimate packet. In these cases, the PDR dropped as low as 0 % (see F1 for 500 ms case in Figure 6). In contrast, with our split buffer, the PDR increased up to 98 % depending on the attack behavior. Notably, the attacker has to send packets with a large number of fragments in short bursts in order to decrease the PDR with our split buffer approach (see N-1 cases in Figure 6). However, in these cases, our packet discard strategy significantly decreases the score of the attack packet once the delayed last fragment is detected as a deviation from the previously observed sending behavior. As a result, attack fragments are purged quickly from the split buffer when a new packet is received.

From these results, we conclude that the buffer reservation attack is viable against an unprotected 6LoWPAN layer even for a tightly resource-constrained attacker. In contrast, our split buffer approach forces an attacker to send short successions of large numbers of fragments during the attack.

6.2 Run-time Performance

To evaluate the run-time performance of our proposed approaches, we first analyzed the cryptographic *per-fragment*

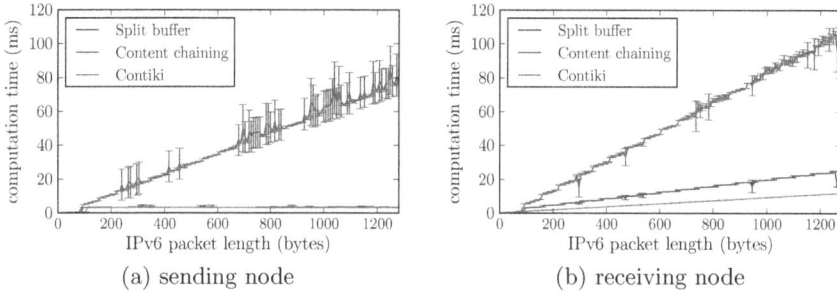

(a) sending node　　　　　　　　(b) receiving node

Figure 7: Processing time at the 6LoWPAN layer. The error bars denote the standard deviation.

Mechanism	ROM	RAM
Contiki	37788	8478
Content chaining	41108 (+8.79%)	9402 (+10.90%)
Split buffer	40052 (+5.99%)	9014 (+6.32%)
All combined	41750 (+10.48%)	9502 (+12.08%)

Table 1: ROM and RAM requirements for our proposed approaches in byte. Numbers in brackets denote added overhead to Contiki.

overhead of the content chaining scheme on a single node. We then examined the computational *per-packet* overheads of the content chaining scheme and the split buffer approach.

For the analysis of the per-fragment overhead, we ran 1000 measurements for fragment payload sizes ranging from 8 to 72 bytes respectively. The average computation time for a single content token, i.e., generation or verification, increases from 0.87 ms to 5.22 ms for growing payload sizes (see Appendix Figure 8). Notably, the processing time for 16 byte inputs increases step-wise compared to 8 byte inputs, although the hash function operates on 16 byte blocks. This is because the hash input has to be extended by an additional block if the length padding information does not fit into the last input block. Moreover, we observed outlier hash operations with run-times above 30 ms and faulty hash digests in about 2 % of our measurements. The number of erroneous operations increased with a growing payload size and thus a rising number of iterative AES operations. During these erroneous operations the hardware interface constantly reported as busy via the *ENC_BUSY* flag.

To evaluate the content chaining and the split buffer overheads, we measured the packet processing time at the 6LoW-PAN layer on a sending and on a receiving node. As we were interested in the expected results without hardware errors, we only considered measurements without erroneous tokens. Specifically, we analyzed the first 10 error-free measurements for each IPv6 packet size ranging up to 1280 bytes.

As shown by the largely overlapping results for the split buffer approach and for the unmodified Contiki in Figure 7(a), the split buffer approach does not impact the performance of a sending node. However, it adds a small overhead of up to 13.19 ms to the packet processing on a receiving node compared to an unmodified Contiki (see Figure 7(b)). The overhead of the content chaining scheme mainly stems from the construction of the content chain at the sending node and the aggregated token verification for all fragments of a packet at the receiving node. As a result, content chaining adds a maximum of 64.22 ms at the sender and of 95.23 ms at the receiver to the packet processing for packets of 1280 bytes. The performance overhead for the content chaining scheme on the receiver side is higher than on the sender side because content chaining also uses the buffer management functionality of the split buffer approach for handling out-of-order packets. Thus, the overall content chaining overhead also includes the majority of the performance overhead of the split buffer approach. Furthermore, the receiving node additionally has to search for unverified out-of-order fragments in the split buffer in case of fragment reordering.

The computational overhead of a forwarder is similar to the overhead on a receiver in case of enhanced route-over routing, and largely resembles the sum of a sender and a receiver for a route-over-based forwarder. A mesh-under-based forwarder is oblivious to packet fragmentation.

To avoid the overhead of the content chaining scheme in network scenarios without a misbehaving node, our scheme can also be implemented with a default-off policy and be switched on on-demand as described in Appendix A.

6.3 Packet Overhead

The *split buffer approach* is a purely local mechanism and, thus, does not require the transmission of additional information. In contrast, the *content chaining scheme* adds 8 bytes per fragment to the overall packet transmission.

To evaluate the packet overhead, we inspected a FRAG1 and a FRAGN with maximum length for standard 6LoW-PAN fragments as well as for fragments containing content tokens with the *wireshark* tool. We then extrapolated the gathered information for higher IPv6 packet sizes.

Notably, unfragmented packets do not contain content token information. Likewise, packets of only 2 fragments result in a token overhead that is as low as 8 bytes because the last FRAGN does not carry token information. Protecting the 6LoWPAN fragmentation of an IPv6 packet of 1280 bytes requires additional 254 bytes. This overhead results from the content tokens and from additional fragments caused by the decreased per-fragment payload space. To put these numbers into perspective, this overhead denotes 11.65 % of the overall packet transmission. This is because an IPv6 packet of 1280 bytes already requires a total transmission of 2180 bytes due to link layer and 6LoWPAN overheads, even without the content chaining scheme.

Overall, the moderate packet overhead increases linearly with the number of fragments in attack scenarios and can otherwise be avoided as described in Appendix A.

6.4 RAM and ROM Overhead

To derive RAM and ROM estimates for our proposed approaches, we analyzed the binary of a simple 6LoWPAN-enabled UDP application for four different configurations with the *msp430-size* tool. While the first binary contains an unmodified 6LoWPAN stack, the other three binaries additionally include i) the content chaining scheme, ii) the split buffer approach, and iii) a combination of both mechanisms.

As shown in Table 1, the ROM overhead of the split buffer amounts to 5.99 % of the Contiki base overhead, whereas content chaining generates 8.79 % overhead. About 530 bytes of the latter result from the implementation of our cryptographic primitive. Content chaining and the split buffer approach use the same buffer management functionality. This is reflected by the low additional overhead for the combined mechanisms compared to the individual overheads. Notably, while not optimized for minimum ROM overhead, our implementation results in less than 4 kB of code.

With respect to RAM, our content chaining scheme and the split buffer approach require 10.90 % and 6.32 % additional memory, respectively, when compared to the Contiki base overhead. For the split buffer approach, this overhead stems from the need to over-provision each buffer slot such that it can hold a 6LoWPAN fragment of maximum length including 6LoWPAN header information, i.e., 81 bytes. As a result, the memory overhead for a reassembly buffer that supports IPv6 packets of 1280 bytes increases by 18.13 %. The remaining overhead results from additional per-packet management information required for our packet discard strategy and for maintenance of the individual buffer slots. This overhead, however, is not only a trade-off for a gain of security. It additionally enables a node to process interleaved legitimate packets during normal operation. Content chaining primarily adds an overhead of 8 bytes for the last verified token to this per-packet management information.

7. RELATED WORK

For our discussion of related work, we distinguish the following two research directions: i) previously identified fragmentation-based attacks and proposed protection mechanisms and ii) existing hash chain schemes.

Packet fragmentation has previously been identify as a potential security risk for today's IP-based communication as well as for 6LoWPAN-enabled networks. Regarding today's IP communication, a large number of fragmentation-based attacks have been identified [40, 7, 8]. However, these attacks commonly focus on deficiencies of the respective IP protocol implementation, e.g., for IDS or firewall evasion, or for DoS purposes [35]. Our work, on the contrary, focuses on inherent design-related issues of the 6LoWPAN layer that emerge when using the 6LoWPAN fragmentation mechanism in resource-constrained network environments.

Recently, Gilad et al. [18] discovered an IP design vulnerability that is based on spoofed fragments with a correctly guessed IP-ID field. While this attack is similar to our identified fragment duplication attack, the authors focus on the exposure of legitimate IP-IDs in today's network environments and do not propose countermeasures like we do.

Incomplete packets have been found to cause vulnerabilities in commodity operation systems and security appliances [22]. We additionally showed that the transmission of large packets stretched over the reassembly timeout allows to maliciously occupy scarce buffer resources. We counter such attacks with our split buffer approach that is inspired by early queueing strategies for congested ATM switches in case of packet-based communication [36, 9].

With regard to 6LoWPAN, the author in [28] claims that implementation deficiencies may enable fragmentation attacks similar to the ones found in IP implementations and that replayed fragmented packets may be a potential security risk. However, in contrast to our work, neither a concrete description of the attacks nor a practical proof is given. Furthermore, the author proposes timestamps or noncryptographic nonces to mitigate replay attacks. None of these approaches help protecting resource-constrained nodes against our identified attacks as an attacker can simply spoof

such information. In [30], the authors analyze the vulnerability of the RPL routing protocol and propose IDS-based countermeasures. Their work is complementary to ours.

Hash chain schemes based on the delayed disclosure of token information such as $\mu TESLA$ [34] have been proposed for the authentication of broadcast messages in resource-constrained environments. While these schemes could also be used to authenticate the sender of fragmented packets, verifying nodes would only be able to authenticate fragments after the delayed token disclosure. Hence, in contrast to our content chaining scheme, verifying nodes would invariably be required to store unauthenticated fragments, even for in-order fragment reception. Moreover, these schemes typically require central coordination or public-key cryptography for hash chain bootstrapping and use long, pre-created hash chains at the sender to compensate the high bootstrapping costs. In contrast, our content chaining scheme does not require special infrastructure or additional cryptography, and keeps the memory overhead at the sender low.

In [29, 15], the authors propose constructions similar to our content chaining scheme for the purpose of secure network programming. However, their approaches require expensive public-key-based operations for the bootstrapping of the hash-chain anchor element, whereas we forgo the RAM, ROM, and CPU overheads of a bootstrapping mechanism. In [12, 25] the above schemes are extended in order to handle packet reordering by employing hash tree-based constructions. These schemes result in considerable token storage requirements at a verifying node for large packets compared to our content chaining scheme.

8. CONCLUSION

In this paper, we analyzed the 6LoWPAN fragmentation mechanism for vulnerabilities at the design level. We focused our analysis on network-internal attackers and an abstract 6LoWPAN network scenario that involves resource-constrained nodes. We revealed two design-level attacks against the 6LoWPAN fragmentation mechanism, the fragment duplication attack and the buffer reservation attack, and showed the susceptibility of 6LoWPAN-enabled nodes with respect to the supported routing mechanisms. The identified attacks are notably cheap allowing an attacker to use tightly resource-constrained nodes for her attacks.

To mitigate these attacks, we propose the content chaining scheme and the split buffer approach with a tailored packet discard strategy. The content chaining scheme allows a node to cryptographically verify that received fragments belong to the same packet on a per-fragment basis. Our split buffer approach fosters direct competition between legitimate senders and an attacker for scarce reassembly buffer resources. In combination with our proposed packet discard strategy, this forces an attacker to invest similar resources for the buffer reservation attack as for a flooding-based attack. Our evaluation confirms the practical existence of the identified attacks and shows that our proposed mechanisms mitigate these at moderate memory and computational costs.

Our work shows that the limited capabilities of resource-constrained nodes and the potentially highly heterogenous resources of IP-enabled devices open new attack vectors for network-internal and network-external attackers. We consider the analysis and the design of secure protocols according to these factors important future work.

9. REFERENCES

[1] A. Becher, Z. Benenson, and M. Dornseif. Tampering with motes: real-world physical attacks on wireless sensor networks. In *Proc. of SPC*, 2006.

[2] C. Bormann. Guidance for Light-Weight Implementations of the Internet Protocol Suite. draft-ietf-lwig-guidance-02 (WiP), 2012.

[3] J. W. Bos, O. Özen, and M. Stam. Efficient hashing using the AES instruction set. In *Proc. of CHES*, 2011.

[4] D. Boyle and T. Newe. Security Protocols for Use with Wireless Sensor Networks: A Survey of Security Architectures. In *Proc. of ICWMC*, 2007.

[5] P. G. Bradford and O. V. Gavrylyako. Hash chains with diminishing ranges for sensors. *International Journal of High Performance Computing and Networking*, 2006.

[6] S. Çamtepe and B. Yener. Combinatorial design of key distribution mechanisms for wireless sensor networks. *Transactions on Networking*, 2007.

[7] CERT. Advisory CA-1996-26 Denial-of-Service Attack via ping. online @ http://www.cert.org/advisories/CA-1996-26.html, 1996.

[8] CERT. Advisory CA-1997-28 IP Denial-of-Service Attacks. online @ http://www.cert.org/advisories/CA-1997-28.html, 1997.

[9] S. Chan, E. Wong, and K. Ko. Fair packet discarding for controlling ABR traffic in ATM networks . *IEEE Transactions on Communications*, 1997.

[10] I. B. Damgård. A design principle for hash functions. In *Proc. of CRYPTO*, 1989.

[11] S. Deering and R. Hinden. Internet Protocol, Version 6 (IPv6) Specification. RFC 2460, 1998.

[12] J. Deng, R. Han, and S. Mishra. Secure code distribution in dynamically programmable wireless sensor networks. In *Proc. of IPSN*, 2006.

[13] A. Dunkels, J. Eriksson, N. Finne, and N. Tsiftes. Powertrace: Network-level power profiling for low-power wireless networks. Technical report, Swedish Institute of Computer Science, 2011.

[14] A. Dunkels, B. Gronvall, and T. Voigt. Contiki – a lightweight and flexible operating system for tiny networked sensors. In *Proc. of IEEE Local Computer Networks*, 2004.

[15] P. K. Dutta, J. W. Hui, D. C. Chu, and D. E. Culler. Securing the deluge Network programming system. In *Proc. of IPSN*, 2006.

[16] L. Eschenauer and V. D. Gligor. A key-management scheme for distributed sensor networks. In *Proc. of ACM CCS*, 2002.

[17] R. Gennaro and P. Rohatgi. How to Sign Digital Streams. 1997.

[18] Y. Gilad and A. Herzberg. Fragmentation considered vulnerable: blindly intercepting and discarding fragments. In *Proc. of USENIX Offensive technologies (WOOT)*, 2011.

[19] K. Hartke and O. Bergmann. Datagram Transport Layer Security in Constrained Environments. draft-hartke-core-codtls-02 (WiP), 2012.

[20] C. Hartung, J. Balasalle, R. Han, C. Hartung, J. Balasalle, and R. Han. Node compromise in sensor networks: The need for secure systems. Technical report, University of Colorado at Boulder, 2005.

[21] T. Heer, O. Garcia-Morchon, R. Hummen, S. Keoh, S. Kumar, and K. Wehrle. Security Challenges in the IP-based Internet of Things. *Springer Wireless Personal Communications Journal*, 2011.

[22] K. Hollis. The Rose Attack. online @ http://seclists.org/bugtraq/2004/Mar/351, 2004.

[23] J. Hui and D. Culler. Extending IP to Low-Power, Wireless Personal Area Networks. *Internet Computing, IEEE*, 2008.

[24] R. Hummen, J. H. Ziegeldorf, H. Shafagh, S. Raza, and K. Wehrle. Towards Viable Certificate-based Authentication for the Internet of Things. In *Proc. of ACM HotWiSec*, 2013.

[25] S. Hyun, P. Ning, A. Liu, and W. Du. Seluge: Secure and dos-resistant code dissemination in wireless sensor networks. In *Proc. of IPSN*, 2008.

[26] IEEE. Part 15.4: wireless medium access control (MAC) and physical layer (PHY) specifications for low-rate wireless personal area networks (WPANs). IEEE 802.15.4-2006, 2006.

[27] E. Kim, D. Kaspar, and J. Vasseur. Design and Application Spaces for IPv6 over Low-Power Wireless Personal Area Networks (6LoWPANs). RFC 6568, 2012.

[28] H. Kim. Protection Against Packet Fragmentation Attacks at 6LoWPAN Adaptation Layer. In *Proc. of ICHIT*, 2008.

[29] P. E. Lanigan and P. Narasimhan. Sluice: Secure dissemination of code updates in sensor networks. In *Proc. of ICDCS*, 2006.

[30] A. Le, J. Loo, A. Lasebae, M. Aiash, and Y. Luo. 6LoWPAN: a study on QoS security threats and countermeasures using intrusion detection system approach. *International Journal of Communication Systems*, 2012.

[31] R. C. Merkle. One way hash functions and DES. In *Proc. of CRYPTO*, 1989.

[32] G. Montenegro, N. Kushalnagar, J. Hui, and D. Culler. Transmission of IPv6 Packets over IEEE 802.15.4 Networks. RFC 4944, 2007.

[33] A. Mpitziopoulos, D. Gavalas, C. Konstantopoulos, and G. Pantziou. A survey on jamming attacks and countermeasures in WSNs. *Communications Surveys & Tutorials, IEEE*, 2009.

[34] A. Perrig, R. Szewczyk, J. D. Tygar, V. Wen, and D. E. Culler. SPINS: security protocols for sensor networks. *Wireless Networks*, 2002.

[35] T. Ptacek and T. Newsham. Insertion, evasion, and denial of service: Eluding network intrusion detection. Technical report, DTIC Document, 1998.

[36] A. Romanow and S. Floyd. Dynamics of TCP traffic over ATM networks. In *Proc. of SIGCOMM*, 1994.

[37] Z. Shelby, K. Hartke, C. Bormann, and B. Frank. Constrained Application Protocol (CoAP). draft-ietf-core-coap-13 (WiP), 2012.

[38] P. Thubert and J. Hui. LoWPAN Fragment Forwarding and Recovery.

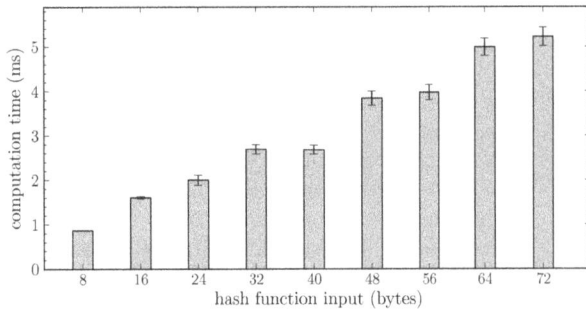

Figure 8: Processing overhead for the generation or the verification of a single content token. The bars represent the mean of the sample for increasing fragment payload sizes. The error bars show the standard error of the mean.

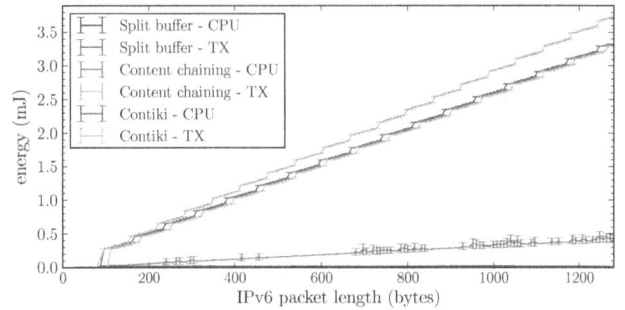

Figure 9: Energy estimation for the computations (CPU) and transmissions (TX) resulting from our proposed defense mechanisms on a sending node. The estimated power consumption for an unmodified Contiki implementation is given as a baseline.

draft-thubert-6lowpan-simple-fragment-recovery-07 (WiP), 2010.

[39] M. Wilhelm, I. Martinovic, J. B. Schmitt, and V. Lenders. Short paper: reactive jamming in wireless networks: how realistic is the threat? In *Proc. of ACM WiSec*, 2011.

[40] G. Ziemba, D. Reed, and P. Traina. Security Considerations for IP Fragment Filtering. RFC 1858, 1995.

APPENDIX

A. ATTACK NOTIFICATION

To enable a default-off policy for our content chaining scheme, we propose a simple detection and notification mechanism for the fragment duplication attack. This mechanism enables the recipient of a packet with duplicate fragments to request the legitimate sender to apply our content chaining scheme to subsequent fragmented packets. This reactive approach allows for a minimal overhead under normal operation, while protecting the network in case of an attack.

A reassembling node can efficiently detect duplicate fragments by comparing the collected payload in the reassembly buffer to the received fragment. If the payload matches, the received fragment can be silently dropped and the reassembly of the packet can proceed normally. However, if the payload differs, the reassembling node sends an ICMP message with a new, dedicated message type to the packet source notifying about the impending attack. Additionally, the node drops the packet with the duplicate fragments.

When the sender of a fragmented packet receives an ICMP message notifying about a fragment duplication attack, it immediately turns on our content chaining scheme for all further fragmented packets it sends. However, it may turn off the content chaining scheme again after a pre-configured timeout (e.g., after 15 minutes). Hence, during the timespan with active content chaining, an attacker is unable to perform the fragment duplication attack. If the attacker still proceeds with the fragment duplication attack after the timeout expired, the recipient of packets with duplicate fragments triggers another ICMP message and the content chaining scheme is activated again.

Spoofing or dropping of a notification message. As the ICMP message that is used to notify the sender about an impending fragment duplication attack is unauthenticated, an off-path attacker Eve could force the resource-constrained nodes in the 6LoWPAN network to turn on the content chaining scheme by sending a spoofed ICMP message. However, the computation and the packet space overheads she generates this way at the legitimate nodes is equal to the overheads of the content chaining scheme without the notification mechanism. Furthermore, the notification mechanism is designed to turn on the content chaining scheme when an attacker is present in the network. If Eve transmits packets for malicious purposes, this clearly is the case.

Likewise, an on-path attacker Mallory could drop legitimate notification messages in order to prevent the content chaining scheme from being activated. However, if her aim is to block legitimate packets by sending duplicate fragments, dropping of traversing packets is a more effective attack.

B. ENERGY EVALUATION

To get an impression of the energy consumption of our proposed mechanisms, we estimated the power required by the CPU during packet processing and by the radio interface during packet transmission on a sending node. For our estimations, we used the energy estimation utility provided by Contiki [13]. We assume the CPU to require 1.8 mA and the radio interface 19.5 mA per second at a supply voltage of 3 V as indicated in the Tmote Sky data sheet.

As shown by the overlapping CPU energy estimates for the split buffer and an unmodified Contiki implementation (see in Figure 9), the computations involved in the split buffer do not impact the energy consumption. The content chaining scheme, however, increases the energy consumption to a maximum of 0.43 mJ for packets of 1280 byte. These results directly correlate to our run-time performance measurements in Section 6.2.

The steps for the energy estimates of the fragment transmissions in Figure 9 depict the additional transmission overhead resulting from further fragments. The slightly shorter step length of the content chaining scheme results from the fact that less IPv6 packet content can be carried per fragment as 8 byte of fragment payload are taken by the content chaining token. As a result, the sending node has to split the IPv6 packet content into additional 6LoWPAN fragments compared to the 6LoWPAN fragmentation procedure without token information.

ETA: Efficient and Tiny and Authentication for Heterogeneous Wireless Systems

Attila Altay Yavuz
University of Pittsburgh
135 N. Bellefield Avenue, Pittsburgh, PA 15260
attila.yavuz@gmail.com

ABSTRACT

Authentication and integrity are vital security services for wireless ubiquitous systems, which require various resource-constrained devices to operate securely and efficiently. Digital signatures are basic cryptographic tools to provide these security services. However, existing digital signatures are not practical for resource-constrained systems (e.g., wireless sensors, RFID-tags). That is, traditional signatures (e.g., RSA, DSA) require expensive operations (e.g., modular exponentiation) that bring high computational cost and power-consumption. Some alternative schemes (e.g., multiple-time signatures, online/offline signatures, pre-computed tokens) are computationally efficient. However, they have large key and signature sizes and therefore are impractical for resource-constrained systems.

In this paper, we develop a new cryptographic scheme called *Efficient and Tiny Authentication (ETA)*, which is especially suitable for resource-constrained devices. That is, *ETA* does not require any expensive operation at the signer side and therefore is more computationally efficient than traditional signatures. Moreover, *ETA* has much smaller private key, signature and public key sizes than that of its counterparts (e.g., multiple-time and online/offline signatures, pre-computed tokens). *ETA* is also fully tolerant to packet loss and does not require time synchronization. All these properties make *ETA* an ideal choice to provide authentication and integrity for heterogeneous systems, in which resource-constrained devices produce publicly verifiable signatures that are verified by resourceful devices (e.g., gateways, laptops, high-end sensors).

Categories and Subject Descriptors

E.3 [**Data Encryption**]: Public key cryptosystems

General Terms

Security

Keywords

Lightweight cryptography; applied cryptography; broadcast authentication; wireless network security

1. INTRODUCTION

Nowadays ubiquitous systems like Internet of Things and Systems (IoTS) and Wireless Sensor Networks (WSNs) are composed of resource-constrained devices (e.g., low-end sensors, smart-cards, RFID-tags). It is vital for such systems to operate securely and efficiently. Hence, it is necessary to provide authentication and integrity services for resource-constrained devices. For instance, guaranteeing the integrity and authentication of financial transactions in a smart-card or RFID-tag is critical for any commercial application. However, this is a challenging task due to the memory, processor, bandwidth and power limitations of these devices.

It is also important to be able to publicly verify the authentication tags produced by resource-constrained devices. This enables any resourceful device (e.g., a laptop or a base station) to publicly audit transactions and system status.

Symmetric cryptography primitives such as Message Authentication Codes (MACs) are computationally efficient and therefore are preferred for resource-constrained devices in small-scale systems. However, such primitives are not scalable for large-distributed systems and are not publicly verifiable [13, 27]. They also cannot achieve the non-repudiation property, which is necessary for various applications (e.g., transportation payment systems, logical/pyhsical access with tiny devices).

Digital signatures rely on public key infrastructures for the signature verification [9]. Hence, they are publicly verifiable, scalable for large systems and achieve the non-repudiation. However, digital signatures also have limitations (discussed in Section 1.1) that prevent them to be practical for resource-constrained devices.

We first briefly discuss some existing alternatives and their limitations. We then present our contribution by summarizing the desirable properties of our scheme. Last, we provide a brief discussion on some prospective application areas of *ETA* and its limitations.

1.1 Related Work

Well-known modular-root based signatures (e.g., RSA [21]) are computationally efficient at the verifier side. However, these schemes require expensive operations[1] at the signer side and have large key/signature sizes. Hence, they are not suitable for computing and transmitting signatures for resource-constrained devices.

Some DLP-based signature schemes (e.g., Schnorr [22], DSA [1]) offer small key and signature sizes with an implementation on Elliptic Curves (ECs) [7]. However, they still require an expensive operation at the signer side. It is possible to eliminate the expensive operation during signature generation via pre-computed cryptographic tokens, which can be generated offline (e.g., DSA with tokens [16]). Despite being computationally efficient, this ap-

---

[1]We refer operations such as modular exponentiation [9], elliptic curve scalar multiplication [7] or pairing [15] as an expensive operation.

Table 1: Private/public key sizes, signature size and signature generation/verification costs of ETA and previous schemes

K-time overhead	Signer			Verifier	
	Private Key Size	Signature Size	Signature Generation Cost	Public Key Size	Signature Verification Cost
ETA	$2\|q\|$	$\|q\|+\kappa$	$2H + Mulq$	$\|q\| + \|H\| \cdot O(K)$	$1.3 \cdot EMul$
HORS	$(t \cdot \kappa)O(K)$	$u \cdot \kappa$	H	$(t \cdot \|H\|)O(K)$	$u \cdot H$
HORSE	$(t \cdot \kappa)O(log_2 K)$	$u \cdot \kappa$	$log_2(K) \cdot H$	$t \cdot \|H\|$	$u \cdot H$
HORS++	$(\kappa \cdot \|H\|) \cdot O(K^2)$	$(\kappa \cdot \|H\|) \cdot O(K)$	H	$\|H\|^2 \cdot O(K^2)$	$u \cdot H$
Online/offline	$2\|n\|$	$2\|n\|$	$H + 0.1 \cdot Muln$	$2\|n\|$	$Expn$
ECDSA	$\|q\|$	$2\|q\|$	$EMul$	$\|q\|$	$1.3 \cdot EMul$
Token-ECDSA	$(\|q\| + \kappa) \cdot O(K)$	$2\|q\|$	$H + 2Mulq$	$\|q\|$	$1.3 \cdot EMul$

(i) K denotes the number of messages to be signed. H and $\|H\|$ denote a cryptographic hash operation and output bit length of H, respectively. $Expn$ and $Emul$ denote a modular exponentiation over modulus n and an ECC scalar multiplication over modulus q, respectively. $Mulq$ and $Muln$ denote a modular multiplication over modulus q and modulus n, respectively. We omit constant number of low-cost operations if there is an expensive operation (e.g., a single H is omitted if there is an $EMul$). We use double-point scalar multiplication for ETA and ECDSA verifications ($1.3 \cdot Emul$ instead of $2 \cdot EMul$). Integers t and u denote the parameters used in HORS and HORSE.

(ii) Given $\kappa = 80$, suggested parameter sizes are as follows: $\|H\| = 160$, $\|q\| = 160$, $\|n\| = 1024$, $t = 256, u = 20$.

(iii) ETA only needs a few low-cost operations for the signature generation, which makes it much more computationally efficient than ECDSA and more efficient than token-ECDSA and HORSE. Note that in our system model (see Section 2), verifiers are resourceful and therefore they can perform $Emuls$. Despite being more computationally efficient than ETA, HORS and HORS++ are impractical for resource-constrained devices due to their extremely high storage and communication overheads (see Table 2).

Table 2: Private/public key and signature sizes of ETA and previous schemes (numerical)

K-time overhead	Signer				Verifier		
	Private Key Size			Signature Size	Public Key Size		
	K=1	$K = 10^2$	$K = 10^4$	$K = 10^4$	K=1	$K = 10^2$	$K = 10^4$
ETA	40 byte	40 byte	40 byte	30 byte	40 byte	1.9 KB	195 KB
HORS	2.5 KB	12 MB	24 MB	200 byte	5 KB	24 MB	48 MB
HORSE	2.5 KB	16 KB	32 KB	200 byte	5 KB	24 MB	48 MB
HORS++	1.5 KB	15 MB	> 1 GB	15 MB	3 KB	30 MB	> 1 GB
Online/Offline	256 byte	256 byte	256 byte	256 byte	256 byte	256 bye	256 byte
ECDSA	20 byte	20 byte	20 byte	40 byte	20 byte	20 byte	20 byte
Token-ECDSA	30 byte	2.9 KB	29 KB	40 byte	20 byte	20 byte	20 byte

ETA has *the smallest signature size* among its counterparts. It also has a small-constant private key, which is significantly smaller than that of its counterparts (see for $K = 10^4$). Notice that, despite being compact, ECDSA requires an expensive operation (i.e., $Emul$) for each data item to be signed and therefore it is much more computationally costly than ETA at the signer side (e.g., the estimated execution time of ECDSA is 1330 μsec, while it is only 4 μsec for ETA, see Section 5 for implementation details).

proach requires storing a pre-computed one-time token for each message to be signed. This incurs a linear storage overhead to the signer side, which is impractical for resource-constrained devices.

Online/offline signatures are generic transformations that can shift expensive signature computations to the offline phase for *any* signature scheme. Previous online/offline signatures (e.g., [6]) have very large key and signature sizes. Recent constructions (e.g., [4,23]) are more space efficient. However, similar to the pre-computed tokens, online/offline signatures also require storing a pre-computed value for each message to be signed. This introduces a linear storage overhead to the signer side.

One-time signatures (OTSs) [10, 20] rely on one-way functions without trapdoors. Hence, they offer a computationally efficient alternative to the traditional signature schemes. One of the most efficient OTSs is Hash-to-Obtain Random Subset (HORS) [20], whose signature generation is as efficient as a single hash computation. However, all these OTSs have extremely large private/public key (e.g., 2.5KB/5KB in HORS) and large signature sizes. Moreover, a private/public key pair can be used only once. This requires distribution of new public keys, which causes further overhead. *Multiple-time signatures* (e.g., HORS++ [19], HORSE [17]), also called K-time signatures, can compute a constant number of signatures under the same private/public key. However, the key and signature sizes of these schemes are larger than that of HORS, which make them prohibitive for resource-constrained devices.

1.2 Our Contribution

In this paper, we develop a new signature scheme, that we call *Efficient and Tiny Authentication (ETA)*, which is ideal for resource-constrained devices. We summarize the desirable properties of ETA below (the representative key sizes are given for the security parameter $\kappa = 80$). Table 1 and Table 2 further compare ETA and its counterparts with respect to various performance metrics.

1) Small Key Sizes: In ETA, regardless of the value of K (i.e., the number of messages to be signed), the size of private key is small-constant (i.e., 320 bytes). This is significantly smaller than that of token-based techniques (e.g., tokens with ECDSA [16]) and online/offline signatures (e.g., [6, 23]) that require $O(K)$ at the signer side. This is more efficient than that of multiple-time signatures (e.g., [17,19,20]) with large key sizes (e.g., 2.5 KB in HORS).

In ETA, the size of public key is also much smaller than that of HORS and its extensions. The size of public key component for each message is $\|H\| = 160$ bits in ETA, whereas it is 5KB in HORS and HORSE ($\|H\|$ denotes the bit length of hash output). The size of K-time public key is $\|H\| \cdot O(K)$ in ETA, whereas it is $\|H\|^2 O(K^2)$ in adaptive-secure HORS++.

2) Highly Compact Signature: In ETA, the size of signature can be as small as 240 bits. This is much more efficient than that of one-time/multiple-time (e.g., 1600-3200 bits in HORS) and online/offline signatures (e.g., 2048 bits [4,23]). It is also more efficient than that of some traditional schemes (e.g., ECDSA, RSA) and token-based alternatives (e.g., 320 bits for ECDSA tokens [16]).

3) Expensive Operation-free Signature Computation: ETA does not require any expensive operation to compute signatures (i.e., one modular addition/multiplication and two hash operations for each message). Hence, ETA is several orders of magnitude more computationally efficient than traditional signature schemes (e.g., ECDSA, RSA) and as efficient as token-based alternatives. It is also comparable to HORS and its multiple-time extensions, while being much more compact than those alternatives.

4) Practical Signing Capability: Existing multiple-time signatures are not practical even for a small K. In ETA, all the signer performance metrics are independent from K, which permits a resource-constrained device to sign very large number of messages without having any performance issue.

5) Immediate Verification and No Synchronization Requirement: Unlike some existing alternatives (e.g., [18, 25]), ETA does not require any time synchronization between the signer and verifier. ETA also achieves immediate verification, which is important for real-time applications (e.g., vehicular networks).

6) Individual Message Verification: ETA allows receivers to verify each individual message independently (unlike some HORS variants such as [17, 25]). This eliminates packet loss, time synchronization and security problems that stem from the interdependency between messages (or their corresponding signatures).

1.3 Applications

ETA is suitable for heterogeneous systems and applications, in which signers are computational, storage and bandwidth limited but verifiers are resourceful. We discuss some illustrative applications, in which ETA can be useful.

• *Increasing the Life Span of WSNs*: Many secure WSN protocols such as clone detection (e.g., [5]), secure code dissemination [8] and secure logging [26] need a signer optimal signature scheme. ETA can substantially increase the life span of WSNs by serving as a building block for such protocols, since it is compact and does not require any expensive operation at the signer side.

Token-based Payment and Access Systems: Nowadays intelligent transportation and mobile systems need mass producible low-cost payment devices [2]. Some examples include toll systems (e.g., E-ZPass), pre-paid systems (e.g., MetroCards in NYC, Pay-as-You-Drive Carsharing), NFC-based mobile payment systems and token-based logical/pyhsical access (e.g., with USBs).

ETA is an ideal primitive for such applications, since it can sign very large number of transactions by incurring near-zero computation, storage and communication overhead to the resource-constrained device. Recall that its counterparts either require an expensive operation for per transaction or incur a linear token storage overhead (which is impractical for such devices).

• *Real-time Authentication in Cyber Physical Systems*: Secure and rapid dissemination of system measurements (e.g., voltage, frequency) is essential to prevent the cascade failures [12] in cyber physical systems like smart-grids. Hence, measurement devices (e.g., phasor measurement units) must sign system readings before multicasting them to the the control centers [14]. Time and security critical nature of these applications require real-time authentication of system readings. ETA is much more practical than its counterparts (e.g., one-time/multiple time schemes [11, 25]) for such applications as shown in Section 5.

• *Limitations*: In ETA, the public key size is linear with respect to K. This requires verifiers to be storage resourceful. Moreover, despite being very efficient during signature generation (online phase), ETA requires expensive operations at its key generation (offline phase). Therefore, the initial key generation must be performed offline before system deployment (similar to online/offline signatures [4] and cryptographic tokens [16]).

Observe that, for our envisioned applications, the signer computational/storage/communication efficiency is much more important than the verifier storage efficiency alone. Furthermore, these applications permit verifiers to be storage resourceful (e.g., base stations in WSNs, control centers in cyber physical systems). Similarly, it is feasible to perform key generation phase offline in these applications. For instance, a key generation center can generate and distribute keys to the devices in payment systems and WSNs. Moreover, ETA's ability to select a very large K without creating a burden at the signer side further remedies these limitations.

2. SYNTAX AND MODELS

Notation. $||$ denotes the concatenation operation. $|r|$ denotes the bit length of variable r. $r \xleftarrow{\$} \mathcal{M}$ denotes that variable r is randomly and uniformly selected from set $\mathcal{M}$. We denote by $\{0,1\}^*$ the set of binary strings of any finite length. H is an ideal cryptographic hash function, which is defined as $H : \{0,1\}^* \to \{0,1\}^{|H|}$, where $|H|$ denotes the output bit length of H. $\mathcal{A}^{\mathcal{O}_0,...,\mathcal{O}_i}(\cdot)$ denotes algorithm $\mathcal{A}$ is provided with oracles $\mathcal{O}_0,...,\mathcal{O}_i$. For example, $\mathcal{A}^{SGN.Sig_{sk}}(\cdot)$ denotes that algorithm $\mathcal{A}$ is provided with a *signing oracle* of signature scheme SGN under a private key sk.

ETA relies on the Schnorr signature scheme [22].

Definition 1 *The Schnorr signature scheme is a tuple of three algorithms (Kg, Sig, Ver) defined as follows:*

- $(y, Y, I) \leftarrow Schnorr.Kg(1^\kappa)$: *The key generation algorithm takes 1^κ as the input. It generates large primes q and $p > q$ such that $q|(p-1)$, and generates a generator α of the subgroup G of order q in $\mathbb{Z}_p^*$. It returns a private/public key pair $(y \xleftarrow{\$} \mathbb{Z}_q^*, Y \leftarrow \alpha^y \bmod p)$ and a system parameter $I \leftarrow (q, p, \alpha)$ as the output.*

- $(s, e) \leftarrow Schnorr.Sig(y, M)$: *The signature generation algorithm takes private key y and a message M as the input. It returns a signature (s, e) as follows:*

$R \leftarrow \alpha^r \bmod p$, $e \leftarrow H(M||R)$, $s \leftarrow (r - e \cdot y) \bmod q$, *where* $r \xleftarrow{\$} \mathbb{Z}_q^*$ *and H is defined as* $H : \{0,1\}^* \to Z_q^*$.

- $b \leftarrow Schnorr.Ver(Y, M, \langle s, e \rangle)$: *The signature verification algorithm takes Y, M and $\langle s, e \rangle$ as the input. It computes $R' \leftarrow Y^e \alpha^s \bmod p$ and returns a bit b, with $b = 1$ meaning valid, if $e = H(M||R')$ and $b = 0$ otherwise.*

Definition 2 *Existential Unforgeability under Chosen Message Attack (EU-CMA) [9] experiment is as follows:*
Experiment $Expt_{SGN}^{EU\text{-}CMA}(\mathcal{A})$

$(sk, PK, I) \leftarrow SGN.Kg(1^\kappa)$, $(M^*, \sigma^*) \leftarrow \mathcal{A}^{SGN.Sig_{sk}(\cdot)}(PK)$,
If $SGN.Ver(PK, M^*, \sigma^*) = 1$ and M^* was not queried, return 1, else, return 0.

The EU-CMA advantage of $\mathcal{A}$ is defined as $Adv_{SGN}^{EU\text{-}CMA}(\mathcal{A}) = Pr[Expt_{SGN}^{EU\text{-}CMA}(\mathcal{A}) = 1]$. The EU-CMA advantage of SGN is defined as $Adv_{SGN}^{EU\text{-}CMA}(t, K) = \max_{\mathcal{A}}\{Adv_{SGN}^{EU\text{-}CMA}(\mathcal{A})\}$, where the maximum is over all $\mathcal{A}$ having time complexity t and making at most K oracle queries.

System Model: There are two types of entities in the system.

(i) *Resource-constrained Signers*: Signers are storage, computational, bandwidth and power limited devices (e.g., wireless sensor, RFID-tag). The objective of ETA is to minimize the cryptographic overhead of signers. (ii) *Resourceful Verifiers:* Storage and computational resourceful verifiers (e.g., a laptop, base station) who can be any (untrusted) entity.

We assume that the key generation/distribution is performed *offline* before deployment. A key generation center generates private/public keys offline and distributes them to the system entities.

Syntax and Security Model: ETA is a multiple-time (K-time) signature scheme, which can sign a pre-determined number of messages under the same public key.

Definition 3 *ETA is comprised of a tuple of three algorithms (Kg, Sig, Ver) defined as follows:*

- *$(sk_0, PK, I) \leftarrow ETA.Kg(1^\kappa, K)$: The key generation algorithm takes the security parameter 1^κ and the maximum number of messages to be signed K as the input. It returns a private/public key pair (sk_0, PK) and a system parameter I as the output, where PK is a vector with $K + 1$ elements.*

- *$\sigma_j \leftarrow ETA.Sig(sk_j, M_j)$: The signature generation algorithm takes the current private key sk_j, $0 \leq j \leq K - 1$ and a message M_j to be signed as the input. It returns a signature σ_j on M_j as the output, and then updates sk_j to sk_{j+1}.*

- *$b \leftarrow ETA.Ver(PK, M_j, \sigma_j)$: The signature verification algorithm takes PK, message M_j and its corresponding signature σ_j, $0 \leq j \leq K - 1$ as the input. It returns a bit b, with $b = 1$ meaning* valid, *and $b = 0$ otherwise.*

The details of ETA algorithms are given in Section 3.

ETA is proven to be (K-time) *Existentially Unforgeable against Chosen Message Attack (EU-CMA)* based on the experiment defined in Definition 4. Adversary $\mathcal{A}$ is provided with two oracles: (i) A *random oracle $RO(.)$* from which $\mathcal{A}$ can request the hash of any message M of her choice up to K' messages. (ii) A signing oracle $ETA.Sig_{sk}(.)$ from which $\mathcal{A}$ can request a ETA signature on any message M of her choice up to K messages.

Definition 4 EU-CMA experiment *for ETA is as follows:*
Experiment $Expt_{ETA}^{EU\text{-}CMA}(\mathcal{A})$

$(sk, PK, I) \leftarrow ETA.Kg(1^\kappa, K)$,

$(M^*, \sigma^*) \leftarrow \mathcal{A}^{RO(.), ETA.Sig_{sk}(.)}(PK)$,

If $ETA.Ver(PK, M^*, \sigma^*) = 1$ and M^* was not queried to $ETA.Sig_{sk}(.)$, return 1, else, return 0.

The EU-CMA-advantage of $\mathcal{A}$ is defined as $Adv_{ETA}^{EU\text{-}CMA}(\mathcal{A}) = Pr[Expt_{ETA}^{EU\text{-}CMA}(\mathcal{A}) = 1]$. The EU-CMA-advantage of ETA is defined as $Adv_{ETA}^{EU\text{-}CMA}(t, K', K) = \max_{\mathcal{A}}\{Adv_{ETA}^{EU\text{-}CMA}(\mathcal{A})\}$, where the maximum is over all $\mathcal{A}$ having time complexity t, making at most K' queries to $RO(.)$ and at most K queries to $ETA.Sig_{sk}(.)$.

3. THE PROPOSED SCHEME

Some DLP-based signatures (e.g., ECDSA [16], Schnorr [22]) can eliminate expensive operations from the signature generation by pre-computing and storing the components (R, r), where $r \overset{\$}{\leftarrow} \mathbb{Z}_q^*$ and $R = \alpha^r \bmod p$. However, this approach incurs linear storage to the signer side (i.e., one token for each message).

It is highly desirable to construct a multiple-time signature scheme, which has a constant signer storage and yet avoids expensive operations. However, this is a challenging task due to the nature of aforementioned schemes. That is, in these schemes, the token R is directly used during the signature computation and therefore its storage cannot be offloaded to the verifier side. This forces a signer either to store or to compute a token for each message.

We outline our strategies that overcome the above limitation.

• *Eliminate R from Signature Generation and Transmission*: We enable signer to compute signatures by relying on random r instead of token R. In contrast to R, random r can be evolved without requiring any expensive operation (e.g., with a hash operation). This offers a small-constant storage at the signer side, since the current r_j can be derived from the previous r_{j-1} efficiently.

In Schnorr, different from ElGamal and ECDSA, not R but the hash of it is used in signature generation as $s \leftarrow r - e \cdot y \bmod q$, where $e \leftarrow H(M||R)$. We mimic R in e by replacing it with a random number $x_j \leftarrow \{0, 1\}^\kappa$ as $e_j \leftarrow H(M_j||j||x_j)$. Our modification preserves the provable security assuming that H is a random oracle (see Theorem 1 in Section 4). This strategy offers small private key (i.e., 320 bits) and signature sizes (i.e., 240 bits).

• *Offload Token Storage to the Verifier Side*: Since we eliminate R from the signature generation, it is possible to store it at the verifier side. Note that R does not disclose r and verifiers are storage resourceful in our system model.

We pre-compute the corresponding R_j of each r_j and give the hash of each R_j to the verifier as $v_j = H(R_j)$ for $j = 0, \ldots, K-1$ before the deployment. This allows a much smaller public key size than that of existing multiple-time signatures. Note that since $\{v_j\}_{j=0}^{K-1}$ are a part of public key, corresponding $\{R_j\}_{j=0}^{K-1}$ are authenticated, despite they are not signed during the signature generation (in contrast to Schnorr).

To verify signatures, we rely on the fact that Schnorr verification algorithm recovers component R_j as $R_j \leftarrow Y^{H(.)} \cdot \alpha^{s_j}$. Given the signature (s_j, x_j, j), the verifier recovers R'_j and checks whether it is the same with the original $R_j \in PK$ as $v_j = H(R'_j)$.

The detailed description of ETA algorithms is given below.

1) $(sk_0, PK, I) \leftarrow ETA.Kg(1^\kappa, K)$:

a) Invoke $(y, Y, \langle q, p, \alpha \rangle) \leftarrow Schnorr.Kg(1^\kappa)$. Set $I \leftarrow (K, q, p, \alpha)$ for ETA.

b) Generate $R_j \leftarrow \alpha^{r_j} \bmod p$, where $r_0 \overset{\$}{\leftarrow} \mathbb{Z}_q^*$ and $r_j \leftarrow H(r_{j-1})$ for $j = 1, \ldots, K - 1$. Generate verification tokens as $v_j \leftarrow H(R_j)$ for $j = 0, \ldots, K - 1$.

c) The private and public key are $sk_0 \leftarrow (y, r_0)$ and $PK \leftarrow (Y, \overrightarrow{v} = v_0, \ldots, v_{K-1})$, respectively.

2) $\sigma_j \leftarrow ETA.Sig(sk_j, M_j)$: Given $sk_j = (y, r_j)$, compute signature σ_j on a message M_j as follows,

a) $e_j \leftarrow H(M_j||j||x_j)$ and $s_j \leftarrow r_j - e_j \cdot y \bmod q$, where $x_j \overset{\$}{\leftarrow} \{0, 1\}^\kappa$. The signature σ_j on M_j is $\sigma_j \leftarrow (s_j, x_j, j)$.

b) Update r_j as $r_{j+1} \leftarrow H(r_j)$ and erase r_j (to save memory).

c) If $j > K - 1$ then return $\perp$ (i.e., the limit on the number of signatures is exceed).

3) $b \leftarrow ETA.Ver(PK, M_j, \sigma_j)$: Recall that $\sigma_j = (s_j, x_j, j)$ and $PK = (Y, \overrightarrow{v})$, where $\overrightarrow{v} = v_0, \ldots, v_{K-1}$. If $j > K - 1$ then return $\perp$. Otherwise, if $v_j = H(R'_j)$ $ETA.Ver$ returns 1, else, it returns 0, where $R'_j \leftarrow Y^{H(M_j||j||x_j)} \cdot \alpha^{s_j}$.

4. SECURITY ANALYSIS

We prove that ETA is a (K-time) *EU-CMA* signature scheme in Theorem 1 (in the random oracle model [3]). We ignore terms that are negligible in terms of κ.

Theorem 1 $Adv_{ETA}^{EU\text{-}CMA}(t, K', K) \leq Adv_{Schnorr}^{EU\text{-}CMA}(t', K)$, where $t' = O(t) + 3K \cdot (O(\kappa^3) + RNG) + K' \cdot RNG$.

Proof: Let $\mathcal{A}$ be a *ETA attacker*. We construct a *Schnorr attacker* $\mathcal{F}$ that uses $\mathcal{A}$ as a sub-routine. That is, we set $(y, Y, \langle q, p, \alpha \rangle) \leftarrow$ *Schnorr.Kg*(1^κ) by Definition 1 and then run the simulator $\mathcal{F}$ by Definition 2 (i.e., *EU-CMA* experiment) as follows:

Algorithm $\mathcal{F}^{Schnorr.Sig_y(.)}(Y)$

- *Setup:* $\mathcal{F}$ keeps three lists $\overrightarrow{\mathcal{M}}$, $\overrightarrow{\mathcal{L}}$, and $\overrightarrow{\mathcal{L'}}$, all initially empty. $\overrightarrow{\mathcal{M}}$ is a message list that records each M_j queried to *ETA.Sig* oracle. $\overrightarrow{\mathcal{L}}[j]$ and $\overrightarrow{\mathcal{L'}}[j]$ record M_j queried to $RO(.)$ oracle and its corresponding $RO(.)$ answer h_j, respectively. $\mathcal{F}$ sets counters $(l' \leftarrow 0, l \leftarrow 0, n \leftarrow 0)$ and continues as follows:

  - $h \leftarrow H\text{-}Sim(M, l, \overrightarrow{\mathcal{L}}, \overrightarrow{\mathcal{L'}})$: $\mathcal{F}$ implements a function $H\text{-}Sim$ to handle $RO(.)$ queries. That is, cryptographic function H is modeled as a random oracle via $H\text{-}Sim$. If $\exists j : M = \overrightarrow{\mathcal{L}}[j]$ then $H\text{-}Sim$ returns $\overrightarrow{\mathcal{L'}}[j]$. Otherwise, it returns $h \xleftarrow{\$} \mathbb{Z}_q^*$ as the answer, assigns $(\overrightarrow{\mathcal{L}}[l] \leftarrow M, \overrightarrow{\mathcal{L'}}[l] \leftarrow h)$ and $l \leftarrow l + 1$.

  - $\mathcal{F}$ creates a simulated *ETA* public key PK as follows:

    (i) Query *Schnorr.Sig*$_y(.)$ on $d_j \xleftarrow{\$} \mathbb{Z}_q^*$ for $j = 0, \ldots, K-1$. *Schnorr.Sig*$_y(.)$ returns a signature (s_j, e_j), where $R_j \leftarrow Y^{e_j} \cdot \alpha^{s_j} \bmod p$ for $j = 0, \ldots, K-1$.

    (ii) Set $PK \leftarrow (Y, \overrightarrow{v})$, where $\overrightarrow{v} = \{H\text{-}Sim(R_j, l, \overrightarrow{\mathcal{L}}, \overrightarrow{\mathcal{L'}})\}_{j=0}^{K-1}$.

- *Execute* $(M^*, \sigma^*) \leftarrow \mathcal{A}^{RO(.), ETA.Sig_{sk}(.)}(PK)$:

  - Queries: $\mathcal{A}$ queries $RO(.)$ and *ETA.Sig* oracles on up to K and K' messages of her choice, respectively. $\mathcal{F}$ handles these queries as follows:

    (i) $\mathcal{A}$ queries $RO(.)$ on a message M. If $l' > K' - 1$ then $\mathcal{F}$ rejects the query (i.e., the query limit is exceeded). Otherwise, $\mathcal{F}$ invokes $h \leftarrow H\text{-}Sim(M, l, \overrightarrow{\mathcal{L}}, \overrightarrow{\mathcal{L'}})$, returns h as the answer and increments $l' \leftarrow l' + 1$.

    (ii) $\mathcal{A}$ queries *ETA.Sig* oracle on a message M_n. If $n > K - 1$ then $\mathcal{F}$ rejects the query (i.e., the query limit is exceeded). Otherwise, $\mathcal{F}$ generates $x_n \xleftarrow{\$} \{0,1\}^\kappa$ and checks if $(M_n||n||x_n) \in \overrightarrow{\mathcal{L}}$. If it holds then $\mathcal{F}$ *aborts* (i.e., the simulation fails). Otherwise, $\mathcal{F}$ simulates the hash output of $M_n||n||x_n$ with e_n by inserting the tuple $(M_n||n||x_n, e_n)$ to $(\overrightarrow{\mathcal{L}}[l] \leftarrow M_n||n||x_n, \overrightarrow{\mathcal{L'}}[l] \leftarrow e_n)$. $\mathcal{F}$ then simulates the *ETA* signature by setting $\sigma_n \leftarrow (s_n, x_n, n)$. $\mathcal{F}$ returns σ_n to $\mathcal{A}$, assigns $\overrightarrow{\mathcal{M}}[n] \leftarrow M_n$ and then increments $(n \leftarrow n + 1, l \leftarrow l + 1)$.

  - Forgery of $\mathcal{A}$: Eventually, $\mathcal{A}$ outputs a forgery on PK as $(M^*, \overline{\sigma^*})$, where $\sigma^* = (s_j^*, x_j^*, j)$, $0 \leq j \leq K-1$. By definition 4, $\mathcal{A}$ wins the *EU-CMA* experiment for *ETA* if *ETA.Ver*$(PK, M^*, \sigma^*) = 1$ and $M^* \notin \overrightarrow{\mathcal{M}}$ hold. If these conditions hold, $\mathcal{A}$ returns 1, else she returns 0.

- Forgery of $\mathcal{F}$: If $\mathcal{A}$ loses in the *EU-CMA* experiment for *ETA*, $\mathcal{F}$ also loses in the *EU-CMA* experiment for *Schnorr*, and therefore $\mathcal{F}$ *aborts* and return 0. Otherwise, if $(M^*||j||x_j^*) \notin \overrightarrow{\mathcal{L}}$ then $\mathcal{F}$ *aborts* and returns 0 (i.e., $\mathcal{A}$ wins the experiment without querying $RO(.)$ oracle).

  Otherwise, given *ETA* forgery $(M^*, \sigma^* = \langle s_j^*, x_j^*, j \rangle)$ on PK, $\mathcal{F}$ continues as follows: $\mathcal{F}$ sets the *Schnorr forgery* on public key Y as $(M^*||j||x_j^*, \gamma^* = \langle s^*, e^* \rangle)$, where $s^* = s_j^*$ and $e^* = \overrightarrow{\mathcal{L'}}[i]$ such that $\exists i, 0 \leq i \leq K' - 1 : (M^*||j||x_j^*) = \overrightarrow{\mathcal{L}}[i]$ and $0 \leq j \leq K-1$. By Definition 2, $\mathcal{F}$ checks if γ^* is a valid Schnorr signature and it has not been queried to *Schnorr.Sig*$_y(.)$. That

is, $R_j = Y^{e^*} \cdot \alpha^{s^*}$ and $(M^*||j||x_j^*) \notin \{d_0, \ldots, d_{K-1}\}$ hold *(Setup phase step-i)*. If these conditions hold then $\mathcal{F}$ wins the *EU-CMA* experiment for *Schnorr* and returns 1. Otherwise, $\mathcal{F}$ loses and returns 0.

Success Probability Analysis: $\mathcal{F}$ succeeds if all below events occur.

- $\overline{E1}$: $\mathcal{F}$ does not abort during the query phase.
- $E2$: $\mathcal{A}$ wins the *EU-CMA* experiment for *ETA*.
- $\overline{E3}$: $\mathcal{F}$ does not abort after $\mathcal{A}$'s forgery.
- *Win*: $\mathcal{F}$ wins the *EU-CMA* experiment for *Schnorr*.
- $Pr[Win] = Pr[\overline{E1}] \cdot Pr[E2|\overline{E1}] \cdot Pr[\overline{E3}|\overline{E1} \wedge E2]$

  - *The probability that event $\overline{E1}$ occurs*: During the query phase, $\mathcal{F}$ aborts if $(M_j||j||x_j) \in \overrightarrow{\mathcal{L}}$, $0 \leq j \leq K-1$ holds, *before* $\mathcal{F}$ inserts $(M_j||j||x_j)$ into $\overrightarrow{\mathcal{L}}$ (i.e., the simulation fails). This occurs if $\mathcal{A}$ guesses the random number x_n and then queries $(M_j||j||x_j)$ to $RO(.)$ *before* querying it to *ETA.Sig*. The probability that this occurs is $\frac{1}{2^\kappa}$, which is negligible in terms of κ. Hence, $Pr[\overline{E1}] = (1 - \frac{1}{2^\kappa}) \approx 1$.

  - *The probability that event $E2$ occurs*: If $\mathcal{F}$ does not abort, $\mathcal{A}$ also does not abort since the simulated view of $\mathcal{A}$ is *indistinguishable* from the real view of $\mathcal{A}$ (see the indistinguishability analysis). Therefore, $Pr[E2|\overline{E1}] = Adv_{ETA}^{EU-CMA}(t, K', K)$.

  - *The probability that event $\overline{E3}$ occurs*: $\mathcal{F}$ does not abort if the following conditions are satisfied:

    (i) $\mathcal{A}$ wins the *EU-CMA* experiment for *ETA* on a message M^* by querying it to $RO(.)$. The probability that $\mathcal{A}$ wins without querying M^* to $RO(.)$ is as difficult as a random guess.

    (ii) $\mathcal{F}$'s forgery is valid and non-trivial. The probability that $(M^*||j||x_j^*) \in (d_0, \ldots, d_{K-1})$ (i.e., $\mathcal{F}$'s forgery is trivial) is $\frac{K}{2^{|q|}}$, which is negligible in terms of κ, where $|q| = 2\kappa$. Since (s^*, e^*) is valid on $(Y, R_j) \in PK$, it is also a valid signature on Schnorr public key Y. Hence, $Pr[\overline{E3}|\overline{E1} \wedge E2] = Adv_{ETA}^{EU-CMA}(t, K', K)$.

    Omitting the terms that are negligible in terms of κ, the upper bound on *EU-CMA-advantage* of *ETA* is as follows:

$$Adv_{ETA}^{EU-CMA}(t, K', K) \leq Adv_{Schnorr}^{EU-CMA}(t', K),$$

In this experiment, the running time of $\mathcal{F}$ is that of $\mathcal{A}$ plus the time it takes to respond K' $RO(.)$ queries and K *ETA.Sig* queries (*RNG* denotes the cost of drawing a random number). $\square$

5. PERFORMANCE ANALYSIS AND COMPARISON

ETA is implemented on an Elliptic Curve (EC) [7], which offers small key and signature sizes.

The private key and signature sizes of *ETA* are constant as $2|q|$ and $|q| + \kappa$, respectively. The public key size is $|q| + |H| \cdot O(K)$.

In *ETA*, the key generation is performed *offline* (before the deployment) by a key generation center, whose cost is $(H + EMul) O(K)$. The signature generation is performed *online* by a resource-constrained device, whose cost is $2H + Mulq + Addq$. The signature verification is performed by a resourceful device, whose cost is $H + 1.3 \cdot EMul$.

Table 1 (see Section 1) compares *ETA* and its counterparts with respect to their storage, communication and computational costs. Table 2 numerically compares the storage/communication overhead of *ETA* with that of its counterparts for the growing K values.

ETA has *the smallest signature size* (i.e., 240 bits) among all of its counterparts. That is, the signature size of *ETA* is 6, 8, 1.3 and orders of magnitudes times smaller than that of **HORS/HORSE**, online/offline signatures, ECDSA/token-ECDSA and HORS++, respectively. *ETA* also has a small-constant private key size (i.e.,

320 bits), which is one of the smallest among all of its counterparts. For instance, the private key size of ETA is *several orders of magnitude smaller* than that of token-ECDSA, HORSE, HORS, HORS++ and online/offline signatures ($K = 10^4$). Similarly, the size of K-time public key is also much smaller than that of HORS, HORSE and HORS++. However, it is larger than that of ECDSA, token-ECDSA and online/offline signatures.

We prototyped ETA on a computer with an Intel(R) Core(TM) i7 Q720 at 1.60GHz CPU and 2GB RAM running Ubuntu 10.10 using MIRACL library [24] (also see item (ii) under Table 1 for the recommended key/parameter sizes used in this comparison). ETA does not require any expensive operation at the signer side and therefore is much more efficient than traditional signature schemes. For instance, the estimated execution time of ETA is 4 μsec, while it is 1330 μsec for ECDSA. ETA is also faster than HORSE (15 μsec) and token-ECDSA (6 μsec), but it is only 3μsec slower than HORS and HORS++.

The signature verification of ETA is more efficient than that of online/offline signature [23] (1774 μsec) and is equally efficient to that of ECDSA and token-ECDSA (1554 μsec). Note that in ETA system model, *verifiers are resourceful* and therefore they can easily perform a double-point scalar multiplication. ETA is less efficient than HORS, HORSE and HORS++, since these schemes only rely on hash functions for the signature verification.

The computational cost of *offline phase* (i.e. the key generation phase) is $O(K)$ expensive operations for ETA, online/offline signatures and token-ECDSA. ECDSA requires a single expensive operation, while HORS, HORSE and HORS++ require $H \cdot O(K)$ operations. We focus on the *online computational efficiency* as it is the most important metric for our envisioned applications.

In summary, ETA *is the only scheme among its counterparts that achieves the signer computational, storage and communication efficiency at the same time.*

6. CONCLUSION

In this paper, we proposed a new signature scheme called ETA, which achieves several desirable properties needed for resource-constrained devices. That is, ETA does not require any expensive operation at the signer side and therefore is much more computationally efficient than traditional PKC-based schemes. ETA has a small-constant private key and signature sizes, which are much more efficient than that of token-based signatures, online/offline signatures and multiple-time signatures. In ETA, the size of K-time public key is much smaller than that of previous multiple-time signatures. Moreover, ETA is packet loss tolerant and it achieves immediate verification property. ETA is also proven to be secure against the adaptive chosen message attacks (in random oracle model). Therefore, ETA is an ideal choice for providing authentication and integrity services for resource-constrained devices.

7. REFERENCES

[1] American Bankers Association. *ANSI X9.62-1998: Public Key Cryptography for the Financial Services Industry: The Elliptic Curve Digital Signature Algorithm (ECDSA)*, 1999.

[2] F. Baldimtsi, G. Hinterwalder, A. Rupp, A. Lysyanskaya, C. Paar, and W. P. Burleson. Pay as you go. In *Proc. of HotPETs*, July 2012.

[3] M. Bellare and P. Rogaway. Random oracles are practical: A paradigm for designing efficient protocols. In *Proceedings of the 1st ACM conference on Computer and Communications Security (CCS '93)*, pages 62–73, NY, USA, 1993. ACM.

[4] D. Catalano, M. D. Raimondo, D. Fiore, and R. Gennaro. Off-line/on-line signatures: Theoretical aspects and experimental results. Public Key Cryptography (PKC), pages 101–120. Springer-Verlag, 2008.

[5] M. Conti, R. D. Pietro, L. V. Mancini, and A. Mei. Distributed detection of clone attacks in wireless sensor networks. *IEEE Trans. on Dependable Secure Compuation*, pages 685–698, 2011.

[6] S. Even, O. Goldreich, and S. Micali. Online/offline digital signatures. In *Proceedings on Advances in Cryptology (CRYPTO '89)*, pages 263–275. Springer-Verlag, 1989.

[7] D. Hankerson, A. Menezes, and S. Vanstone. *Guide to Elliptic Curve Cryptography*. Springer, 2004.

[8] S. Hyun, P. Ning, A. Liu, and W. Du. Seluge: Secure and DoS-resistant code dissemination in wireless sensor networks. In *Proceedings of the 7th international conference on Information processing in sensor networks*, IPSN '08, pages 445–456, Washington, DC, USA, 2008. IEEE Computer Society.

[9] J. Katz and Y. Lindell. *Introduction to Modern Cryptography*. Chapman & Hall/CRC, 2007.

[10] L. Lamport. Constructing digital signatures from a one-way function. Technical report, October 1979.

[11] Q. Li and G. Cao. Multicast authentication in the smart grid with one-time signature. *IEEE Transactions on Smart Grid*, 2(4):686–696, December 2011.

[12] Y. Liu, M. K. Reiter, and P. Ning. False data injection attacks against state estimation in electric power grids. In *ACM Conference on Computer and Communications Security*, pages 21–32, 2009.

[13] J. Lopez. Unleashing public-key cryptography in wireless sensor networks. *Journal of Computer Security*, pages 469–482, Sep. 2006.

[14] Z. Lu, X. Lu, W. Wang, and C. Wang. Review and evaluation of security threats on the communication networks in the smart grid. In *Military Communication Conference (MILCOM)*, November 2010.

[15] M. Mass. Pairing-based cryptography. Master's thesis, Technische Universiteit Eindhoven, 2004.

[16] D. Naccache, D. M'Raïhi, S. Vaudenay, and D. Raphaeli. Can D.S.A. be improved? Complexity trade-offs with the digital signature standard. In *Proceedings of the 13th International Conference on the Theory and Application of Cryptographic Techniques (EUROCRYPT '94)*, pages 77–85, 1994.

[17] W.D. Neumann. HORSE: An extension of an r-time signature scheme with fast signing and verification. In *Information Technology: Coding and Computing, 2004. Proceedings. ITCC 2004. International Conference on*, volume 1, pages 129 – 134 Vol.1, april 2004.

[18] A. Perrig, R. Canetti, D. Song, and D. Tygar. Efficient authentication and signing of multicast streams over lossy channels. In *Proceedings of the IEEE Symposium on Security and Privacy*, May 2000.

[19] J. Pieprzyk, H. Wang, and C. Xing. Multiple-time signature schemes against adaptive chosen message attacks. In *Selected Areas in Cryptography (SAC)*, pages 88–100, 2003.

[20] L. Reyzin and N. Reyzin. Better than BiBa: Short one-time signatures with fast signing and verifying. In *Proceedings of the 7th Australian Conference on Information Security and Privacy (ACIPS '02)*, pages 144–153. Springer-Verlag, 2002.

[21] R.L. Rivest, A. Shamir, and L.A. Adleman. A method for obtaining digital signatures and public-key cryptosystems. *Communications of the ACM*, 21(2):120–126, 1978.

[22] C. Schnorr. Efficient signature generation by smart cards. *Journal of Cryptology*, 4(3):161–174, 1991.

[23] A. Shamir and Y. Tauman. Improved online/offline signature schemes. In *Proceedings of the 21st Annual International Cryptology Conference on Advances in Cryptology*, CRYPTO '01, pages 355–367, London, UK, 2001. Springer-Verlag.

[24] Shamus. Multiprecision integer and rational arithmetic c/c++ library (MIRACL). http://www.shamus.ie/.

[25] Q. Wang, H. Khurana, Y. Huang, and K. Nahrstedt. Time valid one-time signature for time-critical multicast data authentication. In *INFOCOM 2009, IEEE*, April 2009.

[26] A. A. Yavuz and P. Ning. Self-sustaining, efficient and forward-secure cryptographic constructions for unattended wireless sensor networks. *Ad Hoc Networks*, 10(7):1204–1220, 2012.

[27] A. A. Yavuz, P. Ning, and M. K. Reiter. Efficient, compromise resilient and append-only cryptographic schemes for secure audit logging. In *Proceedings of 2012 Financial Cryptography and Data Security (FC 2012)*, March 2012.

GPS+: A Back-end Coupons Identification for Low-Cost RFID

Ethmane El Moustaine, Maryline Laurent
Institut Mines-Télécom, Télécom SudParis, CNRS Samovar UMR 5157
9 rue Charles Fourier, 91011 Evry, France
{Ethmane.Elmoustaine, Maryline.Laurent}@telecom-sudparis.eu

ABSTRACT

Security and privacy for RFID systems are very challenging topics. First, the RFID passive tags prevailing in most of the RFID applications are very limited in processing power, thus making most of the ordinary security mechanisms inappropriate. Second, tags do answer to any reader requests, for this the most innovative RFID proposed protocols are not suitable whether for privacy problems or the high cost of tags.

So far, a variety of public-key identification/authentication protocols have been proposed, but none of them satisfy both the security and privacy requirements within the acceptable restricted resources. Girault described a storage-computation trade-off approach of the famous GPS scheme for low cost RFID tag using t *coupons* stored on tag, but for moderate security level, this approach is still beyond current capabilities of low-cost RFID tags as storage capacity is the most expensive part of the hardware. Moreover, as we demonstrate the GPS scheme cannot be private against active adversary.

In this paper, we present a new private efficient storage-security trade-off of GPS public key scheme for low-cost RFID tags. The ideas are twofold. First, the coupons are stored only on the back-end and not on the tag, so the protocol is private, the number of coupons can be much higher than in Girault's approach, and consumed coupons can be easily replaced with new ones. Second, for authenticating to the reader, the tag only needs simple integer operations, so implementation can be done in less than 1000 gate equivalents (GEs). Our approach takes advantages of the GPS scheme, and is resistant to the classical security attacks including replays, tracking, man in the middle attacks, etc.

Categories and Subject Descriptors

K.6.5 [**Management of Computing and Information System**]: Security and Protection—*Authentication*

General Terms

Security, Design, Theory

Keywords

Lightweight public key Cryptography, RFID, Security, Privacy, GPS scheme, identification

1. INTRODUCTION

Radio Frequency Identification (RFID) is a ubiquitous technology. It is an automatic identification technology using radio frequency and enabling wireless data transmission. It is often referred to as the next technological revolution after the Internet. However, RFID suffers from many security and privacy issues, such as eavesdropping, replay attacks, impersonation attacks, location tracking, etc.

RFID systems are made up of three main components: RFID tag, RFID reader and back-end database (next referred to as back-end for short). Wireless communications between tag and reader are known to be insecure while the communications between the back-end and the reader are assumed as secure. Moreover, the reader and back-end are devices with computing and battery resources, so there are no impediments to secure their in-between communications with strong symmetric or asymmetric key algorithms.

As Mark Weiser already predicted in 1991, one of the main problems against adoption of ubiquitous computing is privacy [16]. The simultaneous provision of privacy and security in low-cost RFID systems does not give other alternatives than designing a lightweight approach based on public key cryptography.

Public key cryptography can achieve a higher security level compared to private key cryptography, while it requires a higher computational overhead [15]. However, in terms of computation complexity and on-tag memory, variants of the GPS public key scheme [9] require less than that required by most private key techniques.

In this paper, we present a new private storage-security trade-off of GPS public key scheme [3] for low-cost RFID tag. The coupons are not stored on the tag as proposed by the other approaches, but they are stored on the back-end (reader or server) as such consumed coupons can be easily replaced with new ones. The direct benefit is that the protocol is private, and the number of coupons can be greatly increased, thus leading to a better security level to denial of service attacks for still low-cost tags. The computation capacities of the tag are limited to one pseudo random number generation and two simple integer operations: one multiplication and one addition. The computation and storage capacity required on the server remain reasonable. This approach is demonstrated to be robust against the classical security and privacy attacks performed over the wireless channel.

This paper is organized as follows. Section II introduces works related to RFID security issues and GPS scheme. Section III then describes our public key GPS scheme-based approach and section IV discusses its robustness to security and privacy threats. Performance issues aspects are also given in section V before our conclusions in section VI.

2. RELATED WORKS

RFID tags cannot implement classical security algorithms such as cryptographic hash functions, AES, DES, RSA, etc. and it is commonly agreed that no more than 2000 gate equivalents (GEs) can be dedicated to security [7]. A variety of secret key protocols have been proposed to support the identification/authentication service and solve some of the security and privacy issues of the RFID systems. Most of them are making use of some secret keys shared between the tag and the back-end database (next referred to as "back-end" for short). To counteract clandestine tracking, some of them are doing key update after each successful authentication session so some randomness is introduced into the message content. However a desynchronization between the tag and the back-end can occur thus leading to a denial of service.

The public key identification/authentication approaches have not been much addressed by researchers although the public key cryptography is more suitable for open system like RFID where users of an RFID tag cannot be identified ahead of time. Only few public key protocols have been proposed in the literature. Peeters et al. [12], Lee et al. [8] propose an RFID authentication protocols based on elliptic curve cryptography, but they require the tag to implement scalar point multiplications in the tag.

In [14], Sekino et al. propose an authentication protocol based on the Niederreiter public key cryptosystem [11]. It is not adapted to low-cost tags, as the tag is required to support hash function, to store large matrix, and to excute matrix operations.

To the best of our knowledge one of the most suitable public key identification protocol for RFID tags is the storage-computation trade-off approach of the famous GPS scheme [3] proposed by Girault [1], but the required storage capacity on the tag for a moderate security level stills beyond current capabilities of low-cost RFID tags.

GPS scheme [3] named after its authors Girault, Poupard and Stern, is a public-key-based zero-knowledge protocol. GPS is now standardized by the international standard ISO/IEC 9798-5 [5], and it is listed in the final NESSIE portfolio [4]. The elliptic curve variant of GPS scheme is illustrated in Figure 1. However, the GPS scheme cannot be private because if the attacker interrogates the tag using the same c he can calculate $yP - X = c(sP)$ which is a constant characterizing each tag.

In [1], Girault proposes a GPS based storage-computation trade-off (next referred to as GPS with coupons for short). As described in Figure 2, this approach consists in storing t coupons of pairs of numbers r_i, x_i (or in storing only x_i if r_i is regenerated on demand locally) for $0 \leq i \leq t - 1$ on the tag with the tag-specific random secret key s. In this case,

the computation by the tag is limited to $y = r_i + s \times c$. However, this adaptation does not support a moderate security level. An attacker can make a denial of service attack by interrogating the tag more than t times because the number of coupons stored on the tag is limited for cost reasons. Storage capacity is the most expensive part of the hardware. Moreover, GPS with coupons has the problem of privacy of GPS.

The GPS with coupons has been widely studied in the literature and different implementations have been proposed [2], [13].

In [10], McLoone et al. propose an implementation that requires only 1000 GEs in less than 150 clock cycles using challenges with a low Hamming weight.

3. THE PROPOSED PROTOCOL

Our approach (GPS+) is a private storage-security trade-off variety of GPS specially designed for low-cost RFID tags. The main advantage of our approach over GPS with coupons is the privacy and the fact that the storage overhead is supported by the back-end and not the tag. Therefore the number of coupons can be hundred times or more the number of stored coupons of the Girault's approach (GPS with coupons) while reducing the cost of the tags. In our approach, tags are only required to implement a deterministic pseudo random number generator (PRNG) with a random secret *seed* (as in GPS with coupons) and two simple integer operations (addition and multiplication), to store one random secret key s.

Note that for each tag, the coupons stored on the back-end are calculated using the same PRNG as on the tag and with the same random secret *seed*.

The reader implements the elliptic curve scalar-point multiplication and stores for each tag its public key $V = -sP$ and its t coupons $r_i P$ for $0 \leq i \leq t - 1$.

As described in Figure 3, upon detecting the tag, the reader generates and sends back to the tag a new challenge $c \in [0, B[$. Upon receiving the reader's challenge, the tag checks that $c \in [0, B[$ and re-generates a new pseudo random value $r_i \in [0, A[$ where $0 \leq i \leq t - 1$, calculates and sends to the reader $y = r_i + s \times c$. As such, the tag only does one pseudo random value generation, one multiplication of two integers, and one addition of two integers.

Note that the tag uses a deterministic pseudo random value generator with a random secret *seed*.

The reader/back-end searches in its database for a public key V satisfying both $yP + cV = (r_i + s \times c)P - c(sP) = r_i P$ and the matching of the resulting $r_i P$ value with a coupon associated to the public key of the tag (the result is in the database DB). If so the tag is identified. Otherwise the back-end/reader aborts the session concluding that the tag is illegitimate. The specifications of GPS state that the parameters are typically set to:

- s is at least 160-bit long (i.e. $S \geq 2^{160}$)

- c is 16, 32, 64-bit long (i.e. $B = 2^{16}$, 2^{32} or 2^{64})

- $\gg$ means "64 or 80-bit more", r_i is a 240, 256 or 288-bit long (i.e. $A = 2^{240}$, 2^{256} or 2^{288})

Note that if a low Hamming weight challenges are used to optimize the implementation, the challenge c can be rather longer, about 860 bits as demonstrated in [10].

```
parameters: (EC, +) an elliptic curve
          P a point of EC
          A, B and S three integers such that $A \gg B.S$
Secret key: $s \in [0, S[$
Public key: $V = -sP$


Tag                                                      Reader/Back-end

choose $r \in_R [0, A[$, compute $X = rP$   $\xrightarrow{X}$     choose $c \in_R [0, B[$

check $c \in [0, B[$                        $\xleftarrow{c}$

compute $y = r_i + s \times c$              $\xrightarrow{y}$   check $y \in [0, A + (B-1) \times (S-1)[$ ,

                                                         and $yP + cV = X$
```

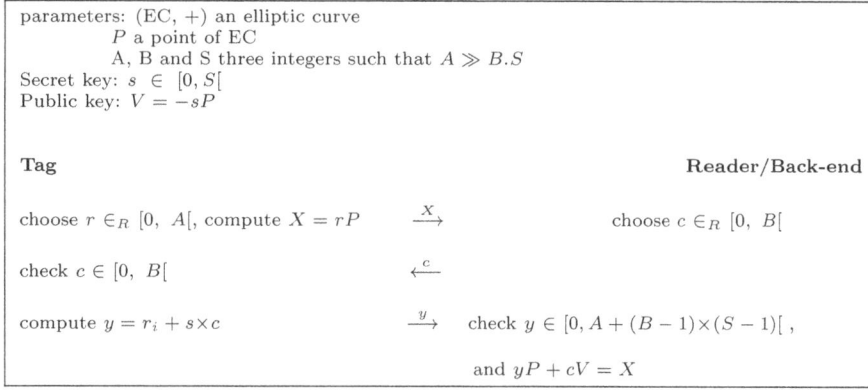

Figure 1: Elliptic curve variant of GPS

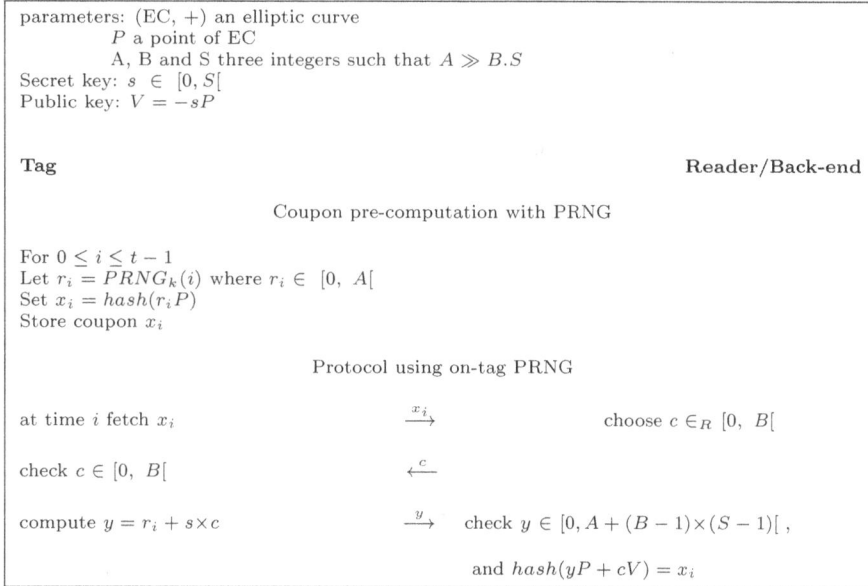

```
parameters: (EC, +) an elliptic curve
          P a point of EC
          A, B and S three integers such that $A \gg B.S$
Secret key: $s \in [0, S[$
Public key: $V = -sP$


Tag                                                      Reader/Back-end

                    Coupon pre-computation with PRNG

For $0 \leq i \leq t - 1$
Let $r_i = PRNG_k(i)$ where $r_i \in [0, A[$
Set $x_i = hash(r_i P)$
Store coupon $x_i$

                    Protocol using on-tag PRNG

at time $i$ fetch $x_i$                     $\xrightarrow{x_i}$     choose $c \in_R [0, B[$

check $c \in [0, B[$                        $\xleftarrow{c}$

compute $y = r_i + s \times c$              $\xrightarrow{y}$   check $y \in [0, A + (B-1) \times (S-1)[$ ,

                                                         and $hash(yP + cV) = x_i$
```

Figure 2: Elliptic curve variant of GPS with coupons

4. SECURITY AND PRIVACY ANALYSIS

In the proposed protocol, we just remove the first message in the GPS scheme, so the security is not weakened.

In the following analysis, we only consider vulnerabilities over the reader-tag channel. The channels between the reader and the back-end are considered as safe as both equipments have computing and battery resources, they are under the same administrative domain and they can implement any security protocols.

4.1 Resistance to replay attacks

For each identification session, reader and tag generates new pseudo-random values c and r_i, respectively; thus making messages randomized (personalized) for each session.

As demonstrated in [3], the probability of impersonation is $1/B^l$ where l is the number of the protocol rounds, and it only depends on the challenge c. A replay attack to the reader is unlikely to occur, as the reader/back-end is assumed implementing a good pseudo-random value generator. As such, they are unlikely to generate the same challenge c. This probability is $1/B$, and to achieve a higher security level the reader can repeat the protocol l times such that

$B^l \geq 2^{80}$. If $B = 2^{32}$, 3 rounds of the protocol achieve a security level higher than 80 bits. However, in many RFID applications, l will often be equal to $l = 1$.

4.2 Resistance to man-in-the-middle attack/ impersonation attacks

For the attacker, the best is experimenting a replay attack on the reader. Indeed, suppose that the attacker wants to be identified as a legitimate tag, he has to produce a valid response y' or to retrieve the tag secret key s. The attacker cannot produce a valid response y' from the previously exchanged messages between the tag and readers because any modification in a previous tag's response will be detected in the attacker's response y' thanks to the challenge c generated by the reader itself as showed in [3]. On the other hand, after GPS security proof the attacker cannot retrieve the secret keys of the tag from the exchanged messages between the tag and the reader. As such, our approach is resistant to man in the middle attacks.

4.3 Privacy support

We suppose that the attacker does not know the tag coupons

```
parameters: (EC, +) an elliptic curve of prime order (160-bit at least)
           P a point of EC
           A, B and S three integers such that A ≫ B.S
Secret key: s ∈ [0, S[
Public key: V = −sP


Tag                                                          Reader/Back-end

                    Coupon pre-computation with PRNG for back-end

                                              For 0 ≤ i ≤ t − 1
                                              Let r_i = PRNG_k(i) where r_i ∈ [0, A[,
                                              Set x_i = r_i P
                                              Store coupon x_i in the data base (DB)

                         Protocol using on-tag PRNG

check c ∈ [0, B[              ←──c              choose c ∈_R [0, B[

at time i re-generate r_i ∈ ]0, A[,

compute y = r_i + s×c         ──y→            check y ∈ [0, A + (B − 1)×(S − 1)[ ,

                                              and yP + cV = r_i P ∈ DB(V)
```

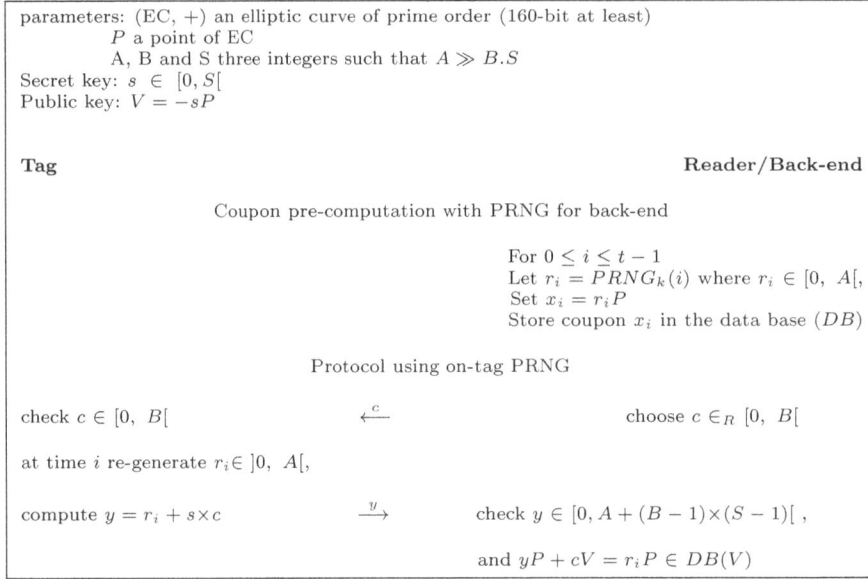

Figure 3: GPS+: A Back-end Coupons Identification for Low-cost RFID

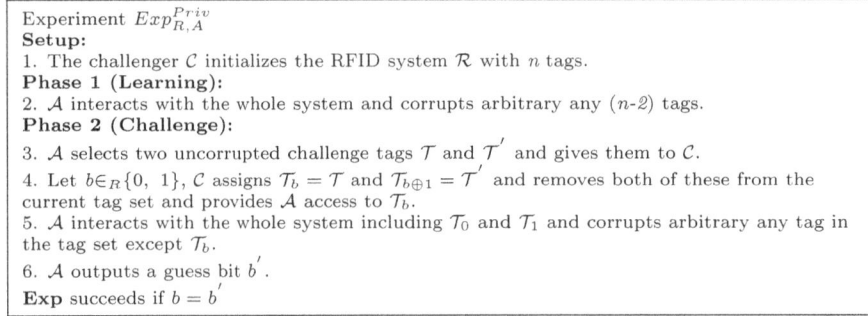

Experiment $Exp_{R,A}^{Priv}$
Setup:
1. The challenger $\mathcal{C}$ initializes the RFID system $\mathcal{R}$ with n tags.
Phase 1 (Learning):
2. $\mathcal{A}$ interacts with the whole system and corrupts arbitrary any (n-2) tags.
Phase 2 (Challenge):
3. $\mathcal{A}$ selects two uncorrupted challenge tags $\mathcal{T}$ and $\mathcal{T}'$ and gives them to $\mathcal{C}$.
4. Let $b \in_R \{0, 1\}$, $\mathcal{C}$ assigns $\mathcal{T}_b = \mathcal{T}$ and $\mathcal{T}_{b \oplus 1} = \mathcal{T}'$ and removes both of these from the current tag set and provides $\mathcal{A}$ access to $\mathcal{T}_b$.
5. $\mathcal{A}$ interacts with the whole system including $\mathcal{T}_0$ and $\mathcal{T}_1$ and corrupts arbitrary any tag in the tag set except $\mathcal{T}_b$.
6. $\mathcal{A}$ outputs a guess bit b'.
Exp succeeds if $b = b'$

Figure 4: Privacy experiment

$r_i P$ for $0 \leq i \leq t - 1$, otherwise the tag's privacy cannot be guaranteed, where the tag public key is known to the attacker.

The goal of the attacker is to obtain information about a tag $\mathcal{T}$ or trace it. To trace $\mathcal{T}$, the attacker must be able to distinguish between $\mathcal{T}$ and another legitimate tag $\mathcal{T}'$. As we have demonstrated in Section 2, the GPS protocol cannot be private, but our solution GPS+ is private as we will demonstrate.

In the following, we introduce a model that is widely adopted for privacy evaluation.

We demonstrate that our protocol guarantees the tag's privacy according to the most commonly used privacy model proposed by Juels and Weis [6].

In this model, an adversary $\mathcal{A}$ cannot make more than a limited number of interactions with the system in each phase (Learning and Challenge, Figure 4), and he cannot communicate and compute more than k overall steps where k is the global security parameter of the RFID system $\mathcal{R}$.

A protocol is private if
$\mid Pr(Exp_{R,A}^{Priv}$ succeeds in guessing $b) - \frac{1}{2} \mid \leq \epsilon(k)$ where $\epsilon(.)$ is a negligible function.

We suppose that the number of coupons stored on the backend is at least twice more than the number of interactions possible to the adversary $\mathcal{A}$, this is a realistic assumption as coupons are stored on the back-end and not on the tag, and the size of each coupon can only be 161 bits (the abscissa of the elliptic curve point and one bit for selecting the ordinate).

In order to demonstrate that GPS+ is private we replace the tag $\mathcal{T}_b$ in phase 2 of the privacy experiment $Exp_{R,A}^{Priv}$ (Figure 4) by a simulator Sim who does not know any secret and we demonstrate that the interaction of $\mathcal{A}$ with Sim will be computationally indistinguishable from an interaction with $\mathcal{T}_b$.

Let X_{Sim} the set of responses of Sim collected by $\mathcal{A}$ in the phase 2, X and X' the set of responses of tags $\mathcal{T}$ and $\mathcal{T}'$, respectively; collected by $\mathcal{A}$ in the phase 1. In order to compromise the privacy (detect the presence of Sim), $\mathcal{A}$ has to find some values $y \in X_{Sim}$ which is invalid for both $\mathcal{T}$ and $\mathcal{T}'$, a necessary condition for this, $\mathcal{A}$ must distinguish between the response of a legitimate tag and a simulator.

According to GPS the message $y = r_i + s×c$ (tag's response in GPS+) is statistically indistinguishable if SB/A is negligible [3]. This means that the communication between $\mathcal{A}$ and the tag $\mathcal{T}_b$ can be simulated. In other words $\mathcal{A}$ cannot distinguish between $\mathcal{T}_b$ and Sim.

So, the probability that $\mathcal{A}$ detects the presence of Sim is negligible. As such, our approach supports privacy.

5. PERFORMANCE EVALUATION

Our proposed protocol is only two steps specially designed for low-cost RFID tags. All the complex tasks such as scalar-point multiplication and coupons storage are done at the server (back-end).

The tag implementing only pseudo random value generator and simple integer operations: one multiplication and one addition.

As our protocol requires fewer resources on the tag than GPS with coupons (no coupons storage on the tag), it can be implemented in less than 1000 GEs using low Hamming weight challenges as demonstrated by McLoone et al. [10]. On the other hand, the reader/back-end implements pseudo random number generator, scalar-point multiplication, and stores one public key and t coupons for each tag.

The identification is asymmetrical as the back-end know the public key V of the tag and not the private key s. The back-end stores for each tag the public key and a set of coupons, apart from the storage overhead of coupons, the cost of identification is almost thus of a conventional symmetric identification because the back-end has to find first a public key V satisfying both $yP + cV = (r_i + s \times c)P - c(sP) = r_iP$ and the matching of the resulting r_iP value with a coupon associated to the public key V. So compared to symmetrical identification, the additional overhead is only to find the coupon r_iP among t coupons that are associated to the public key V which is negligible.

To give an overview of our storage-security trade-off approach, we suppose that the time to interrogate a tag is 300 ms and we show in table I the storage capacity required on the back-end for each tag according to the number of coupons and the time before any security problem (desynchronization, etc.) if the tag is interrogated continuously by an attacker. We note that this time becomes large enough if the number of coupons is important. To increase the number of coupons and to reduce storage overhead on the server, consumed coupons can be deleted and replaced with new coupons (next coupons) computed for example by trusted third party (manufacturer, etc.). Recall that the tag uses a deterministic pseudo random number generator with a random secret $seed$ as in GPS with coupons.

number of coupons	set of tags	storage overhead	tag coupons consumption time
100	100	196,5 KB	30 s
1.000	1.000	19,19 MB	300 s
10.000	10.000	1,87 GB	3000 s
100.000	100.000	187,42 GB	8,33 h
1.000.000	1.000.000	18,3 TB	83,33 h

Table 1: Overview of our storage-security trade-off

6. CONCLUSIONS

In this paper, we propose a private GPS scheme-based public key approach for low cost RFID tags. This solution satisfies the security and privacy requirements for low-cost RFID systems. It benefits from the GPS scheme features like high security level, indistinguishably, and low complexity. It provides remarkable properties such as privacy and low complexity, and resistance to known classical security

attacks. Furthermore, it can be implemented efficiently into low-cost tags in less than 1000 GEs.

7. REFERENCES

[1] M. Girault. Low-size coupons for low-cost ic cards. In *CARDIS*, pages 39–50, 2000.

[2] M. Girault, L. Juniot, and M. Robshaw. The Feasibility of On-the-Tag Public Key Cryptography. In *Workshop on RFID Security – RFIDSec'07*, Malaga, Spain, July 2007.

[3] M. Girault, G. Poupard, and J. Stern. On the fly authentication and signature schemes based on groups of unknown order. *J. Cryptology*, 19(4):463–487, 2006.

[4] F. Inc. IST-1999-12324. final report of european project IST-1999-12324: New european schemes for signatures, integrity, and encryption (NESSIE), April 2004.

[5] ISO/IEC. International standard ISO/IEC 9798 part 5: Mechanisms using zeroknowledge techniques. December 2004.

[6] A. Juels and S. Weis. Authenticating pervasive devices with human protocols. In V. Shoup, editor, *Advances in Cryptology – CRYPTO'05*, volume 3126 of *Lecture Notes in Computer Science*, pages 293–308, Santa Barbara, California, USA, August 2005. IACR, Springer.

[7] A. Juels and S. A. Weis. Authenticating pervasive devices with human protocols. In *CRYPTO*, pages 293–308, 2005.

[8] Y. K. Lee, L. Batina, D. Singelée, and I. Verbauwhede. Low-Cost Untraceable Authentication Protocols for RFID. In S. Wetzel, C. Nita-Rotaru, and F. Stajano, editors, *WiSec'10*, pages 55–64, Hoboken, New Jersey, USA, March 2010. ACM, ACM Press.

[9] J. S. Marc Girault, Guillaume Poupard. On the fly authentication and signature schemes based on groups of unknown order. *J. Cryptology*, 19(4):463–487, 2006.

[10] M. McLoone and M. J. B. Robshaw. New architectures for low-cost public key cryptography on rfid tags. In *ISCAS*, pages 1827–1830, 2007.

[11] H. Niederreiter. Knapsack-type cryptosystems and algebraic coding theory. *Problems Control Inform. Theory/Problemy Upravlen. Teor. Inform.*, 15(2):159–166, 1986.

[12] R. Peeters and J. Hermans. Wide strong private RFID identification based on zero-knowledge. Cryptology ePrint Archive, Report 2012/389, 2012.

[13] A. Poschmann, M. J. B. Robshaw, F. Vater, and C. Paar. Lightweight cryptography and rfid: Tackling the hidden overhead. *TIIS*, 4(2):98–116, 2010.

[14] T. Sekino, Y. Cui, K. Kobara, and H. Imai. Privacy enhanced rfid using quasi-dyadic fix domain shrinking. In *GLOBECOM*, pages 1–5. IEEE, 2010.

[15] S. Vaudenay. On privacy models for rfid. In *ASIACRYPT*, pages 68–87, 2007.

[16] M. Weiser. The computer for the 21st century. *Scientific American*, 265(3):66–75, January 1991.

The Weakness of Integrity Protection for LTE

Teng Wu
Department of Electrical and Computer
Engineering
University of Waterloo
200 University Avenue West
Waterloo, ON N2L3G1
Canada
teng.wu@uwaterloo.ca

Guang Gong
Department of Electrical and Computer
Engineering
University of Waterloo
200 University Avenue West
Waterloo, ON N2L3G1
Canada
ggong@uwaterloo.ca

ABSTRACT

In this paper, we concentrate on the security issues of the integrity protection of LTE. EIA1 and EIA3, two integrity protection algorithms of LTE, are insecure if the initial value (IV) can be repeated twice during the life cycle of an integrity key (IK). Especially for EIA1, because of its linearity, given two valid Message Authentication Codes (MACs) our algorithm can forge up to 2^{32} valid MACs. Thus, the probability of finding a valid MAC is dramatically increased. Although the combination of IV and IK never repeats in the ordinary case, in our well-designed scenario, the attacker can make the same combination occur twice. The duplication provides the opportunity to conduct our linear forgery attack, which may harm the security of communication. To test our linear forgery attack algorithm, we generated two counter check messages and successfully forged the third one. We also examined the attack timing by simulating real communication. From the experimental results, our attack is applicable.

Categories and Subject Descriptors

D.4.6 [**Security and Protection**]: Authentication

General Terms

Security

Keywords

Forgery, MAC, LTE, man-in-the-middle

1. INTRODUCTION

After more than twenty years evolution, cellular system has evolved to the fourth generation. Security issues of the cellular system are attracting more and more attention, because the expenses of attacking the system are much cheaper than before.

Integrity protection can protect messages from being modified. It also can prevent the impersonation attacks. Thus, integrity protection is important in communication, especially in wireless channels. Compared with wired transmission, active eavesdropping in a wireless environment is relatively easy. Without integrity protection, attackers can modify messages transmitted over the air as they wish.

In the public key cryptography domain, integrity is protected by digital signatures; similarly, in symmetric key cryptography, it is protected by MACs. Because computation of digital signatures is inefficient, in some resource-constrained applications, integrity protection is always based on MACs.

A MAC-generating algorithm usually has two components: the underlying cipher and the upper-level structure. An underlying cipher could be the keyed hash function, block cipher or stream cipher. The input messages of the algorithm are allowed to have arbitrary length. Input messages pass through the underlying cipher and become cipher text. This cipher text is assembled by the upper-level structure to get a length-fixed string, which is the output of the MAC. There are many widely used MACs, such as HMAC [7], EMAC [11], XCBC [9], OMAC [17], TMAC [12] and XOR MAC [8].

One threat to integrity protection is forgery attacks. If an attacker wants to impersonate some identities or to modify some messages, he must have the ability to forge the MAC (or the signature in public key cryptography) of some specific data. In this paper, since the cellular communications are protected by MACs, we focus on a forgery attack on MAC. There are basically two ways to forge a MAC: breaking the underlying cipher or bypassing the underlying cipher. The former one is hard to achieve, since the underlying ciphers are usually chosen to be classical ciphers, which have already been proven to be secure in both theory and practice. Thus, attackers always choose the latter one, when they try to forge a MAC. We also chose the latter way, which means our goal is to bypass the underlying ciphers of the cellular system to forge a MAC.

To bypass the underlying cipher, the attacker can apply either a probabilistic method or a structure dependent method. The probabilistic method, for example, the birthday attack, can be launched at most MAC generating algorithms. Although its target is general, the result of a birthday attack is not very significant, because the searching complexity is still exponential. The attack is also not realistic, because in practice, the attacker is not allowed to make that many queries. In contrast, some structure-dependent attacks, which only target on certain MAC-generating algorithms, are more implementable. Our method being able to forge depends on the linear structures of some specific MACs. It is also a kind of structure-dependent method.

1.1 Applications of MAC in LTE

UMTS and LTE are two cellular standards proposed and maintained by 3GPP. UMTS is considered a third-generation cellular communication system. Its successor is called LTE, which is considered to be the fourth generation. As wireless communication systems, the resources of mobile devices are constrained. Thus, MACs are used to protect the integrity of these two systems. In UMTS, the underlying ciphers of MACs are Kasumi [3] and Snow 3G [1]. The former is a block cipher, and the latter is a stream cipher. When UMTS migrated to LTE, AES was standardized. Consequently, Kasumi has been replaced by AES. Besides AES and Snow 3G, a new stream cipher, ZUC [5], has been added into the current standard.

In UMTS, integrity protection algorithms are called UIAs. UIA1 [2] and UIA2 [4] are based on Kasumi and Snow 3G respectively. In LTE, integrity protection algorithms are called EIA. EIA1 is adopted from UIA2, which is based on Snow 3G; EIA2 is based on AES, and EIA3 [4] is based on ZUC. We use EIA in this paper to name these integrity protection methods in accordance with LTE, the latest standard.

1.2 MAC Based On Stream Ciphers and EIA

Wireless communication is widely used in daily life. As a result, stream ciphers are becoming more important than ever before. Originally, most MACs were based on keyed hash functions and block ciphers. Now some researchers have turned to the design of MACs based on stream ciphers.

In the past, the research on stream ciphers based MACs has not received much attention as that of block ciphers or hash functions based MACs. Recently, researchers have realized the importance of MACs using stream ciphers. Some significant work has been presented, such as GMAC [14], Grain-128a [6], EIA1, and EIA3. The underlying cipher of GMAC is a block cipher in counter mode, which makes a block cipher act as a stream cipher. Thus, GMAC can work with other stream ciphers directly. Both EIA1 and EIA3 have adopted some ideas from GMAC with some modifications to form their own MAC-generating algorithms.

It is publicly acknowledged that EIA2 can be considered as a secure MAC-generating algorithm, because it utilizes AES as the underlying cipher and CMAC as the MAC structure. AES is widely used for commercial purposes, and gets many analyses from academic study. AES has proven to be secure so far both theoretically and practically. CMAC is a kind of CBC MAC. Such kinds of MACs have already received many analyses, and proven to be secure.

Unlike EIA2, grounds for trusting EIA1 and EIA3 are not so solid. Since they are not as popular as AES and CBC MAC, the security of these algorithms is still not well studied. To the authors' best knowledge, there has been no significant attack on EIA1 so far, which means it does not evolve after it is created. EIA3 receives more attacks. But only some of them are significant.

1.3 Related Work

There is already some work analyzing the EIA family. The following attacks are two of the most significant.

Cycling Attack [16] can be applied to all polynomial MACs. Since EIA1 is a kind of polynomial MAC, it also suffers such attacks.

ZUC and EIA3 were published later than Snow 3G and EIA1. However, EIA3 has received more analyses than EIA1. The attack [18] proposed by Thomas et al. makes EIA3 change from v1.3 to v1.5. After changing to v1.5, EIA3 does not suffer such attacks any more. However, it is still not immune to our new attack, which will be presented in Section 3.

1.4 Our Contributions

Linear structures of EIA1 and EIA3 enable us to forge a valid MAC if we know two MACs generated by the same IV and IK. On top of such facts, we develop a method that, with known two valid MACs, we can forge up to 2^{32} valid MACs by introducing a coefficient λ. Since we have 2^{32} valid MAC-message pairs, the probability of finding a pair with valid MAC and valid message is quite high. Statistically, there is more than one meaningful pair among those 2^{32} pairs. For brute-force attacks, the probability of finding the valid MAC of a message is $\frac{1}{2^{32}}$. Now the probability that we get a meaningful MAC-message pair is increased to more than $\frac{1}{2^{32}}$. In fact, finding the λ that can generate a meaningful pair is much easier in practice than finding it in theory, because the messages usually have some specific structures. Those structures may shrink the searching space. In addition, such an attack does not aim at only EIA1 and EIA3 but also at some general polynomial MACs.

To prove that our linear forgery attack is doable, we create a scenario from which our attack can be launched. Such a scenario is based on the observation that the authentication of LTE and UMTS is not really mutual, although it is claimed to be a mutual authentication. In Extensible Authentication Protocol - Authentication and Key Agreement (EAP-AKA), only the server sends the challenge to a client. Checking the response of the challenge, the server can authenticate clients. Nevertheless, a client does not challenge the server. Thus, it can only authenticate a server by checking the MAC of the authentication vector. Such a protocol leaves a hole for the replay attack. Ordinarily speaking, the replay attack cannot get anything. However, because of our attack, the replay attack makes the forgery possible.

The rest part of this paper is organized as followings. In Section 2 we introduce MACs and the EIA family. In Section 3, we present some security issues of EIA1 and EIA3, and propose a forgery attack on EIA1 and EIA3. Since the attack makes use of the linearity of MACs, we call it, a *linear forgery attack*. Section 4 describes a scenario from which the linear forgery attack can be launched in practice, and shows some experimental results of our attack.

2. PRELIMINARY

In this section, we introduce the definition of the linear operation. This concept is the vital part of our attack. Generally, most polynomial MACs, such as EIA1 and EIA3, have this linear property.

2.1 Notation, Definition and Data Representation

Table 1 lists all notations used in this paper.

Table 1: Notations used in this paper

Notations	Explanations
$a\|\|b$	Concatenation of a and b.
$(a)_s$	Represent a in a base-s form. Example: $(1001)_2$, $(1FA8D)_{16}$.
$<i>$	The binary representation of i. Example: $<(5)_{10}> = (101)_2$.
L^i	The left shift operator. L^1 is simplified as L. Example: $L(10011011)_2 = (00110110)_2$.
$+$	Bitwise exclusive or.
$\cdot$	Multiplication in finite field. Some times, without ambiguity, we directly write $a \cdot b$ as ab.
$[A]_{i..j}$	The i-th bit to the j-th bit of A.
$\mathbf{M}$	A vector. Example: the message composed of several blocks. $\mathbf{M}=M_0\|\|M_1\|\|\cdots\|\|M_{n-1}$.
$MAC(\mathbf{M})$	The MAC of the message $\mathbf{M}$.

For the purpose of this paper, the *Linear Operation* is defined as

DEFINITION 1. *$F(x)$ represents the operation F applied to x. If the operation F satisfies*

- $F(a + b) = F(a) + F(b)$
- $F(c \cdot a) = cF(a)$,

where a, b and c are from a Galois Field; a, b and $c \cdot a$ $\in$ Domain of F, then F is called linear operation.

In the following part of this paper, all data variables are presented with the most significant bit on the left-hand side. For example, V is a 64-bit integer, $<V> = (V_0V_1 \cdots V_{62}V_{63})_2$, where V_i is the i-th bit of V. V_0 is the most significant bit.

We use the term "package" and "message" to represent the protocol data unit and the information that carried by the package respectively. Package is composed of the header (used for control) and the message. Take the counter check message for example, it is a package of radio resource control (RRC) protocol. The control message sequence number (SQN) is the header, and the values of different counters are the messages. These two parts together are called a *package of the counter check message*.

2.2 CBC-MAC, XOR MAC and GMAC

In wireless communication network, MAC is one of the most important elements to secure the system. CBC-MAC and XOR MAC are two well-known MACs. CBC-MAC comes from the CBC mode of block ciphers. There are a lot of variations of CBC-MAC, such as EMAC [11], XCBC [9], OMAC [17] and TMAC [12]. XOR MAC [8] has two categories, XMACC and XMACR. Compared with CBC-MAC, the structure of XOR-MAC is relatively simple.

GMAC has different design compared with CBC-MAC and XOR MAC, because it is based on the counter mode of the block cipher. GMAC [14] is standardized by NIST in 2007. When it is used for encrypted authentication, it is called GCM. GCM outputs both the encrypted data and MAC. When it is only used to generate MAC without encryption, it is called GMAC. Unlike GCM, GMAC does not output the encrypted data. GMAC is composed of two parts, GCTR and GHASH. GCTR is the counter mode of a block cipher. GHASH is a polynomial hash function. Messages first are encrypted by GCTR, and then passed through GHASH.

2.3 EIA1, EIA2 and EIA3

Because both the underlying cipher and the upper-level structure of EIA2 are proven to be secure, we do not discuss it in this paper. Hence, in this section, we present more details of EIA1 and EIA3 than of EIA2.

2.3.1 EIA1

EIA1 is adopted from UIA2. The underlying cipher is Snow 3G. In the evaluation of EIA1, it is said that EIA1 comes from GMAC. In fact, it only borrows the idea of GHASH. Thus, EIA1 is considered to be a polynomial MAC. Details of EIA1 are shown in Figure 1. At the beginning stage of EIA1, Snow 3G generates a 160-bit key stream. Then this key stream is truncated into P, Q and OTP. P and Q are 64-bit words. OTP is a 32-bit word. In Figure 1, M_i is the i-th block of message. The length of each block is 64 bits.

The mathematical description of EIA1 is given by

$$MAC(\mathbf{M}) = \left[\left(\sum_{i=1}^{n} P^i \cdot M_{n-i} + LENGTH\right) \cdot Q\right]_{0..31} + OTP. \tag{1}$$

Equation 1 can be written as

$$MAC(\mathbf{M}) = \underbrace{\left[\sum_{i=1}^{n} P^i \cdot M_{n-i} \cdot Q\right]_{0..31}}_{part1} + \underbrace{[LENGTH \cdot Q]_{0..31} + OTP}_{part2}.$$

Part 1 is a linear operation of the message. Part 2 is a value only relevant to the length of the message. If the length is fixed, part 2 is a constant. So MAC is computed by a linear combination of one linear block operation and a constant.

2.3.2 EIA2

Kasumi is replaced by AES in the new standard. Accordingly, EIA2 is put into the standard to replace UIA1. The underlying cipher of EIA2 is AES. As a block cipher, AES can use some existing MAC generating algorithms without any changes. CMAC [13] is chosen as the MAC generating algorithm of EIA2. CMAC is a kind of CBC-MAC, which has already been well studied.

2.3.3 EIA3

Originally, ZUC is not in UMTS. It is added to the standard after the system migrates to LTE. The integrity protection based on ZUC is EIA3. The same as EIA1, EIA3 is also claimed to be GMAC. However, Figure 2 suggests that it is much closer to XOR MAC. In Figure 2, $M[i]$ is the i-th bit of message, Z_i is the i-th word of the key stream

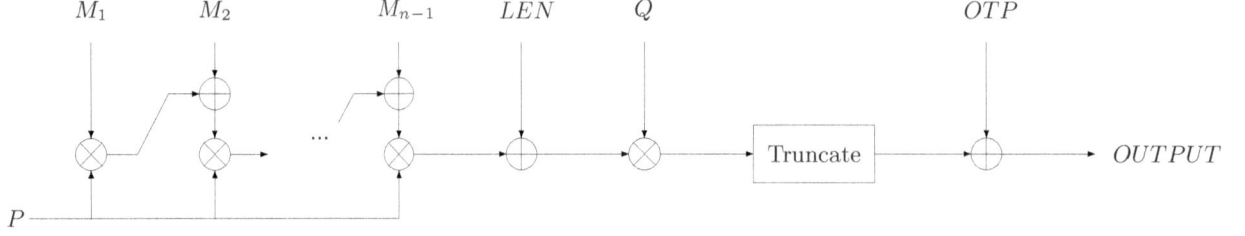

Figure 1: EIA1: based on Snow 3G

generated by ZUC. Z_i starts with the i-th bit of the key stream. EIA3 pads every message with a "1". $Z_{32*(L-1)}$ is the mask to encrypt the intermediate tag. L is equal to $\lceil LENGTH/32 \rceil + 2$.

A mathematical expression of Figure 2 is given by

$$MAC(\mathbf{M}) = \underbrace{\sum_{i=0}^{n-1} M[i]z_i}_{part1} + \underbrace{z_{LENGTH} + z_{32*(L-1)}}_{part2},$$

where $M[i]$ is the i-th bit of message; z_i is the i-th word in the key stream, i.e. $z_i = z[i]||z[i+1]||\cdots||z[i+31]$; $L = \lceil LENGTH/32 \rceil + 2$. The main observation of EIA3 is that Part 1 is a linear operation. This is very straightforward:

$$\sum_i (M[i] + M'[i])z_i = \sum_i (M[i]z_i + M'[i]z_i)$$
$$= \sum_i M[i]z_i + \sum_i M'[i]z_i.$$

Part 2 is a constant if the length is fixed.

2.4 IV Synchronization Mechanism

EIA1, EIA2 and EIA3 synchronize IVs in the same way. As required by the EIA family, $COUNT - I$ and $FRESH$ are input as parameters to form IV of the underlying cipher. $FRESH$ is a random number. $COUNT - I$ is a counter that records how many times IK has already been used so far. It contains two parts, SQN and hyper frame number (HFN). SQN is the sequence number. It is increased after a package is sent. When SQN overflows, HFN is increased. $COUNT - I$ and $FRESH$ together are used to prevent replay attacks.

$FRESH$ and $COUNT - I$ are written in the package to synchronize the transmitter and receiver, which maintain two counters, named $COUNTER_{tx}$ and $COUNTER_{rx}$, respectively. The transmitter uses $COUNTER_{tx}$ as the value of its $COUNT - I$, and then generates key streams. When the receiver receives a package, the value of $COUNT - I$ is compared with the value of $COUNTER_{rx}$. If the value of $COUNT - I$ is greater than $COUNTER_{rx}$, $FRESH$ is used to form IV together with $COUNT - I$. Then the key streams generated by this IV and IK are used to verify the MAC. If the value of $COUNT - I$ is smaller than $COUNTER_{rx}$, this package will be disregarded.

3. SECURITY ISSUE OF EIA1 AND EIA3

We present our work in this section. Compared with other works before, our attack is more practical. We need only two valid MACs to forge another valid MAC.

3.1 Quasi-Linearity Property of EIA1 and EIA3

Let $\mathbf{M} = (M_0, M_1, \cdots, M_{n-1})$, where M_i is a 64-bit vector, treated as an element in $GF(2^{64})$, which is defined by a primitive polynomial $t(x) = x^{64} + x^4 + x^3 + x + 1$. Let α be a root of $t(x)$ in $GF(2^{64})$, and let $\beta = \alpha^{2^{32}+1}$. Then β is a primitive element of $GF(2^{32})$, a subfield of $GF(2^{64})$. The minimal polynomial of α over $GF(2^{32})$ is given by

$$t_1(x) = x^2 + ux + v,$$

where $u = \beta^{17}$, $v = \beta$. Then each element in $GF(2^{64})$ can be represented as $a + b\alpha$, $a, b \in GF(2^{32})$. We define

$$f(\mathbf{M}, P) = \sum_{i=1}^{n} M_{n-i}P^i, for\ P \in GF(2^{64}),\ M_i \in GF(2^{64}).$$

PROPERTY 1. *For any* $\lambda \in GF(2^{32}) \subset GF(2^{64})$,

$$\mathbf{M}_1 = (M_{1,0}, M_{1,1}, \cdots, M_{1,n-1})$$
$$\mathbf{M}_2 = (M_{2,0}, M_{2,1}, \cdots, M_{2,n-1}),$$

then

$$f(\mathbf{M}_1 + \mathbf{M}_2, P) = f(\mathbf{M}_1, P) + f(\mathbf{M}_2, P) \quad (2)$$
$$f(\lambda \mathbf{M}, P) = \lambda f(\mathbf{M}, P). \quad (3)$$

PROOF. According to the definition of $f(\mathbf{M}, P)$, we have

$$f(\mathbf{M}_1 + \mathbf{M}_2, P) = \sum_{i=1}^{n}(M_{1,n-i} + M_{2,n-i})P^i$$
$$= \sum_{i=1}^{n} M_{1,n-i}P^i + \sum_{i=1}^{n} M_{2,n-i}P^i$$
$$= f(\mathbf{M}_1, P) + f(\mathbf{M}_2, P).$$
$$f(\lambda \mathbf{M}, P) = \sum_{i=1}^{n}(\lambda M_{n-i})P^i$$
$$= \lambda \sum_{i=1}^{n} M_{n-i}P^i = \lambda f(\mathbf{M}, P).$$

Thus, the assertions are true. $\square$

We have the MAC of $\mathbf{M}$ generated by EIA1 as

$$MAC(\mathbf{M}) = [Q \cdot f(\mathbf{M}, P)]_{0..31} + [Length \cdot Q]_{0..31} + OTP$$

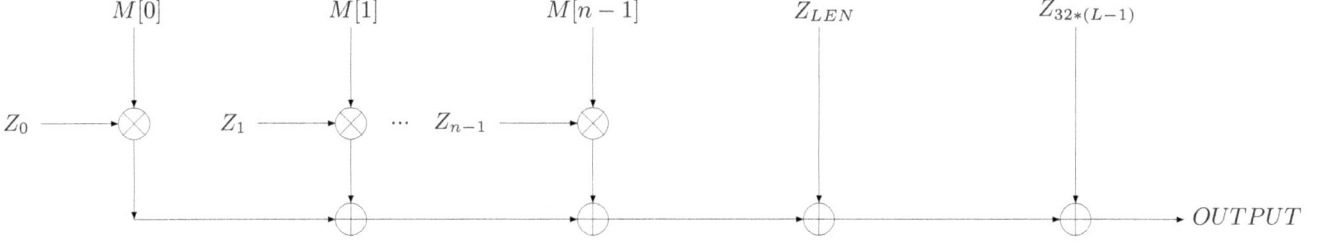

Figure 2: EIA3: based on ZUC

Let $Q \cdot f(\mathbf{M}, P) = a + b\alpha$, $Length \cdot Q = c + d\alpha$, a, b, c, and $d \in GF(2^{32})$. From the MAC generation of EIA1 in Section 2.3.1 and Property 1, the following result follows immediately.

PROPERTY 2. *For any $\lambda \in GF(2^{32})$,*

$$MAC(\mathbf{M}) = a + c + OTP \qquad (4)$$
$$MAC(\lambda\mathbf{M}) = \lambda a + c + OTP. \qquad (5)$$

EIA3 has very similar properties, which can allow attackers to forge one MAC with the probability $1/2$. More seriously, under the chosen plaintext attack, attackers can even recover somes part of the key stream. The details of this attack are presented in the full version of this paper.

3.2 Linear Forgery Attack Algorithm

Assume that we can make three queries to obtain MACs of the messages $\mathbf{M}_i$, for $i = 1, 2, 3$, under the same IV. Let

$$Q \cdot f(\mathbf{M}_i, P) = a_i + b_i\alpha, \ a_i, \ b_i \in GF(2^{32}). \qquad (6)$$

THEOREM 1. *Let (i, j, k) be a permutation of $(1, 2, 3)$. For any $\lambda \in GF(2^{32})$*

$$MAC(\mathbf{M}_{new}) = \lambda(MAC(\mathbf{M}_i) + MAC(\mathbf{M}_j)) + MAC(\mathbf{M}_k) \qquad (7)$$

which is a valid MAC value of the message

$$\mathbf{M}_{new} = \lambda(\mathbf{M}_i + \mathbf{M}_j) + \mathbf{M}_k.$$

PROOF. We give a proof only for $(i, j, k) = (1, 2, 3)$, since the proofs for the other cases are similar. In order to prove (7), we compute the results of both sides of (7). According to Properties 1 and 2

$$MAC(\mathbf{M}_{new}) = \lambda(a_1 + a_2) + a_3 + c + OTP. \qquad (8)$$

On the other hand,

$$\lambda(MAC(\mathbf{M}_1) + MAC(\mathbf{M}_2))) + MAC(\mathbf{M}_3)$$
$$= \lambda(a_1 + c + OTP + a_2 + c + OTP) + a_3 + c + OTP$$
$$= \lambda(a_1 + a_2) + a_3 + c + OTP. \qquad (9)$$

The assertion follows from (8) and (9). $\square$

From (7), we have the following corollary.

COROLLARY 1. *Let (i, j) be a permutation of $(1, 2)$, and $k \in (1, 2)$. For any $\lambda \in GF(2^{32})$, (7) is true. In other words,*

if we have the valid MACs from two queries, then

$$MAC(\lambda(\boldsymbol{M}_1 + \boldsymbol{M}_2) + \boldsymbol{M}_1)$$
$$= \lambda(MAC(\boldsymbol{M}_1) + MAC(\boldsymbol{M}_2)) + MAC(\boldsymbol{M}_1)$$
$$MAC(\lambda(\boldsymbol{M}_1 + \boldsymbol{M}_2) + \boldsymbol{M}_2)$$
$$= \lambda(MAC(\boldsymbol{M}_1) + MAC(\boldsymbol{M}_2)) + MAC(\boldsymbol{M}_2)$$

are valid.

From Corollary 1, we need only two valid MACs to forge a new one. In practice, we can reduce the number of queries by applying Corollary 1. Obtaining two valid MACs generated by the same IV is much easier than obtaining three.

The algorithm to forge a valid MAC by using two known valid MACs is shown in Algorithm 1, where $find\lambda()$ is a function that returns a λ such that either $\lambda(\mathbf{M}_1 + \mathbf{M}_2) + \mathbf{M}_1$ or $\lambda(\mathbf{M}_1 + \mathbf{M}_2) + \mathbf{M}_2$ is a valid message. How to find λ, such that the message is also valid, is discussed in Section 3.3.

Algorithm 1: Linear forgery

Data: two messages $\mathbf{M}_1$, $\mathbf{M}_2$, and the MACs of these two messages $MAC(\mathbf{M}_1)$, $MAC(\mathbf{M}_2)$
Result: one message and its valid MAC
$\lambda = find\lambda()$;
$\mathbf{temp} = \lambda(\mathbf{M}_1 + \mathbf{M}_2)$;
if $\boldsymbol{temp} + \boldsymbol{M}_1$ *is a valid message* **then**
    $\mathbf{M}_{new} = \mathbf{temp} + \mathbf{M}_1$;
    $MAC(\mathbf{M}_{new}) =$
    $\lambda(MAC(\mathbf{M}_1) + MAC(\mathbf{M}_2)) + MAC(\mathbf{M}_1)$;
else
    $\mathbf{M}_{new} = \mathbf{temp} + \mathbf{M}_2$;
    $MAC(\mathbf{M}_{new}) =$
    $\lambda(MAC(\mathbf{M}_1) + MAC(\mathbf{M}_2)) + MAC(\mathbf{M}_2)$;
end
return $MAC(\boldsymbol{M}_{new})$ *and* $\boldsymbol{M}_{new}$;

REMARK 1. *EIA's MAC has 32 bits. An attacker wishes to forge a valid MAC, it is equivalent to him randomly selecting 32 bits; the probability of success is $\frac{1}{2^{32}}$. However, if the attacker can make two queries for obtaining two valid MACs, then he can forge 2^{32} messages with valid MACs. In Section 4, we will demonstrate how the attacker can obtain two valid MACs in practice.*

3.3 How To Find λ

We randomly pick a λ, then we can get a MAC of a message. However, this message may not be a valid message for a protocol. Thus, the problem is how to find a λ that can generate the MAC of a valid message.

Usually in a real environment, it is easier to find λ, because there is a relationship between these two known messages, such as the relationship between two counter check messages. Two counter check messages have very similar structures. Therefore, most bits in the exclusive or of two counter check messages are zeros. We need to consider only very few bits, which are nonzero.

Moreover, even if we cannot find the valid message, our linear forgery attack can still cause some Denial-of-Service (DoS) attacks. Because the MAC is valid, every time the receiver must do the decoding and then finds the message is not well formatted, the computational resource will be occupied by verifying and decoding.

4. APPLICATION

In this section, we design a scenario in which the same IV and IK will occur twice. Following this, our linear forgery attack can be launched to get the valid MAC.

4.1 A Scenario of Fixing IV

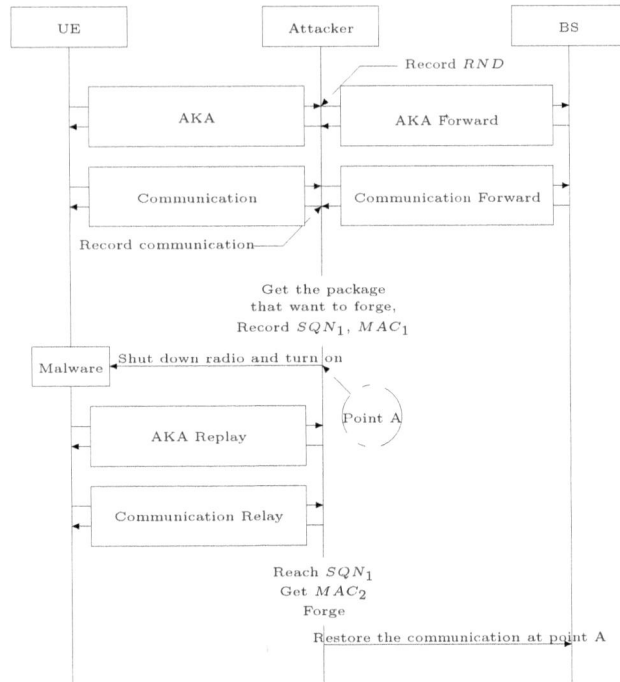

Figure 3: Fixing IV

Figure 3 demonstrates the procedure by which we can get the same SQN together with the same IK. The preconditions are: (1) we can set up the man-in-the-middle (MITM), and (2) there is a malware on the phone that can shut down and turn on the radio. Perez et al. [15] show that Condition (1) is applicable. Condition (2) is also easy to satisfy. We can

choose the Android smart phone to be our target because Android is an open platform, which is popular around the world.

First, the MITM attacker records all user data messages and control messages, including the authentication and key agreement messages. When this attacker observes the package he wants to forge, he shuts down the radio of the victim and then turns it on. The MITM attacker uses the recorded AKA messages to conduct a replay attack. In the AKA protocol, mobile devices are not required to verify whether the random number has been received before or not. They only check the freshness of SQN. However, in some cases, we can make SQN wrap around (the details are shown in full version). Thus, the victim believes it is talking with the real base station. Notice that the EAP-AKA is claimed to be mutually authenticated. The user equipment (UE) proves its identity to an radio network controller (RNC) by replying to the challenge from the RNC. However, since the UE does not send the challenge to the RNC, the RNC can prove itself only by computing the MAC of the authentication vector. The random number in the authentication vector can make sure each authentication vector is unique. However, the UE cannot record all random numbers it received before. This enables the replay attack. Such attack makes the UE accept the fake RNC. Generally, the attacker can get nothing from the replay attack, because he still cannot get the key. But in our case, we do not care about the key. The only thing we care about is the SQN. As long as we get two identical SQNs with the same IK, we can launch our linear forgery attack.

After the victim accepts the random number, it generates the same IK, which is also used in the session suspended by the attacker. When the victim believes the attacker is the real base station, it begins to send packages. The attacker replies with the previously recorded packages. The victim may accept or reject those packages, but it does not matter, because the only target for the attacker is to increase the victim's counter until the SQN reaches the recorded value.

As long as we get the sequence number that we want, the MITM attacker applies our linear forgery attack to forge a valid MAC of the package. This forged package together with the forged MAC will be forwarded to the real base station. Since the MAC will pass the verification, this package will be accepted.

4.2 Counter Check Message

A realistic application of our linear forgery attack is that we can forge the counter check message. In LTE, the integrity of the user plane is not protected. Thus, the counter check message is sent from the RNC to the UE to check the number of data packages that have been transmitted. The RNC includes the most significant s bits of its counter in the counter check message. When the UE receives the counter check message, it compares its own counter with the value included in the counter check message and sends its counter's value back. If the difference is not acceptable, the RNC will release the connection. This procedure is shown in Figure 4. Chen et al. [10] present more details about the counter check message.

We want to forge the counter check message because sometimes the attacker inserts some data into user plane data. If the counter check message is conducted correctly, the RNC will find out the insertion. For example, the MITM attacker

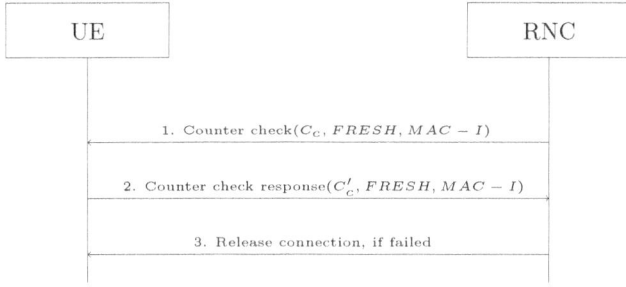

Figure 4: Counter Check Message.

inserts a redirect URL command or advertisement into the web page that the user is browsing. He must expect that RNC cannot detect the insertion. Then he needs to modify the counter check message.

4.3 Launching Attack

We assume that the MAC-I in Figure 4 is generated by EIA1 or EIA3. IVs of EIA1 and EIA3 are composed of two portions. The least significant four bits represent the Radio Resource Controller sequence number (RRC SQN). The other twenty-eight bits represent the HFN. Each time an RRC signal is sent, the RRC SQN is increased by one. If there is an overflow of the RRC SQN, the HFN is increased by one.

Attacking scenario:

1. RNC sends a counter check message to the MITM attacker.

2. The MITM attacker forwards this message to the UE, and gets the reply form the UE with MAC_1.

3. The MITM attacker applies the attack we mentioned above, and gets MAC_2.

4. The MITM attacker forges $\lambda(MAC_1+MAC_2)+MAC_1$ or $MAC_2 + \lambda(MAC_1 + MAC_2)$, then forwards to the real base station.

5. The RNC finds the difference is acceptable, and continues to communicate with the MITM attacker.

6. The MITM attacker can continue to forward messages between the RNC and the UE without being detected by the RNC.

This process creates a forged MAC. In this procedure, there is a drawback, i.e., the connection between the MITM attacker and the RNC may time out during the forgery process. So such attack can forge only the counter check message that is sent not too long after powering up.

4.4 Experimental Results

In order to test our attack, we generate two counter check messages, in which there are two counters, as shown in Table 2. The RRC commands of these two packages are listed in Table 3.

Table 2: Counter Check Messages

Message	Identity	UCounter	DCounter
M_1	10	258	257
	50	260	259
M_2	10	259	258
	50	261	260

Table 3: Counter Check Messages in Hex

Message	RRC command
M_1	0x30 0x27 0xA1 0x25 0xA4 0x23 0xA0 0x21 0xA0 0x1F 0x80 0x01 0x1E 0xA1 0x1A 0x30 0x0B 0x80 0x01 0x0A 0x81 0x02 0x01 0x02 0x82 0x02 0x01 0x01 0x30 0x0B 0x80 0x01 0x32 0x81 0x02 0x01 0x04 0x82 0x02 0x01 0x03
M_2	0x30 0x27 0xA1 0x25 0xA4 0x23 0xA0 0x21 0xA0 0x1F 0x80 0x01 0x1E 0xA1 0x1A 0x30 0x0B 0x80 0x01 0x0A 0x81 0x02 0x01 0x03 0x82 0x02 0x01 0x02 0x30 0x0B 0x80 0x01 0x32 0x81 0x02 0x01 0x05 0x82 0x02 0x01 0x04

4.4.1 Forge Procedure

The xor sum of these two messages is

0x00 0x00 0x00 0x00 0x00 0x00 0x00 0x00
0x00 0x00 0x00 0x00 0x00 0x00 0x00 0x00
0x00 0x00 0x00 0x00 0x00 0x00 0x00 0x01
0x00 0x00 0x00 0x03 0x00 0x00 0x00 0x00
0x00 0x00 0x00 0x00 0x01 0x00 0x00 0x00
0x07

Then chose $\lambda = $0x1B, $\lambda(M_1 + M_2)$ is

0x00 0x00 0x00 0x00 0x00 0x00 0x00 0x00
0x00 0x00 0x00 0x00 0x00 0x00 0x00 0x00
0x00 0x00 0x00 0x00 0x00 0x00 0x00 0x1B
0x00 0x00 0x00 0x2D 0x00 0x00 0x00 0x00
0x00 0x00 0x00 0x00 0x1B 0x00 0x00 0x00
0x41

Finally, we get the message $M_{new} = M_2 + \lambda(M_1 + M_2)$

0x30 0x27 0xA1 0x25 0xA4 0x23 0xA0 0x21
0xA0 0x1F 0x80 0x01 0x1e 0xa1 0x1A 0x30
0x0B 0x80 0x01 0x0A 0x81 0x02 0x01 0x18
0x82 0x02 0x01 0x2F 0x30 0x0B 0x80 0x01
0x32 0x81 0x02 0x01 0x1E 0x82 0x02 0x01
0x45

This message is represented in a binary form. Decoding the binary message, we get the result shown in Figure 6, which demonstrates the xml form of the binary message. From Figure 6, it is obvious that the forged message still contains the counter values of the bearers 10 and 50. The uplink and downlink counter values of the bearer 10 are 577 and 531 respectively. They are increased compared with the real values. The counter values of the bearer 50 are increased as well. The forged value of uplink counter is 274, and the forged downlink counter value is 352. We can also

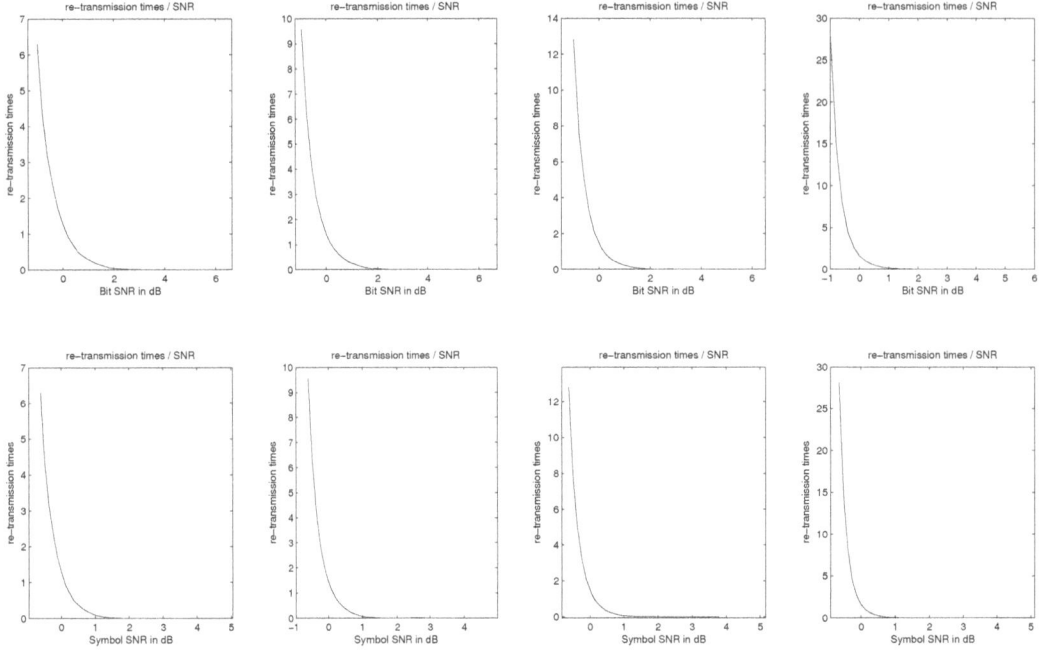

Figure 5: Retransmission times / SNR

verify the MAC of the message $\mathbf{M}_{new}$. $MAC(\mathbf{M}_{new})$ is indeed $MAC_2 + \lambda(MAC_1 + MAC_2)$. Therefore we successfully forge a valid package, in which the value of each counter is increased, and the MAC of the message is valid. This means if the attacker inserts some packages, RNC will not realize that.

```
<DL-DCCH-Message>
 <message>
  <counterCheck>
   <r3>
    <counterCheck-r3>
     <rrc-TransactionIdentifier>30</rrc-TransactionIdentifier>
     <rb-COUNT-C-MSB-InformationList>
      <RB-COUNT-C-MSB-Information>
       <rb-Identity>10</rb-Identity>
       <count-C-MSB-UL>577</count-C-MSB-UL>
       <count-C-MSB-DL>531</count-C-MSB-DL>
      </RB-COUNT-C-MSB-Information>
      <RB-COUNT-C-MSB-Information>
       <rb-Identity>50</rb-Identity>
       <count-C-MSB-UL>274</count-C-MSB-UL>
       <count-C-MSB-DL>352</count-C-MSB-DL>
      </RB-COUNT-C-MSB-Information>
     </rb-COUNT-C-MSB-InformationList>
    </counterCheck-r3>
   </r3>
  </counterCheck>
 </message>
</DL-DCCH-Message>
```

Figure 6: Decoding result

4.4.2 Timing of Attack

Turning the radio off and on usually costs three to six seconds. If RNC does not time out within this range, our attack can be launched. There is no requirement for this time-out duration in the standard. It is decided by the manufacturer. We cannot make a direct test, because analysis of public communications is forbidden by the law. However, since RRC commands are sent in an ARQ fashion, we can

use the retransmitting time to show the time out duration intuitively. We simulate an Additive White Gaussian Noise (AWGN) channel. The modulation scheme is Quaternary Phase Shift Keying (QPSK). The result is shown in Figure 5. Each column is corresponding to a set of coding parameters. The top row shows the bit Signal to Noise Ratio (SNR), while the bottom row represents the symbol SNR.

Figure 5 indicates that when the bit SNR or symbol SNR is around -0.2dB, the base station needs to transmit a message three times on average to ensure that the user can receive that message. If users are in a building, such a situation may occur with high probability. To make sure all users can get services, the time out duration must be relatively long. This may give us a chance to conduct the attack.

5. CONCLUSION AND FUTURE WORK

In this paper, we proposed a method whereby two known valid MAC-message pairs generated by the same IV and IK, we can forge 2^{32} valid MAC-message pairs. In a real environment, we can easily find a meaningful MAC-message pair among those 2^{32} pairs. We also developed an attack that makes the same IV and IK occur twice. This enables our linear forgery attack in practice.

To prevent our linear forgery attack, the structures of EIA1 and EIA3 need to be changed such that either the message is involved in generating the key stream or the MAC is generated in a nonlinear fashion. The problem is how to find a way that can avoid linear structures without compromising efficiency. So far, this issue seems to be a trade-off.

In the next stage of our research, the way to fix IV needs some improvement, for it heavily depends on the timing. If we can find a better way to get the same IV and IK twice, then our attack can launch at any time of communication (not only the time not too long after powering up), and also the failure rate caused by timing out will be reduced.

6. REFERENCES

[1] 3GPP. Specification of the 3GPP Confidentiality and Integrity Algorithms UEA2 & UIA2. Document 2: SNOW 3G Specification. September 2006.

[2] 3GPP. Specification of the 3GPP Confidentiality and Integrity Algorithms. Document 1: f8 and f9 Specification. 2007.

[3] 3GPP. Specification of the 3GPP Confidentiality and Integrity Algorithms. Document 2: Kasumi Specification. June 2007.

[4] 3GPP. Specification of the 3GPP Confidentiality and Integrity Algorithms 128-EEA3 & 128-EIA3. Document 1: 128-EEA3 and 128-EIA3 Specification. January 2011.

[5] 3GPP. Specification of the 3GPP Confidentiality and Integrity Algorithms 128-EEA3 & 128-EIA3. Document 2: ZUC Specification. January 2011.

[6] M. Agren, M. Hell, T. Johansson, and W. Meier. Grain-128a: A New Version of Grain-128 with Optional Authentication. *Int. J. Wire. Mob. Comput.*, 5(1):58–59, December 2011.

[7] M. Bellare, R. Canetti, and H. Krawczyk. Keying Hash Functions for Message Authentication. In *Proceedings of the 16th Annual International Cryptology Conference on Advances in Cryptology*, CRYPTO '96, pages 1–15, London, UK, UK, 1996. Springer-Verlag.

[8] M. Bellare, R. Guérin, and P. Rogaway. XOR MACs: New Methods for Message Authentication Using Finite Pseudorandom Functions. In *Proceedings of the 15th Annual International Cryptology Conference on Advances in Cryptology*, CRYPTO '95, pages 15–28, London, UK, UK, 1995. Springer-Verlag.

[9] J. Black and P. Rogaway. CBC MACs for Arbitrary-Length Messages: The Three-Key Constructions. In *Proceedings of the 20th Annual International Cryptology Conference on Advances in Cryptology*, CRYPTO '00, pages 197–215, London, UK, UK, 2000. Springer-Verlag.

[10] L. Chen and G. Gong. *Communication System Security*, chapter 10, pages 358–359. CRC Press Taylor & Francis Group, 2012.

[11] P. Erez and R. Charles. CBC MAC for Real-Time Data Sources. *J. Cryptology*, 13:315–338, 2000.

[12] K. Kurosawa and T. Iwata. TMAC: Two-Key CBC MAC. In *Proceedings of the 2003 RSA conference on The cryptographers' track*, CT-RSA'03, pages 33–49, Berlin, Heidelberg, 2003. Springer-Verlag.

[13] NIST. Recommendation for Block Cipher Mode of Operation: The CMAC Mode for Authentication. NIST Special Publication 800-38B. 2005.

[14] NIST. Recommendation for Block Cipher Modes of Operation: Galois/Counter Mode (GCM) and GMAC. NIST Special Publication 800-38D. 2007.

[15] D. Perez and J. Pico. A Practical Attack Against GPRS/EDGE/UMTS/HSPA Mobile Data Communications. Presented in Black Hat Conference, 2011.

[16] M.-J. O. Saarine. Cycling Attacks on GCM, GHASH and Other Polynomial MACs and Hashes. In *Proceedings of the Fast Software Encryption 2012*, 2012.

[17] I. Tetsu and K. Kaoru. OMAC: One-Key CBC MAC. In *Pre-proceedings of Fast Software Encryption, FSE 2003*, FSE '03, pages 137–161. Springer-Verlag, 2002.

[18] F. Thomas, G. Henri, R. Jean-Rene, and M. Videau. A Forgery Attack on the Candidate LTE Integrity Algorithm 128-EIA3. *IACR Cryptology ePrint Archive*, 2010:168, 2010.

Perfect Contextual Information Privacy in WSNs under Colluding Eavesdroppers

Alejandro Proaño
Dept. of Electrical and Computer Engineering
University of Arizona
Tucson, AZ, USA
aaproano@ece.arizona.edu

Loukas Lazos
Dept. of Electrical and Computer Engineering
University of Arizona
Tucson, AZ, USA
llazos@ece.arizona.edu

ABSTRACT

We address the problem of preserving contextual information privacy in wireless sensor networks (WSNs). We consider an adversarial network of colluding eavesdroppers that are placed at unknown locations. Eavesdroppers use communication attributes of interest such as packet sizes, inter-packet timings, and unencrypted headers to infer contextual information, including the time and location of events reported by sensors, the sink's position, and the event type. We propose a traffic normalization technique that employs a minimum backbone set of sensors to decorrelate the observable traffic patterns from the real ones. Compared to previous works, our method significantly reduces the communication overhead for normalizing traffic patterns.

Categories and Subject Descriptors

C.2.0 [**Computer - Communication Networks**]: General - Security and Protection

Keywords

Eavesdropping, colluding adversaries, wireless sensor networks, algorithms, security

1. INTRODUCTION

Wireless communications are vulnerable to eavesdropping by anyone equipped with a wireless receiver. When the transmitted information is of sensitive nature, its privacy is protected via cryptographic methods. However, encryption alone cannot prevent the leakage of contextual information such as the location of communicating nodes, the path between the source and the destination, or the time of occurrence of a reported event. Passive eavesdroppers can obtain contextual information by performing traffic analysis using low-level packet identifiers such as packet size and inter-packet timings, even when the contents of the packet remain hidden [4, 6, 9]. Moreover, this information can be

used to launch intelligent attacks of selective and adaptive nature that degrade network performance at low cost [10,13].

In this paper, *we address the problem of preserving the privacy of contextual information in wireless communications.* Though we study this problem in the context of wireless sensor networks (WSNs), our methods are applicable to any static wireless multihop network. We consider an adversary that deploys a network of colluding eavesdroppers at unknown locations within the WSN. The eavesdropping devices can be cheap passive sensors that form an out-of-band collusion network [9, 14]. Eavesdroppers extract communication attributes of interest and centrally process them to derive contextual information.

State-of-the art techniques for hiding contextual information employ bogus transmissions to normalize the eavesdropped transmission patterns [9, 12, 14]. In these schemes, sensors transmit according to a predefined distribution, irrespective of their real traffic profile. Transmissions of real packets conform to the same distribution, thus defeating traffic analysis techniques. However, when the locations of the colluding eavesdroppers are unknown, privacy can be achieved only if all sensors become sources of bogus traffic [9, 12]. In our approach, we significantly reduce the communication overhead by intelligently selecting the bogus sources and loosely coordinating real packet transmissions.

Our Contributions: We propose a resource-efficient traffic normalization scheme that protects contextual information under colluding eavesdroppers. Our scheme achieves perfect privacy while the number of bogus traffic sources is reduced. We map the problem of reducing the bogus traffic sources to the problem of partitioning the WSN into minimum connected dominating sets (MCDSs). Due to the problem complexity, we propose a distributed heuristic algorithm that approximates the WSN partition to MCDSs. We further propose a schedule assignment scheme that reduces packet delay by loosely coordinating transmissions among neighboring sensors.

The remainder of the paper is organized as follows. In Section 2, we present related work. In Section 3, we state our model assumptions. Section 4 presents our traffic normalization scheme. In Section 5, we conduct a performance evaluation and in Section 6, we conclude.

2. RELATED WORK

The problem of hiding contextual information in WSNs has been studied under a local and a global adversary model. Due to space limitations, we focus on the latter model, which is most relevant to our work.

In [9], the authors proposed two traffic normalization methods based on the injection of bogus traffic; periodic collection and source simulation. In periodic collection, each sensor generates bogus packets at a constant rate. To transmit real data, sensors simply substitute dummy packets with real ones. This method prevents colluding eavesdroppers from determining the source of real traffic, the path to the sink, and the sink location, at the expense of significant communication overhead. Our methods achieve the same level of privacy at a considerably lower communication overhead. Source simulation reduces the communication overhead by selecting a subset of sensors as bogus sources, that are chosen to simulate the expected distribution of real events. However, the event distribution must be known apriori.

In [4], the authors proposed a traffic normalization scheme that propagates dummy packets in a probabilistic fashion. A sensor that overhears the transmission of a real packet, forwards a dummy packet to its neighbors with some probability p. The packet is probabilistically flooded in a radius of K hops from the bogus source. Under a global adversary, if an eavesdropper happens to be close to the source or the sink, their location can be inferred.

Besides their overhead, traffic normalization techniques incur unavoidable delay. This is because transmissions of real packets are delayed to conform to predefined transmission patterns. The authors in [12] reduced packet delay by rushing the transmissions of real packets while delaying the transmissions of follow-up dummy packets so that the long-scale traffic statistics are maintained. This approach is not effective when multiple packets need to be transmitted by the same sensor. Moreover, the authors of [1] proved that statistical analysis of the occurred short-long transmission patterns can be used to identify real packets. To address this vulnerability, the authors proposed the generation of fake short-long patterns by introducing dummy events following packets related to real events.

In [11], the number of bogus traffic sources was reduced by constructing a minimum connected dominating set (MCDS) that covers the deployment area. Only the sensors that belong to the MCDS transmit bogus traffic. Sensors that are not part of the MCDS, regulate their transmissions in order to conform to the statistical traffic properties observed by an eavesdropper. Since the eavesdropper's location is unknown, the set of possible eavesdropped rates is inferred via geometric analysis. The scheme in [11] does not address the case of eavesdropper collusion. The method that we present in our present work provides perfect privacy, even if eavesdroppers collude and can eavesdrop on all network communications.

3. SYSTEM AND ADVERSARY MODELS

System Model: We consider a WSN consisting of a set of sensors $\mathcal{V}$. The WSN is organized as a multi-hop mesh topology, which is defined by the sensor communication range and the sensor positions. Sensors are synchronized to a common time reference. Packets are assumed to be re-encrypted on a per-hop basis to prevent eavesdroppers from identifying a packet relayed over multiple hops [8]. Re-encryption is applied to all packet identifiers such as headers at the MAC layer and the payload. Sensors are pre-loaded with secrets that can be used to establish cryptographic keys. Finally, contention management protocols are assumed to conform to the traffic rate assigned to each sensor.

Adversary Model: We assume an unknown number of colluding eavesdroppers to be deployed at unknown locations within the WSN. The set of eavesdroppers observes communication attributes of interest such as the packet sizes, inter-packet times, identity of transmitting nodes (obtained through the unencrypted header fields, or through signal processing techniques). These observations are collectively processed by a central coordinator to extract contextual information. Because the number and positions of the eavesdroppers are unknown, any portion of the WSN communications could be intercepted. In the extreme case, eavesdroppers are able to intercept all packets transmitted in the WSN. This global adversary model is realistic when eavesdropping devices are cheap sensors with similar capabilities to legitimate sensors [4, 9, 12]. Finally, the adversary does not launch active attacks (e.g., jamming, packet modification and injection attacks), or compromise and control any of the sensors in $\mathcal{V}$.

4. RESOURCE-EFFICIENT TRAFFIC NORMALIZATION

In this section, we develop a resource-efficient traffic normalization scheme to prevent the leakage of contextual information. Our scheme consists of two phases: network partition and schedule assignment. First, we motivate our design.

4.1 Design Motivation

Our design is motivated by the excessive communication overhead of state-of-the-art traffic normalization methods. Since the eavesdroppers' locations are unknown, prior methods hide contextual information by normalizing the transmission profiles of all sensors [4, 9, 12]. Moreover, to hide the route to the sink and the sink's location, transmissions between neighboring nodes remain uncoordinated. Lack of coordination can lead to the accumulation of packet delay on a per-hop basis.

In our design, we represent the WSN as a graph $\mathcal{G}(\mathcal{V}, \mathcal{E})$, where $\mathcal{V}$ denotes the set of sensors, and $\mathcal{E}$ the links between them. Set $\mathcal{V}$ is partitioned into disjoint subsets, denoted by $\{\mathcal{D}_1, \mathcal{D}_2, \ldots \mathcal{D}_z\}$. Only one subset is active at any time and active subsets are periodically rotated in a round-robin fashion. Sensors of an active subset are responsible for normalizing the eavesdropped traffic pattern and relaying real packets to their respective destinations. The partition of $\mathcal{V}$ is designed to form special types of subgraphs that satisfy the following principles: (a) all sensors can transmit real traffic without altering their transmission profile; (b) a subset can deliver a packet to any destination; and (c) the number of bogus traffic sources is minimized. Because only a subset of sensors is active at any given time, the communication overhead is drastically reduced.

To decrease the delay in forwarding real packets, we loosely coordinate the sensor transmissions within each $\mathcal{D}_j$, such that the traffic patterns observed by any number of colluding eavesdroppers remains unchanged. We now describe the two phases in detail.

4.2 Phase I: Network Partition

In the first phase, we partition $\mathcal{V}$ to subsets $\{\mathcal{D}_1, \ldots, \mathcal{D}_z\}$. Every subset is active for a fixed time interval. The active subsets are periodically rotated in a round-robin fashion. Sensors of an active subset transmit dummy packets according to a pre-assigned distribution. A sensor with real packets

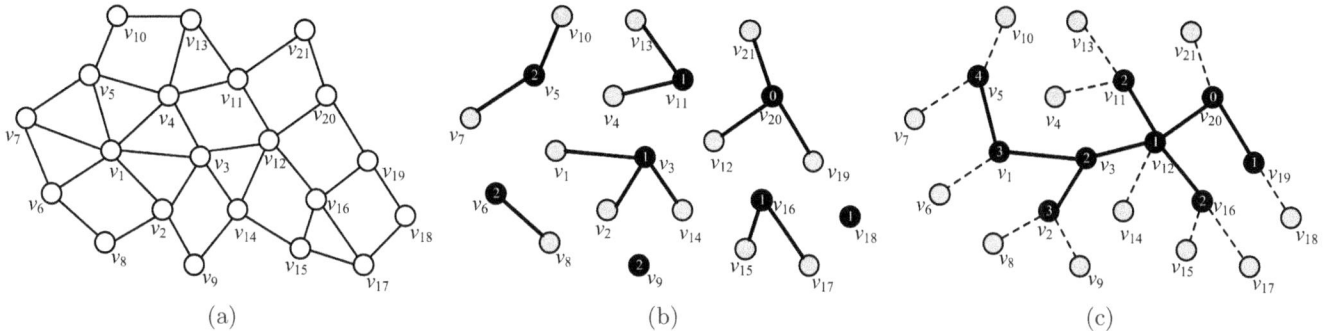

Figure 1: (a) A graph $\mathcal{G}(\mathcal{V},\mathcal{E})$ representing the WSN, (b) a DS generated during Stage 1, (c) an MCDS approximation generated during Stage 2.

for transmission conforms to this distribution by replacing dummy packets with real ones. Thus, transmission of real packets does not alter the traffic patterns observed by eavesdroppers. We reduce the problem of partitioning $\mathcal{V}$, to the problem of *finding disjoint minimum connected dominating sets (MCDSs) that span* $\mathcal{V}$. We now define the MCDS [7].

Minimum Connected Dominating Set: For a graph $\mathcal{G}(\mathcal{V},\mathcal{E})$, a subset $\mathcal{D} \subseteq \mathcal{V}$ is a *dominating set (DS)* if any vertex $u \in \mathcal{V}$ either belongs to $\mathcal{D}$, or is adjacent (within one hop) to some vertex in $\mathcal{D}$. If $\mathcal{D}$ induces a connected subgraph on $\mathcal{G}$, then $\mathcal{D}$ is a *connected dominating set (CDS)*. If $\mathcal{D}$ has the smallest possible cardinality, it forms a minimum connected dominating set (MCDS).

The partition of $\mathcal{V}$ into disjoint MCDSs satisfies properties (a)-(c). Property (a) is satisfied, as the set of MCDSs spans $\mathcal{V}$. Hence, each sensor belongs to one $\mathcal{D}_j$, and is able to transmit real traffic when $\mathcal{D}_j$ becomes active. By design, the transmission profile of an active sensor is not altered when real traffic substitutes bogus traffic. For property (b), a CDS guarantees that any sensor in $\mathcal{V}$ will be either part of $\mathcal{D}_j$ or within one hop from a sensor in $\mathcal{D}_i$. Moreover $\mathcal{D}_j$ is a connected set. Hence, a real packet transmitted by a sensor in $\mathcal{D}_j$ can be forwarded to any sensor in $\mathcal{V}$ using only $\mathcal{D}_j$. Finally, property (c) is satisfied by definition, as an MCDS minimizes the number of active sensors.

A partition of $\mathcal{V}$ into disjoint MCDSs does not always exist for arbitrary graph topologies. The number of disjoint MCDSs is bounded by the minimum vertex-cut size of $\mathcal{G}$. Moreover, determining a single MCDS for arbitrary topologies is known to be an NP-complete problem [5]. In the absence of an MCDS partition guarantee and of a polynomial time algorithm for finding an MCDS, we relax the MCDS partition requirement to allow sensors to be part of more than one MCDSs. We denote the frequency of appearance of a sensor v to any of the z MCDSs as $f(v)$. We aim at finding a set of MCDSs that covers $\mathcal{V}$ and balances between the frequency of appearance, number of MCDSs, and MCDS size. We propose a distributed solution inspired by the heuristic MCDS construction algorithm developed in [3]. Our algorithm computes a set of CDSs $\{\mathcal{D}_1, \mathcal{D}_2, \dots, \mathcal{D}_z\}$, that approximate the partition of $\mathcal{V}$ to MCDSs. We note that the computation and communication overhead for partitioning $\mathcal{V}$ to z CDS is only incurred once during the network initialization. The steps of our algorithm are as follows.

Algorithm 1: MCDS approximation– We generate a $\mathcal{D}_j$ in two stages. We first obtain a DS, and later we expand the DS to a connected graph approximating the MCDS.

For a sensor $v \in \mathcal{V}$, let $m(v)$ be a marker, which can take the values WHITE, BLACK, or GRAY. Let $\mathcal{N}_v$ denote the one-hop neighbors of v, $\delta(v) = |\mathcal{N}_v|$ the degree of v, and $\delta^*(v)$ the *effective degree* of v. Parameter $\delta^*(v)$ is defined as the number of WHITE neighbors of v. Let also $r(v)$ be the rank of v, defined as the order that v changed its marker relative to a leader node. Finally, let $b(v)$ denote the number of higher-ranked BLACK neighbors of v and $f(v)$ the frequency of appearance of v in the CDSs generated thus far. All nodes are initialized to $m(v) =$ WHITE, $\delta^*(v) = \delta(v)$, $b(v) = 0$, $f(v) = 0$, and $r(v) = 0$. The marking process that outputs a DS is as follows.

Stage 1: DS generation

Step 1: A randomly chosen leader v starts the process by changing $m(v)$ to BLACK. Node v becomes a "dominator" and broadcasts $m(v) =$ BLACK, $r(v) = 0$, and $f(v) = 0$.
Step 2: A sensor u with $m(u) =$ WHITE receiving $m(v) =$ BLACK from $v \in \mathcal{N}_u$ is dominated by v. Node u sets $m(u) =$ GRAY, $r(u) = r(v)$, and broadcasts $m(u)$ and $r(u)$.
Step 3: A WHITE sensor v getting $m(u) =$ GRAY from $u \in \mathcal{N}_v$, decreases $\delta^*(v)$ by one, updates the rank to $r(v) = r(u)+1$ if $r(v) \le r(u)$, and broadcasts $\delta^*(v)$, $r(v)$, and $f(v)$.
Step 4: A sensor v changes $m(v)$ to BLACK, if

$$v = \arg\max_{u \in \mathcal{N}_v \cup \{v\}} \left\{ \frac{\delta^*(u)}{\delta^*_{\max}} \times \frac{1}{f(u)+1} \right\},$$

where $\delta^*_{\max} = \max_{u \in \mathcal{N}_v \cup \{v\}} \delta^*(u)$. Node v becomes a "dominator" and broadcasts its new marker value and rank.
Step 5: After receiving the transmission of a BLACK node, a sensor v updates the value of $b(v)$.
Step 6: The marking process is repeated until no sensors are marked as WHITE (i.e., $\delta^*(v) = 0, \forall v \in \mathcal{V}$).

With the termination of Stage 1, all nodes are marked either as BLACK or GRAY, with each GRAY node dominated by a BLACK one. Therefore, the set of BLACK sensors forms a DS. Figure 1(b) shows the DS generated for the graph of Figure 1(a). Since initially $f(v) = 0$ for all sensors, the marking process depends only on $\delta^*(v)$. In our example, v_{20} becomes the leader and broadcasts $m(v)=$BLACK and $r(v) = 0$. Nodes v_{12}, v_{19}, and v_{21} become GRAY and set their rank to zero. In the next iteration, v_3, v_{11}, v_{16}, and v_{18} are added to the DS and change their rank to one. Finally, v_5, v_6, and v_9 are added to the DS and set their rank to two. The network is now partitioned to a set of star subgraphs, where each star consists of a set of GRAY nodes dominated by a BLACK node. The rank of each star increases with its "distance" from the leader node. In Stage 2, we approximate

91

an MCDS by selecting GRAY nodes that connect the stars. The process is as follows.

Stage 2: Approximation of the MCDS

Step 1: Every GRAY node v broadcasts $b(v)$.
Step 2: Leader node v selects GRAY nodes $u \in \mathcal{N}_v$ with

$$u = \arg \max_{\{\mathcal{N}_v, r(v) = r(u)\}} \left\{ \frac{b(u)}{b_{\max}} \times \frac{1}{f(u)+1} \right\},$$

where $b_{\max} = \max_{\{u \in \mathcal{N}_v, r(v) = r(u)\}} b(u)$ and $b(u), b_{max} > 0$. Node u changes its marker to BLACK and its rank $r(u) = r(v) + 1$. Ties are broken arbitrarily.
Step 3: A node $w \in \mathcal{N}_u$ with $m(w) = $ BLACK and $r(w) = r(u)$ becomes dominated by u. Dominated nodes change their rank to $r(w) = r(u) + 1$ and broadcast their new rank. Any GRAY node $v \in \mathcal{N}_w$ overhearing a message from w updates $b(v) = b(v) - 1$ and changes its rank to $r(w)$.
Step 4: A GRAY node v overhearing a rank update message from a BLACK node u with rank $r(u) < r(v)$ changes its dominating node to u and broadcasts $r(u)$ and $b(v)$.
Step 5: The process is iteratively repeated until all GRAY nodes have $b(v) = 0$.
Step 6: If a BLACK node does not dominate at least one other node it changes its marker to GRAY.

At the end of Stage 2, every GRAY node has $b(v) = 0$, i.e., all BLACK nodes of Stage 1 are dominated. Moreover, BLACK nodes are dominated by GRAY nodes of lower rank. Since the process is initiated by the leader node, every BLACK node dominated by the GRAY node gets connected to the leader. This process terminates when all BLACK nodes are dominated. Thus, the resulting subgraph is connected. That is, the set $\mathcal{D} = \{v : m(v) = \text{BLACK}, v \in \mathcal{V}\}$ forms a CDS.

Figure 1(c) depicts the CDS generated after Stage 2. In Step 1, the leader node v_{20} selects GRAY node v_{12} ($b(v_{12}) > b(v_{19}), b(v_{21})$) to connect to the star subgraphs dominated by v_3, v_{11}, and v_{16}. In Step 2, v_{12} becomes BLACK and broadcasts $m(v_{12})$ and $r(v_{12}) = 1$. In Step 3, nodes v_3, v_{11}, and v_{16} change their rank to two and broadcast their new rank. Nodes v_{21} and v_{14} change $b(v_{21}) = b(v_{14}) = 0$ and broadcast their new values. Moreover, v_{14} is now dominated by v_{12} since v_{12} has a lower rank than v_3. In further iterations, nodes v_1 and v_2 change to BLACK to connect v_5, v_6 and v_9, respectively and produce a CDS. In Step 6, the CDS is pruned to eliminate the leaf BLACK nodes v_6, v_9, and v_{18}.

In the last stage, the CDS generation process is repeated to produce another CDS for the partition of $\mathcal{V}$.

Stage 3: CDS Update

Step 1: Increment $f(v)$ by one unit for all nodes in $\mathcal{D}_j$.
Step 2: Repeat Stages 1 and 2 until $f(v) > 0$, $\forall v \in \mathcal{V}$.

In Stages 1 and 2, a sensor v is added to the CDS according to metrics,

$$\frac{\delta^*(v)}{\delta^*_{\max}} \times \frac{1}{f(v)+1} \text{ and } \frac{b(v)}{b_{\max}} \times \frac{1}{f(v)+1},$$

respectively. These metrics are designed to balance between the CDS size and the number of CDSs. By maximizing $\frac{\delta^*(v)}{\delta^*_{\max}}$, we minimize the CDS size in a greedy fashion. Nodes that dominate the maximum fraction of their neighbors are added to the DS. Similarly by maximizing $\frac{b(v)}{b_{\max}}$, nodes that connect the largest fraction of star subgraphs are added to the CDS. On the other hand, $\frac{1}{f(v)+1}$ favors the selection of

nodes that have not been previously included in any CDS. This metric reduces the number of CDSs needed to span $\mathcal{V}$.

The size of each CDSs generated by Algorithm 1 approximates the minimum DS by a factor of eight. Due to space limitations, we provide an inform proof of this claim. We first note that the DSs $\mathcal{D}_j$ generated in Stage 1 are minimal. That is, if a node $v \in \mathcal{D}_j$ changes its color from BLACK to WHITE or GRAY, $\mathcal{D}_j$ no longer forms a DS. This is due to the fact that in Stage 1, a BLACK node only has GRAY neighbors. Hence, a BLACK node that changed its color to GRAY will not be dominated by any other BLACK node. In [2], the authors proved that a minimal DS approximates the minimum DS with an approximation factor of four. In Stage 2, GRAY nodes change their color to BLACK to connect BLACK nodes that belong to the DS. In the worst case scenario, for each BLACK node of the DS, one GRAY node must turn BLACK to connect it to a BLACK node of lower rank (line network topology). Thus, the size of each CDS is at most twice the size of the DS generated in Stage 1. Therefore, the CDS generated by Algorithm 1 is upper-bounded by a factor of eight times the size of the minimum DS.

4.3 Phase II: Schedule Assignment

In the section, we propose the *Deterministic Assignment Scheme (DAS)* for reducing the end-to-end delay of real packets under a fixed communication overhead budget.

In our scheme, time is divided into intervals $I_1, I_2, \dots$ of length T. Only one CDS is active at a given interval. We assume T is sufficiently long to accommodate a number of packets according to the given packet rate, and resolve any contention between active sensors within the same collision domain. The CDSs obtained by Algorithm 1 are periodically activated in a round-robin fashion, allowing all sensors to transmit real data. A CDS $\mathcal{D}_j$ is active in interval I_k, if $j = (k \mod z) + 1$. Sensors of an active $\mathcal{D}_j$ either transmit dummy packets, or replace dummy packets with real ones.

4.3.1 Deterministic Assignment Scheme (DAS)

When sensor transmissions are uncoordinated, the packet relay operation during one interval can be blocked if the next hop completes its transmissions prior to the previous hop. This can be illustrated in the CDS of Figure 1(c). Suppose v_3 wants to send a packet to v_{20} (sink). Assume the each sensor randomly selects to transmit one packet within I_k. If the transmission of v_{12} precedes that of v_3, a real packet p will be relayed one time (from v_3 to v_{12}) during I_k. On the other hand, if v_{12}'s transmission follows the transmission from v_3, p will be relayed twice during I_k, delivering p at v_{20}.

In DAS, transmissions are coordinated to maximize the number of relay operations per I_k. We label an active CDS $\mathcal{D}_j$ as a tree rooted at the sink s. Packets from any sensor in $\mathcal{D}_j$ are delivered to s using shortest path routing on the tree. Hence, a packet originating from a sensor v located at depth $d(v)$ requires $(d(v) - 1)$ relay operations until it is delivered to s. We divide each I_k to subintervals $\{I_k^1, I_k^2, \dots, I_k^\ell\}$ of duration $\frac{T}{\ell}$, where ℓ is the height of the tree. A sensor v at depth $d(v)$ is scheduled to transmit during subinterval $I_k^{\ell - d(v) + 1}$. Formally, DAS implements the following steps.

Algorithm 2: Deterministic Assignment Scheme (DAS)

Step 1: $\mathcal{D}_j$ is labeled as a tree rooted at the sink s.
Step 2: A sensor v located at depth $d(v)$ is labeled with $id_v = (d(v) \mod \ell) + 1$, where ℓ is the height of the tree.

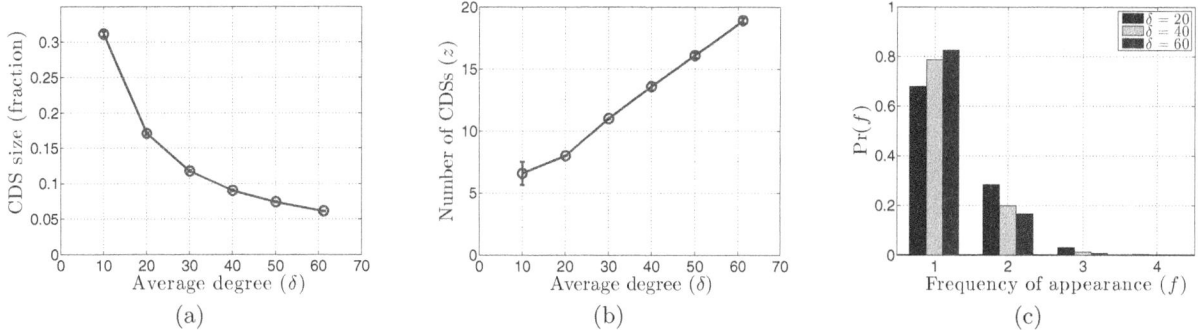

Figure 2: (a) Average CDS size, normalized over the WSN size, as a function of δ, (b) average number of CDSs needed to span $\mathcal{V}$ as a function of δ, (c) empirical probability mass function of f.

Step 3: A sensor with id_v is assigned to transmit one packet in subinterval $I_k^{\ell - id_v}$.

For the CDS $\mathcal{D}_j$ shown in Figure 1(c), suppose that v_{20} is the sink. We first label $\mathcal{D}_j$ as a tree rooted at v_{20}. The label id of each sensor is shown within the circle. Interval I_k is divided into four subintervals (the depth of the tree is $\ell = 4$). Sensor v_5 ($id_5 = 4$) is scheduled to transmit at I_k^1, sensors with $id_v = 3$ are scheduled in I_k^2, and so on.

Note that the sink need not be part of every CDS. If s does not belong to a CDS $\mathcal{D}_j$, any sensor $v \in \mathcal{D}_j$ one-hop away from the source can be selected as the tree root. Such sensor is guaranteed to exist due to the CDS property.

Security Analysis: In DAS, the transmission patterns observed by eavesdroppers are decorrelated from the real traffic pattern. Thus DAS does not reveal the location and time of occurrence of an event. For instance, suppose that a sensor $v \in \mathcal{D}_j$ observed an event $\epsilon(loc, t)$ occurred at time t and location loc. When $\mathcal{D}_j$ becomes active, v transmits packets related to ϵ towards s. Each sensor on the path from v to s will transmit the packets originated by v at random within the designated subintervals according to DAS. These transmissions will not alter the transmission profile observed by any set of colluding eavesdroppers, as real packets substitute dummy ones. Moreover, since hop-by-hop re-encryption is applied at the link layer, copies of the same packet traversing multiple hops remain indistinguishable. Because the adversary cannot distinguish real packets from dummy ones and the transmission pattern is decorrelated from the event pattern, $\epsilon(loc, t)$ is unobservable. However, DAS reveals the sink's location. This is because the sink is the only sensor transmitting during subinterval I_k^1 (all other sensors have an id larger than 1). Hence, DAS may only be adopted when the sink's location must not remain secret.

When the sink's location must be concealed, we use a mechanism that trades communication efficiency for privacy. We label the set $\mathcal{D}_j$ as a tree rooted at a randomly chosen node v. As in the case of DAS, sensor $u \in \mathcal{D}_j$ transmits according to its depth in the tree. To guarantee the delivery of a packet from any sensor to the sink (which differs from the tree root), we divide interval I_k into 2ℓ subintervals. Each sensor $v \in \mathcal{D}_j$ is assigned to transmit one packet in subintervals $I_k^{\ell - id_v}$ and $I_k^{\ell + id_v}$. Based on this schedule, a real packet originating from any sensor v, will reach the randomly selected root by subinterval I_k^ℓ. The real packet will continue its propagation to the rest of the sensors of the tree during subintervals $I_k^{\ell+1}$ to $I_k^{2\ell}$. This mechanism implements

a form of flooding, restricted to the sensors of the CDS. Because the tree root is randomly selected, the transmission schedule cannot be used to infer the sink's location.

5. PERFORMANCE EVALUATION

In this section, we evaluate the performance of our scheme in terms of communication overhead and packet delay. Our simulations were developed using MATLAB 2012. The simulation results are based on 10 independent runs.

5.1 Generation of an MCDS Partition

In this set of experiments, we studied the performance of Algorithm 1 in terms of (a) the average fraction of sensors that belong in a CDS; (b) the number of CDSs needed to span $\mathcal{V}$; and (c) the probability mass function (pmf) of the frequency of appearance. We randomly deployed a WSN within an area of $1,000$ m $\times 1,000$ m and varied the average node degree δ (by increasing the number of sensors). Sensor locations were randomly drawn from a uniform distribution to generate random topologies. We then applied Algorithm 1 and obtained the set of CDSs $\{\mathcal{D}_1, \dots, \mathcal{D}_z\}$ that span $\mathcal{V}$.

Figure 2(a) shows the average fraction of $\mathcal{V}$ that belongs to a CDS as a function of δ. Confidence intervals of 95% are also shown. The CDS size indicates the energy savings compared to prior methods that require all sensors in the WSN to be active at a constant rate [9, 14]. We observe the fraction to be as few as 31% of sensors are active when $\delta = 10$, with less than 7% being active when $\delta = 60$. Figure 2(b) shows the average number of CDSs generated by Algorithm 1 as a function of δ. The value of z is a critical factor for the delay until a CDS that contains the real source becomes active. We observe an almost linear increase of z with δ. We note that z implements a tradeoff between the delay and the communication overhead. A partition of the WSN to fewer CDSs increases the size of each CDS and consequently the number of active sensors. However, less time is required to rotate through each of the CDSs. In Figure 2(c), we show the empirical probability mass function of the frequency of appearance f, which is a measure of the "quality" of the partition of $\mathcal{V}$. We observe that more that 67% of sensors are part of only one CDS, while 95% of the sensors have an f less than four. This indicates that Algorithm 1 favors the creation of CDSs that are disjoint to a large degree, reducing the per-sensor communication overhead.

5.1.1 Communication Overhead and Delay

In the second set of experiments, we compared the performance of DAS with the case where sensor transmissions are

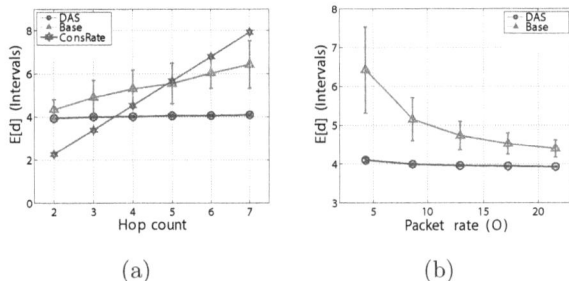

Figure 3: (a) Average delay as a function of the hop count to the sink, (b) average delay as a function of the packet rate.

uncoordinated (referred to as *Base*) and the schemes in [9,12] (referred to as *ConsRate*). In *ConsRate*, perfect contextual privacy is achieved by fixing the packet rate of every sensor. To provide a fair comparison, every scheme was considered under a fixed communication overhead budget. This budget was defined by the number of packets transmitted by all nodes in the network per time interval.

We first considered the end-to-end delay for real packets. Figure 3(a) shows the average end-to-end delay $E[d]$ as a function of the hop count to the sink. The delay is measured in number of intervals until a packet is delivered to the destination. The CDS was rotated per interval. We observe that DAS achieves a constant packet delay irrespective of the hop count to the sink. This is because a real packet always reaches the sink at the end of the interval when the CDS containing the real source becomes active. DAS outperforms the other schemes for a hop count larger than three hops. For shorter hop counts *ConsRate* incurred the lowest delay. This is because for short path lengths, the fixed delay until the corresponding CDS becomes active dominates the overall packet delay. We further observe that DAS introduces a significantly lower delay than the *Base* scheme. Moreover, *Base* has the highest delay variance due to the uncoordinated nature of the real packet relay operation.

We also studied the delay reduction gained by DAS due to the loose coordination of packet transmissions as a function of the average packet rate at each sensor. In Figure 3(b), we compare the packet delay of DAS with the *Base* case. DAS has a fixed delay equal to the CDS rotation delay. On the other hand, in the *Base* scheme the delay decreases with the packet rate. This is primarily due to the reduction of the forwarding delay once a real packet has been transmitted. However, the overall delay is lower-bounded by the delay until a CDS containing the real packet source becomes active.

6. CONCLUSIONS

We addressed the problem of preserving the privacy of contextual information in WSNs under colluding eavesdroppers. We proposed a traffic normalization scheme that significantly reduces the number of bogus traffic sources. This was achieved by partitioning the WSN to a set of CDSs that approximate an MCDS partition. We further reduced the end-to-end packet delay by loosely coordinating the transmissions of sensors within each CDS. We showed that our scheme guarantees the location and time of occurrence privacy of WSN events. Moreover, the end-to-end real packet delay is reduced.

Acknowledgements

This research was supported in part by NSF (under grants CNS-0844111, CNS-1016943, and CNS-1145913) Any opinions, findings, conclusions, or recommendations expressed in this paper are those of the author(s) and do not necessarily reflect the views of the National Science Foundation.

7. REFERENCES

[1] B. Alomair, A. Clark, J. Cuellar, and R. Poovendran. Statistical framework for source anonymity in sensor networks. In *Proc. of the GLOBECOM Conference*, pages 1–6, 2010.

[2] K. Alzoubi, P. Wan, and O. Frieder. New distributed algorithm for connected dominating set in wireless ad hoc networks. In *Proc. of the 35th Annual Hawaii International Conference on System Sciences*, pages 3849–3855, 2002.

[3] X. Cheng and D.-Z. Du. Virtual backbone-based routing in multihop ad hoc wireless networks. Technical report, University of Minnesota, 2002.

[4] J. Deng, R. Han, and S. Mishra. Decorrelating wireless sensor network traffic to inhibit traffic analysis attacks. *Pervasive and Mobile Computing*, pages 159–186, 2006.

[5] M. Garey and D. Johnson. *Computers and intractability*. Freeman San Francisco, CA, 1979.

[6] B. Greenstein, D. McCoy, J. Pang, T. Kohno, S. Seshan, and D. Wetherall. Improving wireless privacy with an identifier-free link layer protocol. In *Proc. of the ACM MobiSys Conference*, pages 40–53, 2008.

[7] J. Gross and J. Yellen. *Handbook of graph theory*. CRC, 2004.

[8] J. Kong and X. Hong. ANODR: anonymous on demand routing with untraceable routes for mobile ad-hoc network. In *Proc. of the MOBIHOC Conference*, pages 291–302, 2003.

[9] K. Mehta, D. Liu, and M. Wright. Location privacy in sensor networks against a global eavesdropper. In *Proc. of the IEEE International Conference on Network Protocols*, pages 314–323, 2007.

[10] A. Proaño and L. Lazos. Packet-hiding methods for preventing selective jamming attacks. *IEEE Transactions on Dependable and Secure Computing*, 9(1):101–114, 2012.

[11] A. Proano and L. Lazos. Hiding contextual information in WSNs. In *Proc. of the IEEE WoWMoM Symposium*, pages 1–6, 2012.

[12] M. Shao, Y. Yang, S. Zhu, and G. Cao. Towards statistically strong source anonymity for sensor networks. In *Proc. of the 27th Conference on Computer Communications*, pages 464–474, 2008.

[13] M. Wilhelm, I. Martinovic, J. Schmitt, and V. Lenders. Short paper: reactive jamming in wireless networks: how realistic is the threat? In *Proc. of the ACM WiSec Conference*, pages 47–52, 2011.

[14] Y. Yang, M. Shao, S. Zhu, B. Urgaonkar, and G. Cao. Towards event source unobservability with minimum network traffic in sensor networks. In *Proc. ACM Wisec*, pages 77–88, 2008.

SnapMe if You Can: Privacy Threats of Other Peoples' Geo-tagged Media and What We Can Do About It

Benjamin Henne
Distributed Computing &
Security Group
Leibniz Universität Hannover
Hannover, Germany
henne@dcsec.uni-
hannover.de

Christian Szongott
Distributed Computing &
Security Group
Leibniz Universität Hannover
Hannover, Germany
szongott@dcsec.uni-
hannover.de

Matthew Smith
Distributed Computing &
Security Group
Leibniz Universität Hannover
Hannover, Germany
smith@dcsec.uni-
hannover.de

ABSTRACT

The amount of media uploaded to the Web is still rapidly expanding. The ease-of-use of modern smartphones in combination with the proliferation of high-speed mobile networks facilitates a culture of spontaneous and often carefree sharing of user-generated content, especially photos and videos. An increasing number of modern devices are capable of embedding location information and other metadata into created content. However, currently there is not much user awareness of possible privacy consequences of such data. While in most cases users upload their own media consciously, the flood of media uploaded by others is so huge that it is almost impossible for users to stay aware of all media that might be relevant to them. Current social network services and photo-sharing sites mainly focus on the privacy of users' own media in terms of access control, but offer few possibilities to deal with privacy implications created by other users' actions. We conducted an online survey with 414 participants. The results show that users would like to get more information about media shared by others. Based on an analysis of prevalent sharing services like Flickr, Facebook, or Google+ and an analysis of metadata of three different sets of crawled photos, we discuss privacy implications and potentials of the emerging trend of (geo-)tagged media. Finally, we present a novel concept on how location information can actually help users to control the flood of potentially infringing or interesting media.

Categories and Subject Descriptors

K.4.1 [**Computer and Society**]: Public Policy Issues—
Privacy

General Terms

Security, Human Factors

Keywords

privacy; social media; awareness; photo-sharing; geo-tagging

1. INTRODUCTION

The amount of media uploaded to the Web is expanding rapidly: For instance, the number of photos uploaded per month on Facebook has risen from 2 billion in 2010 to 6 billion in 2011, and to over 9 billion in 2012 [7, 21]. While privacy issues of shared media are gaining more importance, the focus is still very much on protecting only the content users upload themselves. Increasingly complex privacy settings allow them to decide who is allowed to see what part of their online media. However, the issue of staying on top of what relevant data others are uploading (mostly in good faith) is still very much outside of the control of users. Social network services that allow user tagging usually inform affected users when they were tagged. However, if no such tagging is done by the uploader or a third party, there are currently no mechanisms to inform users of potentially relevant media.

One significant factor causing the flood of shared media is the meteoric rise of the smartphone that allows users to create high-quality photos and videos when and where they want and effortlessly share them on the social Web. In addition to this very visible effect, these devices are changing the landscape of the shared media under the hood as well. Most devices are now capable of gathering location information via GPS or Wi-Fi tracking. This information can be used by location-based services and is often embedded into media created on the device by default. Latest smartphone software additionally integrates facial recognition functions that aim to automatically tag people in photos.

While the privacy issues of location-based services and corresponding privacy-preserving techniques have been discussed at great length, the privacy issues of location information and other metadata embedded into uploaded media have not yet received much attention. There is one very significant difference between these two categories: When using location-based services, like Google Maps, Foursquare, Google Latitude or Facebook Places user mainly affect their own privacy.

While embedded metadata can cause the same privacy implications for the creator of a picture, a critical but often overlooked issue is the fact that location information and other metadata contained in photos or videos can also affect people other than those that upload the media. This is a critical oversight and an issue that will gain importance as the mobile device boom continues.

In this paper, we address the emerging privacy issue that shared media, especially photos, affect other people than their creators themselves. Unlike previous work in this area, we will focus on privacy issues caused by photos uploaded by others and how location information and other metadata can harm, but also help in dealing with the flood of this media and its implications. We analyze the threats in Section 2 and present results of a user survey on the topic of media awareness and privacy in Section 3. Within an analysis of prevalent sharing services in Section 4 we show the complexity and differences in privacy settings of these services. Most of them are limited to access control or are implemented inconsistently and are not well suited to deal with the threats and concerns discussed in this paper. In Section 5 we analyze three sets of photos crawled from the social Web, highlight the amount of shared information and identify problems and trends that show the increasing significance of this topic. We then introduce *SnapMe* a context-based countermeasure in Section 6.

The SnapMe service can assist users in dealing with the flood of media, reducing the risk of privacy events going unnoticed. For this SnapMe leverages the location capabilities of modern devices to create smart privacy zones where users wish to have their privacy protected. Based on co-location checks of these privacy zones and location data of uploaded pictures, the service searches for photos relevant to its users. Additionally, face recognition and a location-based service are used to improve results.

In Section 7 we present related work to this paper. Finally, we conclude this paper in Section 8. A first simulation-based evaluation of the SnapMe service is included in the appendix.

2. PRIVACY THREAT ANALYSIS

A multitude of privacy issues stemming from media on the Web has been researched in the past years. The mere fact of being depicted on the Web can already be a privacy concern for some people. Even being shown at a perfectly harmless place may raise objections for some of them. Even more privacy threats may arise from photos taken in situations where people are intoxicated, do illicit or socially unaccepted things, or are at a place or with someone they would deny to have been with. Research has shown that the fear of such unwanted revealing of information is valid for nearly any other person, i.e. from people outside a person's social circle [1] to people inside a person's social circle including friends and family [3]. Media content may not only harm personal privacy, but can also create financial problems since employers[1], insurance companies[2] and even banks[3] use information from the social Web. An increasing number of people have become cautious about the use of the social Web and the sharing of media. Access control helps to secure people's own media in so far as the implemented

[1]http://www.washingtontimes.com/news/2006/jul/17/20060717-124952-1800r/

[2]http://abclocal.go.com/kabc/story?section=news/consumer&id=8422388

[3]http://www.betabeat.com/2011/12/13/as-banks-start-nosing-around-facebook-and-twitter-the-wrong-friends-might-just-sink-your-credit/

mechanisms meet their requirements. But, still many sites on the Web lack of mechanisms that make people aware of media shared by others, which might threaten their privacy.

2.1 Origin of Damaging Media

Privacy issues can be divided into two categories. Firstly, home-grown problems: Users upload compromising photos of themselves with insufficient protection or forethought, causing damage to their own privacy. A popular example of this category is someone uploading compromising pictures of himself into a public album instead of a private one or onto his Timeline instead of a message. The damage done in these cases is very obvious since the link between the content and the user is direct and the audience (often the peer circle) has direct interest in the content. One special facet of this problem is that the perception of what is considered damaging content by the user can and often does change with time. While this is a serious problem, especially for the Facebook generation, this issue has been well discussed in related work and is not focus of this paper.

Problems created by others fall into the second category: An emerging threat to user privacy originates in media uploaded by others ranging from friends to strangers. This threat is particularly problematic since the harmed person is not involved in the uploading process and thus cannot prevent it from happening and cannot take any effective precautions. Currently there are no countermeasures, except ex post legal ones, to prevent users from uploading potentially damaging content about others. There are two requirements for this kind of privacy threat to take effect: Firstly, to cause harm to a person it must be possible to associate/link the critical media to the person in some way. This link can either be non-technical, such as the person being recognizable in a photo, or technical such as a profile being hyper-linked to a photo or personal information being integrated into image metadata/associated to the image. Secondly, the media in question must contain harmful content for the person linked to it. This can again be non-technical such as being depicted in a compromising way. However, more interestingly it can also be technical. In these cases metadata or associated data causes the harm, such as the time and location indicating that the person has been at an embarrassing location or not where she claimed to have been. Or, it can also simply be an insulting headline or comment attached to a photo.

Since the upload of this type of damaging media cannot be prevented effectively, awareness is the key issue in combating this emerging privacy problem.

2.2 Awareness of Damaging Media

Most popular social network services (SNS) and media-sharing sites allow users to tag objects and people in uploaded media. Media is annotated with names, comments, or is directly linked to user profiles. The direct linking of profiles to pictures was initially met with a great outcry of privacy concerns, since it significantly facilitates gathering information about people and thus enables embarrassing or harmful pictures to be found more easily. It often even broadcasted embarrassing photos to interested parties via notifications. Thus, SNS quickly introduced the option for users to forbid others to link them in media. However, there is also a positive side to this practice of direct linking: The linked person is usually made aware about the linked media and can thus take action to have unwanted content removed

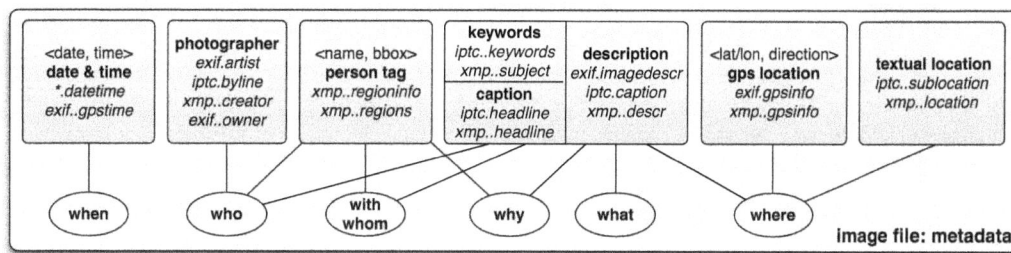

Figure 1: Image metadata and derivable private information (taggable by state-of-the-art tools)

or the visibility of the link restricted [4, 26]. While the privacy mechanisms of current SNS are still limited, hidden and often confusing, once the link is made, the affected users can take action.

A more critical case is the non-linked tagging of photos. In this case a free text tag or comment contains identifying information. However there is no automated mechanism to inform users that they were mentioned in or near a piece of media. The person in question even might not be a member of the service where the media was uploaded. The threat of this kind of tagging significantly differs from the one described above. While the immediate damage is often smaller because no automated notifications are passed to members of the group connected to the user's profile, the threat can remain hidden far longer. The person can remain unaware of this media whereas someone actively looking for information about her can find it via search engines that are capable of indexing the tags attached to the media. Currently the only way to combat this threat is to pro-actively crawl the Web in search of such references (cf. Section 7).

The final category of damaging media does not contain any technical link. Without any link to the person in question this kind of media can only cause harm if someone associated to that person stumbles across it and makes the connection. While the likelihood of causing noticeable harm is smaller, it is still possible. Viral spreading of media did cause serious embarrassment and harm in real world cases. The critical issue in this case is that there is currently no way for a person to pro-actively search for this kind of media to mitigate this threat.

The SnapMe privacy watchdog introduced in Section 6 will address the last two cases and assist users in finding media relevant to them.

2.3 Persistence of Damaging Media

The persistence of media and associated data on the Internet is another privacy concern. There are different expected lifespans. For instance, micro-blog posts or status updates can have a relatively short lifetime in the sense that they are harder to find after a while or are even entirely removed from the visible database [27]. Other content such as images or videos typically remains visible indefinitely. As stated before, without any context information media is not easy to discover and also has little potential to damage a person's privacy. However, the capability to embed metadata into the media (instead of storing it in the provider's database) creates a number of threats that have to be taken into account, as metadata stored in media is just as persistent and replicable as the media itself.

2.4 Threats of Metadata

Metadata stored within media, especially that of photos can include various kinds of context information. It is stored in common formats such as EXIF, IPTC or XMP. Isolated context information is mostly not noteworthy, but linking different information together allows conclusions about people that may harm personal privacy. In the case of embedded metadata different information like who, when, where and what are stored together in the file itself.

While GPS location data raises the most obvious privacy concerns, there is a variety of other meta-information that has to be considered as well. A photo may include textual descriptions of places (location, street, city, state, country) besides GPS coordinates. Some cameras write the artist's name as well as a unique device identifier into media files. An image's caption or description may also include private information. And finally, another new kind of in-file information based on the XMP-based Microsoft Photo Region Schema [20] and Metadata Working Group Regions Schema [18] enables users to tag people with names and bounding boxes directly within an image file. Figure 1 shows the most commonly used metadata, contained information and potentially derivable private information.

There are several ways to add metadata to images. Metadata can be added automatically by digital cameras as well as by many camera phones that typically write EXIF metadata (date, time, artist, camera id, GPS location) into image files. Additional metadata can be added manually or semi-automatically by users via media library and camera software. Location names and GPS coordinates can be added by looking up places on a map or by using (reverse) geo-coding to map between GPS coordinates and location names. Semi-automatic annotation not only exists for location names or geo-tags, but also for first face recognition features that are integrated into media library software, as for instance Google Picasa, Apple iPhoto or Windows Live Gallery. These technologies are converging with new mobile devices as such equipped with Android 4 that integrates face recognition in its API that is also available for third party apps [22], or Apple devices that from iOS 5 heavily integrate Cloud technologies and face detection features based on software acquired from Polar Rose [23]. Finally, apps like SocialCamera [28] bring face recognition to media sharing users. This convergence is particularly noteworthy: If such metadata is directly integrated into the media file, much more care must be taken to protect it. While such metadata is already commonly stored and used in SNS and photo-sharing sites this data is usually only accessible within the service or via its APIs and protected by its access control mechanisms.

Extracting all this separately stored metadata and republishing it together with the media takes conscious effort and thus probably malicious intent. However, downloading a picture with embedded metadata from a site and sending or reposting it somewhere else drags all metadata along, thus exposing it outside of the privacy controls of the hosting service and greatly amplifying the threat posed by the media.

Metadata plays an integral role to privacy for both awareness and damage reasons. Because of this we conducted a survey of the privacy capabilities of current photo-sharing services with a particular focus on metadata handling to exhibit the extent of the situation as presented in Section 4.

3. USER SURVEY

To confirm the assumptions made about the origin of damaging media and user awareness of media, we conducted an online survey. We invited 1,418 people from a mailing list for survey participations at our university. As an incentive, participants were entered into a raffle for two $ 60 vouchers for Amazon. We received 414 valid answers – 53.9% male and 46.1% female. About 25% of the participants had at least one university degree. The average age was 23 ± 4 years. Based on Westin's privacy segmentation index [15], 91.8% of the participants were classified as privacy pragmatists, 6.0% as fundamentalists and 2.2% as unconcerned.

3.1 Threat of Others' Media

Most of our participants seem to have at least a basic level of moral incentive to protect the privacy of people depicted in shared photos. Only 2% of our participants answered that they do *not* think about threats to others *at all* when sharing photos on the Web. The remaining 98% stated that they consider threats to others at least a bit in their sharing decisions. About 61% rated threats to others and threats to themselves with the same value on the 7-point scale from *not at all* to *very much* as criteria for sharing a photo. It is interesting to note that 6.6% of participants weighted threats to others higher than threats to themselves.

Even though most people think about others' privacy when sharing media, they seem to believe that others do not comply to "moral obligation" [4] adequately. We asked our participants about their estimation of the extent of a possible privacy violation by photos shared by different groups on a 7-point scale from *very low* to *very high* (cf. Figure 2 (q17)). Most people rated any possible violation higher than *very low*: Only 1.4% of our participants rated a possible violation across all groups of the sharing people to be *very low*. All participants rated the violation level of photos shared by friends to be the lowest with a mean of 3.64, $sd = 1.85$. The middle level was that of photos shared by friends of friends with 4.69, $sd = 1.66$. The highest estimated violation level was that of media shared by strangers with 5.23 of 7, $sd = 1.95$. Although these may be less likely, participants consider privacy threats by strangers' media to be possible. In fact, 47% of them rated a violation of their privacy by strangers' photos higher than by those of direct and indirect friends.

3.2 Media Awareness

Media shared by non-friends is not easy to discover and might even be impossible to access in some social services. Thus one goal of our survey was to learn about users' awareness of photos on the Web they might be depicted in. When

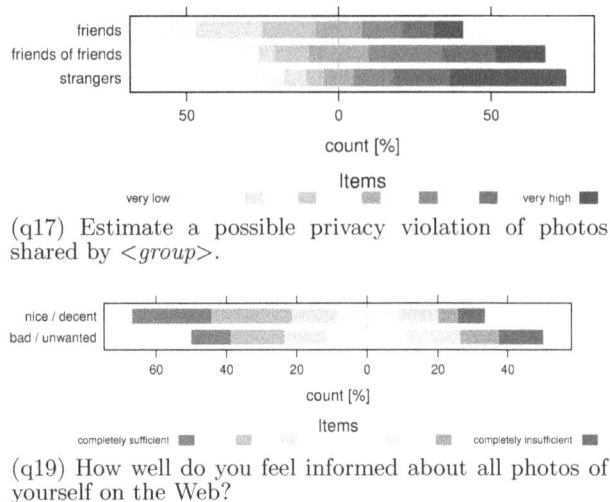

(q17) Estimate a possible privacy violation of photos shared by <*group*>.

(q19) How well do you feel informed about all photos of yourself on the Web?

Figure 2: Extract of online survey results

we asked our participants how they actually get to know about photos they are depicted in, 75% of them answered that they get notifications by automatic email if they are tagged in a photo (94% of these were Facebook users). This kind of information is missing in the case of non-linked tagging or missing tags. It is only valid for the typical profile-linked tag in many SNS. 52% of our participants answered that they get to know about photos of themselves by chance (2% exclusively by chance). 39% of them hear about them in conversations and 30% in friends' messages. 18% are actively looking for photos, 4.6% are informed by messages of non-friends, and 3.4% are not informed at all. We asked the participants how well informed they feel about photos of their own on the Web. On a 7-point likert scale from *completely sufficient* to *completely insufficient* (cf. Figure 2 (q19)) participants answered to feel moderately informed (mean = 3.6, $sd = 1.72$): They answered with a mean of 3.2, $sd = 1.85$ concerning decent photos; when thinking about bad photos, they answered with a mean of 4.0, $sd = 1.85$. In detail, 22% stated that their information about decent photos of themselves was *completely sufficient* and 25% less to *completely insufficient*. In contrast, only 11% stated their information about bad and unwanted photos to be *completely sufficient*, and 39% of the participants asserted that their level of information about bad photos of themselves was *less* to *completely insufficient*. We finally asked our survey participants the hypothetical question whether they would like to use a service that helps them to find relevant photos, which entails the need to screen potential photos. 53.1% of them answered with a clear yes and 41.8% were interested in using it. Only 3.6% argued that costs of screening would overbalance benefits. Others called on uploaders' moral obligation or did not believe that any photos of them were online.

3.3 Discussion

The participants of our survey stated that they realize threats caused by media shared by other people. When asked in which way and how well they are informed about photos they may be depicted in, which could lead to privacy threats to themselves, their answers confirmed that improvements are needed in the area of online media awareness and privacy. Nearly all participants stated that they would be willing to invest time to screen photos if they are informed about them. One participant wrote as final comment that he even would pay a one-time fee for such a service.

4. ANALYSIS OF SERVICE PRIVACY

The following section gives an overview of our privacy analysis of prevalent photo-sharing sites. It focuses on privacy settings (mostly access control) as well as metadata handling.

Flickr provides the most fine-grained privacy/access control settings of all analyzed services. Privacy settings can be made for metadata as well as the image itself. For instance it is possible to configure that an image is publicly viewable, but the metadata is only visible to friends. However, there is a caveat to this feature. With most of the client software used to upload, Flickr extracts the title and description of an image. Additionally, embedded keywords and textual location information (country, state, city) are extracted as image tags. This information is available to all people with access to the image via the website or the API. Flickr stores complete EXIF metadata of all files in its database. Access to EXIF data can be enabled for all people (default) or disabled at once for all images of a Flickr user. The access to GPS data (map view, exact coordinates in HTML source, EXIF data view, API) is only provided if users add images to their Flickr map. Clients like the iPhone Flickr app or Instagram can automatically do this. Access to GPS data can be set separately on a per image base as public (default), restricted to family, friends or contacts, or completely disabled for all photos of a user at once. One particularly interesting feature of Flickr is the geofence. The geofence feature enables users to define privacy regions on a map by putting a pin on it and setting a radius. Access to GPS data of the users' photos inside these regions is only allowed for a restricted set of users (friends, family, contacts).

Flickr allows its users to tag and add people to images. When adding a person (Flickr user), an optional bounding box can be drawn in the image. These actions are allowed to a user's contacts by default, which can be extended to all Flickr members, changed to family and friends or restricted to only the owner. When a user revokes a person tag of himself from an image, no one can add the person to that image again. Flickr resizes uploaded photos and strips off all metadata from resized files. Thus, only the metadata stored in the database and accessed via the website or API, which is protected by the privacy settings, is relevant. However, care has to be taken in the case of Flickr Pro users whose original images including all original metadata (EXIF, IPTC, XMP) are stored. Geo-location data included in these files is not protected by the geofences [12]. Flickr's access control settings are as fine-grained as they are hard to grasp for the average user. Additionally, privacy is handled as an opt-in feature. We analyzed two sets of Flickr data and present the results in Section 5.

Facebook extracts the title and description of an image from its metadata on upload with some clients. All photos are resized after the upload and metadata is stripped off. Facebook officially does not store original photos or original image metadata. Tags (text, person, location) can be added for each image using the Facebook website or any other client using its API. Facebook uses face recognition for friend tagging suggestions based on already tagged friends. Added metadata is stored only in Facebook's database and not in any image file. The access to images is restricted by Facebook's changing, complex and abstruse privacy settings. With given permissions, comments, person tags as well as locations are accessible via the Facebook API.

Picasa Web extracts the image title from metadata when uploading via web or the Picasa application. The image is resized and metadata is stripped off. Picasa Web stores the complete EXIF metadata of each image in its database, making it accessible to everyone who can access the image. Image access is defined on a per-album base: It can be public, restricted to people who know the secret URL, or to the owner. Geo-location data can be protected separately. Images can be downloaded in original size including all metadata. Unlike Flickr Pro, if geo-locations are deactivated for an album, GPS location is removed from the metadata of the downloadable image. Picasa Web allows the tagging of people in images. These tags are stored in Picasa's database and are not included in files as if using Picasa desktop application. Picasa Web allows a two-way synchronization of albums with its desktop application. This synchronization includes person tags.

Google+ integrates Picasa Web for its photos. Every statement made about Picasa Web in the previous paragraph holds for photos in Google+. Access control is extended to enable restricted access for user-selected circles. A preliminary privacy analysis has been done in [16].

Windows Live SkyDrive does very little in the way of metadata processing upon upload. Images can be resized on upload, but the complete original metadata is retained by default. Client applications like Windows Live Photo Gallery can be configured to partially or completely strip metadata from images before uploading. Access control rules can only be set per image; no separate metadata protection is possible. Windows Live SkyDrive is the only online service capable of reading person tags based on the Microsoft Photo Region Schema [20] from files' XMP metadata. These tags can be written to files using Windows Live Photo Gallery; *avPicFaceXmpTagger* [29] can transfer person tags from Picasa's local database into XMP metadata stored within the image.

Locr is a photo-sharing site focused on geo-tagging, so location information is included in most images. All metadata is retained in all images by default. Access control is set on a per image base. Everybody who can access an image can also see its metadata. Locr also offers extensive location-based search options. Geo-data is extracted from uploaded files or set by people on the Locr website. Locr uses reverse geo-coding to add textual location information to images in its database.

Apple iCloud *Photo Stream* allows iOS and OS X users to share photos between different devices and with selected people. When shared via Photo Stream, images are resized and uncommon metadata is removed. Common metadata including location information and the Metadata Working Group Regions Schema information is retained. When sharing images with other Apple users, all this metadata is included; when sharing via a web link only GPS coordinates are removed while textual location information down to addresses or person tags is still included.

Instagram is a service and mobile app that allows posting images like Twitter messages. Resized images stripped off metadata with optional location data are stored by the service. Additionally, Instagram allows posting pictures on different services like Flickr, Facebook, Dropbox, Foursquare etc. Metadata is stored depending on the service. When uploading a photo to for instance Flickr, metadata is stripped off by the app, but title, description as well as location are extracted from the image or can be set by the user. In contrast, the *Hipstamatic* mobile app preserves in-file metadata when uploading images to Flickr.

In summary, metadata and in particular location data handling is dealt with in a bewildering variety of approaches. Some services strip off all metadata, probably to steer clear of privacy issues. This may be the simplest solution to privacy, but also rejects a valuable asset to keep the increasing mass of images under control. Other services retain metadata and even allow complex queries on location data. Every client application has a different privacy scheme: Some use opt-out; some prefer opt-in, while others do not have any options about privacy and metadata at all. Furthermore, when privacy is concerned, usually only GPS coordinates are considered while reverse geo-coded location tags or XMP-based in-file person tagging are neglected. This leaves users in the unenviable position of having to very consciously understand what is being shared and deduce the privacy risk themselves on a service-by-service basis.

5. SURVEY ABOUT METADATA ONLINE

To confirm the growing prevalence of privacy-relevant metadata and location data in particular as well as to judge potential dangers and benefits based on real-world data, we analyzed a set of 20,000 publicly available Flickr images and their metadata[4]. Flickr was chosen as the premiere photo-sharing website as it can be legally crawled, offers the full extent of privacy mechanisms and in general does not remove metadata. We crawled one photo each from 20k random Flickr users, of which 68.8% were Pro users, so the original file could be accessed as well. For the others, only the metadata available via the Flickr API was accessed, including data automatically extracted from EXIF data during upload and data manually added via the website by client applications. 23% of the 20k users denied access to their extracted EXIF data in the Flickr database. We also took a set of 3,000 images by 3k random mobile Flickr users. 46.8% of the mobile users were Pro users; only 2% denied access to EXIF data in the Flickr database.

Figure 3 shows all publicly accessible metadata of both datasets compared to the total number of photos of each set. For each type of privacy related metadata (cf. Figure 1), the figure shows three distinct subsets: Firstly metadata that could be extracted (only) from original images files, secondly EXIF data extracted from images and stored in the Flickr database and thirdly the amount of meta information added by users themselves or their client applications. GPS location data was present in 19% of the 20k picture dataset. In the 3k dataset collected solely from mobile phone users, 34% of the images were supplemented with GPS data. While Flickr hosts many semi-professional DSLR photos, mobile phones are becoming the dominant photo generation tool with the iPhone 4S currently being the most common camera on Flickr [11]. Textual location information like street or city names is currently uncommon on Flickr. However,

---

[4]All datasets were crawled at the end of 2011.

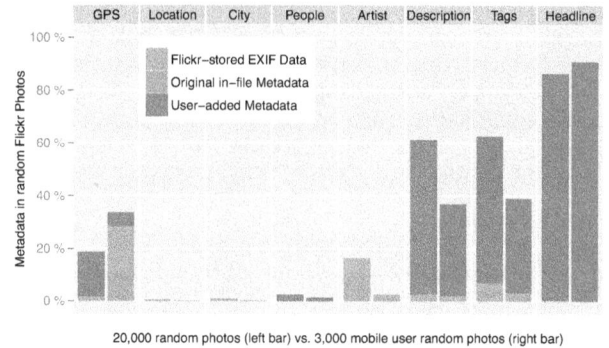

20,000 random photos (left bar) vs. 3,000 mobile user random photos (right bar)

Figure 3: Public privacy-related metadata in 20k random and 3k mobile user photos from Flickr

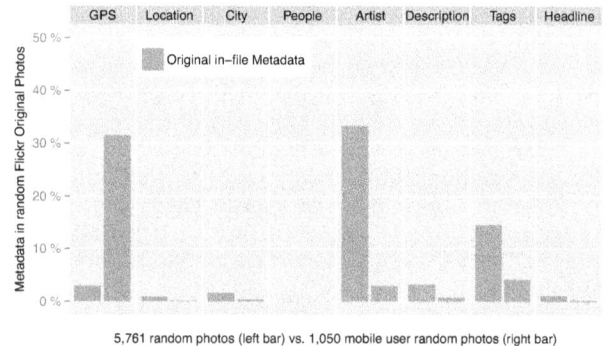

5,761 random photos (left bar) vs. 1,050 mobile user random photos (right bar)

Figure 4: Public privacy-related metadata in 5.7k random and 1k mobile user original Flickr photos

this might change as reverse geo-coding becomes more common in client applications (cf. Locr, Figure 5). In several cases privacy relevant data was stored in miscellaneous tags. Within our sample sets we found telephone numbers, GPS information and different other personal information encoded as plain text in different fields such as artist, title or description. To evaluate the impact on people other than the uploader, we manually counted people in photos with a geo-reference, but no person tag – those are images the possibly concerned people may never be notified about. In the set of 20k images, 16% showed people, having a geo-reference but no person tag in it; in the set of 3k mobile photos we counted 28%.

We further analyzed the subset of images available from Pro users, since these might contain unaltered metadata from the camera. 5761 images from the 20k dataset and 1050 images from the 3k dataset contained in-file metadata. Figure 4 shows the percentage values for the different types of metadata contained in the files. For the other images metadata was either manually removed by the uploader or was never there in the first place. Of the 20k dataset, only 3% of the original photo in-file metadata contained GPS data compared to 32% from the mobile 3k dataset. This shows a clear dominance of mobile devices when it comes to publishing GPS metadata. This itself is unsurprising, since most compact and DSLR cameras do not yet have GPS receivers and only few photographers use external GPS devices together with these cameras – this will change with future

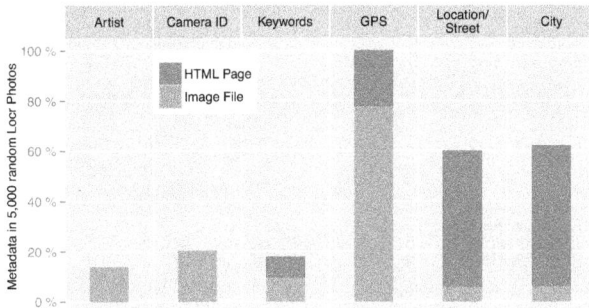

Figure 5: Public privacy-related metadata of 5k random photos from Locr service

cameras. However, since mobile phones are becoming the dominant type of camera concerning the number of published images, it is to be expected that the amount of GPS data available for scrutiny, use and abuse will rise further.

We also collected a 5k dataset of random photos from Locr and analyzed the metadata in the images plus the images' HTML pages built from the Locr database. Figure 5 shows the results from this dataset. Particularly interesting is the higher rate of non-GPS based location information. This trend should be monitored (as reverse geo-coding becomes more popular), since most location stripping mechanisms only strip GPS information and leave text-based tags intact. Additionally, the amount of published camera ids is notable, as these may become part of a unique identifier for a person in the social Web, even if not completely unique as the mobiles' IMEI.

To summarize, one third of the images taken by dominant camera devices contains GPS information. About one third of these images depict people. Summed up, about 10% of the photos may harm other peoples' privacy without their knowledge. We found some users storing private information including phone numbers in metadata – we only talked to their voice mailbox. In-file person tags and textual location data are still only included in few images, but this will change when more applications integrate XMP-based person tags and reverse geo-coding. Android 4 for instance provides a geo-coder API for its apps.

6. SNAPME PRIVACY WATCHDOG

The previous sections discussed privacy threats originating in media uploaded by others. These threats are amplified by the growing amount of metadata – in particular location data – stored within the images. We suggest using one of the contributing factors to this growing threat to actually improve users' privacy by raising awareness of compromising media and enabling them to take actions: We propose leveraging the location tracking capability of modern smartphones to create smart privacy zones in which the users wish to have their privacy protected.

6.1 Design

Previous work by us [25] and by Burghardt et al. [5] attempt to automate manual searches users would otherwise have to do, saving them time. These services are designed as independent value-added services to existing systems and remain close to current SNS practices. The SnapMe watchdog service goes a step further to maximize user benefits. Since our user study showed that the overwhelming majority of our participants were interested in protecting other peoples privacy, we designed a system which would help them to do this. The SnapMe components are integrated into the general upload service of the sharing service to utilize information that could not be accessed by any third party service relying on users' credentials. This integration allows for the watchdog service to act on behalf of any user or user-independent from a higher perspective, i. e. watching for relevant media and notifying users at the time of the upload without any requisite action for users until relevant media has been detected. The proposed watchdog service does not only allow users to keep track of pictures taken in their environment for privacy reasons: it also has a social component that some users may find attractive in-of-itself: e. g. the watchdog service also finds nice/desirable new photos of them.

Figure 6 shows the usage schema of the SnapMe privacy watchdog service: Users registered with the SnapMe watchdog service are marked with R. When a photo is taken and uploaded, the watchdog service may notify the registered users of the event. For user 1 (red), the event would be a false positive since even though the red user was near the location he is not in the photo. User 2 (green) is in the photo and he can evaluate if he objects to the image or not after notification. The non-registered person 3 (blue) will not be notified about the photo even if depicted.

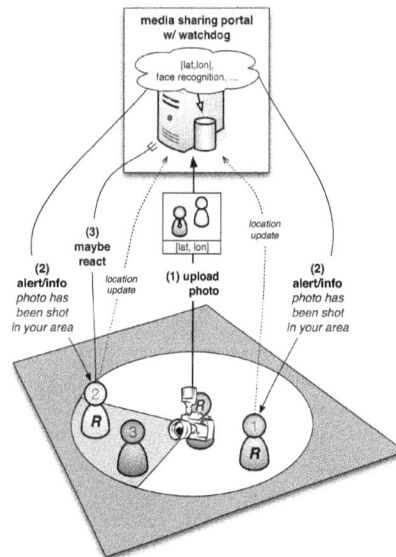

Figure 6: Photo-sharing site with *SnapMe* watchdog service and user notifications

6.1.1 Co-Location

If new media is uploaded to a SNS or a media-sharing site that implements SnapMe, the watchdog service checks if the new content may be relevant to a user. A first indicator for relevance is the question for co-location: SnapMe checks if the location has been marked as private by a user, or if a user was at the time and place where the picture was taken. The location information (GPS coordinates or textual information) of a photo can be embedded in its metadata or sent as context data by the uploading client application. In the latter case, users might select if location information should only be used by the watchdog and thus not be permanently stored afterwards, or if it should be published with the image. For co-location checks, each SnapMe user can define static privacy areas. If a photo is taken in a user's privacy area, he will be notified. In addition to location (coordinates, radius), time constraints are also used to restrict the privacy areas. Using such static areas, users can for instance constantly observe photos taken in the immediate vicinity of their home or business place, or they can observe photos at a one-time event.

The SnapMe watchdog service also features dynamic privacy areas, implemented based on a location-based service (LBS). Compared to static spots on the map, the dynamic privacy areas follow the users. To use the LBS, users install the SnapMe client application on their mobile phones. Every time they want the dynamic area to follow them, they enable the SnapMe LBS client that in turn automatically pushes its location to the service in scheduled intervals until the location updates are paused or stopped. The dynamic approach can be enabled and disabled at will. Thus, it is possible to use it only at special events or places.

Finally, when a new photo with location information is uploaded, SnapMe searches for users in the photo's vicinity and users that have set up matching static areas. To take into account the access control rules of an uploaded photo, those users have to be authorized to see the new media. If both apply for a user, he is notified about the photo and can screen it.

6.1.2 Face Recognition

While the static privacy areas and LBS-based selection of relevant media already reduce the amount of media to be reviewed, there are situations that require an unmanageably high load of photo screening. For example, an event like a festival or protest can easily create such a high amount of pictures taken at a specific time and place that even if only a small privacy radius was set too many pictures would be flagged for viewing. In these cases, face detection and recognition can be used to further reduce or pre-sort the pictures that have to be reviewed by the user. Figure 7 shows how these techniques are integrated into a basic SnapMe photo inspection algorithm: Face detection is used to filter out photos that do not contain faces. If face recognition is enabled for a user, then additionally to a face detection algorithm a face recognition algorithm is executed and the user is only notified about a photo if it is co-located and if his face was recognized. For the LBS-based co-location checks, each user's latest location prior to and directly following the time a photo has been taken are used.

The training data for the face recognition algorithm is of course a very interesting issue: In many cases SNS already have an extensive library of profile pictures that are already in use for face recognition. Allowing a service to use this information for face recognition increases the utility of the watchdog service, but trades information with the provider. This privacy tradeoff must be based on user preferences. Again, the Web 2.0 spirit shows that many people are happy about sharing data as long as they appear in a positive light; so, the acceptance of this tradeoff is not unlikely, but needs to be studied in future work.

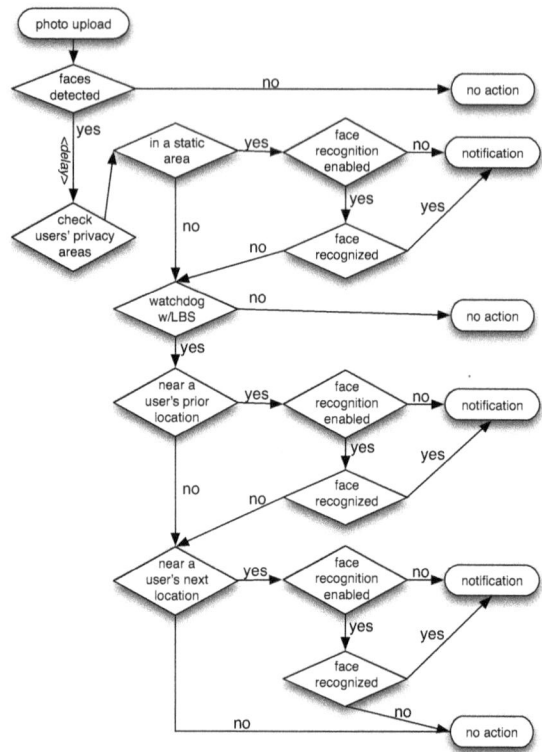

Figure 7: SnapMe photo inspection algorithm

6.1.3 Inference

The watchdog service has to delay the check for each photo for a short time of about some minutes up to about an hour. Firstly, it has to delay the check to consider the time that elapses between taking and uploading an image. Secondly, the use of dynamic privacy areas requires waiting for location updates. SnapMe thus has to cache the location information for the same period of time.

The delaying and caching allows SnapMe to make some inference checking before the location information is dismissed: The location of photos and hence the depicted people may be inferred if any two photos of one photographer are taken within a short time interval.

6.2 Threat Model

The SnapMe privacy watchdog service is designed to leverage the location information of uploaded media. It bases on the assumption that the trends of the rising use of smartphones, the integration of GPS into devices and the mobile taking and uploading of photos develops further.

The service is designed to help users track potentially harmful photos uploaded by others without malicious intent. These uploads include the typical spontaneous or imprudent photos of a person made by friends or strangers in her vicinity. SnapMe has originally been designed to prevent harm, but it also informs its users about nice pictures they would otherwise miss. The watchdog service works on a best-effort approach: It probably cannot detect every single photo of its users on the Web, but strongly facilitates user awareness of relevant data, allowing to minimize the number of hidden dangers posed by unknown media.

If people with malicious intent seriously want to harm a person by publishing comprising media, they can still bypass SnapMe with knowledge about the service details. While SnapMe can help dealing with some of these attacks as well, the protection against malicious intent is currently outside of SnapMe's focus. Malicious people have a huge arsenal of weapons and the watchdog and other services currently cannot defend against methods ranging from slight modifications of metadata to the retouching of photos to create precarious content.

6.3 Privacy Analysis

The privacy watchdog service can help users to reduce the number of relevant media they need to keep an eye on. However, the devil is in the detail since a service like the SnapMe watchdog may also come with its own privacy implications.

When using the static privacy areas, people may want to set their location and time constraints rather broadly to lower privacy concerns. But these constraints have to be limited (number, radius) by the provider, firstly to lower the number of photos that have to be checked and secondly to limit the number of false positive notifications to a manageable amount.

Using the dynamic privacy area based on a LBS may raise location privacy concerns. Users send their current location to the LBS, thus creating a direct correlation between their location and user account. The service in turn uses this information for its checks and can drop it afterwards. The provider does not need to store locations. Users have to trust the provider not to misuse their personal data.

Service users must not obfuscate the dynamic privacy area location information by modifying coordinates, as this would hinder the watchdog from working correctly when searching for relevant media. Users could send a number of fake locations for every real location, making it less easy for the service provider to guess the true location. The service would have to support such different simultaneous locations since otherwise any location interpolation could break. However, this approach does not scale well for at least two reasons: Firstly, it creates a much higher load for the provider. Secondly and more critically, the likelihood of deducing the true location rises anyway when many location updates are sent unless great care is taken in creating complete fake paths and masking the source IP addresses. An imaginable solution to the overall propagation of location information to the service might be enabling the watchdog client only under circumstances when users want their privacy protected. Then again, the service provider could deduce those exact locations that are private to those users. Combining aforementioned solutions, the installed watchdog client could send faked or obfuscated locations in those situation not private to a person and true locations if a person enables his personal watchdog on the client. The service would notice irregularities of locations, but it would be hard to guess which are obfuscated and which are real.

Using the SnapMe watchdog service with its LBS-based features implies that users have to "swap" their location data for the best possible update on uploaded media. Users have to balance their preferences in a privacy-privacy-tradeoff: They surrender parts of their location privacy to the service and gain a higher privacy level concerning photos.

While traditional privacy research aims to preserve user privacy with all its facets and details, this approach goes a step further. By assuming that one cannot protect every detail and aspect in real world deployments, it is up to the user to decide which parts of their personal data are more private than others. The privacy-privacy-tradeoffthat SnapMe users face coincides with user studies [10], which indicate that location privacy is less important to users than other personal data like photos or contact data.

Another concern that people may have about the watchdog service is that the photo notification service may bring along a faster distribution of photos including some unwanted ones. Those concerns may be true, but without the service harmed people may never find out about an incriminating photo and hence could never react.

6.4 Proof-of-Concept Simulation

We evaluated the SnapMe design with a proof-of-concept implementation utilizing simulation to gain basic insights about the service effectiveness and to compare different setups. The upload of virtual photos, the co-location checks and statistics-based face checks were implemented on a real system. Users and photo taking were simulated using the Mobile Security & Privacy (MoSP) simulator [13]. Details on the simulation are described in Appendix A. The best results were achieved with the LBS-based dynamic privacy area with face recognition: With an update interval of 15 minutes and a privacy radius of 20m, users were notified in nearly half of all depictions. Due to their fixed nature, the use of static privacy areas only had a lower effectiveness. The use of face recognition decreased the number of false positive notifications to a more manageable amount.

7. RELATED WORK

PRIMO [5] is a mashup service that searches for photos a user is depicted in by querying different social network services. It uses face recognition based on pre-defined training data and semantic annotations like age or gender to identify persons in published photos. The prototype interfaces to MySpace, Flickr and Facebook to retrieve lists of friends as well as corresponding photos.

In previous work [25] we proposed a co-location based service to search for relevant photos. Based on location traces recorded by their private mobile phones users query different photo-sharing services for photos that may have been shot in their vicinity, setting them up for further review by the user. Based on different incentive models and corresponding privacy and authorization concerns, three different types of the service from sharing-service-integrated to a third party service were proposed.

The two services SocialCamera and Locaccino collect the type of information needed for our privacy watchdog. SocialCamera [28] is a mobile app that detects faces in an image and tries to recognize them with the help of Facebook profile pictures of people that are in the user's list of friends. Recognized people can be tagged automatically and images can be instantly uploaded to Facebook. Locaccino [6] is a Foursquare type application that allows users to upload location-based information into Facebook. These two apps are representative for the willingness of users to share this kind of information on the social Web.

In their work [1], Ahern et al. analyze privacy decisions of mobile users in the photo sharing process. They identify relationships between the location of the photo capture and the corresponding privacy settings. They recommend using context information to help users set up privacy preferences and to increase the user awareness of information aggregation. Another work from Fang and LeFevre [8] focuses on helping users to find appropriate privacy settings in social network services. They present a system by which the user initially only needs to set up a few rules. Through the use of active machine learning algorithms the system helps the user to protect private information based on the individual behavior and taste in the future. In [17] Mannan et al. address the problem, that private user data is not only shared within social network services, but also through personal web pages. In their work they focus on a privacy-enabled web content sharing and utilize existing instant messaging friendship relations to create and enforce access policies.

The three works shown above focus on protecting user privacy based on dangers created by users themselves while sharing media. They do not discuss how users can be protected from other peoples' media. This is prevalent for most of the research work done in this area.

Besmer et al [4] present Restrict Other, which allows a user tagged in a photo to send a request to its owner to hide the linked photo from certain people. This approach also follows the idea that forewarned is forearmed and that creating awareness of critical content is the first step towards the solution of the problem. However the work relies on technical links and as such does not cover the same scope as the privacy watchdog presented in this paper.

CoPE as presented by Squicciarini et al. [26] also takes into account other users' media. They postulate that most of the shared data does not only belong to a single user. Therefore they propose a system to share the ownership of media items and by that strive to establish a collaborative privacy management for shared content. Their prototype is implemented as a Facebook app and is based on game theory, rewarding users that promote co-ownerships of media items. While this work does take into account other users' media, unlike our approach it does not cope with previously unknown and unrelated users.

While not directly dealing with privacy protection, the work of Bekkerman et al. is also of interest: In [2], they focus on the problem of disambiguation of people related web searches. With their presented frameworks they strive to disambiguate web appearances of persons in social network services. One approach is to analyze the link structure of web pages. Another one makes use of a multi-way distributional clustering method. A work with the same aim is presented by Jiang et al. [14]. They use a weighted-graph framework that not only disambiguates people but also automatically tags them. Unlike our approach to find relevant people, both these works do not utilize location information in their disambiguation. However, these approaches are complementary and could be combined.

In [30], Zerr et al. present techniques to automatically detect private pictures by using machine learning. They trained their classification models based on textual tags and visual features extracted from a set of pre-labeled public and private photos. After training they used their system to rank privacy of images and for diversing search results according to the privacy dimension. This approach could be utilized to rank images a SnapMe user has to screen.

To the best of our knowledge there is no existing service that notifies users about possible privacy violations by unknown uploaded pictures as the SnapMe watchdog does.

8. CONCLUSION

In this paper we presented an analysis of the threat to a person's privacy created by other peoples' media-sharing activities. By conducting a user survey we showed that users would like to get more information about relevant media shared by other people and that a majority is interested in protecting other peoples privacy as well. We presented an overview of privacy capabilities and metadata handling of prevalent media-sharing services with respect to their capability of protecting users from other peoples' activities. We also conducted an analysis of privacy related metadata, particularly location data, contained in 28k shared photos from two social services. Based on this, we analyzed the privacy implications and potential of the emerging trend of tagged shared media.

We presented a novel solution on how location information can actually help users to stay in control of the flood of both potentially harmful and interesting media uploaded by others. The SnapMe privacy watchdog service combines different techniques to find media relevant to users and notifies them in time of its upload without any requisite action for that user until relevant media has been detected. To demonstrate the feasibility of our approach we evaluated its basic parameters utilizing simulation. The simulation results quantify the tradeoff between the different components of the system and show that it is possible to be notified of a significant amount of privacy relevant events that otherwise would go unnoticed.

In future work we plan to refine the SnapMe mechanisms to reduce the number of false positive notifications. We plan to evaluate the service with a user study in a real world implementation. Additionally, future work will explore details of the applicability of such privacy-privacy tradeoffs.

9. REFERENCES

[1] S. Ahern, D. Eckles, N. Good, S. King, M. Naaman, and R. Nair. Over-exposed?: privacy patterns and considerations in online and mobile photo sharing. In *Proceedings of the SIGCHI conference on Human factors in computing systems*, pages 357–366, 2007.

[2] R. Bekkerman and A. McCallum. Disambiguating Web appearances of people in a social network. In *Proceedings of the 14th international conference on World Wide Web - WWW '05*, page 463, May 2005.

[3] A. Besmer and H. Lipford. Privacy Perceptions of Photo Sharing in Facebook. In *Proc. SOUPS*, 2008.

[4] A. Besmer and H. Richter Lipford. Moving beyond untagging: photo privacy in a tagged world. In *Proceedings of the 28th international conference on Human factors in computing systems*, CHI '10, 2010.

[5] T. Burghardt, A. Walter, E. Buchmann, and K. Bohm. Primo - towards privacy aware image sharing. In *IEEE/WIC/ACM International Conference on Web Intelligence and Intelligent Agent Technology, 2008. WI-IAT '08.*, volume 3, pages 21 –24, December 2008.

[6] Carnegie Mellons Mobile Commerce Laboratory. Locaccino - a user-controllable location-sharing tool. http://locaccino.org/, 2011.

[7] E. Eldon. New Facebook Statistics Show Big Increase in Content Sharing, Local Business Pages. http://www.insidefacebook.com/2010/02/15/new-

facebook-statistics-show-big-increase-in-content-sharing-local-business-pages/, Februar 2010.

[8] L. Fang and K. LeFevre. Privacy wizards for social networking sites. In *Proceedings of the 19th international conference on World wide web - WWW '10*, page 351, Apr. 2010.

[9] I. R. Fasel and J. R. Movellan. A comparison of face detection algorithms. In *Proceedings of the International Conference on Artificial Neural Networks*, ICANN '02, pages 1325–1332, 2002.

[10] A. P. Felt, S. Egelman, and D. Wagner. I've got 99 problems, but vibration ain't one: A survey of smartphone users' concerns. In *Workshop on Usable Privacy & Security for Mobile Devices, Symposium On Usable Privacy and Security*, SOUPS '12. ACM, 2012.

[11] Flickr. Camera Finder. http://www.flickr.com/cameras, October 2011.

[12] T. Hawk. Is There a Major Security Hole in Flickr's New "Geo-Fences" Feature? http://thomashawk.com/2011/08/is-there-a-major-security-hole-in-flickrs-new-geo-fences-feature.html, August 2011.

[13] B. Henne, C. Szongott, and M. Smith. Towards a mobile security & privacy simulator. In *Open Systems (ICOS), 2011 IEEE Conference on*, Sept. 2011.

[14] L. Jiang, J. Wang, N. An, S. Wang, J. Zhan, and L. Li. Two birds with one stone. In *Proceedings of the 18th international conference on World wide web - WWW '09*, page 1201, Apr. 2009.

[15] P. Kumaraguru and L. F. Cranor. Privacy Indexes: A Survey of Westin's Studies. pages 1–22, Dec. 2005.

[16] S. Mahmood and Y. Desmedt. Poster: Preliminary analysis of google+ privacy. In *Proceedings of the 18th ACM conference on Computer and communications security*, CCS '11, pages 809–811. ACM, 2011.

[17] M. Mannan and P. C. van Oorschot. Privacy-enhanced sharing of personal content on the web. In *Proceeding of the 17th international conference on World Wide Web - WWW '08*, page 487, April 2008.

[18] Metadata Working Group. Guidelines for handling image metadata. version 2.0. http://www.metadataworkinggroup.org/pdf/mwg_guidance.pdf, November 2010.

[19] R. J. Moore. Instagram now adding 130,000 users per week: Analysis. http://techcrunch.com/2011/03/10/instagram-adding-130000-users-per-week/, Mar 2011.

[20] MSDN Library. People tagging overview. http://msdn.microsoft.com/en-us/library/ee719905(v=VS.85).aspx, July 2011.

[21] Onvab. The business of facebook: Facts, users statistics & their usage trends. http://onvab.com/blog/facebook-users-statistics-usage-trends/, 10/2012.

[22] R. Paul. First look: Android 4.0 SDK opens up face recognition APIs. http://arstechnica.com/gadgets/news/2011/10/first-look-android-40-sdk-opens-up-face-recognition-apis.ars, October 2011.

[23] S. Perez. Facial Recognition Comes to iOS 5 via New Developer Tools. http://readwrite.com/2011/07/27/facial_recognition_comes_to_ios_5, 2011.

[24] P. J. Phillips, W. T. Scruggs, A. J. O'Toole, P. J. Flynn, K. W. Bowyer, C. L. Schott, and M. Sharpe. Frvt 2006 and ice 2006 large-scale results, 2007.

[25] M. Smith, B. Henne, C. Szongott, and G. von Voigt. Big data privacy issues in public social media. In *2012 Digital Ecosystems and Technologies Conference (DEST)*, 2012.

[26] A. C. Squicciarini, H. Xu, and X. L. Zhang. Cope: Enabling collaborative privacy management in online social networks. *J. Am. Soc. Inf. Sci. Technol.*, 62(3):521–534, Mar. 2011.

[27] Sysomos Inc. Replies and Retweets on Twitter. http://www.sysomos.com/insidetwitter/engagement/, September 2010.

[28] Viewdle. SocialCamera. http://www.viewdle.com/products/mobile/.

[29] A. Vogel. AvPicFaceXmpTagger 1.7. http://www.anvo-it.de/wiki/avpicfacexmptagger:main, 2010.

[30] S. Zerr, S. Siersdorfer, J. Hare, and E. Demidova. Privacy-aware image classification and search. In *SIGIR '12: Proceedings of the 35th international conference on Research and development in information retrieval*, Aug. 2012.

APPENDIX

A. SNAPME SIMULATION

To gain basic insights about the effectiveness and different parameters of the proposed SnapMe watchdog service, we built a proof-of-concept implementation. We implemented the user registration, the upload mechanism for virtual geo-tagged photos, and the co-location and face checks of the SnapMe service on a real web server. Users and photographing were simulated using the Mobile Security & Privacy simulator [13]. Simulated people moved around, took virtual photos and uploaded them via HTTP to the SnapMe prototype. Additionally, simulated users pushed their location updates to the SnapeMe service. The complete simulation setup is shown in Figure 8.

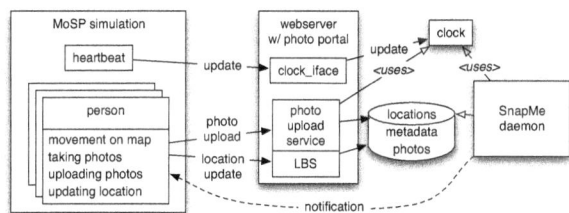

Figure 8: SnapMe prototype and simulation of users and photos

Use of the MoSP simulator enabled us to test the SnapMe prototype in different usage scenarios with different user numbers and configurations. In the presented simulations people move around downtown Chicago. If they pass a coffee shop or specific public places, they randomly stop for a while and move on later. Only SnapMe users and photographers are simulated for efficiency reasons. Other people are not considered in the simulation, since these would neither upload photos to SnapMe nor would they be registered for notifications. Within the simulation, some users take photos and upload them to the SnapMe media-sharing site while others are only consumers of the notification system. Photos are taken with mobile phones with standard lenses having an angle of view of 62° and a photo distance of up to

10 meters. Location updates are made periodically starting at different simulation times. Each run simulated 8 hours of activity.

Based on statistics about Instagram [19] and Facebook [21], simulated photographers take on average one photo every 43 hours. Usage statistics of Instagram as one of the booming media-sharing apps show that in March 2011 the average time between any two uploads made by a single user was 43 hours. Facebook had 1 billion users in September 2012 and about 50 percent of its users logged in on any given day. On average more than 300 million photos were uploaded to Facebook every day. This led to an average between 0.6 and 0.3 photos per person every day, placing the time between any two uploads by a single user to be between 40 and 80 hours. Assuming that predominantly active users upload photos, the Facebook numbers confirm those on Instagram that we also chose for simulation. We also took the fraction of 35% of all users that upload photos at all from the Instagram statistics.

In our setup photos are generated by the simulation each containing information which people are depicted in the photo, how clearly they are depicted, as well as time and GPS location of the photo. People are represented by a unique id. The uniformly distributed clarity of their face determines if a person is recognized by SnapMe or not. A later implementation of the service including real face recognition also could be evaluated using the simulation by generating real pictures with faces from a predefined face gallery.

The simulated face detection decides whether a photo has to be checked by SnapMe or not. Initial Face detection was set very sensitive to avoid missing a photo with people in it. Sensitivity brings along false alarms, but detecting faces in images without any face only increases the number of photos that have to be checked by the service. This slightly influences performance, but not the final privacy results. So we preferred false alarms to misses. Based on [9] we assumed a hit rate of 95% with a false alarm rate lower than 1%. For the following face recognition we assumed a mean false reject rate of 30% and a false accept rate of 1% based on [24]. The SnapMe service uses face recognition to improve person detection and thereby decreases the number of notifications. The false accept rate increases the number of notifications a user has to review. A high false reject rate instead would raise the number of missing notification, thus it should be as low as possible.

A.1 Simulation Results

We evaluated the static privacy area as well as the LBS-based dynamic privacy area with different parameters using the prototypical implementation. In the static setups every user had one static privacy area.

effectiveness in %: notifications per snapped people 5000 people, LBS w/ face recognition average 380 photos with 164 snapped people			
radius / update	5 min.	10 min.	15 min.
10 m	45%	53%	46%
20 m	57%	62%	49%
50 m	78%	60%	54%
100 m	77%	69%	57%

Table 1: 5k users with LBS and face recognition

The best results have been achieved with the LBS-based dynamic privacy area with face recognition as shown in Table 1. The table shows the effectiveness of user notifications (notifications per snapped people) for different radii and location update intervals. The shorter the update interval, the higher the effectiveness. Exceptions for small radii may be caused by blind spots, because we only made co-location checks using the reported locations (cf. Figure 7) and did no interpolation between those. For longer update intervals the effectiveness rises slower with the radius. The LBS-based approach with face comparison is nearly independent from the number of service users (cf. Table 2). The more people are in a photo the more of them are recognized and notified. The service effectiveness only depends on the privacy radius and the location update interval. With an update interval of 15 minutes and a privacy radius of 20m, users were notified for nearly half of all depictions.

effectiveness in %: notified per snapped people (absolut snapped) LBS w/ face recognition, update interval of 5 min. different number of service users (#ppl) with 3 radii				
r/#ppl	500	1k	2k	10k
10 m	67% (3)	100% (2)	74% (34)	27% (677)
20 m	100% (2)	50% (2)	78% (27)	51% (541)
50 m	0% (1)	80% (5)	66% (29)	67% (557)

Table 2: Varying number of user (LBS w/face rec.)

As shown in Table 3, using only static privacy areas only had a low effectiveness. In this case notifications were only sent if users were photographed in their static area, which happened less frequently than if the privacy area followed the walking person. This illustrates that static approaches like Flickr's geofence are not the most effective solutions and are only applicable to hide specific locations from others.

Dynamic and static privacy areas both without the face recognition functionality may only be applicable for areas with few service users. Otherwise the number of notifications rises to a level that makes it difficult for users to review all pictures in a reasonable amount of time as long as there are no other mechanisms that decrease the number or aggregate notifications.

5000 people, location update: 5 min	radius	# photos	photos with people	# depicted persons	# notifications	# notif. per depicted person
static w/ face recognition	50	390	65	165	6	0.04
	100	383	55	215	15	0.07
	250	381	70	192	37	0.19
static w/o face recognition	50	379	51	139	8940	64.3
	100	385	50	146	18190	124.6
	250	380	60	160	57472	359.2
dynamic w/o face recognition	10	386	63	136	1767	13
	20	392	60	181	4346	24
	50	391	61	156	15965	102.3

Table 3: 5k users is different setups

Differential Privacy in Intelligent Transportation Systems

Frank Kargl

University of Ulm & University of Twente
Ulm, Germany & Enschede, Netherlands
frank.kargl@uni-ulm.de

Arik Friedman, Roksana Boreli

NICTA
Sydney, Australia
givenname.surname@nicta.com.au

ABSTRACT

In this paper, we investigate how the concept of differential privacy can be applied to Intelligent Transportation Systems (ITS), focusing on protection of Floating Car Data (FCD) stored and processed in central Traffic Data Centers (TDC). We illustrate an integration of differential privacy with privacy policy languages and policy-enforcement frameworks like the PRECIOSA PeRA architecture. Next, we identify differential privacy mechanisms to be integrated within the policy-enforcement framework and provide guidelines for the calibration of parameters to ensure specific privacy guarantees, while still supporting the level of accuracy required for ITS applications. We also discuss the challenges that the support of user-level differential privacy presents and outline a potential solution. As a result, we show that differential privacy could be put to practical use in ITS to enable strong protection of users' personal data.

Categories and Subject Descriptors

C.2.1 [**Computer-Communications Networks**]: Network Architecture and Design—*Wireless communication*

Keywords

Differential Privacy; Intelligent Transportation Systems; ITS; Privacy

1. INTRODUCTION

Intelligent Transportation Systems (ITS), i.e., the introduction of information and communication technology into transportation systems, and especially vehicles, are generally considered as means to achieve safer, more efficient, and greener road traffic. While some approaches like Car-to-Car communication are still experimental, use of Floating Car Data (FCD) is a more mature ITS technology that is already deployed in the field in many (proprietary) applications. The idea is to turn a vehicle into a mobile sensor that periodically reports its status to a central backend, like a

Traffic Control Center (TCC), by means of a standardized data set, the FCD record. FCD data includes at minimum a timestamp and the vehicle position, but may also include additional data like speed or on-board information from ABS and ESC sensors to detect, e.g., icy roads.

FCD records are used in a variety of applications ranging from fleet management to insurance and tolling applications. Early adopters of FCD include taxi fleets, e.g., in the city of Vienna, where about 2,100 taxis submit FCD records[1], which are then used by the TCC to gain a fine-grained picture of traffic situation on all major roads.

Despite the benefits of ITS and FCD applications, their use also brings concerns that drivers' privacy may be negatively affected. Therefore, FCD records are anonymized in many applications so that they do not contain information that would allow direct identification of specific drivers or vehicles. While this may be a first step towards privacy protection, some identifiers (at least pseudonymous) must still be retained to enable attribution of two successive FCDs to the same car. Otherwise, car counts will not be reliable. As was proposed in previous works [9], a privacy protection mechanism such as k-anonymity may be applied to prevent disclosure of private information. However, this protection can be circumvented, and detailed mining of the FCD database might still reveal a lot of private information about drivers and driving behavior, as shown, e.g., in [15].

The question we want to investigate in this paper is how privacy can be protected more reliably and provably in the context of such data collections in ITS, while still allowing reasonable use for traffic analysis or dedicated applications like road tolling. To this end we focus on differential privacy [6], a formal definition of privacy that allows aggregate analysis while limiting the influence of any particular record on the outcome, typically through the introduction of noise.

Throughout the paper we focus on the following motivating scenarios in ITS. **Scenario 1: identification of traffic conditions** – assessment of traffic conditions, e.g., by calculating the average speed of cars in a certain road segment. Tasks that rely on aggregate information represent the key scenario we would like to accomplish with differential privacy. **Scenario 2: detection of speeding vehicles** – law enforcement agencies who are granted access to FCD databases may be tempted to leverage this access to track and monitor individual drivers. However, this could deter individuals from participating in such schemes. We will show how differential privacy in ITS can mitigate such

---

[1]http://www.wien.gv.at/verkehr/verkehrsmanagement/verkehrslage/projekt.html

privacy breaches. **Scenario 3: eTolling fee calculation** – some applications may nevertheless require access to detailed FCD records, for example, to calculate a road toll based on tracks of journeys. Such applications could be addressed by complementing security mechanisms, beyond differential privacy.

In this paper, we address the challenges in applying differential privacy in practical ITS applications and provide the following contributions:

1) We propose an architecture that integrates differential privacy and additional security mechanisms to provide a comprehensive solution to privacy in ITS.

2) We demonstrate how differentially private mechanisms can be utilized in ITS applications, addressing the accuracy requirements of these applications.

3) We investigate how the privacy parameters can be calibrated within application accuracy requirements, while also considering long-term privacy consequences for the end-user.

2. BACKGROUND AND RELATED WORK

2.1 Privacy Enhancing Technologies in ITS

Protection of private data in ITS has been addressed in the past, often focusing on singular applications and scenarios. As one example, Troncoso et. al. [14] addressed the challenge of privacy-preserving Pay-As-You-Drive (PAYD). Instead of submitting FCD records to the insurance company and having the insurance company calculate the resulting fee, the PriPAYD scheme foresees a trustworthy hardware box installed in the vehicle, which calculates the fee and submits it to the insurance company but without revealing any FCD data. The FCD records are instead given to the driver on USB stick in encrypted form together with a share of the secret key. The second half of the key is given to the insurance company. In case of dispute, both key shares can be combined and the FCD data can be accessed. This way, the driver has full control of the data and can explicitly agree to reveal it to the insurance company.

While many of these approaches achieve the goals of the individual scenario, they have the drawback that they are highly specific and cannot easily be generalized to arbitrary data and arbitrary data processing. Furthermore, the privacy protection relies on the fact that all data processing happens in one On-Board Unit (OBU) and that data leaking from this OBU can be controlled and monitored by the driver. Processing that requires combination of FCD data from different vehicles (e.g., average speed of all vehicles in a given road segment) does not fit into this architecture.

The EU FP7 project PRECIOSA proposed a different approach to privacy preserving data processing in ITS [10, 11]. The *PRECIOSA Privacy-enforcing Runtime Architecture* (PeRA) foresees protection of personal data by augmenting these data with privacy policies and mandatory enforcement of these policies in a distributed system. Whenever personal data are used or communicated, there should also be a policy expressed in the *PRECIOSA Privacy Policy Language* (P3L) that describes the operations allowed on these data. Applications access the data via a dedicated query interface using a SQL-like language called *PRECIOSA Privacy aware Query Language* (PPQL). The *Policy Control Monitor* (PCM) checks the compliance of queries with policies of affected data and either grants or denies access. PeRA is designed to work locally or in a distributed system, the latter case creating a policy enforcement perimeter that can span multiple systems. Within the boundaries of the perimeter, data subjects can rest assured that their personal data are only used in a policy compliant way.

In PeRA, a vehicle transmits data like FCD records together with policies through a confidential communication channel to the importer of a Traffic Control Center. Both data and policy are stored in an encrypted way in the repository and are only accessible via the PCM. PPQL queries can be issued by applications via the Query-API. This approach provides a generic solution to support arbitrary ITS applications, data formats, and operations. It could easily be combined with schemes like PriPAYD to ensure policy compliant data processing in the OBUs and backends.

The concept of differential privacy promises to set hard limits to privacy loss when contributing personal data to a database. However, it has not yet been applied to ITS and its specific applications. In this paper, we will explore how the concept of differential privacy can practically be integrated into the PRECIOSA PeRA framework to provide stronger privacy guarantees for FCD-like applications.

2.2 Differential Privacy

Differential Privacy [6] is a formal definition of privacy that allows computing fairly accurate statistical queries over a database while limiting what can be learned about single records. The privacy protection is obtained by constraining the effect that any single record could have on the outcome of the computation.

DEFINITION 2.1 ((ϵ, δ)-DIFFERENTIAL PRIVACY [5]). *A randomized computation M maintains (ϵ, δ)-differential privacy if for any two multisets A and B with symmetric difference of a single record (i.e., $|A \Delta B| = 1$), and for any possible set of outcomes $S \subseteq Range(M)$,*

$$Pr[M(A) \in S] \leq Pr[M(B) \in S] \cdot \exp(\epsilon) + \delta ,$$

where the probabilities are taken over the randomness of M.

Setting $\delta = 0$ amounts to ϵ-differential privacy.

The ϵ parameter controls the privacy/accuracy tradeoff, as it determines the influence that any particular record in the input could have on the outcome. The δ parameter allows ϵ-differential privacy to be breached in some rare cases.

Differentially private computations can be composed, as shown in [5]: a series of n computations, where computation i is (ϵ_i, δ_i)-differentially private, will result in the worst case in a computation that is ($\sum \epsilon_i, \sum \delta_i$)-differentially private. Therefore, when records enter and leave the database frequently, it is possible to ensure (ϵ, δ)-differential privacy for each record by monitoring the computations performed over the database while the record was in it, and ensuring that the sum of privacy parameters for these computations does not exceed the ϵ and δ bounds.

In this work we focus on *event-level privacy* [8], where the privacy protection is with respect to single records in the database, as in Definition 2.1. In contrast, *user-level privacy* [8] considers the combined effect of all records in the database that pertain to a specific user (or vehicle, in our case). When the number of these records is bounded by c, ϵ-differential event-level privacy amounts to $c \cdot \epsilon$-differential user-level privacy due to composability. In Section 4 we further discuss the user-level privacy.

2.2.1 Privacy Through Perturbation

One of the prevalent methods to achieve differential privacy is the Laplace mechanism [6], in which noise sampled from Laplace distribution is added to the value of a computed function. The probability density function of the Laplace distribution with zero mean and scale b is $f(x) = \frac{1}{2b} e^{-\frac{|x|}{b}}$, and its variance is $2b^2$. The noise is calibrated to the global sensitivity of the function, which is the maximal possible change in the value of the function when a record is added to the database or removed from it.

THEOREM 2.1 (LAPLACE MECHANISM [6]). Let $f : D \to \mathbb{R}^d$ be a function over an arbitrary domain D. Then the computation $M(X) = f(X) + (Laplace(S_G(f)/\epsilon))^d$, where $S_G(f) = \max_{A \triangle B = 1} \|f(A) - f(B)\|_1$, maintains ϵ-differential privacy.

EXAMPLE 2.1. Consider a database of FCD records, where each record includes the speed of a car in km/h. The speed is a number between 0 and 120, and any reported speed outside this range is clamped. Then the following approximations maintain ϵ-differential privacy: 1) Calculating the number of FCD records in the database: $\mathtt{Count(*)} + \mathrm{Laplace}(1/\epsilon)$; 2) Calculating the sum of reported speeds: $\mathtt{Sum(speed)} + \mathrm{Laplace}(120/\epsilon)$; 3) Calculating the average speed of cars: $\frac{\mathtt{Sum(speed)} + \mathrm{Laplace}(240/\epsilon)}{\mathtt{Count(*)} + \mathrm{Laplace}(2/\epsilon)}$. In the last example, we combine two queries, where each query maintains $\frac{\epsilon}{2}$-differential privacy.

3. CHALLENGES IN THE APPLICATION OF DIFFERENTIAL PRIVACY TO ITS

While differential privacy allows to reason formally on the privacy guarantees, it also poses some challenges that may hinder its application in practical systems like ITS.

Computing global-sensitive functions: The `Count`, `Sum` and `Average` functions capture many of the calculations utilized in ITS, and can be evaluated accurately with differential privacy, enabling, e.g., Scenario 1. However, `Max` and `Min` are also valuable functions (e.g., evaluate the speed of the slowest and fastest vehicles in a road section), but have high global sensitivity. Consequently, applying the Laplace mechanism as in Theorem 2.1 to evaluate these functions would provide useless results. We discuss in Section 4.2.1 how techniques relying on local sensitivity [13] can be adapted to overcome this limitation in typical scenarios.

Supporting applications that require precise information: Some applications of ITS require access to precise information. For example, calculating eTolling fees (Scenario 3) is an application, where introduction of noise may be unacceptable as it may result in wrong bills[2]. Noise may also be unacceptable in other applications, such as some safety applications that may have life-and-death consequences. In the scope of this work we focus mainly on applications where noise is acceptable, and even desirable for privacy protection. Other scenarios may be handled through the Controlled Application Environment (CAE), which is part of the existing PRECIOSA framework [4].

Processing time-series data: Differential privacy limits the privacy loss in each query. However, as additional queries are answered by the database, the privacy loss may

accumulate. Since differential privacy maintains composability, it is possible to monitor the overall privacy loss (a worst-case evaluation) and bound it. To address the risk incurred by continuous queries, we describe in Section 4.3.2 an expiry mechanism that ensures that FCD records are removed from the database after participating in a certain amount of queries.

Obtaining user-level privacy: While the privacy loss per FCD record can be monitored and bounded, and thus event-level privacy can be obtained, ensuring user-level privacy is a much more difficult problem. At any point in time, it is possible that multiple FCD records pertaining to the same vehicle (and driver) would be retained in the system and new records that correspond to the same vehicle may be added to the database. Consequently, while differential privacy may prevent an adversary from learning of a specific FCD record that indicates speeding, it does not necessarily prevent from learning that a specific vehicle is frequently speeding. There are theoretical bounds [7] that indicate that such leaks cannot be prevented while still keeping the system usable. However, we use similar arguments in Section 4.3.4 to motivate the choice of ϵ in a way that would quantify this inherent risk.

4. DIFFERENTIAL PRIVACY FOR ITS

In this section, we detail our proposal for a system that enables differentially private use of FCD data for selected ITS applications and services, through an extension of the PRECIOSA PeRA policy enforcement framework.

4.1 System Architecture

The proposed Differential Privacy-enhanced PeRA architecture is shown in Figure 1. For the sake of clarity, we only show the main components relevant to this discussion.

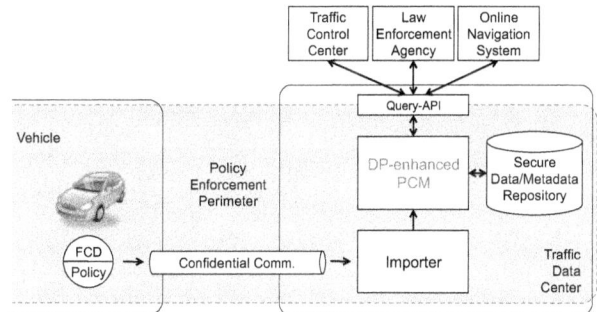

Figure 1: Architecture for enabling the differentially private aggregation of data collected from vehicles in ITS applications.

In line with the existing PeRA architecture, the collection of users' FCD records from the corresponding vehicles is done using a confidential communication channel between the vehicle and the Traffic Data Center (TDC). Collected records are stored in the secure data repository within the TDC. All applications access the FCD data via the Query interface using a set of PPQL queries. As discussed in Section 2, the PRECIOSA P3L policy language already includes the means for expressing, e.g., k-anonymity as a requirement. PPQL enables the formulation of data access queries

---

[2]Though Danezis et. al. [3] proposed a private method for billing, where rebates are issued periodically to compensate for billing errors introduced by differentially private noise.

and the Policy-Control-Monitor (PCM) acts as an enforcement point for privacy control.

The enhancements required to enable differential privacy include the introduction of a DP-Enhanced Policy Control Monitor (DP-enhanced PCM in Figure 1) and the extension of the P3L policy language to enable specifying a set of selected differential privacy parameters, for every FCD or other data record (or set of data records referring to the same event, e.g., position)[3]. These would reflect the level of privacy loss acceptable to the data subject, or as defined by the applicable data protection regulation.

4.2 The Differential-Privacy-enhanced PCM

Differential privacy is suitable for applications that operate on aggregated data, such as the task of assessing traffic conditions outlined in Scenario 1. Such applications access the Traffic Data Center through the Query-API.

In a simple solution, the PCM can use the Laplace mechanism to estimate `Count`, `Sum` and `Average` queries based on their global sensitivity, as was described in Section 2.2. In the next section we demonstrate how additional techniques from the differential privacy literature [13] can be leveraged to evaluate with reasonable accuracy also functions such as `Max` and `Min`, which are frequently used in ITS applications.

4.2.1 Smooth Sensitivity

For some differentially-private computations, the global sensitivity may be too large, and consequently, introducing noise proportional to the global sensitivity would destroy the utility of the computation. For example, the global sensitivity of the max and min functions, computed over values in the range $[0, \Lambda]$, is Λ, and the Laplace mechanism would require adding noise of magnitude Λ/ϵ, consequently destroying utility. To counter this problem, Nissim et. al. [13] proposed adding data-dependent noise. To this end, they defined the *local sensitivity* of a function.

DEFINITION 4.1 (LOCAL SENSITIVITY [13]). *Let* $f : D \to \mathbb{R}^d$ *be a function over an arbitrary domain* D. *The local sensitivity of* f *at point* x *is*

$$LS_f(X) = \max_{Y:d(X,Y)=1} \|f(X) - f(Y)\|_1 \ , \qquad (1)$$

where $d(X,Y)$ *is the distance between datasets.*

Unfortunately, adding noise calibrated to the local sensitivity may still compromise privacy – since the magnitude of noise depends on the data, it becomes a leak channel. To ensure that the magnitude of noise also maintains differential privacy, the concept of *smooth sensitivity* is introduced. While local sensitivity may vary significantly between neighboring datasets, smooth sensitivity changes gradually, and the difference in sensitivity between neighboring datasets is controlled by a parameter β.

DEFINITION 4.2 (SMOOTH SENSITIVITY [13]). *For* $\beta > 0$, *the* β-*smooth sensitivity of* f *at point* x *is*

$$S^*_{f,\beta}(X) = \max_{Y \in D} (LS_f(Y) \cdot exp(-\beta \cdot d(X,Y))) \ . \quad (2)$$

EXAMPLE 4.1. *Let* $X = \{x_1, \ldots, x_n\}$, *where* $0 \leq x_1 \leq \cdots \leq x_n \leq \Lambda$. *The local sensitivity of the function* $f_{\min}(X) = \min(x_1, \ldots, x_n)$ *at point* X *is* $LS_{f_{\min}}(X) = \max(x_1, x_2 - x_1)$.

---

[3] For readability, we will continue our discussion referring just to one FCD record, however other data records or sets of records could be treated the same way.

Nissim et. al. [13] show that the β-*smooth sensitivity of* $f_{\min}$ *at point* X *is:*

$$S^*_{f_{\min},\beta}(X) = \max_{k=0,1,\ldots,n} [\exp(-k\beta) \cdot \max(x_{k+1}, x_{k+2} - x_1)] \ , \quad (3)$$

where $x_k = \Lambda$ *for* $k \geq n$. *Similarly, for* $X = \{x_1, \ldots, x_n\}$, *where* $\Lambda \geq x_1 \geq \cdots \geq x_n \geq 0$, *the* β-*smooth sensitivity of* $f_{\max}$ *at point* X *is:*

$$S^*_{f_{\max},\beta}(X) = \max_{k=0,1,\ldots,n} [\exp(-k\beta) \cdot \max(\Lambda - x_{k+1}, x_{k+2} - x_1)] \ , \quad (4)$$

where $x_k = 0$ *for* $k \geq n$.

Given the β-smooth sensitivity of a function, it is possible to calibrate the noise to obtain a (ϵ, δ)-differentially private output. The following theorem follows from [13]:

THEOREM 4.1 ([13]). *Given* ϵ *and* δ, *set* $\alpha = \epsilon/2$ *and* $\beta = \frac{\epsilon}{2} \cdot \ln(\frac{1}{\delta})$. *Then the computation:*

$$M(X) = f(X) + \text{Laplace}\left(\frac{S^*_{f,\beta}(X)}{\alpha}\right) \quad (5)$$

maintains (ϵ, δ)-*differential privacy.*

EXAMPLE 4.2. *Assume that six cars are stuck in a traffic jam in a road segment, where the speed limit is 90 km/h. Speeds in the FCD database are in the range [0,120]. The cars report the speeds* $\{3, 6, 10, 13, 16, 17\}$. *Evaluating minimum speed with the Laplace mechanism for 1-differential privacy, would require computing* $\min'(X) = 3 + \text{Laplace}(120)$. *In contrast, relaxing the privacy requirement with* $\delta = 0.01$, *for* $(1, 0.01)$-*differential privacy we set* $\alpha = 0.5$ *and* $\beta = 2.3$. *According to Eq. 3,* $S^*_{f_{\min},2.3} = 3$, *hence* $\min'(X) = 3 + \text{Laplace}(6)$ *would still convey that the speed of the slowest car is much lower than expected.*

4.3 Calibrating Privacy Parameters

In this section, we address the calibration of the differential privacy parameters and tracking of privacy loss.

4.3.1 Factors in Parameter Calibration

When a query is executed against the FCD repository, the PCM is required to enforce the privacy policies stated for the affected records. In this process, the following factors should be considered.

Per-application accuracy requirements: ITS applications typically have defined accuracy standards for reporting of selected values. E.g., the Data Quality White Paper [1] published by the U.S. Department of Transport defines the required accuracy of speed reporting for traveller information applications to be in the range of 5-20%. The application requirements represent an upper bound on the variance of the noise introduced by the privacy mechanism for each query, and consequently a lower bound to acceptable values for ϵ and δ.

User-driven privacy settings: The privacy policy attached to each FCD record implies an upper bound on the privacy loss that could be incurred due to participation in queries and correspondingly on the acceptable values for ϵ and δ. As privacy requirements are subjective, acceptable levels of privacy may vary between users. Moreover, future ITS regulations could mandate the default values applicable to all users and all uses of FCD data, e.g., within a specific geographical region.

Affected records: In many functions, the amount of Laplace noise depends only on the privacy parameters, and is not affected by the number of records in the database. Consequently, the relative error may vary depending on the number of queried records. Therefore, to guarantee the required level of data accuracy, the PCM should first verify that enough records participate in the query. In scenarios where a limited number of FCD records are available and / or a lot of queries are issued by applications, there are a number of possible strategies to avoid service disruption due to unavailability of relevant records. These include adapting ϵ to the number of records and based on accuracy demands [16]. In Section 4.3.3 we describe a different approach based on sampling, which is suitable for evaluating average queries.

4.3.2 Managing FCD Lifetimes

The FCD record is the elementary piece of information to which a privacy policy is attached. As noted in Section 2.2, the differential privacy parameter ϵ is composable. If an FCD record participates in a series of queries, where each query q_i is ϵ_i-differentially private, then the overall privacy loss for the FCD record is constrained by $\sum \epsilon_i$. While accuracy requirements imply the acceptable value for ϵ_i in a single query q_i, user-driven privacy settings set a limit on the overall privacy loss $\epsilon = \sum \epsilon_i$ over a period of time.

We assume that FCD records are generated at a constant rate for all vehicles, as is the case with today's systems [1], and that queries are issued at random intervals. We further assume that there is only a limited number of queries during an update interval. To maintain differential privacy for any FCD record in this setting, we rely on two *FCD retention* parameters: *privacy budget* and *expiration time*.

Privacy budget: monitoring a privacy budget is an easy way to ensure that differential privacy requirements are maintained, and was used in frameworks such as PINQ [12] and PDDP [2]. In our architecture, the DP-PCM monitors the privacy budget at the FCD level. Each FCD j has a privacy budget b_j, initially set in the privacy policy attached to the record. For each query q_i, which incurs a privacy loss of at most ϵ_i, the FCD record would participate in the query only if $\epsilon_i \leq b_j$, and consequently the budget will be updated to $b_j \leftarrow (b_j - \epsilon_i)$. If the privacy budget of an FCD record reaches 0, it is removed from the repository.

Expiration time: the privacy policy attached to the FCD record can also state an expiration time, after which the FCD is removed from the repository. Since each vehicle generates new FCD records at a constant rate, the expiration time is critical to ensure that only a limited number of FCD records that originated from the same vehicle reside in the repository at the same time. We will discuss the impact of expiration time on user-level privacy in section 4.3.4.

4.3.3 Example: Evaluating Traffic Conditions

To demonstrate how the PCM can address the accuracy requirements of an ITS application while maintaining privacy constraints, we focus on Scenario 1.

Consider a route guidance application that queries FCD records to determine the average speed on a stretch of road, and accepts a 10 % deviation in the resulting speed. The PCM can use the Laplace mechanism as described in Section 2.2.1, adding Laplace noise to the result up to the acceptable inaccuracy. In addition, the application could also specify a minimum set size for a query. E.g., FCD records

from at least $n = 50$ vehicles on a 1 km road segment would be sufficient to represent the average speed in an accurate way. Then, the PCM can verify before executing the query that enough records are available to answer the query.

Evaluating the number of records: Given a positive number α, sampling a Laplace distribution with scale b would return a number $-\alpha$ or lower with probability at most $0.5 \exp(-\alpha/b)$ (one-sided error). Therefore, to verify that the number of FCD records in a differentially-private count query is at least n, we can set a safety margin α_c, and set $\epsilon_c = \frac{1}{\alpha_c} \ln \frac{1}{2\zeta}$. With probability at least $1 - \zeta$, if the noisy count returns a number greater than $n + \alpha_c$, then there are at least n records in the dataset.

EXAMPLE 4.3. *Assume a safety margin of $\alpha_c = 10$, and set $\zeta = 0.05$. Then, executing a differentially private query with $\epsilon_c = \frac{1}{\alpha_c} \ln \frac{1}{2\zeta} = 0.23$, and obtaining a result of 60 or greater, guarantees with probability at least 0.95 that there are at least 50 FCD records in the database. If any smaller number of records is returned, we abort the query evaluation.*

Executing the *average* query: Once the PCM verifies that there are enough records in the dataset, the actual query can be issued, based on a sample of records with the required size[4]. With probability at most ζ, the two-sided error induced by the Laplace noise with scale b is bounded by $b \ln \frac{1}{\zeta}$. Therefore, the accuracy requirement and the records number bound can be used to derive a bound on the ϵ_s used to evaluate the average speed.

EXAMPLE 4.4. *Assume that there are more than 50 FCD records in the repository, and we would like to evaluate the average speed within 10 % deviation based on a sample of 50 records, where each record holds a value in the range $[0, 120]$. A differentially-private sum query would require Laplace noise of scale $120/\epsilon_s$, and over 50 records, the magnitude of noise added to the sum query should be at most 500. Therefore the PCM should set $\epsilon_s = \frac{120}{500} \ln \frac{1}{\zeta}$. For example, to ensure the bounded deviation with probability 0.95, ϵ_s should be set to at least 0.72.*

Algorithm 1 summarizes the process. For the count evaluation, we take a safety margin α that amounts to 10% of the minimum required record-set size, and the same probability bound ζ as the one used for speed accuracy, but any other reasonable values could be used instead.

4.3.4 Implications for User-Level Privacy Loss

User level privacy, as discussed in Section 2.2, is in general difficult to guarantee when many records are associated with each user, due to the level of noise that would be required in the differentially private functions. However, possible privacy threats can be considered when determining the privacy budget for each FCD record.

As an example, in line with Scenario 2, assume that the police tries to use the system to track down reckless drivers who consistently drive 20 km/h over the speed limit, and

---

[4]In the low-probability case where the noisy evaluation determines there are enough records although their number is below the limit, the query can either be executed on the smaller set, or dummy records with random values can be generated to reach the limit. In either case accuracy will suffer, but privacy would still be maintained.

Algorithm 1: AverageSpeed(P, Λ, n, α_s, ζ)

Input :
$\quad$ P – a road segment for which the average speed
$\quad$ should be evaluated,
$\quad$ Λ – upper bound for speed values,
$\quad$ n – lower bound on number of vehicles to aggregate,
$\quad$ α_s – accuracy bound for speed,
$\quad$ ζ – probability bound for accuracy.

1: $\alpha_c = 0.1n$; $\epsilon_c = \frac{1}{\alpha_c}\ln\frac{1}{2\zeta}.$; $\epsilon_s = \frac{1}{\alpha_s}\ln\frac{1}{\zeta}.$
2: Let RS be the set of all FCD records (one record per vehicle) reported in road segment P, such that for each record r_i with privacy budget b_i, we have $b_i \geq \epsilon_c + \epsilon_s$.
3: $count \leftarrow |RS| + \text{Laplace}(1/\epsilon_c)$.
4: $\forall i \in RS$: $b_i \leftarrow b_i - \epsilon_c$.
5: **if** $count \leq n + \alpha_c$ **then** abort query.
6: Let RS_n be a sample of n records from RS.
7: $avg \leftarrow \left(\text{SumSpeed}(RS_n) + \text{Laplace}(\Lambda/\epsilon_S)\right)/n$.
8: $\forall i \in RS_n$: $b_i \leftarrow b_i - \epsilon_s$.
9: **return** avg.
---

that 2 % of the drivers fall into this category[5]. By querying the system the police aims to conclude that a certain driver is reckless with probability 0.99. From a user u's perspective, it may be desirable to stay "below the radar." Denoting the predicate "u is a reckless driver" with R_u, in differential privacy terms, this could be formulated as follows:

$$\Pr(R_u|DB \cup \text{FCD}_u) \leq \Pr(R_u|DB) \cdot \exp(\epsilon) \ . \quad (6)$$

For any series of queries that maintains ϵ-differential privacy with $\epsilon \leq \ln \frac{\Pr(R_u|DB \cup \text{FCD}_u)}{\Pr(R_u|DB)} \approx 3.9$, the user can avoid being detected by the police.

With respect to this benchmark, it is now possible to interpret the implications of the privacy parameters in terms of the susceptibility of the user to such inferences. For example, if the ϵ per query is 0.01, a new FCD record is generated every 5 minutes and deleted after 5 minutes (so at any time there is only one FCD record in the database per vehicle), an average driving time of one hour each day means that the police would need to monitor the FCD database for more than a month ($\frac{3.9}{0.01 \cdot 12} = 32.5$ days) before it can infer that a certain driver is reckless with high level of confidence. However, the interpretation of the privacy settings in terms of "monitoring period prior to breach" should serve only as a way to roughly judge the implications of different privacy settings in a very restricted scenario, and should not be assumed to reflect a privacy guarantee for a concrete user.

5. CONCLUSION AND FUTURE WORK

In this paper, we have discussed the application of differential privacy to the field of Intelligent Transportation Systems, especially considering the protection of Floating Car Data. As we have shown, event-level differential privacy can be integrated into a policy-enforcement framework

---
[5]According to a report from the U.S. Department of Transport (http://www.nhtsa.gov/staticfiles/nti/pdf/811647.pdf), on limited access highways in the U.S., 20% of drivers exceed the speed limit by more than 10 mph. Although we are not aware of numbers reflecting consistent severe speeding, for the sake of the example we believe our assumptions to be reasonable.

like PRECIOSA PeRA in a straightforward way. We have illustrated how policies could be extended by expiration time and privacy budget parameters to specify and enforce a certain level of differential privacy. Implementing user-level privacy is more challenging and may involve limits to how much data can be stored about any specific vehicle at any time.

6. REFERENCES

[1] AHN, K., RAKHA, H., AND HILL, D. Data quality white paper. Tech. Rep. FHWA-HOP-08-038, U.S. Department of Transportation, Federal Highway Administration, June 2008. Accessed on August 2012.

[2] CHEN, R., REZNICHENKO, A., FRANCIS, P., AND GEHRKE, J. Towards statistical queries over distributed private user data. In *NSDI* (2012).

[3] DANEZIS, G., KOHLWEISS, M., AND RIAL, A. Differentially private billing with rebates. In *Information Hiding* (2011), pp. 148–162.

[4] DIETZEL, S., KOST, M., SCHAUB, F., AND KARGL, F. CANE: A Controlled Application Environment for Privacy Protection in ITS. In *ITST* (2012).

[5] DWORK, C., KENTHAPADI, K., MCSHERRY, F., MIRONOV, I., AND NAOR, M. Our data, ourselves: Privacy via distributed noise generation. In *EUROCRYPT* (2006), pp. 486–503.

[6] DWORK, C., MCSHERRY, F., NISSIM, K., AND SMITH, A. Calibrating noise to sensitivity in private data analysis. In *TCC* (2006), pp. 265–284.

[7] DWORK, C., NAOR, M., PITASSI, T., AND ROTHBLUM, G. N. Differential privacy under continual observation. In *STOC* (2010), pp. 715–724.

[8] DWORK, C., NAOR, M., PITASSI, T., ROTHBLUM, G. N., AND YEKHANIN, S. Pan-private streaming algorithms. In *ICS* (2010), pp. 66–80.

[9] GRUTESER, M., AND GRUNWALD, D. Anonymous usage of location-based services through spatial and temporal cloaking. In *MobiSys* (2003), USENIX.

[10] KARGL, F., DIETZEL, S., SCHAUB, F., AND FREYTAG, J.-C. Enforcing privacy policies in cooperative intelligent transportation systems. In *Mobicom 2009 (Poster Session)* (September 2009).

[11] KARGL, F., SCHAUB, F., AND DIETZEL, S. Mandatory enforcement of privacy policies using trusted computing principles. In *Privacy 2010* (March 2010).

[12] MCSHERRY, F. Privacy integrated queries: an extensible platform for privacy-preserving data analysis. *Commun. ACM 53*, 9 (2010), 89–97.

[13] NISSIM, K., RASKHODNIKOVA, S., AND SMITH, A. Smooth sensitivity and sampling in private data analysis. In *STOC* (2007), pp. 75–84.

[14] TRONCOSO, C., DANEZIS, G., KOSTA, E., BALASCH, J., AND PRENEEL, B. PriPAYD: Privacy-friendly pay-as-you-drive insurance. *IEEE Trans. Dependable Sec. Comput. 8*, 5 (2011), 742–755.

[15] WIEDERSHEIM, B., KARGL, F., MA, Z., AND PAPADIMITRATOS, P. Privacy in inter-vehicular networks: Why simple pseudonym change is not enough. In *WONS* (February 2010).

[16] XIAO, X., BENDER, G., HAY, M., AND GEHRKE, J. iReduct: differential privacy with reduced relative errors. In *SIGMOD Conference* (2011), pp. 229–240.

Private Proximity Testing with an Untrusted Server

Gokay Saldamli, Richard Chow, Hongxia Jin, Bart Knijnenburg
Samsung Research America – Silicon Valley
{gokay.s,richard.chow,hongxia.jin,bart.k}@sisa.samsung.com

ABSTRACT

The privacy of location-based services has gained attention with their increased popularity. To date, citing insufficient privacy demand and inefficient/immature privacy preserving technologies, service providers have not been willing to build private-enhanced systems in which they do not have access to users' location information. However, current practice is likely to change in coming years with increasing privacy awareness and technological advances. For instance, Narayanan et al. recently introduced a fast private equality testing protocol for proximity testing with an untrusted server. In the current work, based on basic notions of geometry and linear algebra, we describe a new three-party protocol for solving the same problem. Our proposed protocol decreases the number of encryptions needed and gives a more efficient solution for private equivalence testing.

Categories and Subject Descriptors

C.2.0 [**Computer-communication Networks**]: General—*Security and protection*

Keywords

Private equality testing, proximity testing, location privacy.

1. INTRODUCTION

Due to recent privacy breaches, end users as well as legal and government entities are increasingly concerned about privacy (for example, see the US hearing on mobile privacy [3] and the Federal Trade Commission report on consumer privacy [13]). Building privacy-aware systems starts with defining a clear data sharing policy that describes the relationship between users and service providers. However, current policies and practices of service providers are biased against privacy since the commercial value of the service providers is built on top of users' (private) information. Therefore, citing insufficient privacy demand and inefficient/immature privacy preserving technologies, most ser-

vice providers are not willing to build truly private systems, as that would compromise their ability to capture user data [1]. Nevertheless, we believe this practice is likely to change in the coming years with increasing privacy awareness and technological advances. In fact, there is much active research in privacy-preserving technologies which make services possible while protecting privacy.

In particular, preserving *location privacy* stands out in this field because location information is one of the most sensitive attributes for user targeting and personalization. In terms of giving better services, location data is collected by service providers and served to third parties for targeted recommendation or advertising, sometimes even without permission [5]. Legal and governmental entities have attempted to regulate service providers (e.g., see the Location Privacy Protection Act of 2011 [2] or the California Location Privacy Act of 2012 [4]), but there is a delicate balance between regulation and service quality and user satisfaction.

The research on location privacy problem has been quite diverse, bringing expertise together from different fields, for example, signal processing and cryptography [9]. The most notable approaches include:

- *geographical masking:* Deterministic noise is added to the location data to hide the original location [18].

- *cloaking:* The user's geographical data are reported at lower resolutions than initially recorded [15].

- *mix zones:* User identities are anonymized by restricting location of identification [7, 14].

- *aggregation:* Users' data are grouped in order to reduce the identification of individuals [8].

- *statistical privacy:* The user's data are obfuscated so that statistical processing is still possible [10, 11].

- *encryption:* The user's location is encrypted. This approach allows measurable security metrics but may require third-parties [26, 23, 21, 20].

The current work studies the use of lightweight encryption methods in comparing private information without leaking it, specifically useful for sharing location between two parties only if the two parties are nearby. In fact, previous work utilizing public-key or homomorphic encryption schemes has the same goals (for instance, see [16, 6, 19]). These methods have high security levels but require intense computations. Hence, they fit well for problems that involve a limited number of users, but designing systems with these methods for millions of users or items is not yet possible.

Recently, Narayanan et al. [22] proposed a fast private equality testing (FPET) protocol with an oblivious server for location privacy. Introducing an oblivious server, their setup is slightly different than the usual private equality testing problem. To be more specific: they assume Chooser-Alice and Sender-Bob have some private values and Alice wants to learn whether Bob's private value and hers are the same, with Alice not learning Bob's values if the values are not the same. A third party (honest but curious) called Server helps them carry out the equality test, but tested values have to be hidden from Server, and Bob should not learn the outcome of the test. Their protocol involves lightweight cryptographic primitives, and hence addresses traditional performance and scalability problems.

Based on basic notions of geometry and linear algebra, we describe a similar three-party protocol for solving the same problem with the same settings, called Vectorial Private Equality Testing (VPET). Our proposed protocol significantly decreases the number of encryptions required and is more efficient than FPET. After briefly reviewing the FPET protocol in the next section, a formal presentation of our new protocol is given in Section 3, which is followed by a discussion of security and efficiency in Sections 4 and conclusion in Section 5.

2. PRIVATE EQUALITY TESTING

Private equality testing (PET) is also known as "comparing information without leaking it". More intuitively, some call it the "Socialist Millionaire Problem". The problem [17] is one in which two millionaires want to determine if their wealth is equal, without disclosing any information about their riches to each other. In fact, it is a variant of the Millionaire's Problem [24, 25] where two millionaires compare their riches to determine who has the most wealth, without disclosing any information about their riches to each other.

We describe the PET problem more formally: let Alice be a Chooser who wants to know whether her private value w_A is equal to the private value w_B of the Sender (Bob). However, Alice should not learn anything else about w_B, while Bob should not learn the outcome of the test.

In fact, various aspects and use cases for the PET problem were considered in a number of studies [12, 6, 26, 22]. The majority of these solutions are two-party protocols using computationally intensive cryptographic tools such as homomorphic, commutative, or public-key encryption. Thus, for real life apps including location sharing in large online social networks, these solutions are currently impractical.

Narayanan et al.'s fast private equality testing (FPET) protocol is a three-party protocol for private location sharing and uses comparably lightweight symmetric-key primitives. FPET requires far less communication and computation compared to prior work. As we solve the same problem, we briefly go through the FPET protocol. Detailed descriptions can be found in the original paper [22].

2.1 Proximity and grid systems

In principle, the neighborhood of a position can be defined by a circle of a given range, and the proximity testing problem is to test whether a contact is inside or outside of the circle. In other words, checking whether a user is in proximity simply means obtaining the location of the user (e.g., via GPS); calculating the distance to this location and testing if the distance is less than the given range. Although this

calculation can give very accurate results, it involves computing trigonometric operations or square roots that can be expensive. Moreover, if the location values are encrypted, the distance calculation could further require a costly homomorphic encryption scheme.

PET can be deployed for private proximity testing without calculating the distance. However, a direct deployment where one compares every single nearby point would be impractical since a neighborhood of a user's location could contain significant number of points. One solution to this problem is to approximate the area of the neighboring circle with some grid cells containing a large number of location points. These cells have unique identifiers and are part of a grid system covering the whole space. In this case, instead of comparing a significant number of exact locations, the user compares the identifiers of the grid cells approximating the neighboring circle.

For instance, in order to use the private equality method for proximity testing, Narayanan et al. [22] introduced an overlapping hexagonal grid system in which the hexagons are labeled by identifiers. In their system, a location lies in at most three overlapping grids. Thus, one runs the protocol three times according to this location labeling scheme. In fact, there are other labeling strategies for better neighborhood coverage (e.g., see [21, 26, 23]), but these increase the number of cells to consider, directly increasing the number of protocol runs. In this work, the method of partitioning the world is not of interest; we assume the most meaningful method is employed for a given application and concentrate on a private equality method for the identifiers.

2.2 FPET protocol

Let $\mathbb{Z}_p$ denote the set $\{0, \cdots, p-1\}$ and the keys k_{ab}, k_a and k_b secretly shared by User A and User B, User A and Server, and User B and Server, respectively. These keys are used by a Pseudo Random Function (PRF) denoted by $E_k(x)$, where k is the PRF key and x is the point at which the function is evaluated. We have the following equations:

$$(k_1, k_2) \Leftarrow E_{k_{ab}}(ctr) \quad \text{and} \quad r \Leftarrow E_{k_b}(ctr), \quad (1)$$

for a counter, ctr incremented once Bob changes location.

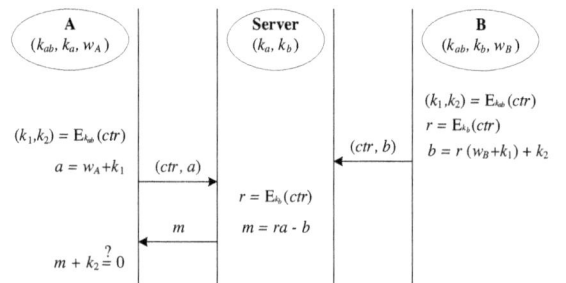

Figure 1: FPET protocol

Assuming that Bob wants to share his proximity of location with Alice, Bob periodically increments ctr by one and computes the values in (1). In order to have exact computations, Bob parses the result so that k_1, k_2 are elements in $\mathbb{Z}_p$ and r is non-zero in $\mathbb{Z}_p$. Bob then computes b and sends (b, ctr) to the Server as seen in Fig. 1.

Alice, wondering whether Bob is nearby, first queries the server to obtain the latest value of ctr for Bob. Alice aborts if the *ctr* received from the server is not fresh; otherwise, Alice computes a and sends (ctr, a) to the Server.

After getting these messages, the Server computes $m = ra - b$ and sends m to Alice. Alice computes $m + k_2 = r(w_A - w_B)$. Notice that if $m + k_2$ is equivalent to 0, Alice learns that Bob is nearby; otherwise, she does not learn the location of Bob.

As the authors claim, showing the security of the protocol is easy using some simple secret sharing arguments. However, the protocol as presented is malleable, i.e. an active attacker may inject/change any of the passed messages and the involving parties would not realize the change in the message. Nevertheless, by deploying some simple mechanisms such as noncing or timestamping, these attacks might be prevented. Therefore, for the sake of presentation simplicity, we follow the Narayanan et. al.'s approach and simply assume that the communication channel is authentic.

The FPET protocol requires one AES computation on Alice and Server, while two are needed on Bob. Moreover, the protocol needs three messages which accounts to 12 bytes per edge, assuming p is 32 bits. Other computations including additions and multiplications are negligible. Compared with two-party protocols, FPET is at least 10 times faster and has 100 times less communication cost [22]. Being a three-party protocol, a server is needed as a third party. This might bring extra complexity compared to two-party setups, but it is a good fit for applications that naturally involve a server.

3. VPET PROTOCOL

We describe a three-party (Chooser-Alice, Sender-Bob, and Server) protocol solving the PET problem with the FPET setting. Our protocol makes use of geometry and linear algebra, particularly vector calculations including the inner product operation.

3.1 VPET Vectorization

The geometry and related representation may vary according to the nature of the values to be tested for equality. For instance, for private location equality testing, the values are 2-tuples (i.e., longitude and latitude) on the earth and not ordered. However, in case of the socialist millionaire problem the values are ordered scalars representing funds in bank accounts.

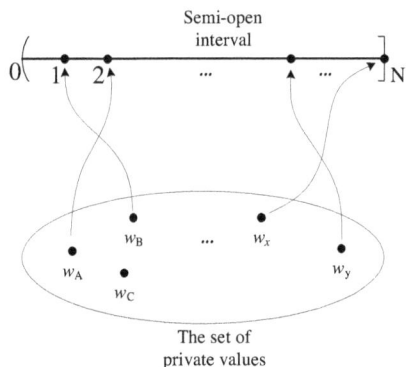

Figure 2: A simple universal map assigning private values to natural numbers.

Without considering the data format, we want to attach

a unique vector for every private value. This vectorization can be done in various ways and could bring up different computational complexities; e.g. the attached vectors can be real or integer valued and might have different arithmetic.

A constructive vectorization starts with a universal map that assigns the private values to a subset of natural numbers. First, we assume the total number of private values N is a multiple of 4 (if not, we may add a few more values and make N a multiple of 4). We enumerate private values in order to have a semi-open interval $(0, N]$ where every integer in this interval represents a private value as seen in Fig. 2.

Next, we embed the line segment into $\mathbb{R}^2$ for vectorization. This may also be done various ways. For example, two ends of the line segment could be bent into a semicircle centered at the origin, and vectors starting from the origin and ending on the semicircle can simply be assigned as private values (see Fig. 3). Note that, with this construction, the angle between consecutive vectors are all equivalent to π/N. Moreover, we do not use the entire circle to exclude the possibility of antipodal vectors.

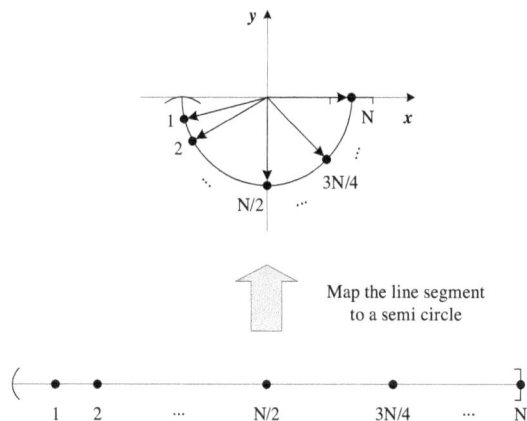

Figure 3: A vectorization using the lower semicircle.

This construction is valid for any finite private value set; however, integer-valued (rather than real-valued) vectors may further increase efficiency. In Section 3.4, we give a suggested vectorization process using only integer values.

Table 1: Notation used for VPET.

N	The number private values; a multiple of 4.
w_A	The vector representing Alice's private value
w_B	The vector representing Bob's private value
u	a random vector perpendicular to w_A
$R(u, \theta)$	An angle preserving map (e.g. rotation)
E	A keyed pseudo random function (PRF)
k	A shared key between A and B
ctr	Counter used for synchronization
θ	Random angle rotating u, computed using E and shared between Alice and Bob.
s	A vector of pseudo random entries computed using E and shared between Alice and Bob.
r	True random value generated by Bob

3.2 VPET Protocol Description

The proposed protocol is based on symmetric key encryption, and hence is fast compared to methods using public-key or homomorphic encryption technologies. Its security assumes the server does not collude with either of the communicating parties. Table 1 gives the notation used in our description of the protocol.

When Alice and Bob declare themselves as friends they set up a shared secret key denoted with k. They also agree on a counter ctr which is initially set to, say, 0. The flow of the protocol is presented in Fig. 4. Assume that Bob wants to share his private value with Alice. Private values are encoded as points on the semi-circle, as described in Section 3.1. Bob increments ctr by one and computes

$$(s,\theta) = E_k(ctr)$$

Bob arranges the output of the PRF, E, in a way that s represents a vector having random entries and θ represents a rotation angle. For simplicity and efficiency, θ should be a multiple of π/N. This can be done simply by choosing a random number from $(0, N]$, and then multiplying by π/N. Next, Bob computes:

$$b = rR(w_B, \theta) + s$$

where r is a non-zero random number, w_B is his private value, and s is taken from the E function. He sends b and ctr to the server.

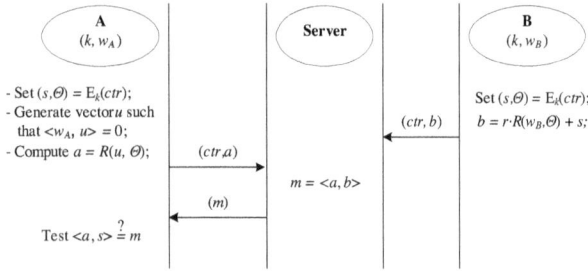

Figure 4: VPET protocol

Alice, in order to compute whether Bob is nearby, first queries the server to obtain the latest value of ctr from Bob. Alice aborts if the ctr received from the server is not fresh, otherwise Alice computes $(s,\theta) = E_k(ctr)$ and a unit vector u perpendicular to w_A (note that there are two such vectors which are antipodal). Alice then blinds u by using the rotation function R, where θ is pseudorandom value by definition. Alice sends $a = R(u,\theta)$ and ctr to the Server.

The Server matches the messages having the same ctr value from Alice and Bob, and performs a single inner product operation giving $m = \langle a, b \rangle$, then sends m to Alice.

$$
\begin{aligned}
m &= \langle a, b \rangle = \langle R(u,\theta), rR(w_B, \theta) + s \rangle \\
&= r\langle R(u,\theta), R(w_B,\theta) \rangle + \langle R(u,\theta), s \rangle \quad (2)
\end{aligned}
$$

Since R is an angle preserving map, $R(w_B, \theta)$ is perpendicular to the vector $R(u,\theta)$ if the private values (vectors) of Bob and Alice are same (i.e., $w_A = w_B$). Notice that in this case the inner product on the left of Eqn. (2) vanishes and only $\langle R(u,\theta), s \rangle$ remains. Alice can compute this value as it does not contain the blinding r. Therefore, Alice computes $\langle R(u,\theta), s \rangle$, and if she finds that $m = \langle R(u,\theta), s \rangle$, she learns that she has the same private vector as Bob.

3.3 Toy Example

In order to describe the process more clearly, we present a toy PET example. Suppose Alice wants to learn whether Bob got the same grade for a math exam. Alice got a grade of $w_A = 70$, located in the lower semicircle in Fig. 5. With the help of a server, Alice learns the grade of Bob, w_B, only if her grade w_A is equal to w_B, learning nothing otherwise. Moreover, the server and Bob learn nothing. Let the grades be $N = 64$ discrete values from the set $[37, 100]$, so in mapping the grade line segment to a semicircle, grade increments are equivalent to $\pi/64$ radians.

We start with shared value generation. If AES is used for pseudorandom generation with key $k = D301E0908128C3B254BDEA4F2C504844$ and $ctr = 3$, the following computation is carried out by two parties after agreeing on the ctr used in the protocol.

$$AES_k(3) = CCD819AB9A623189E540EEEE9006\underline{A1B1}$$

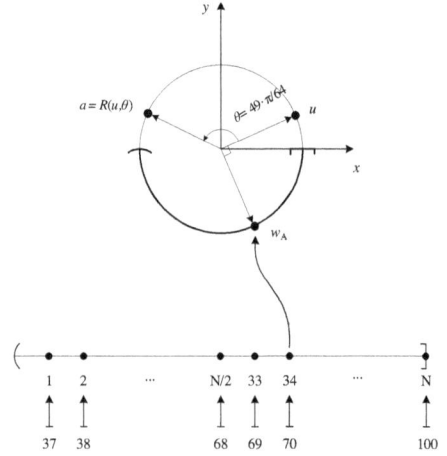

Figure 5: Scalar to vector mapping

Following the pseudo random generation, both Alice and Bob set up the parameters θ and s. This establishment may vary with respect to data used or the security measures. For this example we use the least significant 19-bits of the ciphertext,

$$6A1B1 = \overbrace{\underbrace{110101\,000011}_{s}}^{x \quad\quad y}\underbrace{0110001}_{\theta} \quad (3)$$

and set $\theta = 49$ and $s = (x, y) = (-21, 3)$. Note that, here we use a 6-bit sign-magnitude representation for x and y, however; polar coordinates or floating point representations might also be used.

Assume that Bob also gets 70 from the exam. Bob computes $R(w_B, \theta) = (34 + 49)\pi/64 = 83/64\pi$, and then gets the vector components of $R(w_B, \theta)$ by simply computing $(\cos(83\pi/64), \sin(83\pi/64) \approx (-0.5956, -0.8032)$, where we pick 4-digits of precision. Hence, setting $r = 7$, b is computed as follows:

$$
\begin{aligned}
b &= r \cdot R(w_B, \theta) + s \\
&= 7(-0.5956, -0.8032) + (-3, 21) \\
&\approx (-7.1698, 15.3775)
\end{aligned}
$$

For her part, Alice starts with calculating $a = R(u,\theta)$. The perpendicular vector can be found by simply adding $\pi/2$ to w_A, $u = (w_A + 32)\pi/64 = 66\pi/64$. Hence, $a = u + \theta = (64 + 49)\pi/64 = 115\pi/64$. Alice sends the following a to the Server:

$$\begin{aligned} a &= (\cos(115\pi/64), \sin(115\pi/64)) \\ &\approx (0.8032, -0.5956) \end{aligned}$$

Server takes a and b and computes the inner product and sends the result to Alice

$$\langle a, b \rangle = -7.1698 \cdot 0.8032 + 15.3775 \cdot (-0.5956) = -14.9191.$$

Alice computes

$$\langle a, s \rangle = \langle (0.8032, -0.5956), (-3, 21) \rangle = -14.9191 = \langle a, b \rangle$$

and concludes that Bob got the same grade as her.

3.4 Integer valued Vectorization and VPET

In the previous sections we described an explicit vectorization defined by polar coordinates. However, this vectorization requires subsequent operations on the vectors involving cos and sin functions. Such calculations cause precision errors which makes confirming a match more tricky. Here, we describe another process to vectorize a private value set so that operations are much simpler integer calculations.

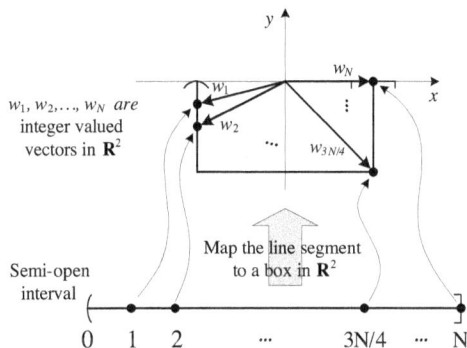

Figure 6: VPET Vectorization Process

The process illustrated in Figure 6 is similar to the one in Sec. 3.1. We let N be a multiple of 4 and map the line segment to a rectangle box in $\mathbb{R}^2$ preserving the distance. Now, the private values lie on the rectangle and we simply assign private values to vectors starting from the origin and ending on these points on the rectangle. Rotation R is defined only on these vectors and their symmetries with respect to the x-axis. There are $2N$ vectors in total and θ is only an integer value in $(0, 2N]$. Notice that this new mapping allows the perpendicular to be easily found, i.e., the perpendicular to $(x, -1)$ is $(1, x)$. Moreover, if r and s are chosen to be integer valued, then the whole protocol can be performed through integer calculations.

Going back to our toy example, Bob having $w_B = (-16, 2)$ should calculate:

$$b = r \cdot R(w_B, \theta) + s = 7(13, 16) + (-3, 21) = (88, 133)$$

On the other side, Alice sends the vector labeled with $w_A + \pi/2 + \theta = 34 + 32 + 49 = 115$ represented by $a = (-16, 13)$ on the regtangle. Server calculates the inner product

$$\langle a, b \rangle = \langle (88, 133), (-16, 13) \rangle = 321$$

and sends the result to Alice, where she computes

$$\langle a, s \rangle = \langle (-16, 13), (-3, 21) \rangle = 321$$

and concludes that Bob got the same grade as herself.

4. SECURITY AND EFFICIENCY

First of all we assume an untrusted server, but a server that is Honest-but-Curious. This assumption is realistic, since violating the honest-but-curious model risks damage to a service provider's reputation. The server correctly routes protocol messages, but does not attempt to create fake users or collaborate with existing users. In particular, for the VPET protocol, server collusion with User B would reveal User A's location.

We see data sub-poenas, hacker break-ins, or inadvertent data release as examples of threats on server-side data. These are the main motivation for the PET type protocols. Threats from users must also be considered, although most of these threats are similar to threats in existing systems.

The security analysis of VPET is quite similar to FPET: both protocols use simple secret-sharing arguments and are based on the indistinguishability principle. We assume the true random number r and the pseudorandom parameters θ and s generated by PRF are cryptographically independent. Moreover, assume that the communication channel is authentic and the server does not collude with either of the communicating parties.

First, Bob does not learn anything since he does not get any messages. Second, Server sees the messages a and b but $a = R(u,\theta)$ and $b = r \cdot R(w_B, \theta) + s$ are blinded by independent variables θ, s, and r. Moreover, the inner product $\langle R(u,\theta), R(w_B, \theta) \rangle$ is implicitly blinded by $\langle R(u,\theta), s \rangle$ in case Server tries to see whether Alice and Bob are at the same place or not. Therefore, the Server cannot deduce anything from these messages. Finally, Alice learns m but if $w_A \neq w_B$ then Alice ends up learning nothing from $r \cdot \langle R(u,\theta), R(w_B,\theta) \rangle$, as she does not know the random value r. However, if Alice colludes with the server she can easily learn Bob's location. Similarly, if Bob colludes with the Server, he can learn Alice's location.

Table 2: Protocol comparison (cost is the number of symmetric encryptions).

Protocol	FPET	VPET
Alice's cost	1	1
Server's cost	1	0
Bob's cost	2	1
Total location data	12 bytes	16 bytes
# of IP packets	3	3

Our protocol decreases the number of encryptions needed for private equality testing (see Table 2). Note that as the number of sharing instances increases, the load on the server might become a bottleneck in deploying private equality sharing. For instance, in a social network with millions of users with hundreds of contacts each, utilizing private sharing could need extensive resources because of the server load. By eliminating the encryption load on the server, VPET gives an efficient solution for private equality sharing. On the other hand, VPET increases the size of the location data slightly (see Table 2) as the vector components have to be transmitted by Bob (in each of three steps, FPET transmits $(4+4+4) = 12$ bytes in total, where VPET transmits $(4+4+8) = 16$ bytes in total). This is a negligible increase in overall transmission cost; for instance, using IP, this would still require 3 IP packets and the additional 4 bytes is small compared to the size of the metadata.

5. CONCLUSION

We present VPET, a new private equality testing protocol using simple geometry and linear algebra. We use the properties of the inner product, basis conversions, and vector operations for efficient and practical equality testing with an oblivious server. The method has equivalent security and transmission costs compared to prior work and yet decreases server load by eliminating the encryption performed on the server. The protocol is a good ingredient for a privacy-preserving location sharing application.

6. REFERENCES

[1] Failcon privacy panel topic: why are location services ignoring these guys? Website, 2010. http://scobleizer.com/2010/10/25/.

[2] Location privacy protection act of 2011. Website, 2011. http://www.franken.senate.gov/.

[3] Senate committee hearing on mobile privacy now underway, watch live. Website, 2011. http://www.engadget.com/2011/05/10/senate-committee-hearing-on-mobile-privacy-/now-underway-watch-1/.

[4] California Location Privacy Act of 2012. https://www.eff.org/cases/california-location-privacy-act-2012.

[5] A. Allan. Got an iPhone or 3G iPad? Apple is recording your moves. O'Reilly Radar: http://radar.oreilly.com/2011/04/apple-location-tracking.html, 2011.

[6] M. J. Atallah and W. Du. Secure multi-party computational geometry. In *Algorithms and Data Structures, 7th International Workshop, WADS 2001*, volume 2125 of *Lecture Notes in Computer Science*, Providence, RI, USA, 2001. Springer.

[7] A. R. Beresford and F. Stajano. Location privacy in pervasive computing. *IEEE Pervasive Computing*, 2(1):46–55, Jan. 2003.

[8] C. Castelluccia. Securing very dynamic groups and data aggregation in wireless sensor networks. In *IEEE International Conference on Mobile Ad-hoc and Sensor Networks (MASS 2007)*, Pisa, Italy, October 2007.

[9] C. D. Cottrill. Location privacy: Who protects? *URISA Journal*, 23(2):49–59, 2011.

[10] S. De Capitani di Vimercati, S. Foresti, S. Paraboschi, G. Pelosi, and P. Samarati. Efficient and private access to outsourced data. In *Proc. of the 31st International Conference on Distributed Computing Systems (ICDCS 2011)*, Minneapolis, MN, June 2011.

[11] C. Dwork. Differential privacy: A survey of results. In *Theory and Applications of Models of Computation, 5th International Conference, TAMC'08*, volume 4978 of *LNCS*, pages 1–19. Springer, 2008.

[12] R. Fagin, M. Naor, and P. Winkler. Comparing information without leaking it. *Communications of the ACM*, 39:77–85, 1996.

[13] Federal Trade Commission. Protecting consumer privacy in an era of rapid change. March 2012. http://ftc.gov/os/2012/03/120326privacyreport.pdf,

[14] J. Freudiger, M. Raya, M. Felegyhazi, P. Papadimitratos, and J. P. Hubaux. Mix-zones for location privacy in vehicular networks. In *Proceedings of the First International Workshop on Wireless Networking for Intelligent Transportation Systems (Win-ITS)*, 2007. QC 20110707.

[15] B. Gedik, K.-L. Wu, P. S. Yu, and L. Liu. Processing moving queries over moving objects using motion-adaptive indexes. *IEEE Trans. on Knowl. and Data Eng.*, 18(5):651–668, May 2006.

[16] C. Hazay and K. Nissim. Efficient set operations in the presence of malicious adversaries. *J. Cryptol.*, 25(3):383–433, July 2012.

[17] M. Jakobsson and M. Yung. Proving without knowing: On oblivious, agnostic and blindolded provers. In *Proceedings of the 16th Annual International Conference on Advances in Cryptology*, CRYPTO '96, pages 186–200, London, UK, 1996. Springer.

[18] M. Leitner and A. Curtis. A first step towards a framework for presenting the location of confidential point data on maps results of an empirical perceptual study. *Journal of Geographical Information Science*, 20(7):813–822, 2006.

[19] H. Lipmaa. Verifiable homomorphic oblivious transfer and private equality test. In *ASIACRYPT'03*, pages 416–433, 2003.

[20] S. Mascetti, C. Bettini, and D. Freni. Longitude: Centralized privacy-preserving computation of users' proximity. In *Proceedings of the 6th VLDB Workshop on Secure Data Management*, SDM'09, pages 142–157, Berlin, Heidelberg, 2009. Springer-Verlag.

[21] S. Mascetti, D. Freni, C. Bettini, X. S. Wang, and S. Jajodia. Privacy in geo-social networks: proximity notification with untrusted service providers and curious buddies. *The VLDB Journal*, 20(4):541–566, Aug. 2011.

[22] A. Narayanan, N. Thiagarajan, M. Lakhani, M. Hamburg, and D. Boneh. Location privacy via private proximity testing. In *Proceedings of the Network and Distributed System Security Symposium (NDSS), San Diego, CA, USA*, 2011.

[23] J. D. Nielsen, J. I. Pagter, and M. B. Stausholm. Location privacy via actively secure private proximity testing. *IEEE Conference on Pervasive Computing and Communications Workshops*, 0:381–386, 2012.

[24] A. C. Yao. Protocols for secure communications. In *Proc. 23rd IEEE Symposium on Foundations of Computer Science (FOCS '82)*, pages 160–164, 1982.

[25] A. C. Yao. How to Generate and Exchange Secrets. In *Proceedings of the 27th Annual Symposium on Foundations of Computer Science (FOCS 1986)*, pages 162–167. IEEE Computer Society, 1986.

[26] G. Zhong, I. Goldberg, and U. Hengartner. Louis, lester and pierre: three protocols for location privacy. In *Proceedings of the 7th international conference on Privacy enhancing technologies*, PET'07, pages 62–76, Berlin, Heidelberg, 2007. Springer-Verlag.

Securing the IP-based
Internet of Things with HIP and DTLS

Oscar Garcia-Morchon, Sye-Loong Keoh, Sandeep S. Kumar, Pedro Moreno-Sanchez
Francisco Vidal-Meca, and Jan Henrik Ziegeldorf
Philips Research Europe, High Tech Campus 34
Eindhoven, The Netherlands
{oscar.garcia, sye.loong.keoh, sandeep.kumar, henrik.ziegeldorf}@philips.com
p.morenosanchez@um.es, francisco.vidal.meca@rwth-aachen.de

ABSTRACT

The IP-based Internet of Things (IoT) refers to the pervasive interaction of smart devices and people enabling new applications by means of new IP protocols such as 6LoWPAN and CoAP. Security is a must, and for that we need a secure architecture in which all device interactions are protected, from joining an IoT network to the secure management of keying materials. However, this is challenging because existing IP security protocols do not offer all required functionalities and typical Internet solutions do not lead to the best performance.

We propose and compare two security architectures providing secure network access, key management and secure communication. The first solution relies on a new variant of the Host Identity Protocol (HIP) based on pre-shared keys (PSK), while the second solution is based on the standard Datagram Transport Layer Security (DTLS). Our evaluation shows that although the HIP solution performs better, the currently limited usage of HIP poses severe limitations. The DTLS architecture allows for easier interaction and interoperability with the Internet, but optimizations are needed due to its performance issues.

Categories and Subject Descriptors

C.2.1 [**Computer-Communications Networks**]: Network Architecture and Design—*network communications*

General Terms

Security; Design; Management; Performance

1. INTRODUCTION

The IP-based Internet of Things (IoT) will enable smart and mobile devices equipped with sensing, acting, and wireless communication capabilities to interact and cooperate with each other in a pervasive way by means of IP connectivity. IP protocols play a key role in this vision since they allow for end-to-end connectivity using standard protocols ensuring that different smart devices can easily communicate with each other in an inexpensive way. Protocols such as IPv6, TCP and HTTP that are commonly used in traditional networks will be complemented by IPv6 over Low power Wireless Personal Area Networks (6LoWPAN) and the Constrained Application Protocol (CoAP) currently in development in IETF. This allows smart and mobile devices used for healthcare monitoring, industrial automation and smart cities to be seamlessly connected to the Internet.

Security and privacy are mandatory requirements for the IP-based IoT in order to ensure its acceptance. The interaction between devices must be regulated during the whole device lifecycle. When a device joins an IoT network, the IoT network has to authorize its joining, such that it is provisioned and configured with the corresponding operational parameters, thus providing *network access*. During operation, devices will interact with each other requiring pairwise session keys, and for that, *key management* is needed.

These security functionalities are the focus of our research, because of two reasons: (i) no standard solution exists yet and the traditional Internet relies on too many different security protocols that are not going to fit on small devices; (ii) device mobility and system scalability are to be considered.

To overcome these issues we propose two security architectures for the IP-based IoT by adapting existing IP security protocols while relying on a single protocol. The first solution is a new variant of the Host Identity Protocol (HIP) that uses pre-shared keys (PSK) while the second solution is based on Datagram Transport Layer Security (DTLS). Beyond the handshakes themselves, we explain how to use them to allow for secure network access, flexible key management and secure communication. We use either the HIP or the DTLS handshake for network access. For key management we integrate both approaches with the Adapted Multimedia Internet KEYing (AMIKEY) protocol and use a polynomial scheme for efficient key management and generation of pairwise keys. With this, our goal is not only to secure a device during its lifecycle but also to analyze which protocol-based architecture would be the best one: the first solution is more efficient but less standard, the second one makes interaction with the Internet easier but performs worse.

The paper is organized as follows. Section 2 overviews the related work and relevant IP protocols. Section 3 details the application scenario and our design goals. In Section 4, we describe our two security architectures that are evaluated in Section 5. Section 6 concludes this paper.

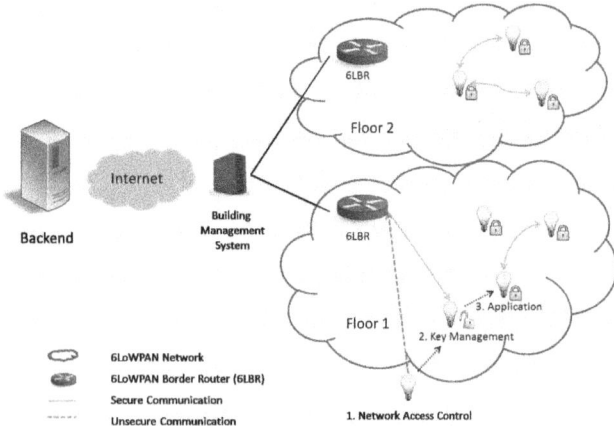

Figure 1: Building Management System Scenario

2. BACKGROUND

Research into security for wireless sensor networks has produced many results. SPINS [7] is a centralized architecture for securing uni- and multicast communication. Sizzle [12], for first time, made feasible the establishment of a secure end-to-end connection to a resource constrained device over the Internet based on HTTPS and Elliptic Curve Cryptography. Liu et al. [14] used polynomial schemes [9] to simplify key agreement in distributed sensor networks. Each node η is assigned a polynomial share $F(\eta, y)$ derived from a secret symmetric bivariate polynomial $F(x, y)$ allowing any pair of devices to establish a common key $F(\eta, \eta') = F(\eta', \eta)$. These solutions are, however, non-standard while the IoT vision requires efficient and standardized solutions. In our work, we extend standard protocols, namely HIP, DTLS, and AMIKEY, to create a secure IP-based IoT.

Host Identity Protocol (HIP) [11] introduces a cryptographic namespace of stable host identities (HIs) between the network and transport layer. Besides enabling mobility and multi-homing, the cryptographic HIs allow for a mutually authenticated Diffie-Hellman key exchange in which the generated symmetric-key usually protects an IPSec association. There are two HIP handshakes: *HIP Base EXchange (BEX)* relies heavily on public-key cryptography while *HIP Diet EXchange (DEX)* [15] defines a lightweight alternative that only relies on a static Diffie-Hellman key exchange being more suitable for constrained devices.

Datagram Transport Layer Security (DTLS) [17] is a datagram-compatible adaptation of TLS that runs on top of UDP. DTLS uses similar messages as defined in TLS including the DTLS handshake to establish a secure unicast link and the DTLS record layer to protect this link. The *DTLS handshake* supports different types of authentication mechanisms, e.g., using a pre-shared key, public-key certificates, and raw public-keys. DTLS is the mandatory standard for protection of CoAP [13] messages.

Adapted Multimedia KEYing (AMIKEY) [1] manages keying materials for securing communications within constrained networks and devices. It is based on MIKEY, a Key Management Protocol intended for use in real-time applications [10]. For this purpose, AMIKEY provides different message exchanges that may be transported directly over UDP and TCP. Essentially, they can be integrated within other protocols, such as HIP and DTLS.

3. SCENARIO AND DESIGN GOALS

Our work targets an *IoT network* running CoAP over 6LoWPAN. Devices in the *IoT network* are mobile or stationary and exhibit tight processing, memory and bandwidth constraints. The *IoT network* is connected to the public Internet through a number of 6LoWPAN border routers (6LBR). Further, we consider a centrally managed scenario in which the devices and services in each *IoT network* are managed by a *domain manager* that is located either within the *IoT network* or in the public Internet. The *domain manager* along with the *IoT network*s it manages is denoted as the *IoT domain*. Figure 1 shows an example of such a scenario – a smart building. Smart devices within the building (e.g., window blinds, lighting devices) form several IoT networks connected to a remote building management system via 6LBR.

This paper has two goals: First, describe how to secure the above scenario during the lifecycle of a device while minimizing the number of (and size of) protocols. Second, discuss what the best approach might be, one based on a more standard protocol such as DTLS or one relying on a less standard protocol combination but with better performance.

In our designs, we assume the Internet Threat Model [16] in which a malicious adversary can read and forge network traffic between devices at any point during transmission, but assume that devices themselves are secure. We consider a simplified lifecycle including three phases that currently lack a standardized (and unified) solution for IoT networks.

- *Secure Network Access* is about how to securely associate a new device to an *IoT network*. An attacker can perform network attacks during this phase, e.g., flooding or using the network for other purposes. The network can also be attacked if secure network access is not present, and thus, only devices that have been authenticated and authorized through a secure network access process should be allowed to communicate within the network. Similarly, devices that leave the IoT network should not be able to access the network with previous access parameters.

- *Key Management* is about how to handle the keys in an *IoT network* used for different purposes. Secure key derivation and management to secure interactions in the IoT domain is required, i.e., different pairwise, group, and network keys. In this way, compromising session keys or a previous derived key will not enable the attackers to obtain information about the currently used keying materials.

- *Secure Communication* is required since the adversary can eavesdrop on traffic in the IoT networkand maliciously modify it. Two communicating parties must establish a pairwise key to assure the confidentiality, integrity and authenticity of the information exchanged.

Due to the resource constrains in IoT devices, our proposed solutions are based on the assumption that a device has been configured with a PSK that is known *a priori* to the domain manager of the IoT domain it wants to join. This assumption is reasonable since a PSK could be embedded and registered during the manufacturing process of a device and the domain manager can retrieve it from a central server. Our solutions can be adapted to work with public-key cryptography when devices become more powerful.

4. DESIGN

In this section, we detail the design of our two solutions to address (i) secure network access, (ii) key management, and (iii) secure communication. The first approach relies on HIP, while the second one uses DTLS. Both of them address the three problems in three corresponding phases. In the following, we first provide a short sketch of the whole system in Section 4.1. Next, we explain each phase individually for the two proposed architectures, including their similarities and differences[1].

4.1 Overview

The first phase accounts for *secure network access*. In our architecture, the network is protected at link layer by means of a symmetric-key (L2 key), which is unknown to the joining device *a priori*. Using its link local address, the joining device authenticates itself to the domain manager of the IoT domain by means of an initial handshake (HIP or DTLS) that is based on a PSK. The PSK is assumed to have been pre-configured in the device (cf. Section 3). On success, the domain manager issues access parameters (L2 key) that would allow the joining device to access the secured IoT network and to receive a routable IPv6 address.

The second phase deals with *key management*. The joining device is provided with keying material to interact with other devices. For pairwise key generation, we make use of the polynomial scheme [9] described in Section 2. Each joining device is issued an individual share of a secret bivariate polynomial by the domain manager along with the access parameters during the network access phase. Using their individual shares, any two devices can then derive pairwise keys based on their identities in a fast and efficient way. In the HIP solution, keys are not used directly, instead they serve as root keying material in the MIKEY key derivation mechanism in order to derive fresh purpose-specific session keys for any pair of devices in the IoT network.

The final phase, *secure communication*, is achieved by protecting the exchange of CoAP messages by means of the DTLS record layer. Keys derived in the key management phase are used to protect the communication links.

4.2 Secure Network Access

We describe the details of the initial HIP and DTLS based handshakes.

4.2.1 HIP based approach

HIP-BEX or DEX as previously described in Section 2 could be used as handshakes in the secure network access phase. However, public-key cryptography is considered to be too expensive for constrained devices. We therefore propose a new variant called HIP-PSK which is more lightweight than HIP-DEX and HIP-BEX.

Figure 2 illustrates the HIP-PSK handshake. It is based on HIP-DEX but removes the ECC Diffie-Hellman method and performs a challenge-response authentication protocol [8] based on a PSK. Thus, instead of computing a static Diffie-Hellman key, the session key is derived by using CMAC [3] as

---

[1]The designs presented in this paper do not include the capability of secure network access in multi-hop networks and secure multicast due to the space limitation. The whole design will be presented in the future. The performance evaluation section shows the results for the prototypes including all functionalities, i.e., also multi-hop and multicast.

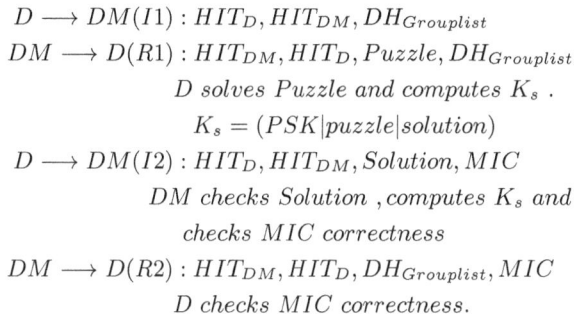

$$D \longrightarrow DM(I1) : HIT_D, HIT_{DM}, DH_{Grouplist}$$
$$DM \longrightarrow D(R1) : HIT_{DM}, HIT_D, Puzzle, DH_{Grouplist}$$
$$D \text{ solves } Puzzle \text{ and computes } K_s .$$
$$K_s = (PSK|puzzle|solution)$$
$$D \longrightarrow DM(I2) : HIT_D, HIT_{DM}, Solution, MIC$$
$$DM \text{ checks } Solution \text{ ,computes } K_s \text{ and}$$
$$\text{checks } MIC \text{ correctness}$$
$$DM \longrightarrow D(R2) : HIT_{DM}, HIT_D, DH_{Grouplist}, MIC$$
$$D \text{ checks } MIC \text{ correctness.}$$

Figure 2: HIP-PSK handshake between a joining device (D) and the domain manager (DM)

$K_s = CMAC(PSK|puzzle|solution)$ with the PSK as the secret input and the puzzle and its solution as nonces guaranteeing freshness. As in HIP-DEX, mutual authentication is achieved by generating and verifying a Message Integrity Code (MIC) on the I2 and R2 messages using CBC-MAC. HIP-PSK also preserves the DoS protection offered by HIP-DEX through the puzzle mechanism.

As in HIP-DEX, HIP-PSK identifies a device by its Host Identity Tag (HIT) computed as a truncation of its Host Identity (HI). The HI is just a random bit string. Alternatively, in the future, a public key could be used for interoperability purposes with HIP-BEX and HIP-DEX.

The details for bootstrapping a new node into the IoT network are the following: The joining device (D), performs the HIP-PSK handshake with the domain manager (DM). HIP-PSK mutually authenticates D and DM and creates a session key K_s. DM then verifies whether D is authorized to join the network. When authorized, DM uses D's HIT HIT_D to generate a polynomial share $F(HIT_D, y)$ from the secret bivariate polynomial $F(x, y)$. DM uses the session key K_s to encrypt and issue the L2 key and polynomial share to D in an additional HIP Update message. D acknowledges the receipt and can now join the secured IoT network. These two last messages are not depicted in Figure 2.

4.2.2 DTLS based approach

In this approach, the DTLS handshake protocol is used instead of HIP-PSK during the secure network access phase. Similar to the HIP approach, our design uses DTLS-PSK [5] instead of public-key based DTLS handshake, because it incurs less overhead and reduces the number of exchanged messages.

The joining device can be authenticated with the domain manager by performing the DTLS-PSK handshake relying on the link-local address. Once the device has been authenticated and authorized for the network, the established DTLS secure channel between the domain manager and the joining device is used to issue the L2 key and polynomial share to the device. At the time of the initial handshake, the joining device has not received an IP address. The domain manager thus generates a polynomial share $F(ID_{D1}, y)$ for the joining device that is bound to a given identifier ID_{D1}. This identifier will be used as the PSK hint in DTLS-PSK [5]. DTLS-PSK provides resiliency against DoS attacks through a cookie mechanism.

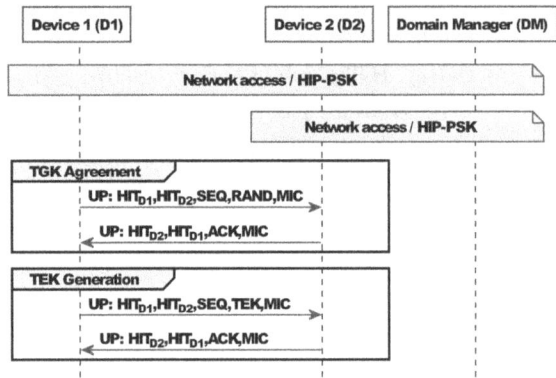

Figure 3: Management of unicast keys in HIP-based approach including TGK and TEK generation

4.3 Key Management

This section describes the details of key management in the HIP and DTLS-based approaches based on the use of the AMIKEY crypto-session bundle (CSB). A CSB is built from some root keying material (the TEK Generation Key (TGK)) and random bits ($RAND$). AMIKEY then defines a lightweight mechanism for the derivation and management of fresh purpose-specific keys, called Traffic Encryption Keys (TEKs), that are used to secure the communication links. We now explain how the root keying material, i.e. the TGK, is obtained, how the CSB is negotiated and set up, and finally how TEKs are requested and generated.

4.3.1 HIP based approach

As described in Section 4.2, each joining device has received its polynomial share linked to its HIT from the domain manager after a successful handshake during the network access phase. Our purpose is to define a method to derive keys for applications, e.g., a CoAP application on a given port. Figure 3 shows how a CSB is set up using root keying material derived from the polynomial share and how fresh TEKs are generated to secure the interactions between a pair of devices $D1$ and $D2$ in an IoT network.

1. $D1$ and $D2$ generate a pairwise key $K_{(D1,D2)}$ by using their polynomial shares and their respective identities: $K_{(D1,D2)} = F(HIT_{D1}, HIT_{D2}) = F(HIT_{D2}, HIT_{D1})$. $K_{(D1,D2)}$ is used as the TGK for the CSB.

2. $D1$ and $D2$ further need to agree on a random value $RAND$. $D1$ provides the random value in a HIP Update message that is protected with a MIC computed with $K_{(D1,D2)}$ using CBC-MAC.

3. $D2$ acknowledges the Update.

If the target application requires mutual authentication, the HIP-PSK handshake (based on $K_{(D1,D2)}$) should be executed between D1 and D2 first using the resulting session key of the HIP handshake as the TGK for the CSB. The $D1$ and $D2$ have now a CSB from which fresh TEKs can be derived upon request in a simple two-step AMIKEY process that we embed in the HIP Update mechanism.

4. $D1$ sends a HIP Update message with the *TEK trigger parameter* to $D2$. The trigger contains the identifier

associated with the new TEK, the desired length and the CSB from which to generate the key.

5. $D2$ generates the new TEK using the TGK and given parameters as inputs to AES-CMAC as specified in AMIKEY/MIKEY [1, 10] and acknowledges the request.

4.3.2 DTLS based approach

As DTLS runs on the transport layer, it can be used directly to protect the applications. The polynomial shares are also used here to provide for fast pairwise key agreement between any pair of devices in the IoT domain. This pairwise key serves as the PSK in DTLS-PSK [5] enabling any two applications running on the devices to derive a session key, equivalent to the TEK in the HIP-approach. In detail:

1. Two applications running on the devices $D1$ and $D2$ start a DTLS-PSK handshake. They exchange their identities ID_{D1} and ID_{D2} as extensions to the first two handshake messages, the *ClientHello* and *ServerHello*.

2. Both devices then generate a pairwise key by using their polynomial shares and their respective identities: $K_{(D1,D2)} = F(ID_{D1}, ID_{D2}) = F(ID_{D2}, ID_{D1})$.

3. The derived key $K_{(D1,D2)}$ is used as the PSK to complete the DTLS-PSK handshake.

4. The result of the DTLS-PSK handshake is a session key (equivalent to a TEK) used to protect the communication link between the two applications on both devices.

4.4 Secure Communication

Once the pairwise session keys have been derived, a secure channel can be created to transport data (CoAP messages) between the devices in the IoT network. For this, we rely on the DTLS record-layer to create a secure transport layer for CoAP.

Any pair of devices in the IoT domain that wish to communicate with each other, establish a CSB and derive a fresh unicast TEK either through HIP or DTLS as described in Section 4.3. The TEK is used in the DTLS record layer (based on AES-CCM) to protect the message exchange between two applications. AES-CCM is an AES mode of operation that defines the use of AES-CBC for MAC generation with AES-CTR for encryption [4]. The CCM counter (corresponding to the DTLS epoch and sequence number fields) are initialized to 0 upon TEK establishment and used in the nonce construction in a standard way.

5. EVALUATION

This section describes the prototype implementation, performance and security properties of our two designs.

5.1 Prototype Implementation

Both prototypes are written in C and run as an application on Contiki OS 2.5 [6]. They have been tested in the Cooja simulator and on Redbee Econotags hardware, which features a 32-bit CPU, 128KB of ROM, 128KB of RAM, and an IEEE 802.15.4 radio with an AES hardware coprocessor.

The prototype of our DTLS-based solution uses the *tinydtls* [2] library with small changes for better performance: separate delivery instead of flight grouping of messages, redesigned retransmission mechanism, and no cookies.

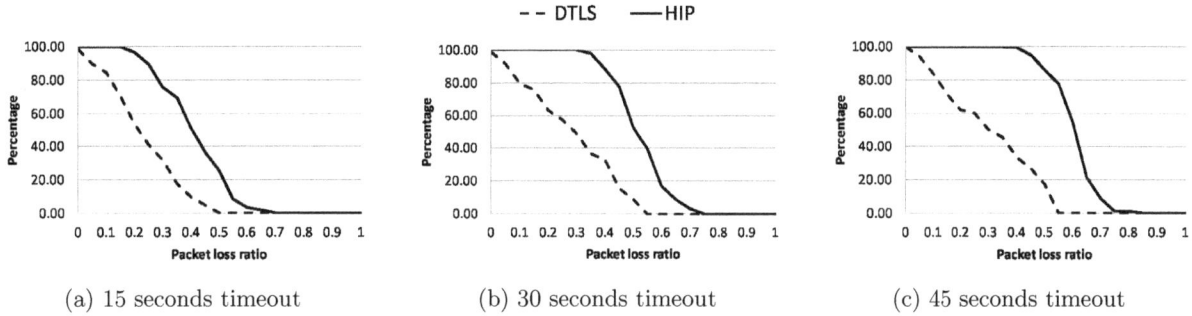

- - DTLS —— HIP

(a) 15 seconds timeout   (b) 30 seconds timeout   (c) 45 seconds timeout

Figure 4: Comparison of the average percentage of successful handshakes (HIP-based and DTLS-based) for different packet loss ratios.

5.2 Performance

This section evaluates the memory and communication overhead of our two proposed solutions.

Table 1: Memory requirements in KB.

	HIP		DTLS	
	ROM	RAM	ROM	RAM
State Machine	5,7	1,2	8,15	1,9
Cryptography	1,7	0	3,3	1,5
Key Management	1,8	0	1	0
DTLS record layer	3,7	0,5	3,7	0,5
Total	**12.9**	**1.7**	**16.15**	**3.9**

Memory needs. Table 1 presents the ROM and RAM consumption of our two designs. The DTLS approach exhibits an overall larger memory footprint because DTLS (i) involves more messages than HIP and (ii) requires SHA2 while HIP-PSK only uses AES-CMAC. However, for key management, HIP adds slightly more overhead, as it requires the definition and handling of new parameter types whereas in DTLS the required information can simply be carried in the payload.

Table 2: Communication overhead for network access and multicast key management.

	HIP	DTLS
Number of messages	6	12
Number of roundtrips	3	4
802.15.4 headers	84 B	168 B
6LoWPAN headers	240 B	480 B
UDP headers	0 B	96 B
Application	304 B	487 B
Total	**628 B**	**1231 B**

Baseline protocol overhead. Table 2 summarizes the required number of round trips, number of messages and the total exchanged bytes for the HIP- and DTLS-based handshake carried out in ideal conditions, i.e. in a network without packet losses. The DTLS handshake is more complex, and thus it involves the exchange of more messages than in HIP. Further, DTLS runs on the transport layer, i.e. UDP, whereas HIP is carried directly over the network layer,

i.e. 6LoWPAN. This directly increases the overhead due to lower layer per-packet protocol headers.

Protocol overhead with packet losses. In a network with packet losses, the solutions will perform worse because security handshakes might fail due to the lost messages. This will increase the delays and lead to a higher protocol overhead affecting how the protocols are used in real-life scenarios. Figure 4 compares the percentage of successful handshakes in both approaches as a function of different timeouts and packet loss ratios. In particular, we observe that the HIP-based approach performs significantly better than the DTLS-based approach in the presence of packet loss. Notably, the HIP-based approach is able to tolerate up to 50 % packet loss and still finish approximately four out of five handshakes within 30 seconds. Note that the distribution of polynomial shares is excluded for the sake of simplicity. Related to this is Figure 5 where we show how an increment in the packet loss ratio leads to a higher delay to perform a successful handshake. Here, the HIP-based approach exhibits a significantly lower delay, since it requires less messages, hence less retransmissions in average.

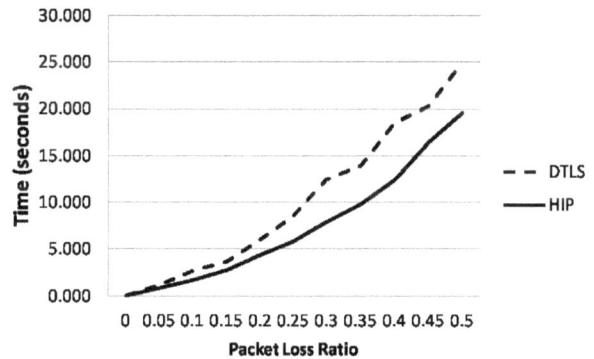

Figure 5: Times to perform network access control and multicast key management, considering only successful handshakes.

5.3 Security Analysis

Both proposed solutions exhibit similar security features. First, the new HIP-PSK handshake relies on a standard challenge-response handshake based on symmetric-keys during which also a session key is derived. The methods are adapted from HIP-DEX, in particular the usage of CMAC for those purposes is recommended by NIST and helps to minimize resource requirements. Second, unicast communications are protected by means of the standard DTLS record layer, and thus, we offer the standard security features, namely confidentiality, authentication, and replay protection by means of counters. Third, network access is performed in a secure way by using the HIP and DTLS handshakes in each of the proposed solutions. Both HIP-PSK and DTLS-PSK provide mutual authentication between the joining device and domain manager. In both cases, corresponding authorization methods at the domain manager – after the authentication of the joining device – ensure that only valid devices access the network. Note that in our design we also include a key at MAC level whose purpose is to ensure that only authorized devices can use the wireless network. This feature is not needed in a single-hop network access case, but we have already included this here because our whole design – not presented here due to space reasons – also addresses multi-hop networks in which this is a must. Fourth, key management is in charge of securing CoAP applications. In the DTLS method, the master key derived from the polynomial share is used as master key for any application, the DTLS handshake is in charge of providing a secure link. In the HIP approach, a key derivation process based on AMIKEY and using standard key derivation functions is applied. This means that in both cases CoAP applications share master keying material and that a rather standard key derivation process is performed to separate the communication channels. In our designs we used PSK due to several reasons including: (i) our interest in comparing the protocol performance without heavier crypto; (ii) public-key crypto still remains a problem in some application areas. However, we fully acknowledge that public-key cryptography simplifies key management.

6. CONCLUSIONS AND FUTURE WORK

This work has shown how to secure the IP-based Internet of Things with focus on (i) secure network access, (ii) key management, and (iii) secure communication. We do this by means of two solutions based on standardized IP protocols, HIP and DTLS, and adapt them to the constraints of low resource devices (bandwidth, memory and CPU). The first approach relies on the less common HIP protocol further combined with the DTLS record layer and leads to good performance; the second one uses the standard DTLS showing worse performance.

Some of the notable contributions of this paper are: (i) the proposal of HIP-PSK, (ii) the usage of both HIP and DTLS for network access; (iii) the integration of AMIKEY into HIP and DTLS for key management; and (iv) the application of polynomial schemes for efficient pairwise key derivation in HIP-PSK and DTLS-PSK.

As a proof of concept, we implemented the two architectures based on HIP and DTLS over Contiki OS running on a Redbee Econotag. The evaluation shows that the HIP based mechanism is smaller in memory footprint (ROM and RAM) compared to the DTLS mechanism. Results of the communication overhead, delay and resiliency to packet loss also show that the HIP based implementation outperforms the DTLS solution.

This work shows that while the broadly used DTLS allows for better interoperability it compromises efficiency. Our less standard solution based on HIP performs better, but currently is not on the standardization roadmap. We hope that our work provides valuable protocol designs and evaluation results (with their pros and cons) which can provide the much needed direction for the standardization effort in IETF to ensure that the best solutions are adopted.

As a next step we plan to further describe how the presented security architectures, based on HIP and DTLS, can easily provide support for secure network access in multi-hop networks and for secure multicast.

7. REFERENCES

[1] R. Alexander and T. Tsao. Adapted Multimedia Internet KEYing (AMIKEY): An extension of Multimedia Internet KEYing Methods for Generic LLN Environments. Internet-draft, IETF, 2012.

[2] O. Bergmann. tinydtls - a basic dtls server template.

[3] L. Chen. Recommendation for key derivation using pseudorandom functions. SP-800-108, Computer Security Division. Information Technology Laboratory. US Department of Commerce, 2009.

[4] M. Dworkin. Recommendation for block cipher modes of operation: The ccm mode for authentication and confidentiality. SP-800-38c, NIST. Technology Administration. US Department of Commerce, 2007.

[5] P. Eronen and H. Tschofenig. Pre-Shared Key Ciphersuites for Transport Layer Security (TLS). RFC 4279 (Proposed Standard), December 2005.

[6] A. Dunkels et al. Contiki - a lightweight and flexible operating system for tiny networked sensors. In *29th Annual IEEE International Conference on Local Computer Networks*, pages 455–462. IEEE, 2004.

[7] A. Perrig et al. Spins: security protocols for sensor networks. *Wireless Networks*, 8(5), 2002.

[8] A.J. Menezes et al. *Handbook of Applied Cryptography*. 5 edition, Aug. 2001.

[9] C. Blundo et al. Perfectly-secure key distribution for dynamic conferences. *Advances in cryptology*, 1993.

[10] J. Arkko et al. MIKEY: Multimedia Internet KEYing. RFC 3830, August 2004. Updated by RFCs 4738, 6309.

[11] R. Moskowitz et al. Host Identity Protocol Version 2 (HIPv2). Internet-draft, IETF, 2012.

[12] V. Gupta et al. Sizzle: A standards-based end-to-end security architecture for the embedded internet. *Pervasive and Mobile Computing*, 1:425–445, 2005.

[13] Z. Shelby et al. Constrained Application Protocol (CoAP). Internet-Draft draft-ietf-core-coap-12, IETF, October 2012.

[14] D. Liu and P. Ning. Establishing pairwise keys in distributed sensor networks. In *Proceedings of the 10th ACM Conference on Computer and Communications Security*, CCS '03. ACM, 2003.

[15] R. Moskowitz. HIP Diet EXchange (DEX). Internet Draft draft-moskowitz-hip-rg-dex-06, IETF, 2012.

[16] E. Rescorla and B. Korver. Guidelines for Writing RFC Text on Security Considerations. RFC 3552 (Best Current Practice), July 2003.

[17] E. Rescorla and N. Modadugu. Datagram Transport Layer Security. RFC 4347, April 2006. Obsoleted by RFC 6347, updated by RFC 5746.

Revisiting Lightweight Authentication Protocols Based on Hard Learning Problems

Panagiotis Rizomiliotis
Dep. of Information and Communication
Systems Engineering
University of the Aegean
Karlovassi, Samos, GR 83200, Greece
prizomil@aegean.gr

Stefanos Gritzalis
Dep. of Information and Communication
Systems Engineering
University of the Aegean
Karlovassi, Samos, GR 83200, Greece
sgritz@aegean.gr

ABSTRACT

At the 2011 Eurocrypt, Kiltz et al., in their best paper price awarded paper, proposed an ultra-lightweight authentication protocol, called $AUTH$. This new protocol is supported by a delegated security proof, against passive and active attacks, based on the conjectured hardness of the Learning Parity with Noise (LPN) problem. However, $AUTH$ has two shortcomings. The security proof does not include man-in-the-middle (MIM) attacks and the communication complexity is high. The weakness against MIM attacks was recently verified as a very efficient key recovery MIM attack was introduced with only linear complexity with respect to the length of the secret key. Regarding the communication overhead, Kiltz et al. proposed a modified version of AUTH where the communication complexity is reduced at the expense of higher storage complexity. This modified protocol was shown to be at least as secure as AUTH.

In this paper, we revisit the security of $AUTH$ and we show, somehow surprisingly, that its communication efficient version is secure against the powerful MIM attacks. This issue was left as an open problem by Kiltz et al. We provide a security proof that is based on the hardness of the LPN problem to support our security analysis.

Categories and Subject Descriptors

E.3 [**Data Encryption**]: Miscellaneous; H.4 [**Information Systems Applications**]: Communications Applications

General Terms

Security

Keywords

RFID authentication protocols, provable security, LPN

1. INTRODUCTION

The last few years, the design of ultra-lightweight authentication protocols has gained a lot of attention. Motivated mainly by the restrictions that the Radio Frequency Identification (RFID) technology imposes on the available resources for security, several protocols have been proposed [2]. Among them, the most promising family of authentication protocols is the family of HB-like protocols that are based on the so-called *Learning Parity with Noise (LPN)* problem.

The LPN problem is an average-case version of the following problem: given a set of noisy binary equations, find a solution that maximally satisfies the equations. In the worst case version LPN is related to the well studied decoding of a random linear code problem that has been proved to be NP-hard by Berlekamp et al. in [3]. Apart from the authentication protocols, several other cryptographic applications, like encryption schemes ([12]), Message Authentication Codes ([19]), string commitment schemes and zero-knowledge proofs ([16]), have been recently introduced based on the LPN problem.

In [15], Juels and Weis proposed HB^+, a symmetric key authentication scheme, inspired by HB ([14]), the work of Hopper and Blum for the secure identification of human beings. The HB^+ has a very simple circuit representation, as it performs only a few dot-product and bit exclusive-or computations. However, the most interesting feature of the protocol is the elegant proof that supports its security analysis. Specifically, in [15], a concrete reduction of the LPN problem to the security of the HB^+ protocol in two attack models was shown. In the first model the attacker is passive and can only eavesdrop the communication between the prover (tag) and the verifier (reader), while in the second model she is active and she can also send queries to the prover. The original proof was further improved in [?], [17].

This security proof does not consider more powerful adversaries that can manipulate messages exchanged between the prover and the verifier. Thus, shortly after the introduction of HB^+, a simple key recovery man-in-the-middle (MIM) attack was proposed ([10]). Motivated by this MIM attack, several variants of HB^+ have been introduced ([5], [6], [7], [8], [11], [21], [22], [26], [28]). However, most of these schemes have been shown to be weak against a MIM attacker.

In this short paper, we will revisit one of these proposals. At the 2011 Eurocrypt, Kiltz et al., in their best paper price awarded paper, proposed an ultra-lightweight authentication protocol, called $AUTH$. This new protocol is supported by a delegated security proof based on the conjectured hardness of the LPN problem against passive and active attacks. To be more precise, they build on a modified version of the LPN problem, the so-called subset LPN problem. $AUTH$ has two shortcomings. Firstly, the security proof does not include MIM attacks and this weakness against MIM attacks was recently verified as a very efficient key recovery MIM attack was introduced with only linear complexity with respect to the length of the secret key [29]. Secondly, $AUTH$ has rather high communication complexity. To cope with the communication overhead, Kiltz et al. proposed a modified version of $AUTH$ in which the communication complexity is reduced at the expense of higher storage complexity. The authors used a technique adapted by Gilbert et al. to enhance the security of HB^+ in [11]. The size of the exchanged messages between the tag and the reader is reduced, while the shared secret key is increased from a vector to a matrix. We will call this protocol $AUTH^\#$. $AUTH^\#$ was shown to be at least as secure as AUTH. However, the evaluation of the resistance of $AUTH^\#$ against MIM attacks was left as an open problem.

In this paper, we revisit the security of $AUTH$, the ultra-lightweight cryptographic protocol for RFID authentication, and we show, somehow surprisingly, that its communication complexity efficient version, $AUTH^\#$, is much more secure. More precisely, we show that this version of $AUTH$ can provably resist against powerful MIM attacks. Our security proof is based on the hardness of the LPN.

1.1 Outline

The paper is organized as follows. In Section 2, we establish the necessary background on the LPN problem, while in Section 3, we present the $AUTH$ and the $AUTH^\#$ authentication protocols. In Section 4, we provide a proof of the security of $AUTH^\#$ against MIM attacks. Finally, conclusions and topics for further research can be found in Section 5.

2. BACKGROUND

2.1 Notation

We try to apply, as possible, the established notation. We use normal, bold and capital bold letters, x, $\boldsymbol{x}$ and $\boldsymbol{M}$ to denote single elements, vectors and matrices, respectively. The Hamming weight $\mathrm{wt}(\boldsymbol{x})$ of a vector $\boldsymbol{x} = [x(0), x(1), \cdots, x(n-1)]$ is the number of nonzero elements and $\boldsymbol{M}^T$ is the transpose of a matrix $\boldsymbol{M}$. Also, $\boldsymbol{0}_m$ denotes the all zero vector of length m and for real numbers $\eta, \psi \in \mathbb{R}$, $]\eta, \psi[= \{x \in \mathbb{R} | \eta < x < \psi\}$. Let $\boldsymbol{a}$ and $\boldsymbol{b}$ be two binary vectors with length l. We use $\boldsymbol{a}_{\downarrow \boldsymbol{b}}$ to denote the subvector of $\boldsymbol{a}$ obtained by deleting all bits of $\boldsymbol{a}$ where $\boldsymbol{b}$ equals 0 (for instance for $\boldsymbol{a} = 10101000$ and $\boldsymbol{b} = 00011010$ we have $\boldsymbol{a}_{\downarrow \boldsymbol{b}} = 010$) and $\boldsymbol{M}_{\downarrow \boldsymbol{b}}$ to denote the submatrix of $\boldsymbol{M}$ obtained by deleting all rows of $\boldsymbol{M}$ where $\boldsymbol{b}$ equals 0. The matrix $\boldsymbol{M}_{\downarrow \boldsymbol{b}}$ can be written as $\boldsymbol{V}(\boldsymbol{b}) \cdot \boldsymbol{M}$, where $\boldsymbol{V}(\boldsymbol{b})$ is a $\mathrm{wt}(\boldsymbol{b}) \times l$ matrix where each row has only one non-zero element.

We use $x \xleftarrow{\$} X$ to denote the assignment to x of a value sampled from the uniform distribution on the finite set X. We use Ber_η to denote the Bernoulli distribution with parameter η, meaning that a bit $\nu \in Ber_\eta$, then $Pr[\nu = 1] = \eta$ and $Pr[\nu = 0] = 1 - \eta$. A vector $\boldsymbol{\nu}$ randomly chosen among all the vectors of length m, such that $\nu(i) \in Ber_\eta$ and

$\eta \in (0, 1/2)$, for $0 \le i \le m-1$, is denoted as $\boldsymbol{\nu} \xleftarrow{\$} Ber(m, \eta)$, while we use $\boldsymbol{b} \xleftarrow{\$} \{0,1\}^k$ to denote a random binary vector $\boldsymbol{b}$ with length k.

An algorithm D is probabilistic polynomial time if D uses some randomness of its logic and for any input the computation of the algorithm terminates in a number of steps that are a polynomial function in the length of the input. Finally, we denote an arbitrary polynomial function of x by $poly(x)$ and by $f(x) = negl(x)$ a function f that is negligible as a function of x, i.e. it vanishes faster than the inverse of any polynomial in x.

2.2 Learning Parity with Noise

The last few years, the *Learning Parity with Noise (LPN)* problem has gained a lot of attention. It appears in two versions, the decisional and the computational one. In [18], it was shown that the two versions are equivalent and depending on the application the most adequate is used. In this paper we use the computational version.

More precisely, for a secret vector $\boldsymbol{x} \in \{0,1\}^l$, we define $\Lambda_{\eta, l}(\boldsymbol{x})$ the distribution over $\{0,1\}^{l+1}$ where a sample is given by

$$(\boldsymbol{r}, \boldsymbol{r}^T \cdot \boldsymbol{x} \oplus \nu)$$

where $\boldsymbol{r} \in \{0,1\}^l$ and $\nu \in Ber_\eta$. We use $\Omega_{\eta, l}(\boldsymbol{x})$ to denote the oracle that outputs samples from the distribution $\Lambda_{\eta, l}(\boldsymbol{x})$. Let U_l denote the uniform distribution over $\{0,1\}^l$. For any $\boldsymbol{x}$, $\Lambda_{\frac{1}{2}, l}(\boldsymbol{x})$ is the same distribution as U_l. The decisional version of the LPN problem is defined as follows.

DEFINITION 1. *The decisional $LPN_{\eta, l}$ problem is (t, q, ϵ)-hard if for any distinguisher D running in time t and making q oracle queries, it holds that,*

$$|Pr\left[\boldsymbol{x} \xleftarrow{\$} \{0,1\}^l : D^{\Omega_{\eta, l}(\boldsymbol{x})}(1^l) = 1\right] -$$
$$Pr\left[D^{U_{l+1}}(1^l) = 1\right]| \le \epsilon.$$

The above description corresponds to the average case LPN problem. In machine learning theory, this problem was introduced by Angluin and Laird [1]. Kearns [20] proved that the class of noisy parity concepts is not learnable within the statistical query model. The worst case version is strongly related to the decoding problem of random linear codes, which is $\mathcal{NP}$-complete [3] and hard to approximate within a factor of 2 [13].

For the average case several studies have been proposed for solving the LPN problem for a constant noise parameter η (for instance see [14], [17], [27]). The most popular algorithm for solving the LPN problem is the BKW algorithm, proposed by Blum, Kalai and Wasserman in [4]. The BKW algorithm was further improved, initially, by Fossorier et al. in [9], and most recently by Levieil and Fouque in [23].

2.3 Subspace and subset Learning Parity with Noise Problems

Several problems have been proposed that are based on the hardness of the LPN problem. In [25], the subspace LWE problem was introduced. The subspace LPN problem is the subspace LWE over a field of size $q = 2$.

Let $\boldsymbol{A}$ be a $l \times l$ binary matrix and $\boldsymbol{b} \in \{0,1\}^l$. We define the distribution,

$$\Gamma_{\eta,l,d}(\boldsymbol{x},\boldsymbol{A},\boldsymbol{b}) = \begin{cases} \bot, & \text{if } \mathrm{rank}(\boldsymbol{A}) < d \\ \Lambda_{\eta,l}(\boldsymbol{A}\boldsymbol{x}\oplus\boldsymbol{b}), & \text{otherwise} \end{cases}$$

and let $\Gamma_{\eta,l,d}(\boldsymbol{x},\cdot,\cdot)$ denote the oracle which on input $\boldsymbol{A}$ and $\boldsymbol{b}$ outputs a sample $\Gamma_{\eta,l,d}(\boldsymbol{x},\boldsymbol{A},\boldsymbol{b})$.

DEFINITION 2. *Let $l,d \in \mathbb{Z}$ where $d \leq l$. The decisional $SLPN_{\eta,l,d}$ problem is (t,q,ϵ)-hard if for every distinguisher D running in time t and making q queries,*

$$|Pr[\boldsymbol{x} \xleftarrow{\$} \{0,1\}^l : D^{\Gamma_{\eta,l,d}(\boldsymbol{x},\cdot,\cdot)} = 1] - Pr[D^{U_{l+1}(\cdot,\cdot)} = 1]| \leq \epsilon,$$

where $U_{l+1}(\cdot,\cdot)$ on input $\boldsymbol{A},\boldsymbol{b}$ outputs a sample of U_{l+1} if $\mathrm{rank}(\boldsymbol{A}) \geq d$ and $\bot$ otherwise.

PROPOSITION 1. *[25] For any $l,d,g \in \mathbb{Z}$ where $d+g \leq l$, if the decisional $LPN_{\eta,d}$ problem is (t,q,ϵ)-hard then the decisional $SLPN_{\eta,l,d}$ problem is (t',q,ϵ')-hard where,*

$$t' = t - poly(l,q)$$
$$\epsilon' = \epsilon + 2q/2^{g+1}.$$

The subset LPN problem ($SLPN^*$) is a weaker version of the $SLPN_{\tau,l,d}$ problem where subsets of the secret $\boldsymbol{x}$ are used. Let $\boldsymbol{v} \in \{0,1\}^l$ and $diag(\boldsymbol{v})$ is the zero matrix with $\boldsymbol{v}$ in the diagonal. We define the distribution,

$$\Gamma_{\eta,l,d}^*(\boldsymbol{x},\boldsymbol{v}) = \Gamma_{\eta,l,d}(\boldsymbol{x},diag(\boldsymbol{v}),\boldsymbol{0}_l)$$
$$= \begin{cases} \bot, & \text{if } \mathrm{rank}(\mathrm{wt}\,\boldsymbol{v}) < d \\ \Lambda_{\eta,l}(\boldsymbol{x}\boldsymbol{v}), & \text{otherwise} \end{cases}$$

From the $\Gamma_{\eta,l,d}^*(\boldsymbol{x},\boldsymbol{v})$ distribution the subset LPN problem is defined as follows.

DEFINITION 3. *Let $l,d \in \mathbb{Z}$ where $d \leq l$. The decisional $SLPN_{\eta,l,d}^*$ problem is (t,q,ϵ)-hard if for every distinguisher D running in time t and making q queries,*

$$|Pr[\boldsymbol{x} \xleftarrow{\$} \{0,1\}^l : D^{\Gamma_{\eta,l,d}^*(\boldsymbol{x},\cdot)} = 1] - Pr[D^{U_{k+1}(\cdot)} = 1]| \leq \epsilon,$$

where $U_{l+1}(\cdot)$ on input $\boldsymbol{v}$, outputs a sample of U_{l+1}, if $\mathrm{wt}(\boldsymbol{v})$, and $\bot$ otherwise.

The security of the $AUTH$ protocol is based on the hardness of the subset LPN problem.

2.4 Definition of security models

We consider three types of attacks: passive, active, and man-in-the-middle attacks.

A passive attacker eavesdrops the communication between a legitimate prover (tag) and the verifier (reader) and then she tries to convince the verifier. An active attacker is more powerful, as she can interrogate a prover for a polynomial number of times and then she interacts with the verifier trying to receive an accept message.

In the man-in-the-middle (MIM) attacks, the attacker can interact with both the prover and the verifier and learn the verifier's decision; *accept* or *reject*. This being the strongest security notion for authentication protocols. It is divided into two phases. In the first phase, the attacker modifies the messages exchanged between the prover and the verifier for q invocations of the protocol, while in the second phase the attacker impersonates the prover. Most of the attacks against HB^+ and its variants are MIM ones.

$$\mathcal{P}(\boldsymbol{x},\eta) \qquad\qquad\qquad \mathcal{V}(\boldsymbol{x},\tau)$$

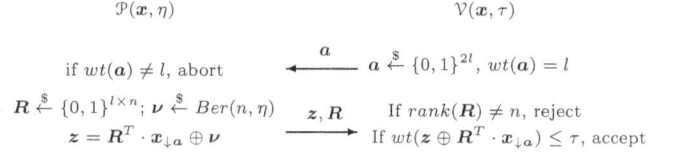

Figure 1: The *AUTH* protocol.

3. THE AUTH AND AUTH# AUTHENTICATION PROTOCOLS

The $AUTH$ protocol is a symmetric key authentication protocol supported by a security proof under the hardness of the subspace LPN problem ([19]). After some initialization phase, the prover $\mathcal{P}$ (the tag) and the verifier $\mathcal{V}$ (the reader) share a secret key $\boldsymbol{x}$ with length $2l$. The basic steps of the protocol go as follows (Fig. 1):

1. The verifier generates a random bit-string $\boldsymbol{a}$ with length $2l$ and sends it to tag $\mathcal{T}$. The Hamming weight of the $\boldsymbol{a}$ must be l.

2. The prover verifies that $\mathrm{wt}(\boldsymbol{a}) = l$ and generates a full rank $l \times n$ random binary matrix $\boldsymbol{R}$ and a bit-string $\boldsymbol{\nu} \in Ber(n,\eta)$. Then, it computes $\boldsymbol{z} = \boldsymbol{R}^T \cdot \boldsymbol{x}_{\downarrow\boldsymbol{a}} \oplus \boldsymbol{\nu}$ and sends to the verifier both $\boldsymbol{z}$ and $\boldsymbol{R}$. If $\mathrm{wt}(\boldsymbol{a}) \neq l$, it aborts the execution of the protocol.

3. The verifier first verifies that the matrix $\boldsymbol{R}$ has rank n and then it accepts if $\mathrm{wt}(\boldsymbol{z} \oplus \boldsymbol{R}^T \cdot \boldsymbol{x}_{\downarrow\boldsymbol{a}}) \leq \tau$, where $n\eta \leq \tau \leq \frac{n}{2}$. If the rank is not correct or the condition is not satisfied, the verifier rejects.

The main disadvantage of AUTH is its extensive communication complexity. In order to reduce this large communication overhead, a trade off between the communication complexity and the key-size was proposed. Actually, they used an idea introduced by Gilbert et al. ([11]) to enhance the security of HB^+. The modified version of the AUTH protocol appears in Fig. 2. We call this modified version $AUTH^\#$. $AUTH^\#$ minimizes the communication complexity, since, instead of sending the $l \times n$ binary matrix $\boldsymbol{R}$, the tag has to send just a l-bit vector $\boldsymbol{r}$. On the other hand, the secret key shared between the verifier and prover increases significantly and a $2l \times n$ matrix $\boldsymbol{X}$ must be stored. The basic steps of the protocol go as follows:

1. The verifier $\mathcal{V}$ generates a random bit-string $\boldsymbol{a}$ with length $2l$, $\mathrm{wt}(\boldsymbol{a}) = l$ and sends it to the prover.

2. The prover verifies that $\mathrm{wt}(\boldsymbol{a}) = l$ and generates a random binary vector $\boldsymbol{r}$ with length l and a bit-string $\boldsymbol{\nu} \in Ber(n,\eta)$. Then, it computes $\boldsymbol{z} = \boldsymbol{r}^T \cdot \boldsymbol{X}_{\downarrow\boldsymbol{a}} \oplus \boldsymbol{\nu}$ and sends to the verifier both $\boldsymbol{z}$ and $\boldsymbol{r}$. If $\boldsymbol{a} \neq l$, it aborts the execution of the protocol.

3. The verifier first verifies that $\mathrm{wt}(\boldsymbol{r}) \neq 0$, otherwise aborts the execution. Then, it accepts if $\mathrm{wt}(\boldsymbol{z} \oplus \boldsymbol{r}^T \cdot \boldsymbol{X}_{\downarrow\boldsymbol{a}}) \leq \tau$, where $n\eta \leq \tau \leq \frac{n}{2}$. Otherwise, the verifier rejects.

In [19], it was proved that AUTH is secure against passive and active attackers given the intractability of the subspace LPN problem. However, recently the very efficient key recovery attack was proposed against AUTH. The attack has

$$\mathcal{P}(\boldsymbol{X}, \eta) \qquad\qquad\qquad \mathcal{V}(\boldsymbol{X}, \tau)$$

if $wt(\boldsymbol{a}) \neq l$, abort $\xleftarrow{\quad \boldsymbol{a} \quad}$ $\boldsymbol{a} \xleftarrow{\$} \{0,1\}^{2l}, wt(\boldsymbol{a}) = l$

$\boldsymbol{r} \xleftarrow{\$} \{0,1\}^l; \boldsymbol{\nu} \xleftarrow{\$} Ber(n, \eta)$ $\xrightarrow{\quad \boldsymbol{z}, \boldsymbol{r} \quad}$ If $\boldsymbol{r} = \boldsymbol{0}_l$, reject

$\boldsymbol{z} = \boldsymbol{r}^T \cdot \boldsymbol{X}_{\downarrow \boldsymbol{a}} \oplus \boldsymbol{\nu}$ $\qquad$ If $wt(\boldsymbol{z} \oplus \boldsymbol{r}^T \cdot \boldsymbol{X}_{\downarrow \boldsymbol{a}}) \leq \tau$, accept

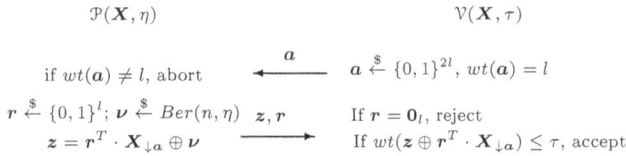

Figure 2: The low communication complexity version of $AUTH$ protocol.

linear complexity with respect to the length of the secret key. In [19], it was also shown that the communication efficient variant, $AUTH^{\#}$, was secure against passive and active attacks. The proof is a trivial application of the methodology followed in [11]. However, it is still an an open problem the evaluation of its resistance against MIM attacks. Next, we show that even when the attacker is able to change some of the responses of the prover, then protocol is secure.

Typically, the false rejection rate P_{FR} of the protocol; i.e. the probability to reject a legitimate tag, equals the probability $wt(\boldsymbol{\nu}) > \tau$ and it is given by

$$P_{FR} = \sum_{i=\tau+1}^{n} \binom{n}{i} \eta^i (1-\eta)^{n-i}.$$

Finally, the false acceptance rate P_{FA}; i.e. the probability to accept a randomly selected response $\boldsymbol{z}$, can be computed as follows:

$$P_{FA} = \sum_{i=0}^{\tau} \binom{n}{i} 2^{-n};$$

i.e. it is equal to the number of binary vectors with length n and Hamming weight at most τ.

4. ON THE SECURITY OF AUTH$^{\#}$

4.1 Definition of security models

We use $\mathcal{V}_{\boldsymbol{X}, \tau}$ to denote the algorithm that it is run by the verifier and $\mathcal{P}_{\boldsymbol{X}, \eta}$ the one run by a legitimate prover. We define two models of security, the $ACT - model$ and the $MIM - model$. In each of the models the adversary runs in two stages. In the first stage she has some interaction with the prover and/or the verifier and in the second she interacts only with the verifier and wins if the verifier returns *accept*. In the $ACT - model$ the active attacker interacts only with an honest prover for a polynomial number of times.

DEFINITION 4. *(ACT-model). In the $ACT - model$ the attack is carried in two phases:*

- **Phase 1.** *The adversary interacts q times with the honest prover.*

- **Phase 2.** *The adversary interacts with the verifier trying to impersonate the prover*

In the $MIM - model$ the attack is carried in two phases and the adversary can manipulate all messages exchanged between the tag and the reader.

DEFINITION 5. *(MIM-model). In the $MIM - model$ the attack is carried in two phases:*

- **Phase 1.** *The adversary interferes for q executions of the protocol. On each execution, the adversary can*

eavesdrop on all messages exchanged between the honest prover and the honest verifier, including the verifier's decision. In addition, she can modify all these messages with the restriction that all the modifications must have been decided before each execution has started.

- **Phase 2.** *The adversary interacts with the verifier trying to impersonate the prover.*

In the $MIM - model$, it is assumed that the attacker cannot decide on the alterations of the exchanged messages during the execution of the protocol. This is the class of the most practical MIM attacks, in which the attacker cannot perform computations on the fly during the execution. This class includes all the MIM attacks that have been proposed so far against LPN-based authentication protocols ([10], [24]).

We define the advantage of an adversary $\mathcal{A}$ against $AUTH^{\#}$ protocol in the $ACT - model$ and the $MIM - model$ as the overhead success probability over the false acceptance probability P_{FA} in impersonating the tag:

$$Adv_{\mathcal{A}}^{ACT}(l, n, \eta, \tau, q) = Pr[\boldsymbol{X} \xleftarrow{\$} \{0,1\}^{(2l,n)}, \mathcal{A}^{\mathcal{P}_{\boldsymbol{X}, \eta}}(1^k) :$$
$$\langle \mathcal{A}, \mathcal{V}_{\boldsymbol{X}, \tau} \rangle = ACC] - P_{FA}.$$

and

$$Adv_{\mathcal{A}}^{MIM}(l, n, \eta, \tau, q) = Pr[\boldsymbol{X} \xleftarrow{\$} \{0,1\}^{(2l,n)},$$
$$\mathcal{A}^{\mathcal{P}_{\boldsymbol{X}, \eta}, \mathcal{V}_{\boldsymbol{X}, \tau}}(1^k) : \langle \mathcal{A}, \mathcal{V}_{\boldsymbol{X}, \tau} \rangle = ACC] - P_{FA}.$$

Proof overview. Mainly we adapt the proof of Theorem 2 in [11]. More precisely, we reduce the security in the MIM-model to the security in the ACT-model. The security in the ACT-model has been already proved in [19]. We will show that if there is an attacker $\mathcal{A}^{\#}$ that can efficiently mount a MIM attack with advantage at least δ against $AUTH^{\#}$, then there is an attacker $\mathcal{A}$ that can mount an active attack. Recall that in the MIM-model, the adversary can modify all the messages exchanged between the reader and the tag. The proof goes as follows.

During the first phase $\mathcal{A}$ has to simulate the tag and the reader for $q^{\#}$ times. As $\mathcal{A}$ has access to an honest tag that it can query freely, there is no difficulty in simulating an honest tag to $\mathcal{A}^{\#}$. The main challenge comes with the task of simulating the honest reader. The strategy that we follow for the reader is easy; the reader accepts the tag only when $\mathcal{A}^{\#}$ does not modify any of the messages.

From the point of view of $\mathcal{A}^{\#}$, the tag is perfectly simulated by $\mathcal{A}$. So the success of the attack depends only on the correct simulation of the reader for $q^{\#}$ executions and the success probability of $\mathcal{A}^{\#}$, i.e. $P_{FA} + \delta$. If p_r is the probability of false simulating the reader (for a single execution), then the overall probability of the attack is given by $(1 - q^{\#} \cdot p_r)(P_{FA} + \delta)$.

LEMMA 1. *[11] Let $\boldsymbol{X}$ be a random $l \times m$ binary matrix and let d be an integer, $1 \leq d \leq \frac{m}{2}$. Then, the probability*

$$p(d) = Pr\left[\min_{\boldsymbol{a} \in \mathbb{F}_2^l, \boldsymbol{a} \neq \boldsymbol{0}_l} (wt(\boldsymbol{a} \cdot \boldsymbol{X})) \leq d\right],$$

is upper bounded by

$$p(d) \leq 2^{-(1 - \frac{l}{m} - H(\frac{d}{m}))},$$

where $H(s) = s \cdot \log_2(\frac{1}{s}) - (1-s) \cdot \log_2(\frac{1}{1-s})$ is the entropy function.

THEOREM 1. *If there is an adversary $\mathcal{A}^{\#}$ that can attack the $AUTH^{\#}$ protocol with parameters (l, n, η, τ) in the MIM-model by modifying $q^{\#}$ protocol executions between the prover and the verifier, with running time $T^{\#}$ and achieving advantage at least $\delta^{\#}$, then, there is an adversary $\mathcal{A}$ that can attack the $AUTH^{\#}$ protocol in the ACT-model with the same parameters by interrogating an honest tag $q^{\#}$ times, with running time at most $T^{\#}$ and with advantage at least $\delta \geq \delta^{\#} - (P_{FA} + \delta^{\#})q^{\#}p_r$, where p_r is a negligible function and P_{FA} is the false acceptance probability.*

PROOF. In the ACT-model, the attacker $\mathcal{A}$ can interrogate a prover. We will show how $\mathcal{A}$ can attack $AUTH^{\#}$ protocol in the ACT-model using the algorithm that the adversary $\mathcal{A}^{\#}$ executes.

During the MIM attack, $\mathcal{A}^{\#}$ is modifying the exchanged messages, and, while, the adversary $\mathcal{A}$ has access to a prover, she has to simulate the behaviour of the verifier. More precisely, her strategy goes as follows.

1. $\mathcal{A}$, simulating the verifier, produces a random bit-string $\boldsymbol{a}$ with length $2l$ and Hamming weight l, and sends it to $\mathcal{A}^{\#}$.

2. $\mathcal{A}^{\#}$ sends $\hat{\boldsymbol{a}} = \boldsymbol{a} \oplus \bar{\boldsymbol{a}}$ to $\mathcal{A}$.

3. $\mathcal{A}$ based on $\hat{\boldsymbol{a}}$ interrogates the prover and sends the produced random binary vector $\boldsymbol{r}$ and the bit-string $\boldsymbol{z}$ to $\mathcal{A}^{\#}$.

4. $\mathcal{A}^{\#}$ produces a new pair $(\hat{\boldsymbol{r}} = \boldsymbol{r} \oplus \bar{\boldsymbol{r}}, \hat{\boldsymbol{z}} = \boldsymbol{z} \oplus \bar{\boldsymbol{z}})$ and sends it to $\mathcal{A}$.

5. $\mathcal{A}$ simulates the verifier as follows. If the triplet $(\bar{\boldsymbol{a}}, \bar{\boldsymbol{r}}, \bar{\boldsymbol{z}})$ is all-zero, the simulated verifier; i.e. $\mathcal{A}$, answers "accept". Otherwise, it rejects.

The previous steps are repeated $q^{\#}$ times. Then, the adversary $\mathcal{A}$ impersonates the prover to a verifier in the ACT-attack, by using the second phase of $\mathcal{A}^{\#}$.

The overall probability p^A of the attack that $\mathcal{A}$ mounts is given by

$$p^A = p_{auth} \cdot (P_{FA} + \delta) \qquad (1)$$

where p_{auth} is the probability of successfully simulating a verifier's behaviour and it depends on the ability of the adversary to simulate the last step; i.e. the acceptance or rejection decision.

Next, we compute p^A. In order for the attack to be successful, the adversary $\mathcal{A}$ must be able to simulate the reader's behavior for $q^{\#}$ consecutive executions of the protocol. Let p_r be the probability to fail in a single execution. Then,

$$p_{auth} = (1 - q^{\#} \cdot p_r). \qquad (2)$$

The probability of false rejecting, when the triplet $(\bar{\boldsymbol{a}}, \bar{\boldsymbol{z}}, \bar{\boldsymbol{r}})$ is all zero, i.e. when $\mathcal{A}^{\#}$ does not modify any of the messages, is P_{FR}. That is, $p_r \geq P_{FR}$.

When $(\bar{\boldsymbol{a}}, \bar{\boldsymbol{z}}, \bar{\boldsymbol{r}}) \neq (\boldsymbol{0}_{2l}, \boldsymbol{0}_n, \boldsymbol{0}_l)$, the probability p_r of false simulating the reader is also defined by the probability that the condition $\mathrm{wt}(\hat{\boldsymbol{z}} \oplus \hat{\boldsymbol{r}}^T \cdot \boldsymbol{X}_{\downarrow a}) \leq \tau$, where $n\eta \leq \tau \leq \frac{n}{2}$, is satisfied. We use $FAIL$ to indicate this event.

The sum $\hat{\boldsymbol{z}} \oplus \hat{\boldsymbol{r}}^T \cdot \boldsymbol{X}_{\downarrow a}$ can be written as

$$\bar{\boldsymbol{z}} \oplus \boldsymbol{r}^T \cdot \boldsymbol{X}_{\downarrow \hat{a}} \oplus \boldsymbol{\nu} \oplus \hat{\boldsymbol{r}}^T \cdot \boldsymbol{X}_{\downarrow a} =$$
$$\bar{\boldsymbol{z}} \oplus \boldsymbol{r}^T \cdot \boldsymbol{V}(\bar{\boldsymbol{a}} \oplus \boldsymbol{a}) \cdot \boldsymbol{X} \oplus \boldsymbol{\nu} \oplus (\bar{\boldsymbol{r}}^T \oplus \boldsymbol{r}^T) \cdot \boldsymbol{V}(\boldsymbol{a}) \cdot \boldsymbol{X} =$$
$$\bar{\boldsymbol{z}} \oplus (\boldsymbol{r}^T \cdot (\boldsymbol{V}(\boldsymbol{a}) \oplus \boldsymbol{V}(\bar{\boldsymbol{a}} \oplus \boldsymbol{a})) \oplus \bar{\boldsymbol{r}}^T \cdot \boldsymbol{V}(\boldsymbol{a})) \cdot \boldsymbol{X} \oplus \boldsymbol{\nu}.$$

Let $\boldsymbol{y}_{\bar{\boldsymbol{a}}, \bar{\boldsymbol{z}}, \bar{\boldsymbol{r}}} = \bar{\boldsymbol{z}} \oplus (\boldsymbol{r}^T \cdot \boldsymbol{DV}_{\boldsymbol{a}}(\bar{\boldsymbol{a}}) \oplus \bar{\boldsymbol{r}}^T \cdot \boldsymbol{V}(\boldsymbol{a})) \cdot \boldsymbol{X}$ and let $\beta_{\bar{\boldsymbol{a}}, \bar{\boldsymbol{z}}, \bar{\boldsymbol{r}}}$ be the Hamming weight of $\boldsymbol{y}_{\bar{\boldsymbol{a}}, \bar{\boldsymbol{z}}, \bar{\boldsymbol{r}}}$. Then, $n - \beta_{\bar{\boldsymbol{a}}, \bar{\boldsymbol{z}}, \bar{\boldsymbol{r}}}$ bits of $\boldsymbol{y}_{\bar{\boldsymbol{a}}, \bar{\boldsymbol{z}}, \bar{\boldsymbol{r}}} \oplus \boldsymbol{\nu}$ follow a Bernoulli distribution of parameter η and the rest $\beta_{\bar{\boldsymbol{a}}, \bar{\boldsymbol{z}}, \bar{\boldsymbol{r}}}$ bits follow a Bernoulli distribution of parameter $1 - \eta$. That is, the Hamming weight $\mathrm{wt}(\boldsymbol{y}_{\bar{\boldsymbol{a}}, \bar{\boldsymbol{z}}, \bar{\boldsymbol{r}}} \oplus \boldsymbol{\nu})$ follows a binomial distribution of expected value $\mu = (n - \beta_{\bar{\boldsymbol{a}}, \bar{\boldsymbol{z}}, \bar{\boldsymbol{r}}})\eta + \beta_{\bar{\boldsymbol{a}}, \bar{\boldsymbol{z}}, \bar{\boldsymbol{r}}}(1 - \eta)$ and variance $\sigma^2 = n\eta(1 - \eta)$.

Since, the expected value is a function of $\beta_{\bar{\boldsymbol{a}}, \bar{\boldsymbol{z}}, \bar{\boldsymbol{r}}}$ we can easily verify that for $\beta_{\bar{\boldsymbol{a}}, \bar{\boldsymbol{z}}, \bar{\boldsymbol{r}}} \geq 1 + \lfloor \frac{\tau - \eta n}{1 - 2\eta} \rfloor$, it holds that $\mu > \tau$. For any $\beta_{\bar{\boldsymbol{a}}, \bar{\boldsymbol{z}}, \bar{\boldsymbol{r}}} \geq 1 + \lfloor \frac{\tau - \eta n}{1 - 2\eta} \rfloor$, any $\boldsymbol{X}$ and $\boldsymbol{\nu}$, it holds that

$$Pr[FAIL] =$$
$$Pr_{\boldsymbol{\nu}}[FAIL | dmin(\boldsymbol{X}) > \beta_{\bar{\boldsymbol{a}}, \bar{\boldsymbol{z}}, \bar{\boldsymbol{r}}}]Pr_{\boldsymbol{X}}[dmin(\boldsymbol{X}) > \beta_{\bar{\boldsymbol{a}}, \bar{\boldsymbol{z}}, \bar{\boldsymbol{r}}}]+$$
$$Pr_{\boldsymbol{\nu}}[FAIL | dmin(\boldsymbol{X}) \leq \beta_{\bar{\boldsymbol{a}}, \bar{\boldsymbol{z}}, \bar{\boldsymbol{r}}}]Pr_{\boldsymbol{X}}[dmin(\boldsymbol{X}) \leq \beta_{\bar{\boldsymbol{a}}, \bar{\boldsymbol{z}}, \bar{\boldsymbol{r}}}]$$

where $dmin(\boldsymbol{X}) = \min_{\boldsymbol{a} \in \mathbb{F}_2^l, \boldsymbol{a} \neq \boldsymbol{0}_l}(\mathrm{wt}(\boldsymbol{a} \cdot \boldsymbol{X}))$.

When, $\mu > \tau$; i.e. $\beta_{\bar{\boldsymbol{a}}, \bar{\boldsymbol{z}}, \bar{\boldsymbol{r}}} \geq 1 + \lfloor \frac{\tau - \eta n}{1 - 2\eta} \rfloor$ from the Chernoff bound we have that $\mathrm{wt}(\boldsymbol{y}_{\bar{\boldsymbol{a}}, \bar{\boldsymbol{z}}, \bar{\boldsymbol{r}}} \oplus \boldsymbol{\nu}) < \tau$ with probability less than $e^{-\frac{(\mu - \tau)^2}{2\mu}}$ and the simulation fails. From the above observation and from Lemma 1, we have that

$$Pr[FAIL] \leq Pr_{\boldsymbol{\nu}}[FAIL | dmin(\boldsymbol{X}) > \beta_{\bar{\boldsymbol{a}}, \bar{\boldsymbol{z}}, \bar{\boldsymbol{r}}}]$$
$$+ Pr_{\boldsymbol{X}}[dmin(\boldsymbol{X}) \leq \beta_{\bar{\boldsymbol{a}}, \bar{\boldsymbol{z}}, \bar{\boldsymbol{r}}}]$$
$$\leq e^{-\frac{(\mu - \tau)^2}{2\mu}} + 2^{-n + 2l + nH(\frac{1 + \lfloor \frac{\tau - \eta n}{1 - 2\eta} \rfloor}{n})}.$$

Similarly to [11], in order to ascertain that the first term is negligible, we define $\hat{d}$ the least integer such that $\mu((\hat{d}) > (1 + c)\tau$ for some $c > 0$ and for all $d \geq \hat{d}$, $e^{-\frac{(\mu - \tau)^2}{2\mu}} \leq e^{-\frac{(c\tau)^2}{2(c + 1)}}$. Also, for practical values of the parameters the exponent of the second term is negative, while the P_{FR} is negligible. Thus, from (1) and (2), the overall probability of the attack is lower bounded by

$$(1 - q^{\#} \cdot (e^{-\frac{(\mu - \tau)^2}{2\mu}} + 2^{-n + 2l + nH(\frac{1 + \lfloor \frac{\tau - \eta n}{1 - 2\eta} \rfloor}{n})}))$$
$$\cdot (P_{FA} + \delta) < p^A.$$

$\square$

From Theorem 1, any efficient attacker achieving a noticeable advantage $\delta^{\#}$ against the $AUTH^{\#}$ protocol in the MIM-model can be turned into an efficient attacker against the same protocol in the ACT-model. However, from [19], this contradicts the hardness assumption of the subspace LPN problem.

5. CONCLUSIONS

The design of lightweight authentication protocols is a challenging task. One of the most recent proposals, $AUTH$, was introduced in 2011 by Kiltz et al., in their Eurocrypt best paper price awarded paper. One of the main advantages of $AUTH$ is the elegant security proof, against passive and active attacks, based on the conjectured hardness of the LPN problem that supports its security analysis. However, due to its high communication complexity, Kiltz et al. presented a variant of $AUTH$ with significant smaller communication overhead, but with higher storage complexity. It was also proved that this variant was at least as secure as $AUTH$.

In this paper, we have revisited the security of $AUTH$ and have shown that its variant is much more secure. More precisely, we showed that it can resist powerful MIM attacks and we provided a security proof based on the hardness of the LPN problem to support our security analysis. However, it remains an interesting open problem the designing of a variant of $AUTH$ that has both small storage and communication complexity.

6. ACKNOWLEDGEMENTS

This research is performed in the framework of the INTERREG III Poseidon project, which is funded by the European Union (80%) and National Funds of Greece and Cyprus (20%).

7. REFERENCES

[1] D. Angluin and P. Laird. Learning from Noisy Examples. *Machine Learning*, vol. 2(4), 1987, pp. 343–370.

[2] G. Avoine. *RFID Security and Privacy Lounge*. The list of papers is available at http://www.avoine.net/rfid/download/bib/bibliography-rfid.pdf.

[3] E. R. Berlekamp, R. J. McEliece, V. Tilborg. On the Inherent Intractability of Certain Coding Problem. *IEEE Transactions on Information Theory*, vol. 24, 1978, pp. 384-386.

[4] A. Blum, A. Kalai, and H. Wasserman. Noise-Tolerant Learning, the Parity Problem, and the Statistical Query Model. Journal of the ACM, vol. 4, 2003, pp. 506-519.

[5] J. Bringer, H. Chabanne, EH. Dottax. HB^{++}: a Lightweight Authentication Protocol Secure against Some Attacks. In Proc. of the IEEE Int. Conference on Pervasive Sevices, Workshop - SecPerU, 2006.

[6] J. Bringer, H. Chabanne. *Trusted-HB*: A Low-Cost Version of HB Secure Against Man-in-the-Middle AttackHB^{++}. *IEEE Transactions on Information Theory*, vol. 54, 2008, pp. 4339-4342.

[7] C. Bosley, K. Haralambiev, A. Nicolosi. HB^N: An HB-like protocol secure against man-in-the-middle attacks. Cryptology ePrint Archive, Report 2011/350 (2011), http://eprint.iacr.org.

[8] D.N. Duc and K. Kim. Securing HB^+ against GRS Man-in-the-Middle Attack. In Proc. of the Symp. on Cryptography and Information Security, 2007.

[9] M.P.C. Fossorier, M.J. Mihaljevic, H. Imai, Y. Cui, and K. Matsuura. A Novel Algorithm for Solving the LPN Problem and its Apllication to Security Evaluation of the HB Protocol for RFID Authentication. Cryptology ePrint Archive, Report 2006/197, http://eprint.iacr.org, 2006.

[10] H. Gilbert, M. Robshaw, and Y. Silbert. An Active Attack against HB^+-a Provable Secure Lightweighted Authentication Protocol. Cryptology ePrint Archive, Report 2005/237, http://eprint.iacr.org, 2005.

[11] H. Gilbert, M. Robshaw, and Y. Silbert. $HB^\#$: Increasing the Security and Efficiency of HB^+. In Proc. of Eurocrypt, Springer LNCS, vol. 4965, 2008, pp. 361–378.

[12] H. Gilbert, M. Robshaw, and Y.Seurin. How to Encrypt with the LPN Problem. In Proc. of ICALP '08, LNCS 5126, 2008, pp. 679-690.

[13] J. Hastad. Some Optimal Inapproximability Results. *J. ACM*, vol. 48 (4), 2001, pp. 798-859.

[14] N.J. Hopper, and M., Blum. Secure Human Identification Protocols. In Proc. of Asiacrypt, Springer LNCS, vol. 2248, 2001, pp. 52–66.

[15] A, Juels, and S.A. Weis. Authenticating Pervasive Devices with Human Protocols. In Proc. of Crypto, Springer LNCS, vol. 3126, 2005, pp. 293–308.

[16] A. Jain, S. Krenn, K. Pietrzak and Aris Tentes. Commitments and Efficient Zero-Knowledge Proofs from Hard Learning Problems. In Proc. of Asiacrypt, Springer LNCS, vol. 7658, 2012, pp. 663–680.

[17] J. Katz, and A. Smith. Analyzing the HB and HB^+ Protocols in the Large Error Case. Cryptology ePrint Archive, Report 2006/326, http://eprint.iacr.org/, 2006.

[18] J. Katz, and J. Shin. Parallel and Concurrent Security of the HB and HB^+ Protocols. *Journal of Cryptology*, vol. 23, 2010, pp. 402–421.

[19] E. Kiltz, K. Pietrzak, D. Cash, A. Jain, and D. Venturi. Efficient Authentication from Hard Learning Problems. In Proc. of Eurocrypt, Springer LNCS, vol. 6632, 2011, pp. 7–26.

[20] M. Kearns. Efficient noise-tolerant learning from statistical queries. In Proc. of the 25th ACM Symposium on Theory of Computing, 1993, pp. 392–401.

[21] X. Leng, K. Mayes, and K. Markantonakis. HP-MP^+: An Improvement on the HB-MP Protocol. In Proc. of the IEEE Int. Conference on RFID 2008, IEEE Press, 2008, pp. 118–124.

[22] J. Munilla, and A. Peinado. HP-MP: A Further Step in the HB-family of Lightweight authentication protocols. *Computer Networks*, Elsevier, vol. 51, 2007, pp. 2262–2267.

[23] E. Levieil, and P.A. Fouque. An improved LPN Algorithm. In Proc. of SCN, Springer LNCS 4116, 2006, pp. 348–359.

[24] K. Ouafi, R. Overbeck, V. Vaudenay. On the Security of $HB^\#$ against a Man-in-the-Middle Attack. In Proc. of Asiacrypt, Springer LNCS, vol. 5350, 2008, pp. 108–124.

[25] K. Pietrzak. Subspace LWE. 2010. Manuscript available at http://homepages.cwi.nl/pietrzak/publications/SLWE.pdf.

[26] S. Piramuthu. HB and Related Lightweight Authentication Protocols for Secure RFID Tag/Reader Authentication. In Proc. of CollECTeR Europe Conference, Basel, Switzerland, 2006.

[27] O. Regev. On Lattices, Learning with Errors, Random Linear Codes, and Cryptography. In Proc. of STOC, ACM, 2005, pp. 84–93.

[28] P. Rizomiliotis. HB-MAC: Improving the Random - $HB^\#$ Authentication Protocol. In Proc. of the 6th International Conference on Trust, Privacy and Security in Digital Business (TrustBus), Springer, LNCS 5695, 2009, pp. 159–168.

[29] P. Rizomiliotis and S. Gritzalis. On the security of AUTH, a provably secure authentication protocol based on the subspace LPN problem. Accepted for publication in the Int. J. of Inform. Security, 2012.

SeDyA: Secure Dynamic Aggregation in VANETs

Rens W. van der Heijden, Stefan Dietzel
University of Ulm
Institute of Distributed Systems
rens.vanderheijden@uni-ulm.de
stefan.dietzel@uni-ulm.de

Frank Kargl
University of Ulm, Inst. of Distributed Systems
University of Twente, DIES Group
frank.kargl@uni-ulm.de

ABSTRACT

In vehicular ad-hoc networks (VANETs), a use case for mobile ad-hoc networks (MANETs), the ultimate goal is to let vehicles communicate using wireless message exchange to provide safety, traffic efficiency, and entertainment applications. Especially traffic efficiency applications benefit from wide-area message dissemination, and aggregation of information is an important tool to reduce bandwidth requirements and enable dissemination in large areas. The core idea is to exchange high quality summaries of the current status rather than forwarding all individual messages. Securing aggregation schemes is important, because they may be used for decisions about traffic management, as well as traffic statistics used in political decisions concerning road safety and availability. The most important challenge for security is that aggregation removes redundancy and the option to directly verify signatures on atomic messages. Existing proposals are limited, because they require roads to be segmented into small fixed-size regions, beyond which aggregation cannot be performed. In this paper, we introduce SeDyA, a scheme that allows more dynamic aggregation compared to existing work, while also providing stronger security guarantees. We evaluate SeDyA against existing proposals to show the benefits in terms of information accuracy, bandwidth usage, and resilience against attacks.

Categories and Subject Descriptors

C.2.0 [**Computer-Communication Networks**]: General—*Security and protection (e.g., firewalls)*; C.2.2 [**Computer-Communication Networks**]: Network Protocols

General Terms

Algorithms; Design; Security

Keywords

VANETs; secure aggregation; multi-hop communication

1. INTRODUCTION

In the past decade, research on vehicular ad-hoc networks (VANETs) has developed from a challenging research application of mobile ad-hoc networks (MANETs) to a viable type of networks that is set to be deployed in the coming decade. A number of standards have been developed, including IEEE 802.11p-2010 Amendment, which was recently incorporated into 802.11-2012 [19], the IEEE 1609 draft standards [20], and ETSI TS 102 637-* [7–9] to facilitate this deployment. These standards provide specifications for on-board units (OBUs) with which vehicles will be equipped and the communication they will use to create a VANET. Communication will potentially be supported by a network of road-side units (RSUs) that provide connection to a back-end infrastructure and the Internet. However, RSU coverage is likely to be sparse, especially on highways. The core research challenges of VANETs include the highly dynamic network topology and short reaction times, which provide strict bandwidth limitations.

For some envisioned applications for VANETs, it suffices to provide high-frequency periodic one-hop communication (beaconing) [8] between vehicles, to exchange current position, speed, and other environmental data. However, many next-generation applications, especially traffic efficiency applications, require multi-hop communication, creating a demand for even more bandwidth. To improve traffic efficiency, knowledge about the area beyond the direct neighborhood of the vehicle is necessary, typically in the order of several kilometers. Such knowledge allows for the detection of traffic jams and alternative routes. However, it is not feasible to simply forward messages generated by beaconing in large areas, due to the well-studied broadcast storm problem [27]. However, applications such as traffic information systems often do not require exact information for their decisions. Information about average speed in certain regions is enough to enable efficient routing. Hence, many authors have proposed to apply in-network aggregation to VANETs [2, 21, 22, 26, 29]. While aggregation can be applied to many different applications, including traffic information systems, road conditions, temperature, and parking spot availability, we will focus on traffic information systems as the main use case for the remainder of the paper. However, we note that our scheme is flexible enough to be applied to other domains, as well. In traffic information systems, aggregation does not only involve the computation of a sum, count or average, but also the dynamic selection of the area over which the aggregate should be computed.

Besides bandwidth efficiency, security is a highly challenging goal for VANETs, due to the potential impact of a successful attack, which could cause crashes or traffic jams [3]. To provide security, the IEEE 1609.2 standard [18] provides certification for each vehicle, requiring signatures on all the messages that are sent. For privacy protection, short-term certificates, typically called pseudonyms, are used. While signatures and certificates provide reasonable sender attribute authentication, hence preventing attacks with commodity hardware, it is likely that insider attackers will be able to extract key material from cars they own and use it for attacks. Therefore, additional security mechanisms are required, both to detect misbehavior of nodes, as well as to maintain a minimum level of information accuracy. For aggregation, security is even more challenging, because vehicles merge information received from several other vehicles and remove redundancy, as well as the original signatures and certificates. Then, only the aggregates information is disseminated further. Thus, unstructured aggregation can allow an attacker to claim his sensor readings are supported by hundreds of vehicles, while these vehicles never existed. Moreover, legitimate nodes may transmit invalid data if they aggregate a set of messages that includes messages from an attacker. Conventional cryptographic signatures, as provided by IEEE 1609.2 [18], cannot solve these two challenges, because they do not foresee advanced cryptographic protocols beyond basic signatures and certificates.

To address these shortcomings, researchers have proposed several new aggregation mechanisms that explicitly address security. Most of these mechanisms are based on a specific underlying aggregation mechanism, which assumes a fixed segmentation of the road. Such aggregation mechanisms are known to not scale well to large area dissemination. Moreover, most existing schemes employ probabilistic counting techniques, such as Flajolet-Martin (FM) sketches [12] as their underlying data structure. Due to the probabilistic nature of the sketches, the resulting schemes that allow aggregation of discrete values still allow attackers to influence aggregated values within certain error bounds.

In this paper, we introduce a new scheme, called Secure Dynamic Aggregation (SeDyA), to provide a stronger and more flexible security mechanism, while still remaining feasible in terms of bandwidth requirements. In particular, our mechanism uses aggregation with flexible road segmentation to allow for good scalability. Like existing schemes, we use secured FM sketches as a basis for our aggregation mechanism, but add to the basic signature-based security mechanisms with a combination of plausibility checks and advanced cryptography, in particular multisignatures and identity-based signatures, to provide stronger guarantees. We evaluate our mechanism against multi-hop beaconing, as well as existing secure aggregation schemes to assess the bandwidth efficiency, accuracy, and security against attacks. In addition, a variant of our scheme that applies only our novel security mechanism, without the security on the FM sketches themselves, provides an improved trade-off between security and bandwidth consumption.

The remainder of the paper is structured as follows: first we introduce the requirements and attacker model in Section 2, followed by a discussion of related work in Section 3. We discuss our new scheme in Section 4 and evaluate it in Section 5. Section 6 concludes the paper with a discussion of open issues and future work.

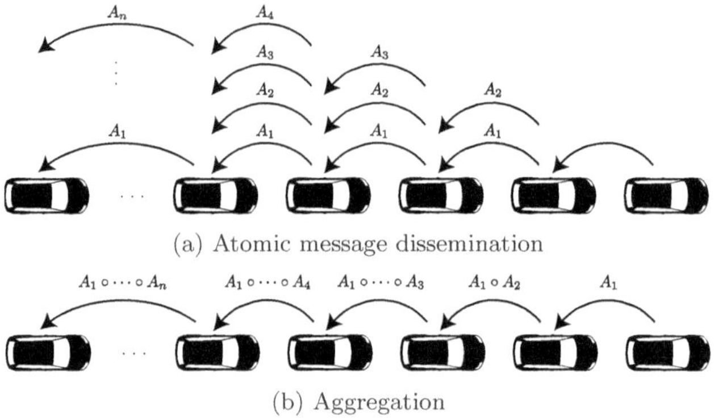

(a) Atomic message dissemination

(b) Aggregation

Figure 1: Comparison of atomic message dissemination and aggregation.

2. SYSTEM MODEL

To foster a better understanding of our proposed security mechanisms, we will introduce a generic model for data aggregation mechanisms, and outline the specific attacker model we assume.

2.1 Network Model

VANETs are a form of mobile ad-hoc network, which is formed by vehicles on the road. Each vehicle is equipped with a wireless communication device operating at 5.9 GHz according to the IEEE 802.11p-2010 Amendment now incorporated in 802.11-2012 [19]. Wireless communication operates similar to the well-known 802.11a-1999 Amendment in 802.11-2012 [19], but on dedicated frequencies. The expected communication range is approximately 250 meters, depending on shadowing effects. Some papers foresee that the networks are supported by roadside infrastructure, which could relay information from and to back-end systems. Additionally, cellular networks could be envisioned to provide additional information dissemination means. However, both kinds of infrastructure are unlikely to be deployed on a large scale. Especially in highway scenarios, the availability of infrastructure is unlikely in the near future. Thus, communication protocols should ideally rely solely on communication between vehicles, possibly spanning multiple hops. The main challenge for multi-hop protocols is to overcome the ephemeral nature of vehicular communication due to high vehicle mobility.

Both periodic single hop data dissemination and multi-hop dissemination of event notifications have recently been standardized in Europe [8, 9] and the US [20]. However, no mechanisms have been standardized yet to support periodic multi-hop dissemination of environmental data, such as current traffic situation. One of the reasons for missing standardization is that simple relaying of information is too bandwidth inefficient to be used for large scale dissemination of messages, as shown in Figure 1.

Hence, a number of protocols have been proposed (cf. Section 3.1) that employ information aggregation. The basic idea is simple: whenever a vehicle receives information from a neighbor, it decides whether the received information can be merged with already known information, and only the merged information is disseminated further. Further vehicles do the same, resulting in an aggregated view of a stretch of road. For instance, a traffic information system will form

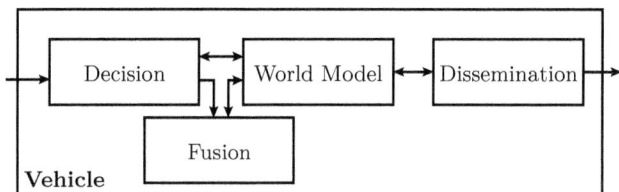

Figure 2: Basic architecture model of aggregation mechanisms.

aggregates like "there is a traffic jam from kilometer X to kilometer Y on road Z", which can be encoded efficiently and disseminated in large regions.

Due to the lack of infrastructure, the aggregated view is created as a collaborative effort of all vehicles on the road, and each vehicle operates as an equal peer. While variations are possible, most aggregation mechanisms follow the basic structure shown in Figure 2 [4]. Four main components define the aggregation process. A vehicle receives information from local sensors or remote vehicles.

- The **decision** component compares newly-received information with other information already contained in the *world model* and decides whether two items of information are similar enough to be aggregated.

- The **fusion** component takes two items of information and merges them to form an aggregate, for instance by averaging the speed contained.

- The **world model** represents a vehicle's current knowledge about the surroundings.

- Periodically, the **dissemination** component selects a, possibly further summarized, subset of the *world model* for dissemination to other nodes.

The main challenge of the different components is to create a bandwidth-efficient yet non-biased summary of the real world situation that is still accurate enough to support applications such as advanced navigation systems.

2.2 Attacker Model

We assume an insider attacker that aims to disrupt traffic, typically for his own gain, by creating false traffic jam reports or hiding real traffic jams. Besides such targeted attacks, we also assume attackers interested in disrupting the aggregation mechanisms, exploiting denial of service attacks, and reducing the quality of disseminated information by as much as possible.

We assume that the entire implementation of the aggregation mechanism and all its parameters are known to the attacker, allowing the computation of a maximum impact given sufficient information about other vehicles on the road. Moreover, the attacker has complete control of the OBU of any vehicle he physically owns, including key material. However, the attacker cannot obtain private key material from other cars remotely. Like most other works, we do not consider denial of service attacks that simply disrupt the network by jamming globally, or by violating MAC protocol parameters to prevent any transmissions from occurring.

Although the attacker may control multiple vehicles, it is assumed that the majority of the network participants is

honest. Due to the relatively large aggregation areas that SeDyA will use, this assumption also implies that the aggregation area contains an honest majority. The assumption of an honest majority is motivated by the price of a vehicle. The attacker cannot obtain many pseudonyms for the same vehicle in the same timespan, because we assume that the pseudonym mechanism is secure. This means that the pseudonyms provided to the vehicle are bounded to use in a relatively short timespan, on the order of a few minutes.

However, for location privacy, it has been shown that it is not sufficient to have one pseudonym at any time. To solve this issue, pseudonym exchange schemes like Mix-Zones [15] use several pseudonyms in the exchange period, requiring a period in which an attacker can have more than one pseudonym. There are many pseudonym mechanisms in related work, but a discussion of these is out of scope for this paper. For the purpose of this work, we assume that the attacker is bounded to at most three pseudonyms at any time, which allows for the complete overlap of two pseudonym change periods.

3. RELATED WORK

Data aggregation is a well-researched field with applications in different domains that require data summarization to achieve higher efficiency. Before the advent of VANETs, aggregation has been widely researched in the domain of wireless sensor networks (WSNs). Several mechanisms for data aggregation [11] and corresponding integrity protection mechanisms [25] have been proposed. In contrast to VANETs, WSNs typically consist of low-cost sensor devices, which are deployed in the field to collect a set of data. Data aggregation is used to combine data from several nodes, which is then forwarded to central back-end systems. Aggregation mechanisms developed for WSNs are typically not applicable to VANETs, because they assume limited node mobility, hierarchic structures, and few data sinks interested in the aggregation result.

Hence, a different set of mechanisms, which is optimized for high node mobility and a large set of interested vehicles, has been proposed for aggregating data in VANETs. We will first survey a number of aggregation schemes that do not consider security specifically to highlight a number of challenges when designing an efficient aggregation schemes for VANETs. Following, we introduce existing work on secure aggregation.

3.1 VANET Aggregation

One of the first mechanisms proposed is SOTIS [29]. Here, the road is divided into segments of fixed size, which correspond to wireless range. All vehicles continuously send beacons containing their current velocity, and beacons from the same segment are combined by calculating the average speed. Only the average speed per segment is disseminated in a larger area. While SOTIS reduces the communication overhead by only forwarding averages per segment, it does not scale to larger areas [26]. Since segment size is fixed, the total bandwidth reduction achieved is constant, as opposed to schemes which use larger segment sizes to disseminate information about areas further away. Moreover, fixed segments can fail to correctly depict small traffic jams at segment borders, and waste bandwidth in case of larger phenomena spanning multiple segments, which could be aggregated further. Hence, newer aggregation schemes use flex-

Old sketch: | 1 | 0 | 0 | 1 |

$H(c_1) =$ | 0 | 0 | 1 | 0 |

$H(c_2) =$ | 0 | 0 | 0 | 1 | **OR**

| 1 | 0 | 1 | 1 | $\rightsquigarrow \#elements = 2^2/\rho$

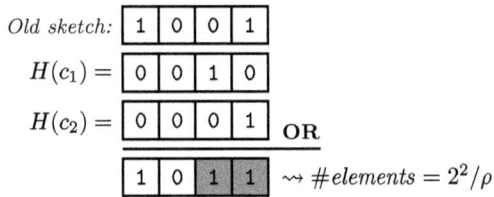

Figure 3: Two vehicles with identities c_1, c_2 add their count to an existing sketch. The initial sequence of uninterrupted ones, here $l = 2$, is used to approximate the total number of values in the sketch.

ible road segmentation [2, 3, 22] to better adapt to the real situation on the road.

The aforementioned schemes suffer from the problem of duplicate reports. If a vehicle's velocity is counted multiple times in the same aggregate, the resulting average will be biased. Lochert et al. [21] propose to use an enhanced version of Flajolet Martin (FM) sketches [12] for aggregating the number of free parking spots in certain city regions with automatic duplicate elimination. We will shortly introduce the FM sketch data structure, because several secure aggregation mechanisms build on the idea of securing FM sketches against malicious manipulation.

Originally used for large databases, FM sketches are a counting method that can be used to approximate the total number of distinct unique values with low storage overhead. The data structure used is a sequence of m bits, which is initially set to all zeros. To increase the sketch's count by 1, a hash $H(c_i)$ of an element's unique id c_i is calculated where H is a geometric hash function. Using a geometric hash function, the probability that the output is n is $1/2^n$. Then, the $h(i)$-th bit in the sketch is set to 1, as shown in Figure 3. Because the hash is based on the unique id, the same element can be added multiple times without setting more than one bit to 1. The number of elements in the sketch can be approximated using the length l of the initial uninterrupted sequence of one bits in the sketch as:

$$\#elements = 2^l/\rho$$

with $\rho \approx 0.775351$. To increase the approximation's accuracy, multiple sketches can be used and their results averaged. The approach using multiple sketches is called Probabilistic Counting with Stochastic Averaging (PCSA) [12].

In addition to simple counts, sketches can be adapted to represent sums and averages. Due to their efficient representation of duplicate insensitive merging functions, they have been widely adopted for aggregation schemes.

3.2 Secure Aggregation

The aggregation mechanisms discussed so far do not consider security, most notably integrity, of aggregates explicitly. Hence, it is possible for a malicious attacker to modify the reported values. Raya et al. [24] propose a security mechanism that can be applied to fixed segments aggregation schemes, such as SOTIS. Once all vehicles within a segment have agreed on an average value, the goal is to secure the average against further modification. The paper discusses three different signature schemes to achieve integrity protection. Simply attaching all participating vehicles' signatures as a list achieves the lowest computational overhead

while requiring a lot of bandwidth to accommodate all signatures. Onion signatures, meaning that each vehicle resigns the aggregate's existing signature instead of adding its own signature to a list, reduce the bandwidth usage. On the other hand, they increase the computational overhead, because each receiving vehicle needs to re-calculate all signatures. The biggest limitation of the scheme is that it can only be applied to fixed segments aggregation, which has been proven to not scale well [26].

Instead of first agreeing on an aggregate value and then signing it, Dietzel et al. [5] add signed atomic reports, which have been used in an aggregate's calculation, as attestation meta-data. This meta-data is then used to verify that the atomic values have been correctly aggregated. To save bandwidth, only a subset of of all atomic values is added. The subset is chosen such that the atomic reports' locations are equally distributed throughout the aggregate area. Still, the resulting mechanism can only provide a heuristic for the aggregates' correctness. Because not all signed atomic reports are added, an attacker can still modify the resulting aggregate within certain limits. However, the scheme can be applied to arbitrary dynamic aggregation schemes, because no fixed values need to be agreed before calculating the attestation meta-data.

Picconi et al. [23] follow a different idea for achieving probabilistic integrity protection. Their scheme borrows from interactive commitment schemes where a sender calculates the aggregated value and a cryptographic commitment on the value and sends both to the receiver who then asks for one or more atomic reports used in the calculation. To avoid interactivity, Picconi et al. use trusted hardware in the sender's vehicle, which challenges the commitments instead of the actual receiver. In case of misuse, the role of the trusted hardware is to guarantee that a proof of the failed challenge is sent to allow other vehicles to detect the attacker. Beyond the reliance on trusted hardware, one of the scheme's problems is that it can still only provide a probabilistic detection of attackers.

A first attempt to secure an FM-sketch-based aggregation scheme is made by Garofalakis et al. [13]. The authors only consider aggregation schemes, which count the number of witnesses of a binary event, such as "there is an accident at position X." Each vehicle that agrees to the event hashes its id into the corresponding FM sketch. For each bit set to 1, the vehicle id, bit position in the sketch, and a signature on these values is added as proof data. Hence the sketch is protected against inflation by an attacker. However, a signature and corresponding certificate have to be kept for each bit set to 1. To protect against deflation, a second sketch is kept that represents the inverse count c', that is, $c' = N - c$ if c is the actual count and N the maximum expected value.

Similarly, [17] counts witnesses on binary events, but reduces the overhead of signatures and certificates needed. Instead of FM sketches, z-smallest is used as underlying probabilistic counting method. The idea of z-smallest is that, given n elements uniformly distributed between 0 and 1, the z-smallest element gives an approximation of n by calculating z/c where c is the value of the z-smallest element. To protect against inflation, the authors exploit the fact that it is difficult for an attacker to forge z signed reports with a value smaller than or equal to c.

Han et al. present SAS [16], a scheme to protect average

values rather than binary events. The authors assume that the underlying aggregation scheme uses fixed segments to calculate the average values. Moreover, the authors assume that vehicle only report aggregates to traffic management centers (TMC), which allows then to use symmetric cryptography and shared keys between each car and the TMC. As counting method, FM sketches are used. However, a number of enhancements over Garofalakis' scheme are proposed.

For inflation protection, again signatures for each 1 bit are generated. However, the signatures on those 1 bits that are part of the initial uninterrupted sequence of ones are merged. Only the signatures of the additional 1 bits are kept separate. To protect against deflation, the authors propose to use a hash chain that represents the length of the initial sequence of one bits. Because the hash cannot be inverted, attackers cannot remove bits from the sequence. The size of the hash chain is constant, as is the combined signature on the initial sequence of ones. Only the extra signatures on one bits, which are not part of the initial sequence, use extra space.

SAS eliminates the limitation to binary events and offers a number of bandwidth improvements over other secure aggregation schemes. In addition, SAS provides good protection against inflation and deflation of sketch values. However, the scheme only allows central infrastructure to check the integrity of aggregates. More importantly, the scheme is limited to fixed segments and consequently suffers from limited scalability.

3.3 Open Issues

Having discussed related work, we now point out some key remaining issues, which are the goals that our scheme, SeDyA will aim to solve. The challenge is twofold: dynamic aggregation and secure aggregation.

First, an issue that all secure aggregation schemes have in common is that they are bounded to a predefined aggregation area. This area is defined either completely in advance (fixed segments) or determined when the aggregate is first generated and not changed afterwards. On the other hand, when we examine most VANET aggregation schemes that do not consider security, we observe that the aggregation is performed in a dynamically defined region. This key distinction prohibits secure aggregation schemes from achieving the same accuracy and scalability as aggregation schemes that do not consider security.

Second, there is the challenge of security. A core challenge of secure aggregation is that a receiver of an aggregate must be able to verify that the aggregation process was preformed correctly. One solution that allows such verification is the set of mechanisms applied by SAS [16]. However, SAS does not fit our requirements, as our goal is to directly disseminate information in the network rather than reporting summarized to a centralized back-end. Thus, the goal for SeDyA is to provide a way to allow for dynamic aggregation and address the requirements of typical VANET aggregation scenarios, as discussed in the previous sections. Beyond dynamic segments, the removal of the TMC allows for potential vulnerabilities, as the hash chains that SAS uses can no longer be employed. This is due to the fact that the hash chains are bound to the usage of a central authority (the TMC). We will address these vulnerabilities in the construction of our new scheme, SeDyA.

4. SEDYA

SeDyA provides improvements on related work in two important areas: security and dynamic aggregation. The mechanism is divided into three phases, as shown in Figure 4: the *aggregation phase* (Phase 1) where vehicles collaboratively agree on an aggregate, the *finalization phase* (Phase 2) where additional signatures are added to attest aggregate correctness, and the *dissemination phase* (Phase 3) where the finalized aggregates are disseminated in larger regions. We will provide a short overview of all three phases before explaining each in more detail.

For example, consider a traffic jam of 6 kilometers length. The traffic jam can be described by indicating the 6 km section as the area and providing an aggregated average speed that is close to zero. Because the length of the traffic jam is unknown to the first vehicles initiating the aggregation process, the *aggregation phase* allows a dynamic description of the area. The aggregation phase ends when a minimum duration has expired and a node detects the edge of an area of homogeneous speed. This allows for accurate aggregates with relatively low standard deviation between the contained atomic speed values and a good approximation of the real world situation. The aggregation phase also contains an optional security mechanism, which simplifies earlier detection of malicious aggregates at the cost of additional bandwidth usage.

When the edge of a homogeneous speed area is detected, the *finalization phase* is initiated, which provides additional security: in this phase, the aggregate message is transmitted from the node on the edge back through the aggregation area. As the message is forwarded, each forwarding node checks the contents of the message against its observed values. If the message is accurate, the forwarder attaches a certificate and signature and forwards the message; otherwise, it discards the message. Instead of using regular ECDSA signatures and public key certificates, an identity-based multisignature scheme is used to reduce the signature size. In order to focus on the conceptual aspects of our scheme, we omit details of the cryptographic mechanisms used in this section. However, the underlying cryptography is discussed in Appendix A.

Eventually, the forwarding process will reach the other end of the aggregation area. Once a node detects that its location is not contained within the aggregation area, it initiates the *dissemination phase*. In this phase, the message is disseminated to potentially interested vehicles further away. Each receiving node can compare the amount of attached certificates to the estimated amount of participants to determine the quality of the aggregate. In addition, SeDyA's multisignature provides a method to verify the amount of vehicles that agree with the aggregate. This can be compared to the amount of vehicles that the aggregation mechanism indicates, when performing plausibility checks and the resolution of conflicting information.

4.1 Aggregation Phase

In the aggregation phase, the goal is to agree on the contents of the aggregate. That is, the aggregation area must be determined, and the average speed must be measured. First, a vehicle (e.g., v_1 in Figure 4) encodes the average speed into an aggregate A_1. In addition, the – initially small – aggregation area needs to be encoded. We use two types of probabilistic counting here: FM sketches and LC sketches.

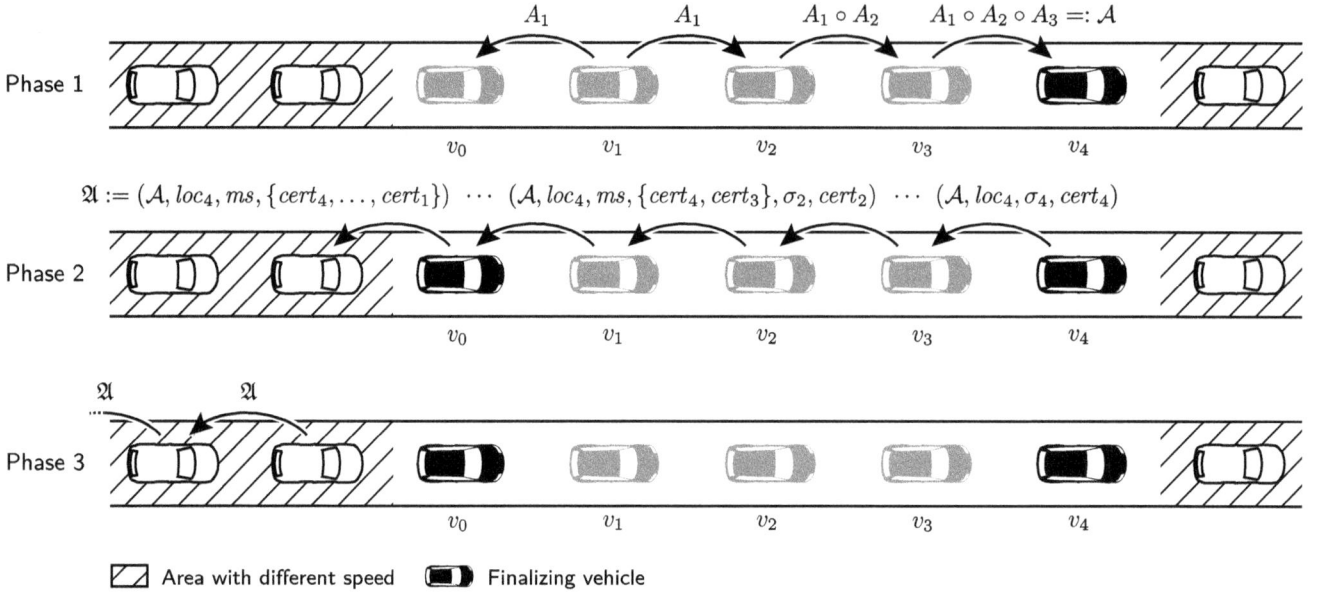

Figure 4: This figure illustrates the different phases of SeDyA. In particular, it shows the aggregate-and-verify steps in the first and second phase.

LC sketches are a variation on FM sketches that is designed to be more accurate, but they cannot represent large values as well as FM sketches can. Their functionality is described and compared to FM sketches in [10]. We use LC sketches to count the amount of participants in the aggregation area, while we use FM sketches to describe the bounds of the area and to describe the measured average speed itself. We apply PCSA over several FM sketches to increase the accuracy of FM sketches at the cost of additional hash operations and bandwidth.

Encoding the average speed is a straight-forward application of FM sketches. To create a dynamic aggregation scheme based on FM sketches, we eliminate fixed segments used by related work and encode the aggregation area as follows. First, the road is marked with a set of fixed *reference points*. Instead of storing its absolute location, each vehicle calculates its relative distance to the last passed reference point, marked as d_i in Figure 5a. We use an FM sketch to store the average of these relative distances as the center of the aggregate's area (C in the figure). Thus, when v_1 creates an aggregate A_1, it will approximate $R_1 - v_1$; when v_2 receives A_1, it computes a new aggregate that will approximate $avg(R_1 - v_1, R_2 - v_2)$, and so on. Whenever a vehicle contributes to the aggregate, the average center is updated by adding its relative distance to the FM sketch and increasing the vehicle count. In the aggregation phase, the area is not yet well-defined, because it can still expand. Once a vehicle determines that it is on the edge of an area of homogeneous speed, it enters the finalization phase, and the area is defined by the location of this edge vehicle (F in Figure 5b) and the center of the area C. The assumption here is that, due to the homogeneity of the vehicles' speeds, the vehicle distribution is mostly uniform within the aggregate area.

During the aggregation phase, aggregates are disseminated as follows. Whenever a new aggregate is started, it is broadcast to direct neighbors. Each vehicle will merge the received aggregates with that of other vehicles and its own observations, by merging the associated FM and LC sketches. The vehicles then forward the message after a random waiting time. This forwarding is bounded, to protect against a typical broadcast storm problem. The amount of messages a node may transmit every second is set to a contention-dependent value between 1 and 10, inversely dependent on the size of the neighbor table. At each merge operation, each vehicle also checks whether it is at the edge of an aggregation area. To detect this, we use the vehicle's sensors and neighbor table to detect a change in average speeds. Each vehicle compares its velocity with the neighbors in front and

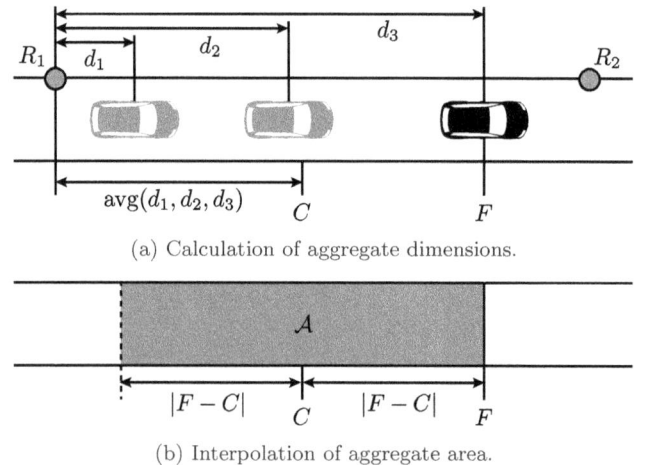

(a) Calculation of aggregate dimensions.

(b) Interpolation of aggregate area.

Figure 5: During the aggregation process, the aggregate's center C is calculated as the average relative distance between participating vehicles and a reference point (R_1). Assuming uniform distribution of vehicles, the aggregate area $\mathcal{A}$ is interpolated as $2 \cdot |F - C|$ where F is the position of the finalizing vehicle.

behind. If the speed difference is higher than a threshold, the aggregate area edge is reached, and the aggregate can be finalized.

Before we discuss the finalization phase in detail, we address the security of the aggregation phase. We consider the security mechanism to be optional in this phase: while it improves the scheme through early detection of malicious aggregates, the main security contribution of SeDyA is in the next phase (the *finalization phase*). The security mechanism in the aggregation phase is inspired by two mechanisms from related work; SAS [16] and AM-FM sketches [13]. We create a signature for each bit in the FM or LC sketch when the bit is set to 1. The values signed are the position of the respective bit, the current time slot, and the reference point with respect to which the added distance is computed. Note that the reference point is the same for every participating vehicle, a prerequisite for the signatures to be verifiable by other vehicles. This approach to securing FM sketches is comparable to the mechanism that SAS and AM-FM apply.

As previously discussed, we use FM sketches to compute averages: this means that each vehicle can sign more than one bit in each sketch. This is analogous to the method used in SAS [16]. In AM-FM sketches [13], this mechanism is not used; instead, multiple rounds of communication are executed to produce an average. As multiple rounds of communication are not feasible in our setting, because we do not aggregate towards a single sink node, using FM sketches to compute averages is the only feasible approach.

However, this approach leaves an open attack vector. An attacker can exploit the fact that he possesses key material to modify an aggregate as desired and then sign the necessary bits. Such manipulation is also possible when attacking schemes in related work, but more challenging in proposals that use a central authority. Although this attack on SeDyA can often be detected and traced back to the attacker through plausibility checks, an intelligent attacker could use the approach to influence the output while remaining hidden. Therefore, we provide an additional security mechanism in our finalization phase. The key difference is that the mechanism in the finalization phase only operates after the aggregation phase; an attack will thus be detected later than by using security in the aggregation phase. We note that due to the significant overhead involved in the security mechanism in the aggregation phase, it may be more efficient to only employ the security mechanism in the finalization phase, despite the delayed detection of an attack.

4.2 Finalization Phase

In this phase, the goal is to provide vehicles within the aggregation area the opportunity to verify the contents of the aggregate. The first step is to produce the finalized aggregate message. This message consists of the aggregate from the previous phase, the current location of the finalizing vehicle, a signature, and a certificate. The finalization location is stored to allow receivers of the message to reconstruct the aggregation area, as described in the previous section. The signature of the finalizing vehicle signifies that it agrees with the contents of the aggregate. Here "agreeing" means that the contained average does not deviate by more than a predefined threshold from the vehicle's own speed readings. The certificate is attached to enable receivers to verify the signature.

Because more than one vehicle may detect the edge of the

aggregate at the same time, there is a short waiting time followed by a repetition of the same message to agree on one finalized aggregate to be disseminated further. After waiting, the finalized message of the vehicle with the lowest ID is selected for further dissemination. At this point, vehicles within the aggregation area may initiate the forwarding protocol. The aim of this protocol is to allow each vehicle to add its signature and certificate to the finalized message. Due to the high bandwidth requirements for implementing this signature process naively, we apply multisignatures to reduce the amount of signatures. Using multisignatures, several signatures can be merged, thus saving bandwidth. However, a list of all corresponding certificates is still necessary to verify the signers' attributes. Hence, we use identity-based signatures to reduce public key size compared to regular certificates. The details of these cryptographic mechanisms are discussed in Appendix A.

When a vehicle receives a signed message in the finalization phase, it will only forward it if it is a correct and legitimate message. This removes the majority of errors and naive attacks. The decision of whether the message is legitimate is a purely data-centric one: the vehicle checks whether it is inside the aggregation area, and then compares its own sensor readings with the contents of the message. If the message aligns with the vehicle's sensor readings, the vehicle signs the message and forwards it. To protect the network from a broadcast storm problem, we bound the maximum amount of messages a vehicle is allowed to broadcast using the same mechanism as discussed in the previous phase. Due to this redundant, flooding-based message dissemination, it is unlikely that an attacker can selectively drop messages to prevent the transmission of an aggregate.

Because of the non-deterministic forwarding mechanism, it is possible that a vehicle receives multiple versions of the aggregate at the same time, but with different, overlapping sets of previous signers. The overlap can be detected because of the attached identity sets, but the multisignature cannot be decomposed. Hence, the merging vehicle would need to create a message with duplicate identities in the resulting identity set. To resolve this, we introduce a specific message format, where each forwarded message consists of the aggregate, the multisignature, the set of certificates, and a separately-kept signature and certificate of the current sender. This allows to resolve duplicate identities when merging aggregates; details of the mechanism are discussed in Appendix B.

Finally, we stress that the goal of the finalization phase is not just to produce a valid multisignature. The multisignature also transfers the information that sufficient nodes in the aggregation area agree with the aggregated data. This fact allows the receiver of such a multisigned message to estimate the amount of vehicles that agree with the message, compared to the amount of participants in the aggregation process. We thus exploit the properties of digital signatures to transfer information about the results of plausibility checks of the signers. As a result, we can transfer trust in the aggregate correctness over multiple hops, allowing for dynamic and large aggregation areas.

4.3 Dissemination Phase

Finally, the dissemination phase is where finalized aggregates are distributed throughout the network. The dissemination phase is initiated when a message reaches the op-

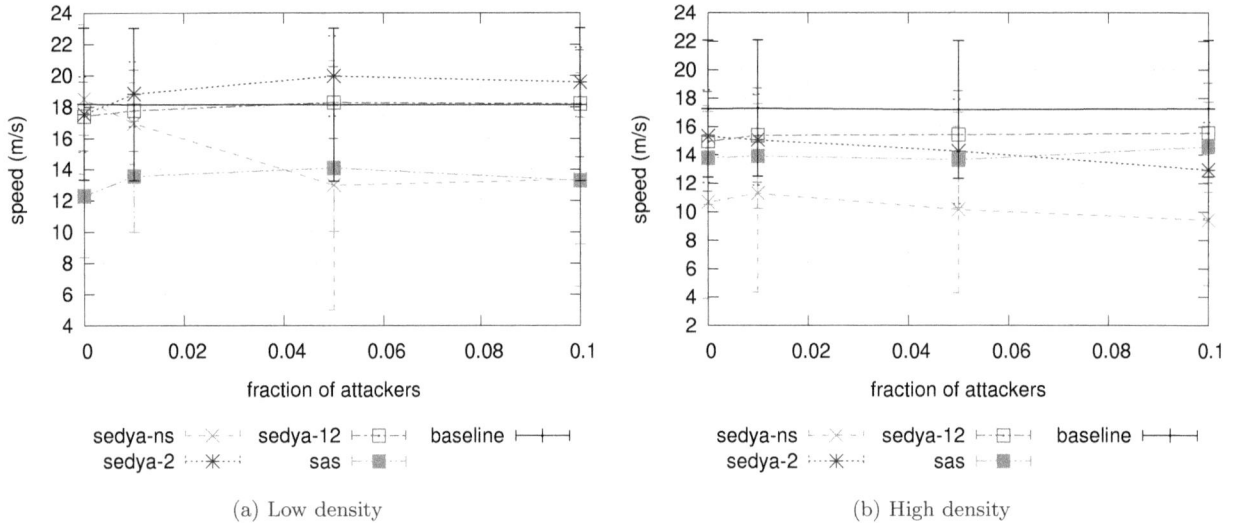

(a) Low density	(b) High density

Figure 6: This figure shows the average speed over one period for the low and high density scenarios, with a varied amount of attackers, for each of the studied schemes.

posite edge of the aggregation area, as shown in Figure 4. Once, the aggregate is received by vehicles that are not in the area, they switch to the dissemination phase. Each vehicle simply forwards the messages it receives using any generic efficient dissemination scheme, such as advanced adaptive geocast [1]. To compute the confidence value for a message, receivers determine the amount of participants in the multisignature and compare it to the reported amount of participants in the aggregation process, as described in the previous section. This check allows receivers to identify whether a message can be trusted for navigation decisions. It is possible to build more complex systems to resolve conflicts in data, as has been done in previous work, for instance, using the generic aggregation model from [5]. For such mechanisms, SeDyA provides reliable input concerning the amount of participants.

Finally, we note that the computed confidence values, in addition to data quality metrics, can be used to determine forwarding priorities. This provides a way to allow high quality messages to persist in the network, while low quality messages will be filtered. We consider developing and analyzing a dissemination scheme that uses such metrics in detail an interesting challenge for future work.

5. EVALUATION

In this section, we aim to show that SeDyA achieves the intended goals and is feasible to implement. To do this, we have implemented and simulated SeDyA in a network simulator. The implementation is available from the authors on request.

5.1 Simulation

The network simulator we use is an updated version of the JiST/SWANS simulator [28], which includes enhancements by the University of Ulm, published in 2008, which we have maintained internally[1].

Our simulations are performed on a highway setting, with

two lanes and a length of 5 kilometers. Initially, the vehicles are randomly distributed over the road; each vehicle uses a simple mobility model that contains car-following and overtaking. This model is based on the highway model in the Ulm JiST/SWANS distribution. The density of the network is either high (800 vehicles) or low (200 vehicles). The beacon frequency is set to 10 Hz with a beacon size of 209 byte (including security overhead). We use default IP and 802.11p implementations available in the JiST/SWANS distribution. On the physical layer, we use the Ray-Leigh fading model, the TwoRay path loss model, a transmit power of 10.9 dB and an additive noise model. The mobility model is configured for a maximum speed of 36 m/s with an acceleration capability of 1 m/s (5 m/s for breaking) and a target average of 25 m/s. Our simulation ran over a period of 20 seconds, with a single 9 Mbit/s channel for all communication, including beacons.

SeDyA itself is configured to use 64 bits for the LC sketch that counts participants, 4 sketches of 8 bits each for describing the location and 8 sketches of the same size for describing the payload (i.e., the speed measured). In all of these, the MD5 hash is used to hash elements; for PCSA, a second, independent hash is required, for which we use SHA-1. Both of these are implemented using the standard Java API. We remark that it is not required for the security of our scheme that these hashes are cryptographically secure, so MD5 and SHA-1 are sufficient. To simulate pseudonyms, we require that each period in which an aggregate is produced, the vehicle in question signs with a different identifier.

5.2 Attack Implementation

To prove that our scheme works, we need to show that the it is secure. Proving security against external attackers is typically done using security proofs. However, we rely on existing mechanisms that have been shown to be secure against these types of attacks. The novel part of our scheme is designed to protect specifically against insider attacks. In

---

[1]Due to licensing constraints, this source code is not public;

please contact us if you require the source code for your research.

such an attack, a node possess the necessary key material to be a legitimate participant. An attacker typically has control over his own vehicle, and can therefore always manipulate the sensor readings, even when a perfect hardware security module is used. Hence, we concentrate on attacks where malicious nodes try to influence the aggregation mechanism by crafting spoofed aggregates.

For these reasons, we have chosen to provide an attack implementation that abstracts from any particular attack on the scheme. Instead, we provide the attacker with tools to selectively increase or decrease aggregates. In this paper, we show our analysis for the most interesting attack: creating a fake traffic jam. We argue that the converse, attempting to hide a traffic jam, can be prevented analogously. Moreover, active safety mechanisms implemented using CAM messages can prevent accidents in case traffic jams are hidden by an attacker.

We typically assume that a single insider attacker tries to influence the aggregation result. The reason is that attacks on aggregation are especially interesting for attackers that only possess a single or few vehicles. Due to the aggregation, high impact can be achieved with little resources. In addition, we simulate scenarios where multiple independent attackers try to achieve the same goal. These attackers are randomly selected from the set of all vehicles.

Finally, we note that the attacker is assumed to be aware of the security mechanism and plausibility mechanisms in SeDyA. He will thus attempt to maximize the impact of an attack by working within the bounds of the plausibility mechanism that SeDyA provides.

5.3 Results

In this section, we present the results of our evaluation. As metric for performance, we use the speed measured by the aggregation mechanism in addition to a plot with the real speed that the mobility model provides. The speed values produced by the mobility model are the baseline: the best possible scheme would produce exactly this graph. By examining the average speed produced, in addition to the standard deviation, we can see how the scheme performs in the presence of attackers. The standard deviation is particularly important: a high standard deviation compared to the baseline indicates that a significant amount of messages contains higher or lower values. This means that either the aggregation mechanism is particularly poor for some messages, or the attacker is successful for some messages. In either of these cases, the scheme creates uncertainty for the receiver: she can no longer rely on a message, because there is no way to verify which message is good and which message is bad.

We have varied our simulations over several parameters: the density of the network, type of security mechanism and fraction of attackers. The density of the network is varied by setting the amount of vehicles on the road to 200 (for a density of 0.04 vehicles/meter) or 800 (for a density of 0.16 vehicles/meter).

To assess SeDyA's performance, we compare SeDyA against an improved version of SAS [16]. We take SAS as a baseline and add plausibility checks similar to those used in SeDyA's finalization phase (cf. Section 4.2) in order to make the two schemes more comparable. Also for comparability, we exchanged SAS' symmetric cryptography mechanisms with asymmetric ones, as suggested by the SAS authors in the

original paper. The corresponding graphs are labeled `sas`. For SeDyA itself, we evaluate a version of the underlying aggregation mechanism without SeDyA's security added (labeled `sedya-ns`), the full version of SeDyA (`sedya-12`), and a variant where only the security mechanisms presented in the finalization phase are used (`sedya-2`). In addition, we note that `sedya-ns` is not equipped with plausibility checks, while all others (including `sas`) are.

In addition to these communication-based schemes, our graphs include the baseline, which is supplied by the mobility model. The baseline indicates the natural variation in speed on the road; an ideal scheme should result in similar aggregated values. Finally, the fraction of attackers indicates the fraction of vehicles on the road that is malicious, which is varied from 0% to indicate no attacks at all, to 10%, which indicates a high amount of attackers. This fraction is computed over all simulated vehicles: it is possible for the attackers to form a local majority, as the attacker-controlled vehicles are randomly selected from all participants.

In the graphs in Figures 6a and 6b, we show the results for a low and a high density network. Each set of parameters has been repeated eight times. In particular for the low density setting, performance of SeDyA is good: showing an average speed of 17.4 ± 2.2, 17.5 ± 2.5 and 18.5 ± 4.8 m/s for the case without attackers using `sedya-12`, `sedya-2` and `sedya-ns`, respectively. SAS, on the other hand, underestimates the speed at 12.3 ± 3.9 m/s. The `sedya-ns` line clearly shows the impact of an attack, even with few attackers. Both the lower speed and the increased standard deviation indicate that the impact of attacks varies per message, creating high uncertainty for the receiver of any such message. On the other hand, we note that plausibility checks for all the schemes with security avoid high-impact attacks, as in both Figure 6a and Figure 6b, the graphs remain consistent as the amount of attackers increases. We note that for `sedya-12` and `sedya-2`, the standard deviation for each of these points is between 2.0 and 3.7 m/s; for Figure 6a, the standard deviation is lower, while for 6b all standard deviations are greater than 3. The cause for these standard deviations is in part due to the inherent inaccuracy of FM sketches, and due to the natural variation in the traffic, some vehicles will not have accurate information. We note that in SAS, the standard deviation for Figure 6a is significantly higher than SeDyA's, with values between 3.5 and 4m/s, while for 6b it is comparable. SeDyA's reduced performance in the high density scenario can be explained by the fact that the determination of edges is based on average speed reported by neighbors. This leads to larger aggregation areas for higher densities, which has an impact on accuracy.

In addition to the performance of the security mechanism, it is important to consider the bandwidth usage. Thus, we have computed the bandwidth consumed by the aggregation mechanism per vehicle per second in bytes. Clearly, SAS is more efficient in terms of bandwidth due to the flooding-based mechanisms that SeDyA employs in the finalization and dissemination phases. However, SeDyA's overhead is still reasonable due to the fact that our simulations were bounded to a single channel of 9 Mbit/s. In reality, the assignment of channels is still unclear; however, [6] defines 6 Mbit/s for safety applications and an additional two channels (6 Mbit/s and 12 Mbit/s) for traffic efficiency applications. Thus, SeDyA's bandwidth usage, which is on the

(a) Low density

(b) High density

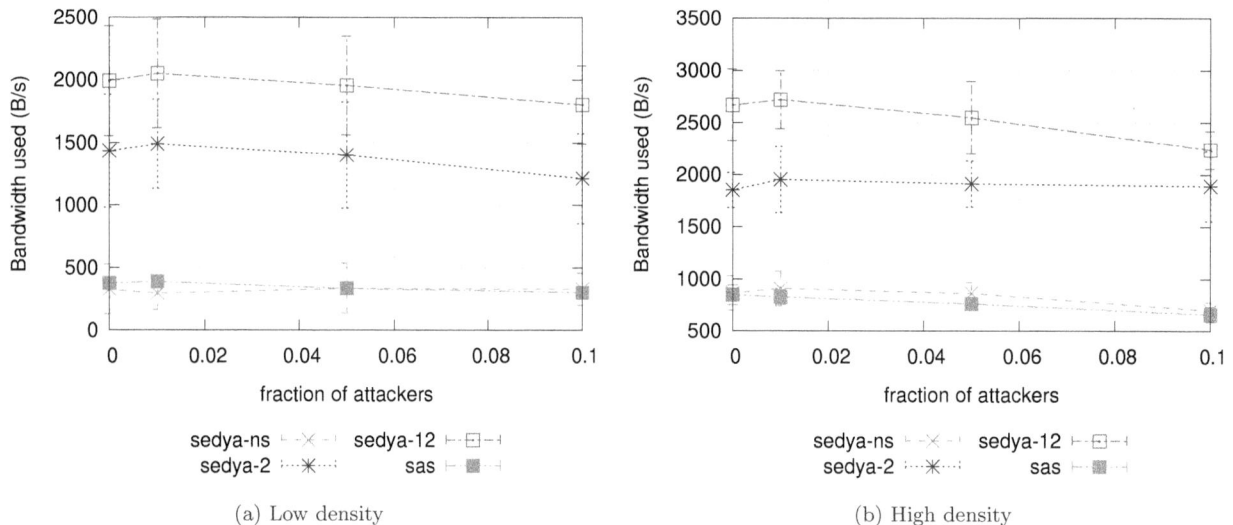

Figure 7: This figure shows the consumed bandwidth in bytes per vehicle per second in the low and high density scenarios.

order of 2 kilobytes per second per vehicle in a high density scenario, is well within the limits of the available bandwidth. When no spatial re-use is assumed, that is, if all vehicles are in each others transmission range, it would take 750 vehicles to use the complete 12 Mbit/s channel. Thus, we conclude that SeDyA is capable of producing improved results, particularly in low density settings, while maintaining a reasonable level of bandwidth consumption.

We remark that our bandwidth requirements are indeed much higher than those claimed by other secure aggregation schemes. However, our scheme is generally applicable to aggregation mechanism, rather than restricting the aggregation process to a particular implementation. In addition to this increased flexibility, we show our scheme with secure FM sketches (`sedya-12`), which provides a higher level of security in exchange for higher bandwidth requirements. Therefore, we illustrate that there is a fundamental trade-off between security and bandwidth efficiency, a trade-off that we have explored using the different schemes we have implemented.

Although the very high level of security that `sedya-12` provides is not always the best trade-off, we claim that it provides a good insight of what is fundamentally possible when faced with insider attackers. `Sedya-2` provides a more balanced trade-off, sacrificing some security for increased efficiency, while still providing good performance. As noted, future work should focus on applying our security mechanism to other aggregation schemes, and the incorporation of SeDyA's multisignature information into a misbehavior detection scheme.

6. CONCLUSION

We have introduced a new secure aggregation scheme that satisfies the unique requirements posed by VANETs. Our main improvements over related work are supporting dynamic aggregation mechanisms and offering additional security guarantees that prevent insider attacks from influencing the aggregation results. Our simulations show that the impact of generically implemented attacks is significantly less than for related work. However, our scheme's band-

width usage is higher. We also show that our scheme is feasible, despite the increased bandwidth requirements, by running it in a limited bandwidth channel, which also accommodates secure beaconing. Currently, we are further exploring the design space and trade-offs between high levels of protection and bandwidth efficiency. We envision that adaptive schemes could dynamically react to changing network situations and adapt the security overhead accordingly. Moreover, we are investigating a more general attack detection framework that uses information offered by schemes like SeDyA to detect misbehaving nodes in VANETs. In addition, we believe it is possible to apply SeDyA to existing aggregation mechanisms, by replacing the sketches in our first phase with the existing mechanisms. Thus, SeDyA can provide a generically applicable method for securing aggregation schemes that permit finalization and dependable aggregates. Our results using a custom-tailored aggregation scheme show that dynamic aggregation and proper security mechanisms can be combined and significantly reduce the attack vectors for insider attackers.

7. REFERENCES

[1] B. Bako, F. Kargl, E. Schoch, and M. Weber. Advanced adaptive gossiping using 2-hop neighborhood information. In *Global Telecommunications Conference, 2008. IEEE GLOBECOM 2008.*, pages 1–6. IEEE, 2008.

[2] S. Dashtinezhad, T. Nadeem, B. Dorohonceanu, C. Borcea, P. Kang, and L. Iftode. TrafficView: a driver assistant device for traffic monitoring based on car-to-car communication. In *Vehicular Technology Conference, 2004. VTC 2004-Spring. 2004 IEEE 59th*, volume 5, pages 2946 – 2950 Vol.5, may 2004.

[3] S. Dietzel, B. Bako, E. Schoch, and F. Kargl. A fuzzy logic based approach for structure-free aggregation in vehicular ad-hoc networks. In *Proceedings of the sixth ACM international workshop on VehiculAr InterNETworking - VANET '09*, page 79, 2009.

[4] S. Dietzel, F. Kargl, G. Heijenk, and S. Florian. On the poptential of generic modeling for vanet data

aggregation protocols. In *IEEE Vehicular Networking Conference (VNC)*, pages 78–85, 2010.

[5] S. Dietzel, E. Schoch, B. Konings, M. Weber, and F. Kargl. Resilient secure aggregation for vehicular networks. *IEEE Network*, 24(1):26, 2010.

[6] ETSI. Intelligent Transport Systems (ITS); European profile standard for the physical and medium access control layer of Intelligent Transport Systems operating in the 5 GHz frequency band, November 2009. ETSI ES 202 663, final draft, version 1.1.0.

[7] ETSI. Intelligent Transport Systems (ITS); Vehicular Communications; Basic Set of Applications; Part 2: Functional Requirements, 2010.

[8] ETSI. Intelligent Transport Systems (ITS); Vehicular Communications; Basic Set of Applications; Part 2: Specification of Cooperative Awareness Basic Service, 2010.

[9] ETSI. Intelligent Transport Systems (ITS); Vehicular Communications; Basic Set of Applications; Part 3: Specifications of Decentralized Environmental Notification Basic Service, 2011.

[10] Y.-C. Fan and A. L. Chen. Efficient and robust sensor data aggregation using linear counting sketches. In *2008 IEEE International Symposium on Parallel and Distributed Processing*, page 1, 2008.

[11] E. Fasolo, M. Rossi, J. Widmer, and M. Zorzi. In-network aggregation techniques for wireless sensor networks: a survey. *IEEE Wireless Communications*, 14(2):70, 2007.

[12] P. Flajolet and G. N. Martin. Probabilistic counting algorithms for data base applications. *Journal of Computer and System Sciences*, 31:182–209, October 1985.

[13] M. Garofalakis, J. M. Hellerstein, and P. Maniatis. Proof sketches: Verifiable in-network aggregation. In *2007 IEEE 23rd International Conference on Data Engineering*, page 996, 2007.

[14] C. Gentry and Z. Ramzan. Identity-based aggregate signatures. In *Public Key Cryptography - PKC 2006, 9th International Conference on Theory and Practice of Public-Key Cryptography*, pages 257–273, 2006.

[15] M. Gerlach. Assessing and improving privacy in VANETs. *ESCAR Embedded Security in Cars*, 2006.

[16] Q. Han, S. Du, D. Ren, and H. Zhu. SAS: A secure data aggregation scheme in vehicular sensing networks. In *2010 IEEE International Conference on Communications*, page 1, 2010.

[17] H.-C. Hsiao, A. Studer, R. Dubey, E. Shi, and A. Perrig. Efficient and secure threshold-based event validation for VANETs. In *WISEC'11*, pages 163–174, 2011.

[18] IEEE. IEEE trial-use standard for wireless access in vehicular environments - security services for applications and management messages, 2006. IEEE Std 1609.2-2006.

[19] IEEE. IEEE Standard for Information technology–Telecommunications and information exchange between systems Local and metropolitan area networks–Specific requirements Part 11: Wireless LAN Medium Access Control (MAC) and Physical Layer (PHY) Specifications. IEEE Std 802.11-2012 (Revision of IEEE Std 802.11-2007), March 29 2012.

[20] IEEE. IEEE Trial-Use Standard for Wireless Access in Vehicular Environments (WAVE). IEEE Standard 1069.*, 2012.

[21] C. Lochert, B. Scheuermann, and M. Mauve. Probabilistic aggregation for data dissemination in VANETs. In *Proceedings of the fourth ACM international workshop on Vehicular ad hoc networks - VANET '07*, page 1, 2007.

[22] T. Nadeem, S. Dashtinezhad, C. Liao, and L. Iftode. Trafficview: a scalable traffic monitoring system. In *Mobile Data Management, 2004. Proceedings. 2004 IEEE International Conference on*, pages 13 – 26, 2004.

[23] F. Picconi, N. Ravi, M. Gruteser, and L. Iftode. Probabilistic validation of aggregated data in vehicular ad-hoc networks. In *Proceedings of the 3rd international workshop on Vehicular ad hoc networks - VANET '06*, page 76, New York, New York, USA, 2006. ACM Press.

[24] M. Raya, A. Aziz, and J.-P. Hubaux. Efficient secure aggregation in VANETs. In *Proceedings of the 3rd international workshop on Vehicular ad hoc networks*, VANET '06, pages 67–75. ACM, 2006.

[25] Y. Sang, H. Shen, Y. Inoguchi, Y. Tan, and N. Xiong. Secure data aggregation in wireless sensor networks: A survey. In *PDCAT'06*, pages 315–320, 2006.

[26] B. Scheuermann, C. Lochert, J. Rybicki, and M. Mauve. A fundamental scalability criterion for data aggregation in VANETs. In *Proceedings of the 15th annual international conference on Mobile computing and networking - MobiCom '09*, page 285, 2009.

[27] Y.-c. Tseng, S.-Y. Ni, Y.-s. Chen, and J.-P. Sheu. The Broadcast Storm Problem in a Mobile Ad Hoc Network. *Wireless Networks*, 8(2-3):153–167, 2002.

[28] University of Ulm. vanet.info simulation portal. http://www.vanet.info/?q=node/11.

[29] L. Wischoff, A. Ebner, H. Rohling, M. Lott, and R. Halfmann. SOTIS - a self-organizing traffic information system. In *Vehicular Technology Conference, 2003. VTC 2003-Spring. The 57th IEEE Semiannual*, volume 4, pages 2442 – 2446 vol.4, april 2003.

APPENDIX

A. CRYPTOGRAPHY IN SEDYA

In this appendix, we briefly discuss the cryptographic primitives required to implement our scheme with additional bandwidth efficiency. First we review different types of signatures, followed by a discussion of multisignatures and aggregate signatures.

Signature A digital signature is a cryptographic primitive that guarantees a it was generated using a message M and a private key sk, when verified with the corresponding public key pk. The private key is held only by the signer, while the public key is available to anyone. The signature can then be verified by anyone that has possession of a legitimate copy of the corresponding public key.

Multisignature Given a group of n participants, each with their own key-pair (pk_i, sk_i) a multisignature on a message M can be generated by any subset of participants L by computing this multisignature from the individual signature

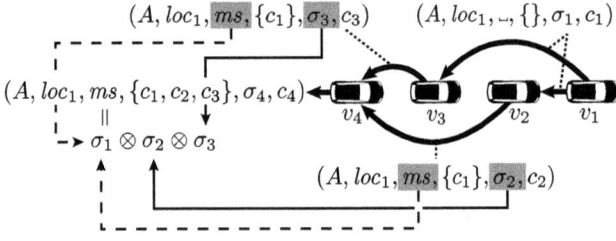

Figure 8: Vehicle v_4 receives multisignatures with overlapping content from v_2 and v_3. Because signatures are only embedded with 1 vehicle delay, v_4 can use the separately received σ_2 and σ_3 together with $ms (= \sigma_1)$ to construct a duplicate-free multisignature representing $\sigma_1, \sigma_2, \sigma_3$.

that each participant in L generates on M. Then, the signature can be verified by using all the public keys of the participants in L and computing from them the public key of L.

Aggregate Signature Given a group of n distinct participants and n messages, an aggregate signature is a signature that can be computed from the signatures that are generated by each participant on its own message. The necessary inputs for verification are the n messages, the n public keys and the aggregate signature.

In [14], an identity-based multisignature scheme is developed, is used as a first step to construct an identity-based aggregate signature scheme, which we will also briefly discuss. Although signing is relatively cheap, verification requires three pairing operations and n point additions (for n participants) and it is not possible to work with multiple CAs in the same multisignature. The identity-based multisignature scheme is defined as follows (we refer interested readers to the original work for details):

Setup Generate two groups G_1, G_2 of prime order q and a pairing $e : G_1 \times G_1 \rightarrow G_2$, a generator P of G_1, a master secret key $s \in \mathbb{Z}/q\mathbb{Z}$ and master public key sP, and two hash functions H_1, H_2 that both map text to G_1.

Key generation Given identity I, compute the key-pair with public key $H_1(I)$ and secret key $sH_1(I)$.

Sign Choose random $r_i \in \mathbb{Z}/q\mathbb{Z}$ and compute the signature as $(r_i H_2(m) + s H_1(I), r_i P)$.

Aggregation To aggregate signatures, simply perform point addition for all the individual signatures; $(\sum_i S_i, \sum_i T_i)$ for signatures written as (S_i, T_i), given that all signatures are on the same message m.

Verify Given a multisignature (S_n, T_n), verify that $e(S_n, P) = e(T_n, H_2(m))e(sP, \sum_i P_i)$.

Note that a strong requirement for the usage of multisignatures is that the message is the same. The goal of aggregate signatures is to side-step this requirement. However,

aggregate signatures are not a viable alternative to multisignatures, at least for SeDyA, because they require all messages to be included. Nevertheless, they can be used to provide compression of signatures in SAS [16], as also noted by the original work. Thus, we briefly show the identity-based aggregate signature scheme that [14] presents:

Setup Generate two groups G_1, G_2 of prime order q and a pairing $e : G_1 \times G_1 \rightarrow G_2$, a generator P of G_1, a master secret key $s \in \mathbb{Z}/q\mathbb{Z}$ and master public key sP, two hash functions H_1, H_2 that both map text to G_1, and a hash function H_3 that maps text to $\mathbb{Z}/q\mathbb{Z}$.

Key generation Given identity I, the CA will compute two key-pairs with public keys $P_{i,0} = H_1(I_i, 0)$, $P_{i,1} = H_1(I_i, 1)$ and private keys $sH_1(I, 0)$, $sH_1(I, 1)$ respectively.

Sign Determine a w that has not been used before, which will be the same for each message. It need not be random, only unique to this signing procedure. Compute $H_2(w)$, $c_i = H_3(m_i, I_i, w)$ and generate random $r_i \in \mathbb{Z}/q\mathbb{Z}$; then the signature is $(w, r_i H_2(w) + s P_{i,0} + c_i s P_{i,1}, r_i P)$.

Aggregation To aggregate signatures, simply perform point addition for all the individual signatures; $(w, \sum_i S_i, \sum_i T_i)$ for signatures written as (w, S_i, T_i). It is not allowed to aggregate signatures with different w.

Verify Given a multisignature (w, S_n, T_n), verify that $e(S_n, P) = e(T_n, H_2(w))e(sP, \sum_i P_{i,0} + \sum_i P_{i,1})$.

B. MULTISIGNATURE COMPUTATION

Instead of using just the multisignature as signature for a message, SeDyA includes a multisignature and a regular signature while the message is in the finalization phase. The reason for this is shown in Figure 8, which shows four vehicles with comparable speeds. In this figure, v_1 is the finalizing vehicle: it generates the finalized message A and adds its location, loc_1, to the message, in addition to his signature and certificate. When both v_2 and v_3 receive this message, both of them set the multisignature to the signature of v_1, σ_1 and put the certificate c_1 into the certificate list. Then, both vehicles create their own signature and forward the message in the same format. Now, v_4 receives both messages and computes a new multisignature as $\sigma_1 \otimes \sigma_2 \otimes \sigma_3$. However, if v_2 and v_3 had directly computed their multisignatures as $\sigma_1 \otimes \sigma_2$ and $\sigma_1 \otimes \sigma_3$, then v_4 would have computed $(\sigma_1 \otimes \sigma_2) \oplus (\sigma_1 \otimes \sigma_3)$. To see why this is problematic, note that the certificate set needs to explicitly specify the set of certificates that were included. If a counter is used for this purpose, then an attacker has the opportunity to perform a denial of service attack on the verification process, because this process involves the multiplication of all certificates. Thus, we include the necessary certificates explicitly; repeated certificates therefore lead to additional computational cost.

Adversarial Testing of Wireless Routing Implementations

Md. Endadul Hoque
Purdue University
mhoque@cs.purdue.edu

Hyojeong Lee
Purdue University
hyojlee@cs.purdue.edu

Rahul Potharaju
Purdue University
rpothara@cs.purdue.edu

Charles E. Killian
Purdue University
Google Inc.
ckillian@cs.purdue.edu

Cristina Nita-Rotaru
Purdue University
crisn@cs.purdue.edu

ABSTRACT

We focus on automated adversarial testing of real-world implementations of wireless routing protocols. We extend an existing platform, Turret, designed for general distributed systems, to address the specifics of wireless routing protocols. Specifically, we add functionality to differentiate routing messages from data messages and support wireless specific attacks such as blackhole and wormhole, or routing attacks such as replay attacks. The extended platform, Turret-W, uses a network emulator to create reproducible network conditions and virtualization to run unmodified binaries of wireless protocol implementations. Using the platform on publicly available implementations of two representative routing protocols we (re-)discovered 14 attacks and 3 bugs.

Categories and Subject Descriptors

C.2.1 [**Network Architecture and Design**]: Wireless communication; C.2.m [**Miscellaneous**]: Security

General Terms

Design, Security

Keywords

Adversarial testing; Virtualization; Wireless; Routing

1. INTRODUCTION

Mobile ad-hoc networks allow a set of wireless nodes to communicate with each other without any central infrastructure. As traditional routing protocols do not perform well in a constrained environment such as wireless networks, significant work has been put into designing routing protocols for wireless networks. Examples include proactive protocols such as DSDV [32], reactive protocols such as AODV [33] and DSR [21], or hybrid protocols such as DST [34]. Moreover, due to increased threats that exist in wireless networks,

several secure routing protocols have been designed. Examples include: SAODV [38], ODSBR [11], ARAN [36], Ariadne [20]. Several of these protocols such as AODV, ARAN were implemented and are available from public repositories.

Given the importance of routing as a fundamental component of wireless networks, many protocols have been subjected to model checking the design [12] and testing the simulator-based implementation [6,39]. While model checking greatly helps to verify the validity of the model, real-world implementations often bring vulnerabilities that model abstraction does not capture. Also, while simulators provide easier and simpler ways to describe a protocol, the simplicity sacrifices some aspects of realism such as interaction with the operating system components.

Recent research [23,24] showed the importance of performing adversarial testing where software is tested beyond just basic functionality by examining edge cases, boundary conditions, and ultimately conducting destructive testing. Adversarial testing makes protocols more robust to arbitrary and extreme conditions and can lead to discovering vulnerabilities in implementations, many of which might have not occurred in simulator-based implementations.

Previous work related to wireless routing implementations has focused exclusively on performance comparison across protocols [14, 16], or using test-beds to investigate properties of the network stack such as performance of TCP in multihop ad hoc networks [13, 16].

In this paper, we focus on adversarial testing of real-world implementations of wireless routing protocols. We consider attacks and failures that are created through manipulation of protocol messages and are specific to wireless routing protocols, having a global impact on the protocol performance. We use an experimentation environment that allows binaries to run in their native operating systems while limiting the impact of noise and interference on the system performance. Our aim is to automatically discover potential attacks by exploiting vulnerabilities in protocol design and implementation. Our contributions are:

- We extend an existing platform, Turret [23], designed for general distributed systems, to address the specifics of wireless routing protocols. The platform uses a network emulator to create reproducible network conditions and virtualization to run unmodified binaries of wireless protocol implementations. The platform requires the user to provide a description of the protocol messages and corresponding performance metrics. We present Turret-W, an extension of an existing platform, that includes a wireless malicious proxy that differentiates routing messages from data messages and supports wireless specific attacks such as blackhole and wormhole, or routing attacks such as replay attacks. Our approach is cost effective as compared to the hardware and manpower costs required by the approaches in [14].

- We demonstrate attack discovery with Turret-W using detailed case studies on two representative wireless routing protocols: a reactive protocol AODV, and a secure reactive protocol ARAN, whose implementations we obtained from public repositories. We found 1 new and 7 known attacks in AODV, and 6 known attacks in ARAN, for a total of 14 attacks. While most of attacks we found are protocol level attacks, one attack in AODV was solely an implementation level attack and such an attack could have been discovered by testing the actual implementation under adversarial environments.

- We show that Turret-W also helps to find bugs because it implicitly supports a testing environment that is realistic and controllable. Bugs are different from attacks in that they can cause performance degradation in benign executions. We discovered 3 bugs in target implementations, 2 in AODV and 1 in ARAN.

The rest of the paper is organized as follows. §2 provides an overview of the platform we use in this paper, §3 describes our methodology and presents two case studies. §4 describes related work and §5 summarizes the paper.

2. PLATFORM OVERVIEW

Our goal is to test real-world implementations, where the network conditions can be reproducible and also isolated from outside world interference. Platforms that leverage virtualization and some form of emulation fit this profile. We selected Turret [23] because it provides these capabilities. However, Turret is not designed for wireless networks, but assumes more general message-passing systems. We first give an overview of Turret, the platform that we built on, and then describe how we extended it to support wireless routing protocols. We refer to Turret with our extension as Turret-W.

2.1 Turret Overview

Turret is a platform for performance attack discovery on unmodified distributed system binaries running in realistic environments. Turret uses virtualization to run arbitrary operating systems and applications, and network emulation to connect these virtualized hosts in a realistic network setting. Turret requires a description of the external API of the message protocol, and a set of metrics that capture the application performance of the system.

Specifically, Turret uses KVM [17] virtualization techniques, to allow several virtual machines, each acting as an individual node, to run on the same physical host. Each node can then run an application and communicate with other nodes through an emulated network, achieved through NS-3 [7] [1]. Turret maps each virtualized node with a shadow node inside NS-3 by creating a *Tap-Bridge* connection between the virtualized node and its corresponding end node.

[1]Note that Emulab [4], MobiNet [26], Orbit [9] could also conceptually replace NS3. Emulab with fixed wireless provides more realism. However, the approach provides less reproducible results because of unwanted disturbance on the wireless channel and requires a separate implementation of the malicious version of the tested routing protocol.

A controller bootstraps the system, (i.e. starts NS-3 and runs application binaries inside the virtual machines), and instructs nodes whether they will act as benign or malicious. A node marked as malicious will then be equipped with a *malicious proxy* which is a part of the Tap-Bridge layer on the NS-3 shadow node. This malicious proxy takes care of intercepting packets and modifying them according to an *attack strategy*. To make this possible, the malicious proxy requires the user to provide it with a list of *messages formats* that the protocol relies on. This way, different fields inside a message can be automatically modified based on the selected attack strategy.

Action	Action Description	Parameter
Drop	Drops a message	Drop probability
Delaying	Injects a delay before it sends a message	Delay amount
Duplicating	Sends the same message several times instead of sending only one copy	Number of duplicated copies
Diverting	Sends the message to a random node instead of its intended destination	None

Table 1: Message delivery actions in Turret

Action	Action Description	Parameter
LieValue	Changes the value of the field with a specified value	The new value
LieAdd	Adds some amount to the value of the field	The amount to add
LieSub	Subtracts some amount from the value of the field	The amount to subtract
LieMult	Multiplies some amount to the value of the field	The amount to multiply
LieRandom	Modifies the value with a random value in the valid range of the type of the field	None

Table 2: Message lying actions in Turret

Turret supports two types of malicious actions: *Message Delivery Actions* which affect when and where a message is delivered (see Table 1) and *Message Lying Actions* which affect the contents of a message (see Table 2).

2.2 Turret Limitations for Wireless Routing

Turret does not focus on routing protocols or on wireless networks and suffers from the following limitations:

Distinguishing between control plane and data plane: While Turret can inject attacks and faults into any message-oriented protocol, it does not differentiate the data layer from the routing layer. As Turret injects attacks based on message types, it can not inject attacks on packets in bulk data transfer application such as ftp. In the case of routing, many data plane attacks or degradation in performance can be amplified if the routing mechanism is disrupted and it does not matter if the application is a message oriented application or a bulk data transfer application. For wireless networks, the separation is needed to support basic attacks such as *blackhole* in which an attacker will drop all data packets but participate in the routing algorithm correctly.

Replaying packets: Turret does not provide the functionality to replay packets. When replaying a packet, an attacker records another node's valid control messages and resends them later to other benign nodes via legitimate channels. This causes other nodes to add incorrect routes to their routing table. Such attacks can be used to impersonate a specific node or simply to disturb the routing plane.

Establishing side-channels (wormholes): Turret does not support colluding attacks because finding an effective colluding strategy in distributed systems results in a state space explosion. However, an attack specific to wireless networks that requires coordination between two attackers and shown to be very detrimental is the wormhole attack where two colluding adversaries cooperate by tunneling packets between each other to create a shortcut in the network.

2.3 Turret-W Overview

We modified Turret to address the above limitations. The new platform, Turret-W, is shown in Figure 1. The *controller* is the core component that drives the system. It generates a topology file for the network emulation layer using a configuration file provided by the user that specifies various parameters such as the network topology, number of nodes, number of malicious nodes. Then it starts the virtual machines and binds each of them to the underlying network emulation layer. It then loads the routing service (e.g., AODV) at the routing layer and instantiates the application (e.g., *iperf*) at the application layer. It accepts the list of attack strategies created by the strategy generator and injects them into the malicious proxy. Finally, it collects log messages that are useful in post-processing for estimating application performance running on top of the routing protocol.

Wireless network emulation: We configured the network emulator, NS-3, to emulate WiFi links. Like in Turret, virtual machines operate on top of a *network emulation layer* provided by NS-3. In principle, each virtual machine acts as an individual node in the system and is connected to a corresponding shadow node inside NS-3 using a Tap-Bridge connection. A Tap-Bridge connection makes an NS-3 net device appear as a local device inside the virtual machine thereby allowing the virtual machine to use the underlying net device as if it were its own net device for WiFi transmission. The network emulation layer creates a virtual multihop wireless environment to transmit packets from a source to a destination virtual machine.

Supporting wireless routing specific actions: We modified the malicious proxy to differentiate between packets originating from the routing layer and the application layer. We achieve this by utilizing the port number to differentiate between application and routing related packets. This makes it possible to implement a blackhole attack wherein a malicious node acts benign at the routing layer but selectively/entirely drops packets originating from the application layer. We implement the wormhole attack as follows: the private channel is implemented as part of the malicious proxy inside NS-3 and hence the virtual machines are agnostic about the channel. According to the routing protocol, if a node does not hear back from one of its neighbors, the node considers the route to the neighbor as a stale route. Therefore, to convince one colluding node that the other colluding node is its neighbor, we allow the colluding nodes to send their beacon messages (e.g., *hello*) to each other. However, to prevent a colluding node at one end from assuming that the benign neighbors of the other colluding node are its neighbors, the colluding nodes do not forward the beacon messages received from their benign neighbors over the private channel. All other routing protocol messages are forwarded by the colluding nodes over the private channel so that they can perform the wormhole attack in the route discovery process. As a result, Turret-W supports all the malicious actions presented in in Tables 1, 2, and 3.

Attack strategy generation: The strategy generator is responsible for generating a list of attack strategies that a protocol-under-testing should be tested against.

These sequence of strategies are generated based on the malicious actions given in Tables 1 and 2 along with a value that decides the severity of that action. To support the additional wireless specific attacks listed in Table 3, we have extended Turret's basic set of malicious actions with replay, blackhole and wormhole attacks.

Action	Action Description	Parameter
Replay	Records valid control messages from a node and resends them to other benign neighbors	None
Blackhole	Drops all data packets but participates in the routing protocol correctly	None
Wormhole	Creates a wormhole between two colluding nodes and tunnels packets between each other	Types of the control messages to be tunnelled
Wormhole with blackhole	Creates a wormhole between two colluding nodes and tunnels routing packets between each other, but drops all data packets	Types of the control messages to be tunnelled

Table 3: Malicious actions added by Turret-W

3. EXPERIMENTAL RESULTS

We demonstrate our platform on real-world implementations of two well known on-demand wireless routing protocols: AODV [33], and ARAN [36], whose implementations we obtained from [1, 2]. In this section, we describe our experimental setup and discuss our case studies on these protocols. We summarize all the attacks and bugs reported by Turret-W in Table 4.

3.1 Experimental Setup

All our experiments are performed on a Dual-Quad core Intel(R) Xeon(R) CPU E5410@2.33GHz with 8 GB RAM host machine. We use Ubuntu 10.04.4 LTS to serve as the host OS. In all experiments, we use 12 VMs, each allocated 128 MB RAM. For AODV, we use Debian 6.0.5 with Linux Kernel 2.6.32-5-686 as the guest OS. One of the advantages of our platform is that it allows us to support exactly the environment for which a binary was compiled. For instance, since ARAN requires an older kernel, we use Fedora Core 1 with Linux kernel 2.4.22-1.2115.nptl.i386 as the guest OS.

Our emulated network is a multihop wireless adhoc network. For the 802.11 MAC layer, we used 802.11a with bit rate of 6 Mbps. We performed our experiments using a static grid topology. As an application, we run *iperf* [5], a network benchmarking tool, on the VMs. In all the experiments, the performance of the application we report are averaged over ten runs.

As we are interested in performance attacks, we obtain a performance baseline using *benign testing*, where we randomly select pairs of source and destination nodes and transfer a stream of UDP packets between them for 30 seconds. Since we do not intend to stress the protocol or its implementation, we select a lower data rate of 128 Kbps so that the impact of attacks can be easily observed – a low packet delivery ratio implies an attack [11, 31].

In *adversarial testing*, we randomly select malicious nodes and inject malicious strategies during the entire execution. For all attack strategies applied to routing messages, the malicious node drops application packets with a probability of p (= 0.3 in our case). The purpose of using $p = 0.3$ is two-fold: 1) it helps differentiate if the effect of the attack is route discovery failure or packet dropping, and 2) since

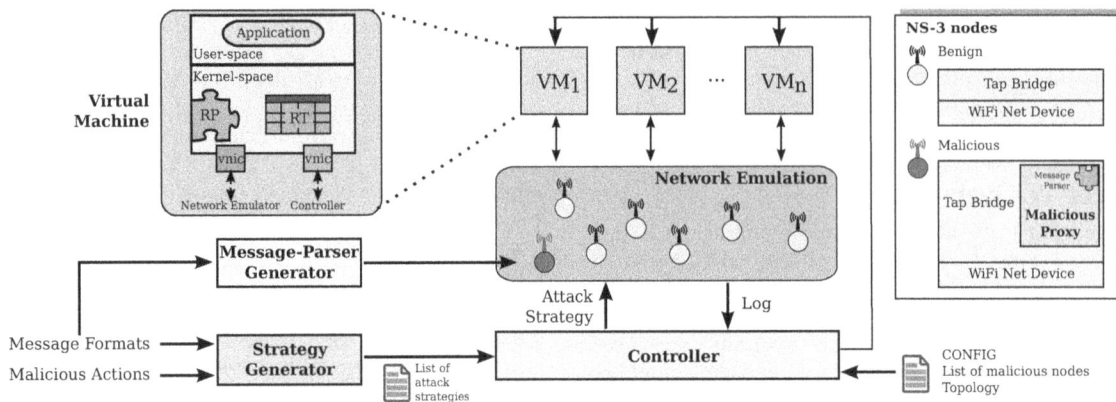

Figure 1: Turret-W (RP:Routing Protocol, RT:Routing Table and VNIC:Virtual Network Interface Card)

we focus on the insider attackers, $p = 0.3$ characterizes the less aggressive behavior which is typical behavior of insider attackers. We vary the total number of adversaries in the network from 1 to 4 exhibiting a homogeneous behavior, i.e., we inject the same attack strategy to each malicious node.

To demonstrate the effect of blackhole attacks and wormhole attacks, we perform experiments with three different configurations of adversaries: blackhole with one adversary, blackhole with two adversaries, combination of wormhole and blackhole with two colluding adversaries. When a blackhole attack strategy is injected, an adversary participates benignly in the routing protocols but drops 100% of application packets. The effect of wormhole is the most noticeable in terms of application performance when we combine the blackhole strategy with the wormhole strategy. Note that except the cases of blackhole and/or wormhole attacks, we use the packet dropping probability $p = 0.3$ in all other malicious executions as mentioned above.

As a performance metric, we use *packet delivery ratio* (PDR), i.e., a ratio of the total number of packets received by the receiver to the total number of packets sent by the sender. PDR is easy to measure and does not require any instrumentation in routing protocol implementations. We plan to investigate other security metrics for future work. We formally define an attack as follows:

Definition 1 - Performance Attack: When the difference in the performance metric for a benign execution to that of a malicious execution exceeds by a threshold, δ, we say that the attack strategy has resulted in a successful attack.

Here, δ is a system parameter and is dependent on the system-under-test. In our experiment we decide to choose 0.2 as our threshold.

One of the by-products of using our system is the ability to find bugs that have an impact on performance. We define a bug as follows:

Definition 2 - Performance Bug: An implementation-level fault that produces a degradation of performance during a benign execution.

3.2 Case Study: AODV

We now describe how we used Turret-W to test AODV [33]. We omit the graphs due to lack of space.

Implementation used: We use AODV-UU-0.9.6 implementation, publicly available from [1] that is RFC 3561 [10] compliant. The AODV-UU consists of two components — a loadable kernel module (kaodv) and a user space daemon process (aodvd). The kaodv module intercepts and handles network packets by registering hooks with the Linux kernel

using the netfilter framework. We use the default values for configuration related parameters presented in [1, 10].

During the benign testing (to obtain a baseline to compare against) of AODV-UU, we discovered two unknown implementation bugs caused by a subtle interplay between the AODV-UU code and the kernel.

Bug 1. Kernel interaction order. In an attempt to measure TCP streaming performance between a source and a destination that is multiple hops away, we observed that packets were not getting delivered in the benign case. We investigated the cause and identified this bug. By design, whenever an application sends a packet with a destination to which the route is either invalid or unavailable, kaodv is supposed to hold the packet and notify aodvd to perform a route discovery. After finishing the route discovery, aodvd should notify the kernel to update the routing table and the koadv module to release the withheld packet.

In the implementation, however, the order of notification upon completion of a route discovery was incorrect, i.e., in the reverse order. This bug could not have been discovered if we had not attempted to measure TCP streaming performance where the first packet, *i.e.*, SYN packet is crucial to establish the connection. However, we observed packet loss when initially using UDP, but like others, we attributed this to the lossy behavior of UDP inside the wireless channel. We fix the bug by reversing the order of the two notifications.

Bug 2. Route packets harder. In the process of obtaining a baseline using a network benchmark tool *iperf*, we observed performance degradation over time despite the route being available and valid in the routing table.

When the kernel transport layer hands-over any locally generated packet to the IP layer, kaodv receives the control of the packet via a hook registered with *netfilter*. So, kaodv is responsible for returning a value to netfilter so that netfilter can decide what to do – accept/drop/ignore the packet or call the hook again.

When kaodv receives the control for a packet and already has a valid route, kaodv notifies netfilter to continue processing the packet by returning NF_ACCEPT. On receiving NF_ACCEPT, netfilter sends the packet down the network stack without performing any further iptables tests [19]. As a result, netfilter does not send the packet to the correct next hop node on the route to the destination. We fix this bug by invoking ip_route_me_harder() inside kaodv before returning NF_ACCEPT.

Attack causing crashing. We discovered an implementation attack that can cause all neighbors of a malicious node to crash. When a malicious proxy modifies an RREQ

Protocol Impl.	Discovery Type	Name	Description
AODV-UU 0.9.6 [1] Updated: April 13, 2011	Attack*	Lie RREQ type 2	Lie about RREQ message type by setting to 2 (RREP) (causes crashing)
	Attack [35]	Lie RERR type 1	Lie about RERR message type by setting to 1 (RREQ)
	Attack [35,38]	Lie RREP hop 0	Lie about the hop count in route response to be 0
	Attack [35]	LieAdd RREQ reqid 10	Increment the route request id of route request by 10
	Attack [35,38]	LieAdd RREP destsq 10	Increment the destination sequence number of route response by 10
	Attack [35,38]	Replay RREP	Replay both route response and hello messages
	Attack [38]	Blackhole	Drop all data packets
	Attack [35,38]	Wormhole + Blackhole	Colluding malicious nodes drop all data packets
	Bug*	Kernel interaction order	Notifies the two components about the route discovery in a wrong order
	Bug*	Route packets harder	Returning NF_ACCEPT from hooks causes Netfilter not to check iptables
ARAND 0.3.2 [2] Updated: Jan 31, 2003	Attack [25]	Drop RDP 100%	Drop each route request message
	Attack [25]	Delay REP 2s	Delay forwarding of route response message by 2 seconds
	Attack [25]	Divert REP	Divert route response message
	Attack [25]	Drop ERR 100%	Drop route error message
	Attack [11]	Blackhole	Drop all data packets
	Attack [11]	Wormhole + Blackhole	Colluding malicious nodes drop all data packets
	Bug*	Wrong postal address	Intermediate nodes forward REP to the source instead of the next hop

Table 4: Attacks and bugs (re-)discovered by Turret-W. Attacks/bugs with (*) means newly discovered.

message to be an RREP, a recipient processes this altered RREQ message as an RREP message. The base RREP message (20 bytes) is smaller in length than a base RREQ message (24 bytes) [10]. Therefore, a recipient of the malformed RREQ assumes it as an RREP with extensions [10] which causes AODV-UU to crash. The malicious node becomes successful in exploiting an *integer overflow* and *buffer overflow* vulnerability in the AODV-UU code. This is solely an implementation level attack that cannot be discovered in a benign environment, however it can cause significant damage. It can be fixed by cautious type and boundary checking with an awareness of possible adversarial attempts.

Attacks caused by basic malicious actions. We rediscovered several attacks on AODV-UU based on message delivery and lying actions which decrease the PDR below the accepted threshold. Note that the AODV protocol itself is susceptible to these attacks [35,38]. In case of our benign experiments, we observe a 98% PDR.

Replay RREP. By replaying a RREP message received from a node, an adversary can fool its benign neighbors to believe that they are the neighbors of the node which, in reality, are at least two hops away. This attack is more damaging than others because replaying the periodic hello messages causes these pseudo-links never to expire. We observe the PDR drops from 75% to 17% as the number of adversaries increases.

LieAdd RREP destsq. Whenever a node receives a control packet from another node with the destination sequence number higher than what it has in its routing table, the node selects the route via this other node. A malicious node can add a positive value with the destination sequence number of an RREP message which causes the recipient to select the route through the malicious node. We observe that this attack results in at most 56% drop in PDR.

LieAdd RREQ reqid. Each RREQ message is uniquely identified by the *request identifier* in conjunction with the originator's IP. For each new route request, the request identifier is incremented by one. No node ever responds to an older RREQ message. A malicious node tricks the destination to respond to an RREQ with a future request identifier so that the source will be left with only one available route, i.e., through the malicious node. We observe that this attack causes the PDR to drop from 78% to 62% with the increase in the number of adversaries.

Blackhole/wormhole attacks. To evaluate the performance of AODV-UU in the presence of a blackhole attacker, we allow an intermediate malicious node to drop all the data packets. Later, we introduce an additional blackhole node that can collude with the other blackhole node via a private channel to perform a wormhole attack. The PDR drops from 70% to 50% with the increase in blackhole nodes, whereas the PDR drops to 40% in case of the wormhole attack.

3.3 Case Study: ARAN

We now describe how we used Turret-W to test ARAN [36]. **Implementation used:** We rely on the implementation of ARAN called arand-0.3.2, publicly available from [2]. This user space routing daemon built for Linux kernel 2.4 relies on the Ad hoc Support Library (ASL) [3]. For the cryptographic functionalities, it uses OpenSSL [8]. We use the default values for parameters as used in [2].

Bug. Wrong postal address. We discovered an implementation bug during the benign experiments in the setting of a multi-hop wireless network. By design, a route discovery request should be flooded via broadcast and the response should be delivered via unicast following the reverse path. However, in the implementation, upon receiving a response, an intermediate node tries to send the packet directly to the source node instead of its correct next hop node. When the intermediate node is not a direct neighbor of the source node, the route discovery will fail. We fix this bug by using the correct next hop address.

Attacks caused by message forwarding actions. We rediscovered several attacks on arand based on malicious delivery that have intense impact on the application performance. Note that the ARAN protocol itself is susceptible to these attacks [11,25]. In case of our benign experiments, we observe 99% PDR.

Divert REP and Drop ERR: By diverting a route reply (REP) message and by dropping a route error (ERR) message, a malicious node can cause the most damage among these attacks. Both these messages are sent via unicast by design, and therefore, if an intermediate malicious node drops or diverts these messages , the upstream nodes on the route remain unaware of the on-going attack. Four malicious nodes can drop the PDR to below 30% by diverting REP messages and to 40% by dropping ERR messages.

Drop RDP. An intermediate malicious node can drop a route discovery (RDP) message instead of re-broadcasting. However, this attack causes a slow decrease in PDR because every intermediate node re-broadcasts the RDP packet and therefore, even if a malicious node does not forward the RDP, the destination eventually receives the RDP message from other benign node(s). The PDR can drop below 60% with the increase in adversaries.

Blackhole/wormhole attacks. We evaluate `arand` in the presence of blackhole/wormhole attackers in the network. In the presence of one blackhole attacker, the PDR drops to 80%. Adding another blackhole node drops the PDR to 42%. However, when two blackhole nodes collude with each other to perform a wormhole attack, the PDR drops to 28%.

4. RELATED WORK

Model-checking [18, 28] and theoretical verification [29] techniques have been used to verify the correctness of models or designs. While theoretical techniques have been helpful to show the correctness of the model, protocols and some characteristics of the environment, the high-level descriptions often do not exactly match the real-world implementations. Many vulnerabilities are introduced during the process of implementation due to the complexity and performance optimization that was not originally considered during the design phase. Hence, verifying correctness and robustness of implementations is another type of challenge [29].

To that end, exploration based model checking techniques [30] and systematic fault injections [15, 27] test real implementations under various conditions. However these works do not consider adversarial environments.

There have been some recent effort on finding attacks automatically in implementations [22–24, 37]. Kothari et al. [22] automatically find attacks that manipulate control flow by modifying messages using static analysis. Stanojevic et al. [37] automatically search for gullibility in two-party protocols by dropping packets and modifying packet headers. Lee et al. [24] automatically discover performance attacks caused by insiders in distributed systems without requiring implementation modification. However [22] requires a priori knowledge about vulnerability. All these works except [23] require the implementation to be written in specific languages.

Turret [23] is closely related to our work as it tries to find attacks on unmodified distributed system implementations in realistic environments. However, it focuses on general distributed systems, and does not consider wireless environments. We extend Turret for wireless environments and add features to support wireless routing protocols such as separation of the data layer from the routing layer so that attackers can independently attack both the layers, and side channels that attackers can use to collude.

5. CONCLUSION

Given the importance of routing as a fundamental component of wireless networks, it is critical to subject their implementations to adversarial testing before deployment. To aid developers in this task, we develop Turret-W, an adversarial testing platform for wireless routing protocol implementations with minimal physical resources. We demonstrate our system by evaluating publicly available real world implementations of AODV and ARAN. In total, we (re-)discovered 14 adversarial attacks capable of either crashing the benign nodes or deteriorating their performance by disrupting the routing service and 3 implementation bugs that have an impact on the application performance.

6. REFERENCES

[1] AODV-UU. http://sourceforge.net/projects/aodvuu/.
[2] ARAN. http://prisms.cs.umass.edu/arand/.
[3] ASL. http://sourceforge.net/projects/aslib/.
[4] Emulab - network emulation testbed. http://www.emulab.net/.
[5] Iperf. http://sourceforge.net/projects/iperf/.
[6] Network Simulator 2. http://www.isi.edu/nsnam/ns/.
[7] Network Simulator 3. http://www.nsnam.org/.
[8] OpenSSL toolkit. http://www.openssl.org/.
[9] Orbit. http://www.orbit-lab.org.
[10] RFC 3561. http://tools.ietf.org/html/rfc3561.
[11] B. Awerbuch, R. Curtmola, D. Holmer, C. Nita-Rotaru, and H. Rubens. ODSBR: An on-demand secure byzantine resilient routing protocol for wireless ad hoc networks. *TISSEC*, 2008.
[12] F. De Renesse and A. Aghvami. Formal verification of ad-hoc routing protocols using spin model checker. In *IEEE Melecon*, 2004.
[13] S. M. ElRakabawy and C. Lindemann. A practical adaptive pacing scheme for TCP in multihop wireless n/ws. *ToN*, 2011.
[14] R. S. Gray, D. Kotz, C. Newport, N. Dubrovsky, A. Fiske, J. Liu, C. Masone, S. McGrath, and Y. Yuan. Outdoor experimental comparison of four ad hoc routing algorithms. In *Procs. of MSWiM*, 2004.
[15] H. S. Gunawi, T. Do, P. Joshi, P. Alvaro, J. M. Hellerstein, A. C. Arpaci-Dusseau, R. H. Arpaci-Dusseau, K. Sen, and D. Borthakur. Fate and Destini: a framework for cloud recovery testing. In *NSDI*, 2011.
[16] A. Gupta, I. Wormsbecker, and C. Wilhainson. Experimental evaluation of TCP performance in multi-hop wireless ad hoc networks. In *Mascots*, 2004.
[17] I. Habib. Virtualization with kvm. *Linux Journal*, 2008.
[18] G. Holzmann. The model checker spin. *Software Engineering, IEEE Transactions on*, 23(5):279–295, 1997.
[19] N. Horman. Understanding and programming with netlink sockets. http://www.smacked.org/docs/netlink.pdf, 2004.
[20] Y. Hu, A. Perrig, and D. Johnson. Ariadne: A secure on-demand routing protocol for ad hoc networks. *WN*, 2005.
[21] D. Johnson and D. Maltz. Dynamic source routing in ad hoc wireless networks. *Mobile computing*, pages 153–181, 1996.
[22] N. Kothari, R. Mahajan, T. Millstein, R. Govindan, and M. Musuvathi. Finding protocol manipulation attacks. *Sigcomm CCR*, 2011.
[23] H. Lee, C. Killian, C. Nita-Rotaru, and J. Seibert. A Platform for Finding Attacks in Unmodified Implementations of Intrusion Tolerant Systems. *Poster at OSDI*, 2012.
[24] H. Lee, J. Seibert, C. Killian, and C. Nita-Rotaru. Gatling: Automatic attack discovery in large-scale distributed systems. *NDSS*, 2012.
[25] Q. Li, M. Zhao, J. Walker, Y.-C. Hu, A. Perrig, and W. Trappe. SEAR: A secure efficient ad hoc on demand routing protocol for wireless networks. *Security Comm. Networks*, 2(4):325–340, 2009.
[26] P. Mahadevan, A. Rodriguez, D. Becker, and A. Vahdat. Mobinet: a scalable emulation infrastructure for ad hoc and wireless networks. *Sigmobile CCR*, 2006.
[27] P. Marinescu and G. Candea. Efficient testing of recovery code using fault injection. *ACM ToCS*, 2011.
[28] K. McMillan. Symbolic model checking: an approach to the state explosion problem. Technical report, 1992.
[29] M. Musuvathi, D. Engler, et al. Model checking large network protocol implementations. In *NSDI*, 2004.
[30] M. Musuvathi, D. Park, A. Chou, D. Engler, and D. Dill. CMC: Pragmatic approach to model checking real code. *Sigops*, 2002.
[31] S. Paris, C. Nita-Rotaru, F. Martignon, and A. Capone. Efw: A cross-layer metric for reliable routing in wireless mesh networks with selfish participants. In *Infocom*, 2011.
[32] C. Perkins and P. Bhagwat. Highly dynamic DSDV for mobile computers. *ACM Sigcomm CCR*, 1994.
[33] C. E. Perkins and E. M. Royer. Ad-hoc On-Demand Distance Vector Routing. In *IEEE Mcsa*, 1997.
[34] S. Radhakrishnan, G. Racherla, C. Sekharan, N. Rao, and S. Batsell. Dst-a routing protocol for ad hoc networks using distributed spanning trees. In *IEEE WCNC*, 1999.
[35] K. Sanzgiri, B. Dahill, B. Levine, C. Shields, and E. Belding-Royer. A secure routing protocol for ad hoc networks. In *IEEE ICNP*, 2002.
[36] K. Sanzgiri, D. LaFlamme, B. Dahill, B. Levine, C. Shields, and E. Belding-Royer. Authenticated routing for ad hoc networks. *IEEE JSAC*, 2005.
[37] M. Stanojevic, R. Mahajan, T. Millstein, and M. Musuvathi. Can you fool me? towards automatically checking protocol gullibility. In *HotNets*, 2008.
[38] M. G. Zapata and N. Asokan. Securing ad hoc routing protocols. In *ACM WiSE*, 2002.
[39] X. Zeng, R. Bagrodia, and M. Gerla. Glomosim: a library for parallel simulation of large wireless networks. *Sigsim*, 1998.

Entropy Harvesting from Physical Sensors

Christine Hennebert
CEA/LETI
christine.hennebert@cea.fr

Hicham Hossayni
CEA/LETI
hicham.hossayni@cea.fr

Cédric Lauradoux
INRIA
cedric.lauradoux@inria.fr

ABSTRACT

Finding entropy sources is a major issue to design non-deterministic random generators for headless devices. Our goal is to evaluate a collection of sensors (*e.g.* thermometer, accelerometer, magnetometer) as potential sources of entropy. A challenge in the analysis of these sources is the estimation of min-entropy. We have followed the NIST recommendations to obtain pessimistic estimations from the dataset collected during our campaign of experiments. The most interesting sensors of our study are: the accelerometer, the magnetometer, the vibration sensor and the internal clock. Contrary to previous results, we observe far less entropy than it was expected before. Other sensors which measures phenomena with high inertia such as the temperature or air pressure provide very little entropy.

Categories and Subject Descriptors

G.3 [**Probability and statistics**]: Random number generation; K.6.5 [**Security and Protection**]: [Miscellaneous]

General Terms

Non-deterministic random bit generator

Keywords

Entropy sources, *min-entropy* estimators and sensors.

1. INTRODUCTION

CONTEXT – Non-deterministic random bit generators (NDR-BGs) play an important role on security: key generation, nonces or masking to name a few. The most common failure in the design of an NDRGB is related to the entropy sources ([11, 13] for instance). The use of bad entropy sources have ruined the security of many systems [14]. All the standards including the RFC 4086 [8], the German BSI report AIS 20 and 31 [1], the NIST report SP800-133 [4] emphasize the

need to analyze the sources and to have as many as possible. On headless devices, entropy sources are scarce. We evaluate the main available resources: physical sensors.

CONTRIBUTION – We address two key-points in the design of an NDRBG. The first point is the evaluation of the physical sensors. We plug different sensors on a node and measure physical quantities (*e.g.* acceleration, temperature or air pressure) and on-MCU features (clocks desynchronization) during three experiments. We focused on three modes of operation to evaluate the influence of the environment: stability, dynamic and attack mode (saturation).

The second point addressed is the discussion on *min-entropy* evaluation. The performance of an entropy source is quantified by *min-entropy*. We consider that in many works the analysis of this quantity is not rigorous. In previous results, the *min-entropy* computed for an observation of the source can be overestimated because the data sequence is biaised. To handle this problem, we have followed the process recommended by the NIST. To our knowledge, we are the first to do so.

OUTLINE – The paper is organized as follows. In Section 2, we provide the definitions related to (N)DRBGs and entropy. The related work of this paper is then given (Section 3). We describe the platforms in Section 4 and the experiments performed to carry out our research in Section 5. The analysis of these datasets is made in Section 6 for the NIST tools and in Section 7 for our results.

2. DEFINITIONS

Before going any further, we provide four definitions to remove any ambiguity in the acronyms or terms used in the rest of the paper. We have chosen to follow the notations used by the NIST in its latest recommendations [5, 6].

Definition 1 (Deterministic RBGs)
A deterministic random bit generator (DRBG) produces a pseudo-random sequence of bits from an initial secret value called a seed (and, perhaps additional input). A DRBG is often called a pseudo-random bit (or number) generator.

Definition 2 (Non-deterministic RBGs)
A non-deterministic random bit generator (NDRBG) produces (when working properly) outputs that have full entropy. Also called a true random bit (or number) generator in the literature.

Definition 3 (Entropy sources)
An entropy source provides random bitstring. There is no

assumption that the bitstring are output in accordance with the uniform distribution.

Two types of entropy sources are distinguished in practice: *dedicated and opportunistic sources.*

DEDICATED SOURCES – These sources are designed only to produce randomness and are often included in secure embedded systems (*e.g. smartcards*). Many hardware sources have been proposed: semiconductor thermal noise [15] or ring oscillators [12]. More examples can be found in the proceedings of CHES. These sources can be very efficient and benefit from tamper resistance features. However, dedicated sources can be very expensive in term of cost or integration time. *We consider the case in which adding dedicated hardware is not affordable.*

OPPORTUNISTIC SOURCES – Because dedicated sources are not always an option, designers have started to divert some peripherals from their primary purpose to seek entropy. The first proposition made was to exploit hard-drive timing [7]. Since then many other sources have been exploited: mouse, keyboard stroke [16], performance counters. These specific sources are found in commodity computers but they are not available on headless devices (which leads to attacks against the Linux random number generator [13]). *On these devices, physical sensors are a main option for entropy.*

To compare entropy sources, we need to have some metrics for security and efficiency. The Shannon entropy and the min-entropy are the main metrics for security.

Definition 4 (Entropy)
Let S be a source of entropy which produces values over $\mathcal{X}$. We associate to these outputs a random variable X and x denotes a value sampled from X. The Shannon entropy and min-entropy for the source are given by:

$$H(X) = -\sum_x \mathbf{Pr}[X = x] \log_2 \mathbf{Pr}[X = x], \quad (1)$$

$$H_\infty(X) = -\log_2(\max_x \mathbf{Pr}[X = x]). \quad (2)$$

The Shannon entropy (Equation 1) was the first obvious choice to study sources. However, the Shannon entropy fails to measure the capability of an adversary. In [2], the authors show that min-entropy (Equation 2) is a better metric. It is also the metric recommended by the NIST [6].

Remark 1 *Determining the min-entropy of a source requires the prior knowledge of the corresponding probability distribution. The term min-entropy of a source S is used in an ambiguous way in many papers. Let us define the random variable Y associated to an observation of S. This observation of S is defined as a set of n values sampled from S and belonging to the set $\mathcal{Y} \subseteq \mathcal{X}$. At a first sight, computing $H_\infty(Y)$ can not be considered as a good estimator of $H_\infty(X)$.*

There are two solutions to solve this deadlock. First, we are able to make a complete or unbiased observation. It means that $\mathcal{Y} = \mathcal{X}$ and n is large enough such that the probability distribution of X and Y are very closed. Both conditions seem hard to achieve. Second, we use an estimator of $H_\infty(X)$ based on the observation Y. We denote this value $\mathbf{Est}_{H_\infty}(Y)$ or $\mathbf{Est}(Y)$ for short. To our knowledge, we are the first to speak of this problem. We have chosen the second solution. The NIST has proposed several tools to evaluate $\mathbf{Est}(Y)$. They are described in Section 6.

In the rest of the paper, we use the notation $H_\infty(Y)$ to denote the biased min-entropy estimator, $\mathbf{Est}(Y)$ a better estimator and $H(Y)$ an estimator of Shannon entropy.

3. RELATED WORK

Many propositions of NDRBGs and entropy sources exist for systems having strong user interactions (Fortuna [9]). Unfortunately, they are not portable to headless devices. Three works are particularly close to our paper: TinyRNG from Francillon and Castelluccia [10], the evaluation of iPhone's sensor by Lauradoux, Ponge and Roeck [17] and the work of Voris, Saxena and T. Halevi [19] published at WiSec 2011.

TinyRNG [10] is a modification of Fortuna [9] suitable for sensor nodes. In their work, the authors of [10] have investigated the opportunity to use channel errors as a source of randomness. Their motivation was that bit errors during communications are difficult to observe/predict and manipulate if certain conditions are met on the signal quality. However, they did not estimate how much entropy can be extracted from this source. We have considered other sources of entropy and provide an estimation of min-entropy.

In [17], Lauradoux, Ponge and Roeck have proposed a new online estimator for Shannon entropy. The different estimators for Shannon entropy are compared on datasets produced by the different sensors (GPS, accelerometer and compass) available on an iPhone. Compared to our work, they focused on the health test needed to check the correctness of sources rather than the estimation of min-entropy.

Voris, Saxena and T. Halevi [19] have proposed a complete design of NDRBG for RFID. Their work implements a proposition of Barak and S. Halevi [3] at CHES 2003 using an accelerometer and a randomness extractor based on Toeplitz matrix. The authors of [19] have conducted an intensive campaign of experiments using a WISP node and a smartphone to evaluate the performance of an accelerometer as entropy source. Their work is very closed to ours but two criticisms can be made. First, they have computed $H_\infty(Y)$ which can lead to an overestimation of min-entropy without giving any information on their observation size and diversity. Despite working on different hardware than in [19], our results for $H_\infty(Y)$ are of the same order. However, the results obtained using $\mathbf{Est}(Y)$ are different (see Section 7).

Second, the authors of [19] have adapted a design built for dedicated sources to an opportunistic source. Their NDRBG follows the advices of Barak and S. Halevi [3] under which it is mandatory that the adversary cannot influence the entropy source. Unfortunately, Voris, Saxena and T. Halevi [19] were unable to prove that an adversary cannot influence an accelerometer. Our opportunistic sources do not need such a proof because we rely on the security model of Barak and S. Halevi [2]. They also discuss the merits (Figure 3 in [19]) of other sensors which are considered in our work. However, their discussions are only qualitative without experiments to support them.

4. PLATFORM DESCRIPTION

4.1 Hardware

Our experiments are based on two sensor nodes: eZ430-RF2500 and Zolertia Z1 platform. The main characteristics of our sensors are summarized in Table 1.

TEXAS INSTRUMENT eZ430-RF2500 – The eZ430-RF2500

is a development platform proposed by Texas Instrument. It uses the MSP430F22x4 processor paired with the CC2500 multi-channel RF transceiver designed for low-power wireless applications. The MSP430 integrates two independent clocks: VLO (very-low-frequency oscillator) and DCO (digitally controlled oscillator). The shift between these two clocks is exploited for entropy generation by TI in [20].

ZOLERTIA Z1 (ULTRA-LOW POWER 16-BIT MCU 16MHz) – It is a low power wireless module compliant with IEEE 802.15.4 and Zigbee protocols. Its core architecture is based upon the MSP430 micro-controller and a TI CC2420 radio transceiver. Two sensors are embedded on the Z1: (a) a digital temperature sensor with $\pm 0.5°C$ accuracy (in the range $25°C$ to $85°C$), and (b) a digital accelerometer (3-Axis).

In addition to the internal thermometer and accelerometer, the Z1 node supports external sensors: two analog sensors (Phidget) and one digital sensor (Ziglet) can be used concurrently. We describe below the Phidget sensors used.

Vibration Sensor: it embeds a piezoelectric transducer. As the transducer shifts from the mechanical neutral axis, bending creates strain within the piezoelectric element and generates voltage.

Magnetic sensor: this is a ratiometric Hall-effect sensor which provides a voltage output that is proportional to the applied magnetic field.

Motion Sensor: it detects changes in infrared radiation that occurs when there is an object/person's movement with a different temperature from the surroundings. This sensor is also characterized by a narrow sensing area.

Gas pressure sensor: it measures absolute gas pressure from 20 to 250 kPa (2.9 to 36.3 psi) with a maximum error of $\pm 1.5\%$. It is suitable for measuring vacuum, or atmospheric pressure; it can also be used as a crude barometer.

Temperature & humidity sensor: it measures the relative humidity (RH) from 10% to 95% with a typical error of $\pm 2\%$RH at 55% RH. It also measures ambient temperature in the range of -30°C to +80°C with a typical error of $\pm 0.75°C$ in the range 0°C to 80°C.

4.2 Software

For the Z1 nodes, we used the Contiki OS to develop the experiment programs. For the eZ430-RF2500, we used IAR Embedded Workbench Kickstart to build and debug embedded applications for MSP430.

5. EXPERIMENTS

We place the aforementioned sensors in different scenarios and collect data. Below, we describe the different experiments made. We used a sampling frequency of 50Hz.

5.1 Stability Mode

Our experiments have started with retrieving sensor data in a stable situation without any influence, which significantly reduces the abrupt changes in the measured phenomenas. In these experiments, we set the sensors in stable conditions and away of all disturbances for a day.

5.2 Saturation Mode

We try to put the sensors in extreme conditions. These experiments differ from one sensor to another, depending on the phenomenon being measured, and which can sometimes be very difficult to implement. We have not attempt to saturate the humidity and the gas pressure sensors because the conditions needed can damage the circuitry of the node.

ACCELEROMETER – To saturate the accelerometer, we have used a playground roundabout.

VIBRATION SENSOR – We used a Power Plate found in a gym club which can generate from 10 to 50 vibrations per second.

TEMPERATURE SENSOR – There are two ways to saturate a temperature sensor, the first one, by increasing the temperature, but in the absence of a precise temperature controller we may burn the circuit. The second solution is to lower the temperature. We used the second solution and put the temperature sensor in the freezer at -18°C.

MOTION SENSOR – The principle of this sensor is to detect temperature differences between an object in its environment. To saturate this sensor, we stick it on a human body in such way that it detects only the body heat.

MAGNETIC SENSOR – We put it between two magnets.

5.3 Dynamic mode

In addition to the stability and saturation mode, we made several measurements in other situations. For instance, we used all the sensors during a road trip. The road we took going through the top of some mountains up to 1176 meters above sea level, which can vary the atmospheric pressure and the temperature of the car. Curves and shift provided a good challenge for the accelerometers. And vibration made the old car a good ground for experimenting vibration sensor.

6. NIST METHODOLOGY

Recently, the NIST has released a first draft of recommendations in order to evaluate both the entropy source model and the meaning of entropy [6]. In [6], the NIST gives several definitions and tests applied to data in order to evaluate the amount of entropy provided by the sources. These tests are not to be mistaken with the NIST statistical tests suite [18] which are made to test if a bit-string is conformed or not to some characteristics of an independent and identically distributed (i.i.d.) sequence. Applying the NIST statistical tests suite to the data produced by our sensors is not particularly significant because it is quite obvious that sequences obtained are biased.

The methodology [6] described to estimate min-entropy is different if the probability distribution of S is known or not. If the source S is i.i.d., some specific tests are applied. Otherwise, five tests are used to compute different statistic on the digitized samples and to provide information about the structure of the data. Their application to non-i.i.d. data will produce an underestimation of the contained min-entropy. No information is assumed on the probability distribution associated to the output of S. Each test reveals information about the unknown distribution given a statistical measurement. The entropy is estimated by minimizing over this set of distribution. For the following tests, a confidence level of 95% is considered.

The tests are based on the property that the expectation of a random value in the probability domain is analogous to the mean of a statistical set in the statistical domain. So, finding the probability distribution that characterizes the data structure is to find the parameter that the probability is equal to the most pessimistic estimate of the mean of the observed series that is the low-bound of the confidence interval. There are five tests: the frequency test ($\mathbf{Est}_1(Y)$),

Sensor	Output type	Accuracy	Sensibility	Max. Consump.	Operating Temp. Min, Max
Vibration	Ratiometric	—	—	400 μA	-20, 70 °C
Magnetic	Ratiometric	± 0.5%	[0, 1000] (Gauss)	2 mA	-20, 85 °C
Motion	Ratiometric	—	5 meters	15 μA	-20, 85 °C
Gas pressure	Ratiometric	± 1.5 %	20,250 Kilopascals	5 mA	0, 85 °C
Temperature	Ratiometric	± 2 °C	[−40, 100] °C	3.9 mA	-30, 80 °C
Humidity	Ratiometric	± 5 %	10%, 90% RH	3.9 mA	-30, 80 °C

Table 1: Summary of sensors characteristics.

the collision test ($\mathbf{Est}_2(Y)$), the partial collection ($\mathbf{Est}_3(Y)$), the compression test ($\mathbf{Est}_4(Y)$) and the Markov test.

The five previous tests operate on relatively small values. This limitation is clearly an issue for the analyst who needs to study sequence of large values. Due to this problem, we were unsuccessful with the Markov test because our data were too large to have a reasonable computation time.

7. ANALYSIS

During our campaign, we were unable to gather the same amount of data in all experiments. We succeed to gather large amount of data (near a million) in the stability mode but not in the dynamic one (near 50000). The explanation for this discrepancy is the complexity to maintain the sensors into the dynamic state or the necessity to preserve their integrity. It results that our confidence level is higher in the stability mode than in the saturation/dynamic mode.

Another important element of our analysis is the number of distinct states observed during the data collection. The maximum and minimum values are given for each sensor in Table 2. This table shows that our dataset are biased except for the TI's clocks desynchronization for which we observe almost all the possible states. For opportunistic sources, we believe that biased observations cannot be avoided. It emphasizes the need to have pessimistic min-entropy estimators.

	Distinct states	
Sensor	min.	max.
Accelerometer.X	1	489
Accelerometer.Y	1	1024
Accelerometer.Z	11	11
Internal thermometer	5	622
External thermometer	175	697
Vibration	44	1018
Magnetic	9	216
Gas pressure	48	92
Motion	227	1815
Humidity	229	238
TI's clocks	63843	63843

Table 2: The min-max number of states by sensors.

COMPARISON OF MIN ENTROPY TOOLS – In Figure 1, we have compared the different measures of min-entropy and Shannon entropy for an experiment with the vibration sensor. The four estimators are more pessimist than $H_\infty(Y)$. The frequency estimation $\mathbf{Est}_1(Y)$ is the closest to the biased $H_\infty(Y)$ value due to its nature. The most pessimist

tests are the partial collection test $\mathbf{Est}_3(Y)$ and the compression test $\mathbf{Est}_4(Y)$. These trends are verified for all the sensors.

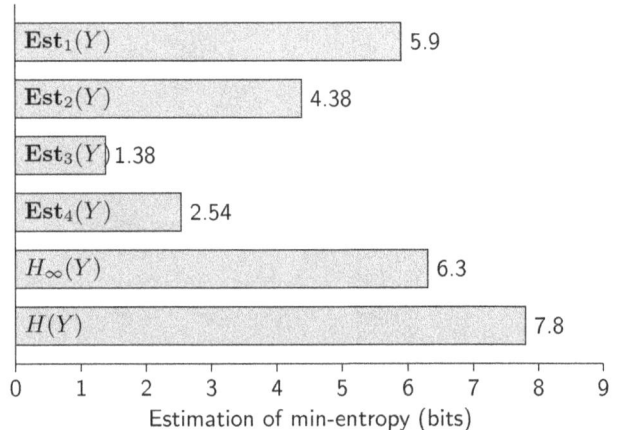

Figure 1: Comparison of the estimators for the vibration sensor on a power plate at 30Hz.

In Figure 2, we have validated the estimators on sensors for which min-entropy results can be (almost) predicted. As a witness, we have used the internal and external thermometers. Temperature is a physical phenomenon with high inertia. In the stable mode, both sensors have measured the lab temperature. Due to the precision limits of the sensors, there is only a few variation in the outputs of both sensors. Therefore, all the entropy can only come from the characteristics of the electronic components. The biased estimator $H_\infty(Y)$ evaluates that there is 3.55 bits of min-entropy for the internal sensor. This value looks excessive. The NIST estimators $\mathbf{Est}_3(Y)$ and $\mathbf{Est}_4(Y)$ considered that the min-entropy is very close to zero.

The case of the external thermometer is a bit different. The input signal of the analog-to-digital converter (ADC) is more noisy for the external thermometer than for the internal one. The wires and impedances connecting the sensor to the ADC are responsible for this noise. As a consequence, the estimation is higher with the external sensors.

To synthesize the results of all the estimators, we have considered the minimum of the four NIST estimators:

$$\mathbf{Est}(Y) = \min(\mathbf{Est}_1(Y), \mathbf{Est}_2(Y), \mathbf{Est}_3(Y), \mathbf{Est}_4(Y)).$$

GENERAL COMMENTS – The results obtained for all the sensors are given in Table 3. In all the modes, the sensors measuring physical phenomena with a high inertia have very little min-entropy. This is the case for the thermometer, the gas pressure and the humidity. The most interesting sources

Sensor	Stability mode		Dynamic mode		Saturation mode		Entropy behaviour under adversarial control
	$\mathbf{Est}(Y)$	$H_\infty(Y)$	$\mathbf{Est}(Y)$	$H_\infty(Y)$	$\mathbf{Est}(Y)$	$H_\infty(Y)$	
Internal thermometer	≈ 0	≈ 0	0	0	0.00	0.00	Nullify entropy
External thermometer	0.05	4.79			0.00	5.75	Nullify entropy
Humidity	0.16	6.21					Unknown
Gas Pressure	0.29	2.66					Unknown
Magnetic	0.62	2.98			0.02	0.94	Decrease entropy
Motion	0.11	5.5			0.65	5.51	Inconclusive
Vibration	0.17	2.87	0.48	6.12	3.02	7.85	Increase entropy
Accelerometer.X	0.22	1.38	1.34	4.90	0.00	0.00	Nullify entropy
Accelerometer.Y	0.42	1.05	2.39	7.75	0,00	0.00	Nullify entropy
Accelerometer.Z	0.36	1.72	2.83	6.15	0.33	0.70	Decrease entropy
TI's clocks	1	14,29					Unknown

Table 3: Summary of results.

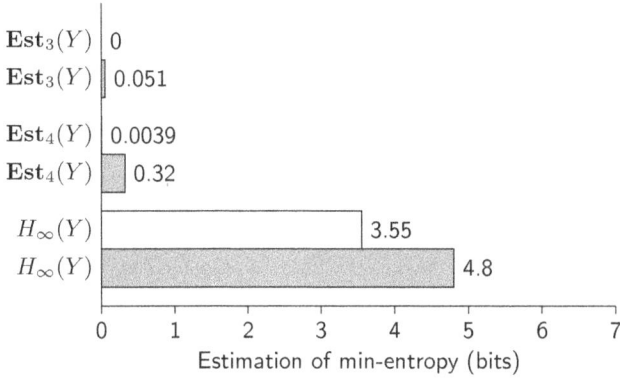

Figure 2: Observations for the internal (white) and external (grey) temperature sensor.

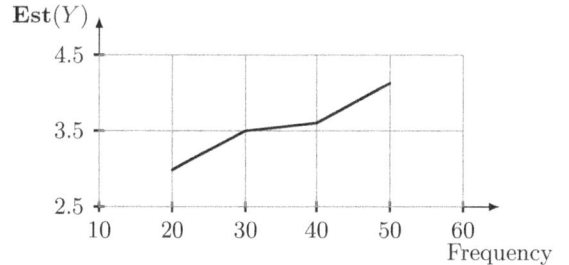

Figure 3: Effect of the power plate frequency on the vibration sensor.

of entropy are the vibration sensor, the accelerometer, the TI's clocks and the magnetic sensor. For these sensors, the mode has a significant impact on the estimation.

MAGNETIC SENSOR – In the stability mode, the best sources of min-entropy are the TI's clocks and the magnetic sensor. In our lab condition, it appears that the background noise is enough to create variation on the sensor.

Despite the small number of states in the output space, the magnetic sensor returns 0.62 bits of min-entropy. To saturate the magnetic sensor, we had put it between two magnets. At the first time, the magnets were in the same direction ($\mathbf{Est}(Y) = 0.37$), and at a second time we reversed the direction of one magnet. We were able to almost saturate the sensor in this second time ($\mathbf{Est}(Y) = 0.02$), by forcing it to return only nine values for all samples. This allows us to conclude that the magnetic sensor can be reliable as a source of entropy if we define a threshold value for the values output by the sensor. This health test can detect saturation attacks or Faraday cage attacks.

VIBRATION SENSOR – In the stability mode, the results for the vibration sensor are not very promising. They are better for the dynamic mode (see Table 3). Our attempt to saturate this sensor on a power plate was a failure. We never saturate the sensor and increase significantly its min-entropy. The Figure 3 shows the effect of the power plate working frequency on the min-entropy. This sensor seems particularly difficult to saturate and our power plate does

not allow us to increase the frequency in order to find the saturation point. The best option for an adversary is to keep the sensor in the stability mode.

ACCELEROMETER – This sensor provides little entropy in the stability mode, this is reflected by the relatively limited number of states observed. The dynamic mode is quite better but we obtain very good results on the power plate. The min-entropy is almost greater to one bit even if the frequency seems to affect the axis differently.

In the saturation mode, a playground roundabout was enough to saturate two axis (X and Y) of the accelerometer but, as it turns on a plan surface, it fails to saturate Z (0.7 bits of min-entropy). The results of this experiments are given in Table 3.

We have also manually manipulated the accelerometer with a lasso to generate an helical trajectory. The results with this last test are not included in these tables as the datasets collected did not fulfill the requirements to pass the NIST min-entropy tests. Surprisingly, the results show that we can completely saturate the X and Y axes, forcing them to return the maximum acceleration value (156), and because of irregularities in our movements, the Z axis returning few values with more dominant maximum value.

8. CONCLUSION

Our study has shown that the best candidates to produce entropy are the accelerometer, the vibration sensor and the magnetic sensor. On this point, our results agree with the work of Voris, Saxena and T. Halevi [19]. But we have shown that the amount of min-entropy collected is clearly overestimated in [19].

Generating a cryptographic key requires to collect many samples from opportunistic sources. The study of these sources is very challenging as they have unknown probability distribution and we can only obtain biased observations of their realization. The NIST estimators have improved the situation but many open problems are still around. We would like to mention the need to have a conditional mean entropy estimator. Indeed, using different sources to feed an NDRBG requires to know the correlation between the sources. In most of the NDRBGs used in practice, the sources are assumed to be independent. However, this hypothesis seems incorrect and a conditional mean entropy estimator would be a precious tool to improve sources analysis.

Acknowledgement

This research work was supported by the ANR VERSO ARESA2 project and the European FP7 project BUTLER, under contract no. 287901.

9. REFERENCES

[1] Functionality classes and evaluation methodology for physical random number generators. Technical Report AIS 31, Bonn, Germany, September 2001. http://tinyurl.com/9wv8dtj.

[2] B. Barak and S. Halevi. A model and architecture for pseudo-random generation with applications to /dev/random. In *ACM Conference on Computer and Communications Security - CCS 2005*, pages 203–212, Alexandria, VA, USA, November 2005. ACM.

[3] B. Barak, R. Shaltiel, and E. Tromer. True Random Number Generators Secure in a Changing Environment. In *Cryptographic Hardware and Embedded Systems - CHES 2003*, Lecture Notes in Computer Science 2779, pages 166–180, Cologne, Germany, September 2003. Springer.

[4] E. Barker and A. Roginsky. Recommendation for Cryptographic Key Generation. NIST Special Publication 800-133, July 2011.

[5] E. Barker and A. Roginsky. Recommendation for Random Bit Generator (RBG) Constructions. Draft NIST Special Publication 800-90C, Aug. 2012.

[6] E. Barker and A. Roginsky. Recommendation for the Entropy Sources Used for Random Bit Generation. Draft NIST Special Publication 800-90B, Jan. 2012.

[7] D. Davis, R. Ihaka, and P. Fenstermacher. Cryptographic Randomness from Air Turbulence in Disk Drives. In *Advances in Cryptology - CRYPTO '94*, Lecture Notes in Computer Science 839, pages 114–120, Santa Barbara, CA, USA, August 1994. Springer.

[8] D. Eastlake, J. Schiller, and S. Crocker. Randomness Requirements for Security, June 2005. RFC 4086.

[9] N. Ferguson and B. Schneier. *Practical Cryptography*. 2003.

[10] A. Francillon and C. Castelluccia. TinyRNG, A Cryptographic Random Number Generator for Wireless Sensor Network Nodes. In *Symposium on Modeling and Optimization in Mobile, Ad Hoc, and Wireless Networks, IEEE WiOpt 2007*, Limassol, Cyprus, April 2007. IEEE.

[11] I. Goldberg and D. Wagner. Randomness and the Netscape browser. *Dr. Dobbs Journal*, Jannuary 1996.

[12] J. D. Golic. New Methods for Digital Generation and Postprocessing of Random Data. *IEEE Trans. Computers*, 55(10):1217–1229, 2006.

[13] Z. Gutterman, B. Pinkas, and T. Reinman. Analysis of the Linux Random Number Generator. In *IEEE Symposium on Security and Privacy - S&P 2006*, pages 371–385, Berkeley, CA, USA, May 2006. IEEE Computer Society.

[14] N. Heninger, Z. Durumeric, E. Wustrow, and J. A. Halderman. Mining Your Ps and Qs: Detection of WidespreadWeak Keys in Network Devices. In *USENIX Security Symposium*, pages 205–219, Bellevue, WA, USA, July 2012. USENIX.

[15] B. Jun and P. Kocher. The Intel random number generator. *Cryptography Research Inc. white paper*, 1999.

[16] J. Kelsey. Entropy sources. In *NIST RNG Workshop*, July 2004. http://tinyurl.com/6bkbwbm.

[17] C. Lauradoux, J. Ponge, and A. Roeck. Online Entropy Estimation for Non-Binary Sources and Applications on iPhone. Rapport de recherche RR-7663, INRIA, June 2011.

[18] A. Rukhin and al. A statistical test suite for random and pseudorandom number generators for cryptographic applications, 2001.

[19] J. Voris, N. Saxena, and T. Halevi. Accelerometers and randomness: perfect together. In *Fourth ACM Conference on Wireless Network Security - WISEC 2011*, pages 115–126, Hamburg, Germany, June 2011.

[20] L. Westlund. Random Number Generation Using the MSP430. Technical Report SLAA338, Texas Instruments, 2006.

ASK-BAN: Authenticated Secret Key Extraction Utilizing Channel Characteristics for Body Area Networks

Lu Shi, Jiawei Yuan, Shucheng Yu
Department of Computer Science
University of Arkansas at Little Rock
Little Rock, AR 72204
{lxshi, jxyuan, sxyu1}@ualr.edu

Ming Li
Department of Computer Science
Utah State University
Logan, UT 84322
ming.li@usu.edu

ABSTRACT

Recently there has been an increasing interest on bootstrapping security for wireless networks merely using physical layer characteristics. In particular, the focus has been on two fundamental security issues - device authentication and secret key extraction. While most existing works emphasize on tackling the two issues separately, it remains an open problem to simultaneously achieve device authentication and fast secret key extraction merely using wireless physical layer characteristics, without the help of advanced hardware or out-of-band channel.

In this paper, for the first time, we answer this open problem in the setting of Wireless Body Area Networks (BANs). We propose ASK-BAN, a lightweight fast authenticated secret key extraction scheme for intra-BAN communication. Our scheme neither introduces any advanced hardware nor relies on out-of-band channels. To perform device authentication and fast secret key extraction at the same time, we exploit the heterogeneous channel characteristics among the collection of on-body channels during body motion. Specifically, with simple body movements, channel variations between line-of-sight on-body devices are relatively stable while those for non-line-of-sight devices are unstable. ASK-BAN utilizes the relatively static channels for device authentication and the dynamic ones for secret key generation. On one hand, ASK-BAN achieves authentication through multi-hop stable channels, which greatly reduces the false positive rate as compared to existing work. On the other hand, based on dynamic channels, the key extraction process between two on-body devices with multi-hop relay nodes is modeled as a max-flow problem, and a novel collaborative secret key generation algorithm is introduced to maximize the key generation rate. Extensive real-world experiments on low-end COTS sensor devices validate that ASK-BAN has a high secret key generation rate while being able to authenticate body devices effectively.

Categories and Subject Descriptors

C.2.0 [**General**]: Security and Protection; C.2.1 [**Network Architecture and Design**]: Wireless Communication

General Terms

Security, Design

Keywords

Wireless Body Area Network; Sensor; Authenticated Key Generation; RSS; Physical Layer

1. INTRODUCTION

Secure wireless communications have been more imperative than ever with increasing prevalence of wireless devices. Among the others, two most fundamental issues for secure wireless communications are device authentication and secret key extraction. Over years research in this area has shifted its attention to bootstrapping security for wireless communications merely based on physical layer characteristics. Such a fact is mainly caused by increasing concerns on drawbacks of applying conventional public and symmetric-key techniques in wireless networks: pre-loading secret keys on heterogeneous wireless devices is less practical; wireless devices are more likely subject to physical compromise attacks; cryptographic primitives for authentication and key distribution are expensive for many wireless applications; most cryptographic primitives assume computation boundary of attackers; so on and so forth. Bootstrapping security from physical layer characteristics can eliminate the complex process of key distribution and the computational assumptions, and thus is believed to offer better efficiency and security for wireless networks.

Existing literature in this direction mainly utilizes three types of physical layer characteristics for bootstrapping security: *advanced hardware* [1, 5, 3, 20, 21], *out-of-band communication channels* [17], and *wireless channel measurements* [28, 14, 25]. The first two approaches both assume the availability of additional resources. Information measured or extracted from the advanced hardware (e.g., multiple antenna) and auxiliary out-of-band (OOB) channels (e.g., ambient radio channels) is used for device authentication [5, 3, 22, 11] or secret key generation [1], or both of them together [27]. However, in ubiquitous environments, wireless devices, especially commercial-off-the-shelf (COTS) ones, are usually constrained in hardware configuration. System stack requires extra modifications to meet the configuration. And OOB communication channels are not always available. The third approach, wireless channel measurements, bootstraps security by only measuring wireless communication channels (e.g., RSS). With minimal requirements on the wireless system, wireless channel measurement based approach is promising in bootstrapping security for wireless devices in ubiquitous environments. Particularly, practical systems often require device authentication and secret key generation to be fulfilled simultaneously. To our best knowledge, there is no such work that is able to simultaneously provide effective device authentication and fast secret key extraction simply by wireless channel measurements.

In this paper, we answer this open problem in the setting of wireless Body Area Networks (BANs) and propose a lightweight, body movement-aided authenticated secret key extraction scheme for intra-BAN communication, namely ASK-BAN. ASK-BAN does not assume the existence of any advanced hardware or out-of-band communication channel with nodes in BAN, ensuring that it can be widely applied to COTS devices. In ASK-BAN, device authentication and secret key extraction are simultaneously achieved only based on measurements of the communication channels between the BAN nodes. Device authentication guarantees that all the sensor devices to communicate with the CU are on the same body of one person, which utilizes *relatively stable channels* between on-body devices and distinguishes them from off-body devices that have *remarkably unstable channels* to the CU due to simple body movements that the patient artificially performed during the process. Concurrently, secret keys are extracted between every authenticated on-body device and the CU, utilizing their *relatively unstable channels*. ASK-BAN is designed based on our two important observations of channel characteristics when the patient is conducting some simple body movements: 1) channels between CU and on-body sensors (OBSs) deployed in line-of-sight (LOS) vicinity tend to be much more stable than OBSs deployed in non-line-of-slight (NLOS) locations. However, channels between off-body devices and CU experience much severer fluctuations than on-body channels, whether the OBSs are LOS or NLOS to CU. 2) Authentication is transitive in BANs. That is, if node A believes node B is on-body and the CU believes node A is on-body, it is safe for CU to believe that node B is on-body.

Specifically, between every on-body sensor and CU, ASK-BAN attempts to find two types of channels – multi-hop stable channels and multi-hop unstable channels – for authentication and key extraction respectively. Using the transitivity property, an on-body sensor is accepted if it has at least one relatively stable multi-hop channel to the CU. Along the multi-hop unstable channels between one sensor and the CU, a secret key is generated with the help of relay nodes. That is, pairwise keys between relay nodes are utilized to maximize the key generation rate and entropy of the final secret key between that sensor and the CU, in terms of number of bits.

Our experiments on real sensor devices show that both stable and unstable multi-hop on-body channels are very easy to be created in practice. Our scheme is shown to be able to simultaneously provide node authentication and secret key extraction with a high key rate.

Our Contribution The contribution of this paper can be summarized as follows.

- To our best knowledge, ASK-BAN is the first scheme that provides authenticated secret key extraction using only wireless channel measurements.

- In most scenarios, combined with body movements, ASK-BAN greatly reduces false positive rate through the multi-hop authentication scheme.

- ASK-BAN introduces a novel collaborative secret key extraction scheme with multi-hop relay nodes based on the max-flow algorithm, which can find application in other wireless systems.

The rest of the paper is organized as follows. Section 2 gives an overview of related work. Section 3 defines the problem as well as the system model. We illustrate our observations of unique BAN channel characteristics in Section 4, which is followed by the detailed description of ASK-BAN in Section 5. Section 6 evaluates and discusses our experimental and simulation results of implementing ASK-BAN on real sensors. We conclude this paper in Section 7.

2. RELATED WORK

In this section, we review existing research on non-crypto key generation and authentication schemes based on physical layer characteristics.

First, we note that using a *non-wireless* channel and under some constrained scenarios, it is easy to simultaneously achieve secret key generation and device authentication. Existing works in this direction are mainly *biometric-based* and *motion-based*. Using physiological signals, many schemes have been proposed to measure and compare physiological information collected by the sensors, such as electrocardiogram (ECG), photoplethysmogram (PPG), iris and fingerprint, to assist authentication and key establishment without a priori distribution of keying material. For authentication, [22] introduced a security mechanism using biometric traits as the authentication identity. And [11] presented a light-weight secure access control scheme for implanted medical devices (IMDs) during emergencies, utilizing basic biometric information or iris data to prevent unauthorized access. For key generation combined with authentication, schemes in [27, 32, 29, 30] established physiological data-based keys between devices for verification. However, the major drawback of biometric-based techniques is that the biometrics derived from physiological features are usually accompanied with high degrees of noise and variability inherently present in the signals. Also it is difficult to guarantee consistent physiological signals measurements with same accuracy for every sensors located in different positions. Moreover, not all the physiological parameters have the same level of entropy for key generation. According to [4], for example, heart rate is not a good choice because its level of entropy is not satisfactory. Given the above issues, their applications are limited.

For motion-based key generation, [1] established a secure connection between two devices by shaking them together and generating a key from the measured acceleration data by appropriate signal. For authentication, [5] used information extracted from companion accelerometers and coherence measurements to determine whether the devices are on the same body. For authentication with key extraction, schemes in [20, 21] exploited the same movement patterns when shaking devices together for authentication, and generated shared secret keys based on the the measured acceleration data in the shaking process. But similar to biometric-based ones, these schemes require specialized sensing hardware and human participation, which is demanding for COTS devices.

On the other hand, using *wireless channel* for authentication and/or key generation has been of great interest recently. For device authentication, wireless channel has been used to determine *device proximity*. Cai et al. in [3] utilized multiple antenna to perform ad hoc pairing of nearby wireless devices, in which the proximity of the sender can be implied by the difference between the received signal strengths (RSS) measured by distinct antennas on the receiver. Similar schemes also include Amigo [28] and Ensemble [14] which perform proximity-based authentication of physically co-located/closely placed devices, using channel measurement-based signatures or variations in RSS. Recently, Shi et al.[25] proposed BANA, basing the lightweight authentication scheme on RSS measurements only. By artificially introducing body motions or channel disturbance, BANA authenticates on-body devices due to relatively stable channels to CU compared to those from off-body attackers to on-body BAN nodes. However, BANA only considers authentication for LOS on-body devices, which is limited in some sensor deployments. For key generation, several seminal works were proposed, including Mathur et. al. [18] and Jana et. al. [13]. Along this direction, one of the key research topics is to improve the key generation rate. Lai et. al. [15] exploited random channels associated with relay nodes in the wireless network as additional random sources for key generation. Note that [15] was only concerned about key generation between two nodes with one-hop relay nodes.

Nevertheless, it has been demanding to realize authentication and key generation at the same time using wireless channel alone, primarily due to a dilemma: authentication usually requires proximity, while fast key generation requires channel fading that proximity cannot provide. Take BANA in [25] for example, since its authentication process does not result in a credential and hence is "memoryless", it is difficult to derive an authenticated secret key extraction scheme by straightforward combination of BANA with existing key extraction techniques. Alternatively, directly utilizing channels between BAN sensors and the CU would result in a very low key generation rate, because these channels are too stable to carry high entropy for key extraction. For only RSS-based solutions, fast key extraction and device authentication seem to be two conflicting objectives due to the gap between their distinct requirements on channel stability.

In this paper, we address the challenge and take a step forward for achieving effective authentication and fast key generation concurrently only based on wireless channels in BAN. Unlike previous work, our scheme, ASK-BAN, does not require advance hardware for physical layer characteristic measurement, nor does it rely on any auxiliary OOB channel. Since wireless channel characteristics can be measured by most COTS devices, ASK-BAN can be easily applied in a wide range of applications.

3. PROBLEM DEFINITION

3.1 System Model and Assumptions

In our system, the wireless BAN is composed of n sensors and one control unit (CU). Worn on the body surface of a patient, these sensors measure physiological signals (e.g., heart rate, blood pressure, etc.) of the patient and transmit the collected data to the CU. As COTS sensors, they are resource-constrained with limited energy supply, memory space and computation capabilities. CU is worn on body or placed near the body with close physical proximity, i.e. with a distance of smaller than 1 meters to each of the on-body sensors, responsible for aggregating and/or processing the received data, and relaying the data to caregivers, physicians, emergency services and even medical researchers locally or remotely. CU could be a hand-held device such as smart phone or PDA.

All the devices in the BAN are able to communicate over wireless channels (e.g., Bluetooth, ZigBee, WiFi, etc.) directly to each other through their radio interfaces. Neither advanced hardware (e.g., multiple antenna, accelerometer, GPS) nor out-of-band channel is considered to exist with the sensors. We assume that the relative positions between the BAN nodes are static during the security bootstrapping process with body movements. Extensive existing research work has shown that, coherent signal observations located greater than half wavelength away from two communicating wireless devices are typically not correlated. In this paper, we place every node at least half wavelength (for Zigbee radios it is approximately 12.5cm) away from each other to ensure uncorrelated wireless channels.

Note that body movements are involved during the running of our proposed protocol. Considering some patients have limited moving capability, we introduce several easily-done body movement options in our experiments:(1) slowly walking at random; (2) slowly rotating by sitting on a spinning chair; and (3) sitting on a rolling wheelchair, which is moved back-and-forth along a straight line with the help of caregiver.

3.2 Attack Model

In this paper, at least one attacker node is present in the system. Multiple attacker nodes may exist and collude with each other with advanced hardware. Attacker locations could be either line-of-sight (LOS) or non-line-of-sight (NLOS) to the BAN user and the legitimate devices including sensors and the CU. Following existing proximity-based authentication schemes [3, 28, 14], our primary goal for authentication is to differentiate on-body BAN devices, whether LOS or NLOS to the CU, from those off the body. Thus we assume attacker devices are deployed off-body. In other words, we do not consider attacks wherein malicious devices are placed on the patient's body. But the distance between the attacker and the patient could vary largely in a wide range, e.g., from 1 or 2 meters to tens of meters.

Among different attack scenarios, we are mainly concerned about impersonation attack, in which attacker devices attempt to pretend to be a legitimate on-body sensor or the CU in order to join the BAN, thereby constructing a shared secret key with CU or sensors for the purpose of launching further attacks during communication. Attackers are aware of the deployed security mechanisms, transmission technology, and the technical specs of the sensors and the CU. They are able to fabricate physical addresses (e.g. MAC address), eavesdrop the wireless channel, replay or inject false data, and transmit packets with varying power. Beyond the above capabilities, attackers may be knowledge-able about the wireless channel environment surrounding the BAN. For instance, an attacker may have investigated the location in advance and measured the signal propagation models in that location using his own devices. Also, historical data collected in previous BAN activities might be used by the attacker for prediction of the path loss of the channel between himself and a legitimate node.

Note that we do not consider jamming and Denial-of-Service (DoS) attacks in this paper. Furthermore, CU is assumed to be not compromised. As the CU could be a hand-held device such as a smartphone, advances in existing techniques for mobile security can applied to safeguard the CU, which are out of the scope of this paper.

3.3 Design Requirements

The main goal of our design is to achieve authenticated key generation, i.e., efficiently establishing a shared secret key between each legitimate on-body sensor and CU while the system effectively differentiating valid sensors/CU from off-body attacker nodes, thereby securing future communication. Our scheme is designed for application scenarios such as setting up on-body devices at home, in hospital, or even during moving.

In addition, the authenticated key generation scheme is expected to have following properties: (1) Lightweight: our scheme shall not involve expensive operations on on-body devices that are resource-constrained ; (2) Usability: common users, such as patients, do not have to get involved in complicated setup and use of the BAN. Instead, *Plug-n-play* is a preferred usability goal; (3) Fast authentication and key extraction: applying our scheme would not put the patient's life at risk in emergency scenarios; this requires that our scheme shall be able to authenticate the nodes and extract keys of satisfying length in a fairly short time period; (4) Compatibility: our scheme shall be compatible with *commercial-off-the-shelf* (COTS) sensors and does not require additional hardware or changing the existing system stack; (5) Reliability: our scheme shall work under various types of scenarios with desirable accuracy.

4. CHANNEL CHARACTERISTICS WITH BODY MOTIONS IN BAN

To bridge the gap between fast secret key extraction and device authentication, in this paper we made some significant observations of special channel characteristics along with body motions in BAN. These new findings lead to a solution to the dilemma mentioned above and build the basis of our authenticated key extraction scheme. For brevity, in the following part we use *on-body channel* to denote the communication channel wherein both transceivers located on the same human body or one of them is in close vicinity of the body (i.e. the CU). And *off-body channel* is referred to as the channel wherein one transceiver on/close to body and the other off-body at a distance away.

4.1 Distinct RSS Variations among On-Body Channels with Body Motions

Previous work [7, 6, 16, 25] has shown that in a BAN, there exists significant differences of RSS variation profiles between on-body and off-body channels when body motions are involved. In this paper, we claim that, with body motions, the channel variations among distinct on-body channels may differ notably even if the on-body sensors remain

in relatively static positions to each other, but the variation for all these on-body channels are still relatively stable compared to those for off-body channels. That is, depending on different positions of the on-body devices including sensors and the CU, some on-body channels, especially those NLOS to each other, may experience more dramatic variations than other on-body ones over time in terms of both amplitude and changing rate. But considering both the on-body channels and off-body channel together, off-body channels prominently display much larger RSS fluctuations than those of all the on-body channels.

Experimental Evidence. To validate our claim, on-body channel measurements were carried out in time domain, with the test setup consisting of six Crossbow TelosB motes (TRP 2400). TelosB motes have the same hardware configuration as many COTS medical sensors [26] such as ECG and EMG devices. TelosB platform includes an IEEE 802.15.4 radio with an integrated antenna, a low-power MCU with extended memory and an optional sensor suite. As shown in Fig. 1(a), five of these devices are configured as on-body sensors, placed on: chest (S_1), left abdominal side (S_2), right side of the back (S_3, S_4), and high centered back (S_5). The remaining one (S_6) works as the CU and is located closed to other sensors. The CU is fixed to a wooden pole carried by the subject in front of him. During the experiment care was taken to ensure that CU was keeping relatively stationary to all the on-body sensors during the experiment. All the on-body sensors were all fixed at their respective locations in the whole process.

The subject performed easily-done body movements we suggested in Section III, including walking randomly, slowly rotating on a chair, and moving back-and-forth on a rolling wheelchair. Note that substantial body motions should be avoided to keep on-body sensors and the CU relatively stationary to each other. Each activity lasted for one minute. We measured the first two types of movements in a small office and a medium office. The third one was performed in a large corridor inside a college building.

During the experiment, all the devices, including the CU, broadcast messages in the round-robin fashion. When one device is broadcasting, others measure the RSS received from that device, respectively. Each round takes 200ms, i.e., each node obtains 5 RSS measurements for every other node per second. Fig. 2 shows the RSS measurements in one of these settings. From the figure, we can clearly see that for channels to the CU, S_4 and S_5 (cf. the bottom two waveforms in the top figure) obviously exhibit larger RSS variations than others. Note that during the experiment S_4 and S_5 were put on the back of the subject while the CU faces the front of subject's body. For channels to S_4, only S_3 and S_5 are relatively stable (cf. the top two waveforms in the bottom figure) and all the other channels undergo relatively high variations.

We classified these channels by their average RSS variations (ARVs) into two groups by existing classification algorithm. Here, ARVs are calculated by averaging the RSS variations between consecutive measurements. Then letting all the sensors authenticate one another in the system directly, sensors with ARV in the group of smaller ARV mean value are accepted; otherwise rejected. Based on the authentication result, we can draw a graph indicating the relation of whether or not one sensor accepts the other one as an authenticated on-body sensor. If two sensors accept each

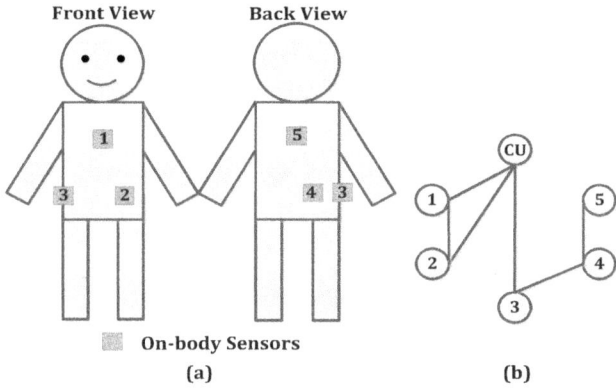

Figure 1: (a) Sensor deployment on the body; (b) Sensor Trust Relationship Topology.

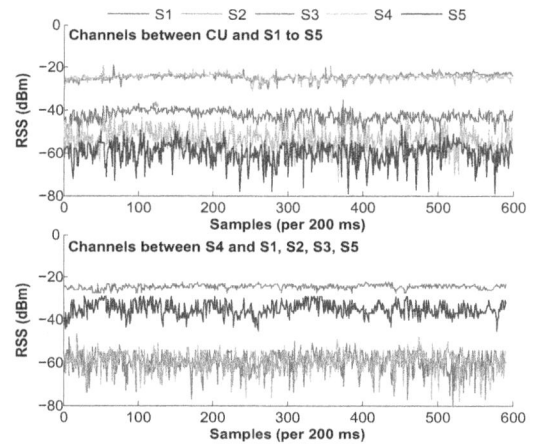

Figure 2: RSS variations among on-body channels.

other, we say that they have a *trust relationship*, or they *trust* each other, for which we draw a solid line between them. In this way, we obtain a trust relationship graph of all the tested sensors in our experiment, as shown in Fig. 1(b). Note that, a trust relationship between two sensors is established if and only if both of them accept each other. If one sensor accepts the other but conversely the other rejects, no trust relationship is established between them.

From Fig. 1(b), we can see that S_1, S_2 and S_3 are accepted by the CU while S_4 and S_5 are rejected. And S_3 and S_5 are accepted by S_4 while remaining ones are rejected. Considering trust relationship transitivity, i.e. multi-hop trust relationship, we noticed the following phenomenon illustrated in Fig. 1(b): for any pair of on-body sensors, at least one multi-hop path of trust relationship can be found between them, producing actually a connected graph. Our experiment shows that such a connected graph can be easily achieved by strategically deploying a few extra on-body sensors to serve as "hubs". For example, if we attach one sensors to the arm as the "hub", most of the on-body sensors around it would be connected through this "hub" – a one-hop trust path exists between each of them and the nearby "hub". If the CU is placed at LOS locations to these "hubs", channels between the "hubs" and the CU tend to be stable and trust paths between them can be easily found, thereby interconnecting the CU and BAN nodes by trust paths through "hubs". With a few "hub" sensor nodes, the authentication range is able to cover the whole body.

To sum up, our observation includes two prominent characteristics of on-body channels while body motions are involved:

(1) On-Body channels exhibit obviously different variations. For example, in Fig. 2, S_1, S_2 and S_3 have stable RSS values with small fluctuations for their channels to the CU, while channels from S_4 and S_5 to CU obviously experiencing larger RSS variations. For channels to S_4, only S_3 and S_5 have stable RSS values; all the other nodes display highly variable RSS values. For other experimental settings the similar phenomenon is observed again. As a close approximation to the actual channel property, especially for heterogeneous devices, fluctuation of RSS values reflects the variations of the channel.

(2) Channels between LOS on-body devices tend to be much more stable than those in NLOS locations. This is clearly shown in Fig. 2. For example, S_3 has much stabler RSSs for its channel to S_4 than other nodes.

S_5 is somewhat stabler than S_1, S_2, and the CU. Recall that in this sensor placement, S_4 and S_5 are both on the back of the subject and in clear LOS locations to each other. S_3 is deployed very close to S_4 with clear LOS. S_1, S_2 and the CU are all on the front side of the subject.

4.2 Theoretical Explanation

As we know, direct path loss, multi-path, shadowing and other interference all play important roles in radio wave propagation. The instantaneous received signal strength is a sum of many components coming from different directions due to severe reflection of the transmitted signal reaching the receiver. And the time-variant on-body propagation channels are more complicated because of the effects of the human body. According to [23], on-body signal propagation is mainly composed of a creeping wave diffracted from the human tissue and trapped along the body surface. For different positions on the body, the received signals are further affected by human movements, device placement and surrounding environment. [9] has shown that for on-body channels, the distance between transmitter and receiver has weak correlation to the path loss since shadowing effect has more influence due to different body shapes. Besides, [8] points out that both voluntary and involuntary movements also cause shadowing affecting the line-of-sight.

Line-of-sight channels. Although radio propagation over on-body channels are affected by many factors, it is well understood that the direct path (DP) plays a dominant role among all the factors if the devices are at very close range. Unsurprisingly, the fading of these channels remains relatively stable as long as the devices are kept static at their positions.

Non-line-of-sight channels. As the line-of-sight was obstructed due to the device placement, or body movements break the line-of-sight, fading of NLOS channels is more unpredictable. For BANs, the channel fading is also affected by creeping wave diffracted from the human tissue and trapped along the body surface. Therefore, NLOS channels tend to be more fluctuating in terms of both aptitude and rate.

5. MAIN DESIGN OF ASK-BAN

In this section, we present the main design of our authenticated key extraction scheme ASK-BAN, which is based on the channel characteristics along with body motions. ASK-BAN focuses on fast shared secret key generation between each valid on-body sensor and the CU during the authentication process.

5.1 Overview

Based on wireless channel measurements, we find that there exists a paradox of achieving effective authentication and efficient key extraction simultaneously. That is, while requiring stable channels in terms of RSS variations for authentication, the system needs unstable channels to provide more randomness and higher entropy for fast key generation. To resolves this situation, according to our channel characteristic observations in Section 4, ASK-BAN introduces a "double-win" strategy by involving easily-performed body movements and utilizing different RSS variations not only between on-body and off-body channels but also among on-body channels themselves.

To be specific, ASK-BAN provides multi-hop authentication between the CU and on-body sensors with the help of trusted sensors as relay nodes. We claim that *the trust relationship is transitive*. For example, while channel between sensor A and sensor C experiencing larger RSS fluctuations, if RSS variations between sensor A and sensor B and that between sensor B and sensor C are both stable, i.e. A trusts B and B trusts C, then A can trust C with high confidence, and A-B-C is a *trust path* between A and C. Therefore, ASK-BAN asks the verifying nodes (being on-body sensors or the CU) to check whether or not there exists a trust path, possibly including multiple hops, to reach the suspect node. The existence of such a path indicates that the suspect node can be accepted safely. False positive rate in ASK-BAN is expected to be remarkably reduced in many circumstances even with sparsely distributed sensor placement.

For secret key extraction, the main challenge is to achieve a high key generation rate during the authentication process. To this end, between each on-body sensor and the CU, ASK-BAN exploits possible multi-hop paths that exhibit relatively large RSS variations. Different from the collaborative key generation in [15], ASK-BAN utilizes multi-hop relay nodes between the sensor and the CU. It is easy to verify that multi-hop relay solution is secure as long as the on-body devices are deployed half wavelength away from each other for channel independence.

5.2 The ASK-BAN Protocol

The details of our authenticated secret key generation scheme can be described as the following steps:

(1) *Pairwise Key Generation:* A shared secret key will be generated for each pair of sensors in the system, denoted as k_{ij} between sensor S_i and sensor S_j ($k_{ij} = k_{ji}$). In our experiments, we choose Adaptive Secret Bit Generation (ASBG) [13] to build the shared secret key from RSS measurements, which utilizes a modified version of Mathur's quantizer [19] in conjunction with Cascade's information reconciliation [2] and privacy amplification based on leftover hash lemma [12]. Different from ASBG, ASK-BAN generates $(n+1)^2$ pairwise keys among $n+1$ nodes including the CU. Naive application of ASBG will result in $(n+1)^2$ rounds of key generation, which is unacceptably inefficient. To tackle this issue, ASK-BAN proposes a time division duplex (TDD) method to aggregate the communication. That is, nodes in ASK-BAN broadcast messages in the round robin fashion. In this way, we generate $(n+1)^2$ pairwise secret keys with an increase of the time complexity by the

---

Algorithm 1: Initial Authentication

for $i = 1$ **to** $n + 1$ **do**
  S_i broadcasts a hello message $M = (x; t_0; t)$;
  for $j = 1$ **to** $n + 1$ *And* $j \neq i$ **do**
    S_j responses after $x + t_{rj}/1000$ seconds, keeps repeating every t ms for t_0 seconds; S_i measures RSS and calculates S_j's ARV_j;
  end
  S_i performs classification on all the ARVs;
  for $j = 1$ **to** $n + 1$ *And* $j \neq i$ **do**
    if S_j *is valid* **then** S_i records (S_j, T);
    else S_i records (S_j, F);
  end
  S_i constructs a trust table based on all the records;
end

---

order of n than in ASBG, assuming that each broadcast is efficient.

(2) *Initial Authentication:* As shown in Algorithm 1, CU first broadcasts a hello message $M = (x, t_0, t)$ using a certain transmission power P_{tx} to the sensors around it, requesting responses after x seconds, where x is a random number picked by CU. Upon receiving the hello message, each responding sensor randomly chooses a small number t_r and broadcasts it, indicating the starting time for sending response message. CU collects and checks all the t_r values, ensuring no duplicated ones exist to avoid further transmission collision. And then all the responding sensors broadcast response messages m in the TDD manner as scheduled, repeatedly every t milliseconds and last for t_0 seconds. During the t_0 seconds, each node, including the CU, measures the RSS value of each received message. It is important to note that t is required to be no less than the channel coherence time. Alternatively, this kind of RSS measurements can also be done in parallel with the pairwise key generation stage (step 1) to save time. In that case, the response messages m become whatever messages transmitted in step 1. To avoid confusion, we define this process here as a separate step.

After having collected the RSSs from all the responding sensors, each node calculates the average RSS variation (ARV) for all the other nodes. Applying classification algorithm to these ARV values, they will be partitioned into two groups, where one group has a smaller mean of ARVs and the other group has a larger one. According to the classification result, the CU accepts the sensors whose ARVs belong to the group with smaller ARV mean and rejects the remaining ones in the other group. In this way, each BAN node authenticates all the other nodes. To help further communication, each node records its accept/rejection decision with corresponding node IDs into a table, showing its trust relationship with other sensors in the system by T (accept) or F (rejection).

(3) *Authenticated Secret Capacity Broadcast:* To construct a key between itself and the CU, each sensor broadcasts the secret capacities of channels between itself and others, and obtains the weighted capacity topology of the whole system. For the convenience of presentation, in this paper we define "secret capacity" as the number of bits with each pairwise secret key generated in step 1. In the rest of the paper we alternatively call it "secret capacity" or just "capacity" for brevity. ASK-BAN performs authentication along with broadcasting capacity information since previous authentication is memoryless.

Algorithm 2: Authenticated Secret Capacity Broadcast

---

for $i = 1$ **to** $n + 1$ **do**
  for $j = 1$ **to** $n + 1$ *And* $j \neq i$ **do**
    S_i broadcasts a secret capacity message
    $M_{ij} = (ID_i, ID_j, T/F, C_{ij})$;
    each sensor than S_i stores M_{ij}, measures RSS;
  end
end
for *each node* S_i **do**
  set trusted group $TG \leftarrow \emptyset$;
  compute ARV for all the other sensors;;
  perform classification on all the ARVs;
  for $j = 1$ **to** $n + 1$ *And* $j \neq i$ **do**
    if S_j *is valid* **then** $TG \leftarrow TG \cup \{S_j\}$;
  end
  set $VG \leftarrow TG$;
  while $VG \neq \emptyset$ **do**
    for *each* $S_j \in VG$ **do**
      for $k = 1$ **to** $n + 1$ *And* $S_k \notin TG$ **do**
        if M_{jk} indicates T **then** $TG \leftarrow TG \cup \{S_k\}$,
        $VG \leftarrow VG \cup \{S_k\}$;
      end
    end
    $VG \leftarrow VG \backslash \{S_j\}$;
  end
  save TG as the trust table;
  construct a security capacity topology based on all the
  capacity messages of nodes in TG;
end

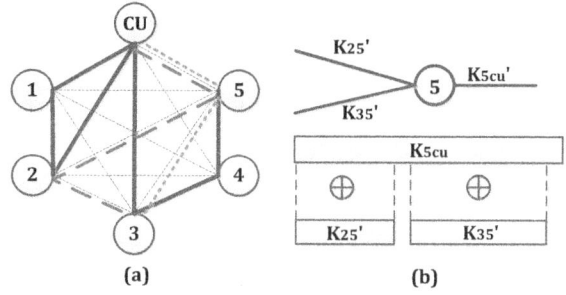

Figure 3: (a) Max-flow path from Sensor 3 to CU; (b) Max-flow multi-path merging scenario.

Algorithm 3: Key Aggregation Broadcast

---

each node runs the max-flow algorithm for source and CU
with the secrecy capacity graph;
for *each node* S_j *other than source and CU* **do**
  for *each max-flow path* P_x *that* S_j *belongs to* **do**
    determine the keys k'_{ij} and k'_{jk} from k_{ij} and k_{jk} for
    neighbor S_i and S_k respectively;
    broadcast $M_{xj} = (j, p_{ij}, p_{jk}, k'_{ij} \oplus k'_{jk})$;
    // p_{ij}/p_{jk} are positions of k'_{ij}/k'_{jk} in k_{ij}/k_{jk}.
    source and CU store M_{xj} if j is trusted;
  end
end
for *each max-flow path* P_x **do**
  source and CU derive a shared key k_x using M_{xj}'s;
end
source and CU derive the final shared key as the
concatenation of k_x's;

To be specific, a sensor node S_k broadcasts a capacity message $(ID_k, ID_l, T/F, C_{kl})$ which contains the ID of the endpoints of the channel with each of its neighbors, say S_l, the trust relationship learned from previous steps, and the channel secret capacity C. Sensors that receive capacity messages store the messages in the buffer temporarily. Meanwhile, each of S_k's neighbors measures the channel, collects RSS values and calculate S_k's ARV for later authentication. This means that broadcasting the capacity messages shall last for t_0 seconds similar to step 2, in the TDD manner with possibly repeated broadcasting of the messages.

For a single node, it assumes there is a null trusted group at the beginning. After all sensors broadcasting own capacity messages and getting capacity messages from others, each one performs classification on the collected ARVs, producing two groups with different ARV mean values, and then adds the sensors whose ARV values are in the group with smaller ARV mean into the trusted group. The nodes in the trusted groups are believed to be authenticated or trusted. The capacity messages of trusted neighbors will be processed to find the nodes that are trusted by these neighbors, i.e., those with a T in the neighbors' capacity message. These newly found node are added into the trusted group if they were not there. This process is recursively executed until all the nodes with a trust path to the node are added to the trusted group. At the end of this phase, each node will has the knowledge of all the channel secret capacity information as well as the set of trusted neighbors. An undirected weighted graph, depicting the capacity topology, can be derived based on the capacity messages, with the weight of each edge representing the secret capacity on the channel. Algorithm 2 summarizes the above process.

(4) Deciding Maximum Entropy: Based on the capacity topology, we would like to know the maximum size of secret key, in terms of bit number, that can be delivered from each

sensor to CU based on the channels with different secret capacities. In the above authentication process, CU might directly accept some on-body sensors, whose channels to CU are stable with a low-entropy key between each of them and CU. Even if the sensor is authenticated by multi-hop authentication, it cannot guarantee that the direct unstable channel between itself and CU has the highest entropy. The task is thus to find out the maximum secret key that can be obtained between a sensor and CU according to the capacity topology. We realize that this actually becomes a generalization of single-source single-sink maximum-flow problem[24]. Therefore, each sensor node runs the maximum-flow algorithm on the topology to find the path(s) through which the entropy of the key information transmitted from itself to CU can be maximized.

(5) Key Aggregation Broadcast: After finding the max-flow path(s) between a source sensor node and the destination node CU, the sensor node securely exchanges its derived secret key along the path(s) to help itself and CU extract a shared maximum secret key. For this purpose, each of the remaining sensors on the path(s), except the source sensor and the destination node CU, broadcasts the XORed value of the keys shared with its previous-hop and next-hop sensors in turn. For example, in Fig.3, if there is a max-flow path 3-2-5-CU between node 3 and the CU, the intermediate nodes 2 and 5 shall broadcast $k_{23} \oplus k_{25}$ and $k_{25} \oplus k_{5CU}$ respectively. Note that if these two keys are not of the same length, the longer one will be truncated for the XOR operation. Only having the knowledge of its shared key k_{23} with node 2, node 3 derives the shared key k_{25} from the broadcast message $k_{23} \oplus k_{25}$ thereby obtaining k_{5CU} from $k_{25} \oplus k_{5CU}$. In the similar way, CU can derive the shared key k_{23} and k_{25}

based on the broadcast messages. On this max-flow path, the shared secret key between node 3 and CU will be either k_{23} or k_{5CU}, which is truncated to the length of the shorter one of k_{23} and k_{5CU}. If there exists other max-flow paths between node 3 and CU, a key will be obtained on each of these paths following the above process, and all these keys are concatenated to form the ultimate shared secret key between node 3 and CU.

It is noteworthy that this kind of multi-hop relay does not result in losing the entropy of the ultimate shared key, which can be easily proven using the method in [15]. During the broadcast process, when the source sensor and the CU receive such a key aggregation message, they will refer to the trust table stored in the memory to decide whether or not to accept the content of the message. The broadcast message is accepted if and only if its sender is in the trust table. Based on the information obtained from every accepted broadcast messages, with the key possessed by itself, this sensor gets all the shared keys along the path(s) by XOR operations if all the nodes on the path(s) are trusted. The final secret key shared with the CU is the concatenation of the shared keys from individual max-flow paths.

Here, special attention shall be paid to the case of *merging* or *splitting* of the paths on the nodes, i.e., there might be two or more flows merging into one flow on a node or one flow splitting into two or multiple flows. As the topology graph is undirected, *merging* and *splitting* are essentially the same. Take Fig.3 for example, there are two max-flow paths 3-2-5-CU and 3-5-CU, which join at node 5. As the secret key bits extracted from different paths are required to be independent to each other, we shall guarantee no overlapped bits used by the XORed value sent from node 5 for the two paths. Specifically, in the two messages $k'_{25} \oplus k'_{5CU}$ and $k'_{35} \oplus k''_{5CU}$ for the two paths respectively, k'_{5CU} and k''_{5CU} shall be non-overlapped segments of k_{5CU}, where k'_{25} and k'_{35} are bits drawn from k_{25} and k_{35} separately. Therefore, besides the XORed value of neighboring keys, the broadcast message also needs to include the bit segment starting position of each key used by the XOR operation. That is, k'_{5CU} may start from bit position P_1 in k_{5CU} with length L_1 and k''_{5CU} with position P_2 length L_2, where $P_1 + L_1 \leq P_2$. Note that the lengths L_1 and L_2 are not necessary to be included in the broadcast message, because the receiving nodes are able to derive this information for each path after running the same max-flow algorithm. But the broadcast message shall point out which max-flow path the message is for, i.e. the message $k'_{25} \oplus k'_{5CU}$ is for path 3-2-5-CU. In implementation, such information can be represented using bit maps to save space. The above processing method is also applicable for n-to-1 merge where $n > 2$. More generally, it can be easily applied to n-to-m cases wherein both merging and splitting happen on the same node. Algorithm 3 describes the process combining both step (4) and (5).

5.3 Security Analysis

Node Authentication: With artificially introduced simple body movements, we assume that for one-hop authentication, i.e. direct authentication between two devices without any relay node, off-body devices have a very low probability, denoted as p, to falsely get accepted by on-body devices. In ASK-BAN, for authentication with the help of k-hop relay nodes, the chance of off-body devices being accepted mistakenly increases from p to kp. However, $p \leq 1$ and

p is generally very low. Moreover, a BAN device can find at least one multi-hop trust path to the CU with the help of "hubs". In practice a patient will not wear too many BAN devices on-body, implying that the value of k shall be small in real-world applications (e.g., $k \leq 3$ for Fig. 1(b)). Thus, with small kp, off-body devices actually do not get more chances to be authenticated. From another perspective, due to extra nodes for relaying purpose in ASK-BAN, every legitimate on-body sensor has more opportunities to be accepted. Therefore, multi-hop authentication would not result in a significant false positive rate in reality.

Secrecy of the Extracted Key: As attackers are off-body and their channels to on-body devices are uncorrelated to on-body channels, they are not able to derive the secret key bits generated by on-body nodes. It is remarkable that, in step 5 we did not impose any requirement for RSS-based authentication, while step 1 does not have this problem since step 2 can actually be integrated with step 1 without affecting the system performance. We claim that this does not compromise our security goals. The reason is that what the attacker is able to achieve at best is to broadcast messages in the name of an legitimate node. However, broadcasting XORed value of his own keys or other random strings does not give the attacker more chances of obtaining the secret keys shared between on-body sensors, nor does it reduce the entropy of the final shared key derived by on-body sensors. In fact, this kind of behavior can only cause denial-of-service attack, which is out of the scope of our security goals.

Also we point out that broadcasting XORed values of each node dose not cause losing entropy of the shared key on each max-flow path between this node and the CU as discussed in [15], nor does it result in losing entropy of another on-body node's shared secret key as long as the on-body channels are not correlated with off-body channels. Moreover, the secret keys shared by different on-body sensors and the CU are not required to be independent since they trust each other.

Man-In-The-Middle Attacks: As stated in [10], signal-based key generation schemes are vulnerable to Man-in-the-Middle (MITM) attacks only by using off-the-shelf hardware. Although Eve is positioned at least half a wavelength away from Alice and Bob and Eve only has uncorrelated estimates between Alice and Bob, MITM attack can be launched by impersonating both Alice and Both and injecting Eve's own information during the channel response estimation (the quantization phase, specifically), which is used by Alice and Bob as part of their secret key.In this way, the secret key generated by Alice and Bob might be revealed partially. Although it seems practical, this kind of attack does not work well on our ASK-BAN system. Since ASK-BAN performs authentication and key generation at the same time, attackers has few chances to pass authentication and get involved in the key extraction phase. In addition, in ASK-BAN, the secret key between Alice and Bob is generated not only based on the physical properties between themselves, but also based on the channel measurements of relaying devices between them. In this case, it is difficult for attackers to guess all bits of the final secret key.

Beam-forming Attacks: Theoretically, a powerful faraway attacker might form a special beam using advanced devices, such as directional antenna, attempting to produce relatively stable channels to on-body devices and finally get accepted by the system mistakenly. As analyzed in [3], we believe this type of attack is hard to implement in practice, if not completely impossible. In particular, the width of the main lobe beam is inversely proportional to the antenna array size. To successfully launch this attack, a large antenna is required since the distance between every two on-body devices is no more than 1 to 2 meters. In most of the real-life scenarios, larger antenna array will probably raise suspicion. For NLOS antenna arrays, it is more difficult to perform such an attack since attackers cannot accurately direct the antennas toward the patient who is conducting random body movements during the whole process. Furthermore, multipath effects caused by walls and indoor objects will also distort the intended beam.

5.4 Discussion

Node deployment: In ASK-BAN, BAN nodes are required to be strategically deployed such that there exists both stable (trust) path(s) and unstable path(s) between every sensor to the CU. In practice, this is easy to achieve. For example, we can either place several in front of the body and others on the back/side of the body, or attach a few extra COTS sensors to arms/legs to serve as "hubs", keeping their relatively close to each other to guarantee the existence of LOS and NLOS channels. Our experiments show that this kind of placement is effective and easy-to-use for ordinary users.

It is important to note that not all of these device are medical sensors in some circumstances, i.e., a few of them might be the extra nodes specially introduced to help with authentication and/or secret key extraction. It is not necessary for extra nodes to have the capability of measuring physiological features; they could be general devices with the basic communication and forwarding ability. Therefore, using extra nodes would not increase the costs greatly.

In addition, as ASK-BAN relies less on the strict relative positioning between the CU and each BAN sensor, it relaxes the requirements for patient on controlling body movements.

Scalability: Number of nodes may impact the performance of our scheme since ASK-BAN utilizes TDD for message broadcasting in most steps. A large number of nodes would result in a long duration for each round of TDD, thereby probably causing some pairs of nodes to measure RSSs in different coherence time periods for their shared channel. Key generation rate will also be affected due to the high error rate for RSS measurements between the pairs. To eliminate these potential effects, we can either limit the number of sensor nodes to a reasonable range, or force the nodes to cluster into groups of fixed size. For example, while node 1 just measuring nodes $\{1, 2, \cdots, k\}$, node 2 measures nodes $\{2, \cdots, k, k+1\}$, $\cdots$, and node n measures nodes $\{n, 1, 2, \cdots, k-1\}$. The corresponding algorithm will be the same except that each node needs to set the secret capacity of nodes outside its set as 0.

6. EVALUATION

To evaluate our scheme ASK-BAN, experiments are conducted under different settings. The evaluation mainly focuses on two aspects - effectiveness of node authentication and efficiency of secret key extraction. During the experiments we considered various factors that may affect the performance of the scheme, including room size, type of patient's body motion, placement of the on-body nodes as well as differences between subjects.

6.1 Experimental Setup

Experiments were conducted on Crossbow TelosB motes (TRP2400) which are all equipped with IEEE 802.15.4 radio. We used ten TelosB motes in experiments: eight motes as on-body sensors, one as the CU and one as the off-body attacker. During the experiments, the realtime RSSs measured by the motes are sent to the computer for analysis and simulation. We also varied the ratio of number of on-body sensors to that of off-body attackers. For device authentication, we emphasize on the effectiveness of differentiating on-body motes from off-body nodes with our multi-hop authentication scheme. For secret key extraction, our major concern is the key generation rate between sensors and the CU.

To show the advantage of ASK-BAN, we compare its authentication performance with BANA. Experiments were conducted in three locations - a small room, a medium-sized room and a relatively large corridor in a college building. Three subjects, two males and one female, are involved in the experiments. The following body movements are studied: walking randomly; sitting-and-rotating, i.e. subject who acts as a patient sits on a chair and the chair is rotated; sitting-and-rolling, i.e. subject who acts as a patient sits on a chair with wheels and is moved around by another subject. These movements are easy to self-perform or accomplished with help for patients even with limited moving capability. On each subject, according to the usual positions of COTS on-body sensors in real applications, the mote placement location includes chest, arms, back, waist, thighs. It is important to note that we do not have stringent requirements on the movements. For example, for walking randomly, subjects are allowed to walk normally rather than walking specially slowly. Also, the CU is not required to be placed at a strictly fixed location. The CU can either be put away from the body or be hang on the body. In addition, on-body sensors can be placed on both the back and the front of the body without affecting the performance of ASK-BAN. Note that BANA only tested the cases wherein sensors are all placed in front of the body and facing CU.

6.2 Results and Evaluation

6.2.1 Node Authentication

For node authentication alone, we conducted 18 experiments with the random combination of the following factors - experiment location, type of body movement, mote placement and subject. In the experiments we sample the collected RSSs every 200ms. For some of the scenarios, we varied the ratio of on-body sensors to off-body attackers from 7:1 to 5:3. Note that in the settings of large corridor, the attacker is either static or following behind the rolling wheelchair, while in the settings of small room and medium room the attacker is static inside/outside the room. The results are shown in Table 1 with comparison to BANA.

From the table, we can see that the overall false positive rate for the 18 experiments in ASK-BAN is almost 16 times less than that of BANA, reducing from 37.72% to 1.75%. Such a dramatic difference can mainly be explained by the flexible sensor placement in the experiments. With sensors in the front, on the back, on the side or other positions of the body, some of them do not have LOS channels to the CU directly. As BANA was designed for direct LOS on-body device authentication, it is not surprising that its false

	ASK-BAN	BANA
Small	2.38%	38.10%
Medium	2.70%	37.84%
Corridor	0%	37.14%
Sitting-and-rotating	0%	34.29%
Sitting-and-rolling	0%	37.14%
Walking	4.55%	40.91%
Subject 1	3.13%	31.25%
Subject 2	2.27%	40.91%
Subject 3	0%	39.47%
overall	1.75%	37.72%

Table 1: The false positive rates for ASK-BAN and BANA.

positive rate obviously increases to an extent that is not affordable since on-body sensor with NLOS channels to the CU would probably be rejected.

Interestingly, the false positives in ASK-BAN mainly happened in the small room and medium room scenarios, as well as the walking scenarios. This can be partially explained by the fact that small rooms and medium rooms tend to have more severe multipath effect due to the close distance from the patient to the walls while the patient is randomly walking. In these experiments, the false negative rate of ASK-BAN under different on-body to off-body node ratios remains 0, which is the same as in BANA.

6.2.2 Authenticated Secret Key Extraction

Our experiments also validate the efficiency of ASK-BAN in terms of secret key extraction rate. In our experiments, to obtain the precise key generation rate, we lasted the key extraction process for 30 seconds during the authentication. Based on the 30-second measurements, we calculate the final key generation rate (bps) as the total number of generated secret key bits during this process divided by 30. In these tests, we tried to maximize the number of RSSs measured per second, thereby expecting the maximal value of the secret key extraction rate. Using TelosB motes (TRP2400) we found that during each round of TDD broadcast, a transmission time (t) of $6ms$ for each mote results in a near-perfect packet delivery ratio (PDR). When the time t is reduced to $4ms$, the PDR dramatically decreases by up to 30%. If t is long, the measured channels might be more independent due to short coherence time, from which the key entropy benefits, while the key generation rate is reduced. If t is short, the key generation rate is increased but with lower entropy. To balance the tradeoff and find a propriate value of t, we tried both $5ms$ and $6ms$ in the experiments.

As mentioned Section 5.2, ASK-BAN adopts the ASBG scheme for pairwise key generation, which uses a modified version of Mathur's quantization. For Mathur's quantization, two thresholds $q-$ and $q+$ are used such that RSS values within $[q-, q+]$ are dropped, where $q- = mean - \alpha * std\_deviation$ and $q- = mean + \alpha * std\_deviation$, $0 < \alpha < 1$. The *Range* of remained RSS values is divided into M intervals and then for each RSS value $\lfloor \frac{Range}{M} \rfloor$ bits can be extracted. During this process, appropriate choice of quantization parameters is critical to the final secret key rate. In particular, the quantization thresholds and intervals play important roles. Lower quantization thresholds and less intervals would produce more bits, but possibly with higher bit error rate as well as lower entropy.

In our experiments, we varied the parameters and attempted to find the best ones for future reference. For this

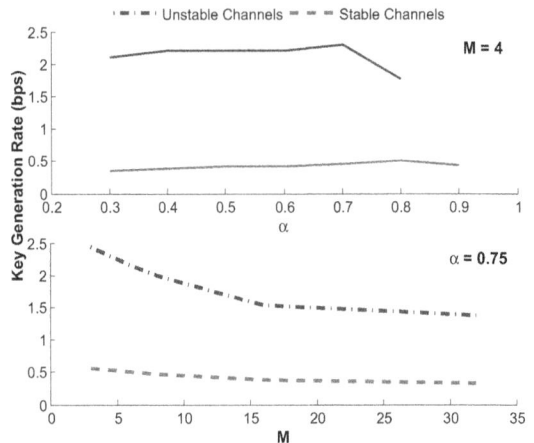

Figure 4: Secret key rate versus quantization thresholds and intervals, based on single channel.

purpose, we picked RSS serials for the measured channels, including relatively stable channels and unstable ones. And then we tried to extract secret keys based on single channel, using the ASBG scheme with varying α and M respectively. Experimental results show that $\alpha = 0.7$ and $M = 4$ result in best key generation rate in general as shown in Fig.4. Therefore, in the rest of our experiments, we stick to these values for α and M.

Key Generation Rate of ASK-BAN: Based on our experiments with eight on-body devices, Fig. 5 presents results of small room scenario and corridor scenario. Specifically, in small room scenario sitting-and-rotating was performed by the three subjects respectively, while in corridor scenario sitting-and-rolling was performed, with similar configuration of sensor placements. We found that ASK-BAN is able to achieve an average secret key rate of $7.29bps$ in the corridor if $t = 6ms$ for each node. For the small room scenarios, while the corresponding rate is about $8.03bps$ with $t = 5ms$, for $t = 6ms$ setting it is also about $8.03bps$. Therefore, to generate a 128-bit key, ASK-BAN only needs $15.9s$ in small room scenarios and $17.5s$ in corridor scenarios, which outperforms other candidate solutions we considered for BANs. On the other hand, if we utilize the direct channel to the CU for each node to extract the secret key, the average bit rates are about $1.04bps$, $0.90bps$, and $0.94bps$ for the settings of corridor-$6ms$, small room-$5ms$, and small room-$5ms$, respectively. This means that ASK-BAN boosts the secret bit rate for about 8 times than that if using direct channels to the CU. Note that for $t = 5ms$ and $t = 6ms$, the final key rate is comparable.

We also applied the collaborative secret key generation method suggested by Lai et al.[15] for comparison. To collaboratively generate the shared secret key between one sensor and the CU, all the available sensor nodes are selected as one-hop relay nodes. That is, multiple paths, each with one relay node chosen from other nodes are built between that sensor and the CU. The comparative results are shown in Fig. 5. From this figure, it is easy to see that ASK-BAN is about 2 to 4 times faster than one-hop relay method. Meanwhile, we noticed that the secret key bit rates in small rooms are slightly larger than those in the corridor on average.

In summary, along with node authentication, ASK-BAN is able to achieve up to 9bps for a single node. To update keys over time, instead of regenerating authenticated key from

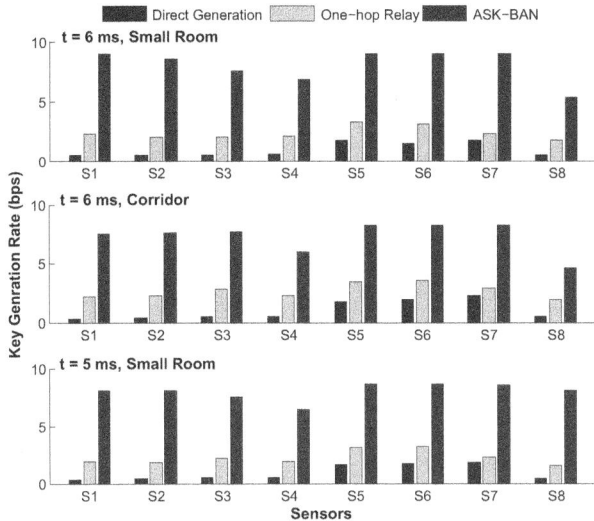

Figure 5: Comparison of secret key rate of ASK-BAN utilizing max-flow algorithm, one-hop relay method, and direct generation.

scratch, a complementary mechanism [31] can be combined with our scheme, which utilizes dynamic secrets extracted from real-time communication to update the system secret by XOR operation.

Secrecy of On-body Channel: To measure the secrecy of the on-body channels, we evaluated the mutual information between on-body channels and off-body channels. Assume A and B are two on-body nodes and C is the off-body attacker. When A is broadcasting, the RSSs measured by B are denoted as RSS_{AB} and those by the attacker are RSS_{AC}. And RSS_{BA} and RSS_{BC} represent the corresponding values by A and C respectively when B is broadcasting. We use mutual information $I(RSS_{AB}; RSS_{AC})$ and $I(RSS_{BA}; RSS_{BC})$ to estimate the channel dependencies for AB-AC and BA-BC separately. $I(RSS_{AB}; RSS_{BA})$ is also used to estimate the dependency between channels AB and BA. We selected channels on the max-flow paths and examined the above channel dependency values. Results show that mutual information between on-body channels and off-body channels is less than 0.5 on average for 6 to 7 bits RSS measurements, indicating good independence between on-body channels and off-body channels. And the mutual information for RSSs measured by the two endpoints for each channel is around 1 on average. Endpoints that measure the channel in consecutive time slots exhibit higher dependency than those measured in more distributed time slots.

7. CONCLUSIONS

In this paper, for the first time we propose ASK-BAN, a lightweight authenticated secret key extraction protocol for BAN only based on wireless channel measurements. We observed that the heterogeneous channel qualities among the collection of on-body channels - those between line-of-sight (LOS) on-body devices are relatively stable while those for non-line-of-sight (NLOS) devices are more dynamic. By utilizing this channel property, we solved the self-contradictory paradox of achieving effective node authentication and fast secret key extraction simultaneously. Our multi-hop authentication scheme in ASK-BAN can significantly reduce the false positive rate compared to previous work. To maximize the secret key generation rate, we combine the multi-hop authentication scheme with a novel collaborative secret key extraction solution based on the max-flow algorithm. Experimental and simulation results show that ASK-BAN is able to provide accurate node authentication while achieving fast secret key extraction.

For future work, we would like to further improve the secret key extraction rate. Moreover, it is interesting to provide protection against strong attacks such as attackers with directional antenna.

8. REFERENCES

[1] D. Bichler, G. Stromberg, M. Huemer, and M. Löw. Key generation based on acceleration data of shaking processes. In *Proceedings of the 9th international conference on Ubiquitous computing*, UbiComp'07, Berlin, Heidelberg, 2007. Springer-Verlag.

[2] G. Brassard and L. Salvail. Secret-key reconciliation by public discussion. In T. Helleseth, editor, *Advances in Cryptology at EUROCRYPT at 93*, volume 765 of *Lecture Notes in Computer Science*, pages 410–423. Springer Berlin / Heidelberg, 1994.

[3] L. Cai, K. Zeng, H. Chen, and P. Mohapatra. Good neighbor: Ad hoc pairing of nearby wireless devices by multiple antennas. In *Proceedings of the Network and Distributed System Security Symposium, NDSS 2011, San Diego, California, USA, 6th February - 9th February 2011*. The Internet Society, 2011.

[4] S. Cherukuri, K. Venkatasubramanian, and S. Gupta. Biosec: a biometric based approach for securing communication in wireless networks of biosensors implanted in the human body. In *Parallel Processing Workshops, 2003. Proceedings. 2003 International Conference on*, pages 432 – 439, oct. 2003.

[5] C. Cornelius and D. Kotz. Recognizing whether sensors are on the same body. In *Proceedings of the 9th international conference on Pervasive computing*, Pervasive'11, Berlin, Heidelberg, 2011. Springer-Verlag.

[6] S. Cotton, A. McKernan, A. Ali, and W. Scanlon. An experimental study on the impact of human body shadowing in off-body communications channels at 2.45 ghz. In *Antennas and Propagation (EUCAP), Proceedings of the 5th European Conference on*, pages 3133 –3137, april 2011.

[7] S. Cotton, A. McKernan, and W. Scanlon. Received signal characteristics of outdoor body-to-body communications channels at 2.45 ghz. In *Antennas and Propagation Conference (LAPC), 2011 Loughborough*, pages 1 –4, nov. 2011.

[8] F. Di Franco, C. Tachtatzis, B. Graham, M. Bykowski, D. Tracey, N. Timmons, and J. Morrison. The effect of body shape and gender on wireless body area network on-body channels. In *Antennas and Propagation (MECAP), 2010 IEEE Middle East Conference on*, pages 1 –3, oct. 2010.

[9] F. Di Franco, C. Tachtatzis, B. Graham, D. Tracey, N. Timmons, and J. Morrison. On-body to on-body channel characterization. In *Sensors, 2011 IEEE*, pages 908 –911, oct. 2011.

[10] S. Eberz, M. Strohmeier, M. Wilhelm, and I. Martinovic. A practical man-in-the-middle attack on signal-based key generation protocols. In *Computer Security - ESORICS 2012*, volume 7459 of *Lecture Notes in Computer Science*, pages 235–252. Springer Berlin Heidelberg, 2012.

[11] X. Hei and X. Du. Biometric-based two-level secure access control for implantable medical devices during

emergencies. In *INFOCOM, 2011 Proceedings IEEE*, pages 346 –350, april 2011.

[12] R. Impagliazzo, L. A. Levin, and M. Luby. Pseudo-random generation from one-way functions. In *Proceedings of the twenty-first annual ACM symposium on Theory of computing*, STOC '89, pages 12–24, New York, NY, USA, 1989. ACM.

[13] S. Jana, S. N. Premnath, M. Clark, S. K. Kasera, N. Patwari, and S. V. Krishnamurthy. On the effectiveness of secret key extraction from wireless signal strength in real environments. In *Proceedings of the 15th annual international conference on Mobile computing and networking*, MobiCom '09, pages 321–332, New York, NY, USA, 2009. ACM.

[14] A. Kalamandeen, A. Scannell, E. de Lara, A. Sheth, and A. LaMarca. Ensemble: cooperative proximity-based authentication. In *Proceedings of the 8th international conference on Mobile systems, applications, and services*, MobiSys '10, pages 331–344, New York, NY, USA, 2010. ACM.

[15] L. Lai, Y. Liang, and W. Du. Phy-based cooperative key generation in wireless networks. In *Communication, Control, and Computing (Allerton), 2011 49th Annual Allerton Conference on*, pages 662 –669, sept. 2011.

[16] B. Latré, B. Braem, I. Moerman, C. Blondia, and P. Demeester. A survey on wireless body area networks. *Wirel. Netw.*, 17(1):1–18, Jan. 2011.

[17] S. Mathur, R. Miller, A. Varshavsky, W. Trappe, and N. Mandayam. Proximate: proximity-based secure pairing using ambient wireless signals. In *Proceedings of the 9th international conference on Mobile systems, applications, and services*, MobiSys '11, pages 211–224, New York, NY, USA, 2011. ACM.

[18] S. Mathur, W. Trappe, N. Mandayam, C. Ye, and A. Reznik. Radio-telepathy: extracting a secret key from an unauthenticated wireless channel. In *Proceedings of the 14th ACM international conference on Mobile computing and networking*, pages 128–139. ACM, 2008.

[19] S. Mathur, W. Trappe, N. Mandayam, C. Ye, and A. Reznik. Radio-telepathy: extracting a secret key from an unauthenticated wireless channel. In *Proceedings of the 14th ACM international conference on Mobile computing and networking*, MobiCom '08, pages 128–139, New York, NY, USA, 2008. ACM.

[20] R. Mayrohofer and H. Gellersen. Shake well before use: authentication based on accelerometer data. In *Proceedings of the 5th international conference on Pervasive computing*, Pervasive'07, Berlin, Heidelberg, 2007. Springer-Verlag.

[21] R. Mayrohofer and H. Gellersen. Shake well before use: Intuitive and secure pairing of mobile devices. *Mobile Computing, IEEE Transactions on*, 8(6), june 2009.

[22] C. Poon, Y.-T. Zhang, and S.-D. Bao. A novel biometrics method to secure wireless body area sensor networks for telemedicine and m-health. *Communications Magazine, IEEE*, 44(4):73 – 81, april 2006.

[23] J. Ryckaert, P. De Doncker, R. Meys, A. de Le Hoye, and S. Donnay. Channel model for wireless communication around human body. *Electronics Letters*, 40(9):543 – 544, april 2004.

[24] F. Shahrokhi and D. W. Matula. The maximum concurrent flow problem. *J. ACM*, 37(2):318–334, Apr. 1990.

[25] L. Shi, M. Li, S. Yu, and J. Yuan. Bana: body area network authentication exploiting channel characteristics. In *Proceedings of the fifth ACM conference on Security and Privacy in Wireless and Mobile Networks*, WISEC '12, pages 27–38, New York, NY, USA, 2012. ACM.

[26] Shimmer. http://www.shimmer-research.com/p/sensor-and-modules.

[27] K. Singh and V. Muthukkumarasamy. Authenticated key establishment protocols for a home health care system. In *Intelligent Sensors, Sensor Networks and Information, 2007. ISSNIP 2007. 3rd International Conference on*, pages 353 –358, dec. 2007.

[28] A. Varshavsky, A. Scannell, A. LaMarca, and E. De Lara. Amigo: proximity-based authentication of mobile devices. In *Proceedings of the 9th international conference on Ubiquitous computing*, UbiComp '07, pages 253–270, Berlin, Heidelberg, 2007. Springer-Verlag.

[29] K. Venkatasubramanian, A. Banerjee, and S. Gupta. Pska: Usable and secure key agreement scheme for body area networks. *Information Technology in Biomedicine, IEEE Transactions on*, 14(1):60 –68, jan. 2010.

[30] K. K. Venkatasubramanian and S. K. S. Gupta. Physiological value-based efficient usable security solutions for body sensor networks. *ACM Trans. Sen. Netw.*, 6(4):31:1–31:36, July 2010.

[31] S. Xiao, W. Gong, and D. Towsley. Secure wireless communication with dynamic secrets. In *INFOCOM, 2010 Proceedings IEEE*, pages 1–9, march 2010.

[32] F. Xu, Z. Qin, C. Tan, B. Wang, and Q. Li. Imdguard: Securing implantable medical devices with the external wearable guardian. In *INFOCOM, 2011 Proceedings IEEE*, pages 1862 –1870, april 2011.

Chorus: Scalable In-band Trust Establishment for Multiple Constrained Devices over the Insecure Wireless Channel

Yantian Hou, Ming Li
Department of Computer Science
Utah State University
Logan, UT 84322
houyantian@gmail.com,
ming.li@usu.edu

Joshua D. Guttman
Department of Computer Science
Worcester Polytechnic Institute
Worcester, MA 01609
guttman@cs.wpi.edu

ABSTRACT

Secure initial trust establishment for multiple resource constrained devices is a fundamental issue underlying wireless networks. A number of protocols have been proposed for secure key deployment among nodes without prior shared secrets (ad hoc), however so far most of them rely on secure out-of-band (OOB) channels (e.g., audio, visual) which either only work with a small number of devices or require auxiliary hardware. In this paper, for the first time, we design a solution that enables secure initialization of a group of wireless devices, which works merely within the wireless band. Our proposed solution is based on a novel physical-layer primitive for authenticated string comparison over the insecure wireless channel, called *Chorus*, which simultaneously compares the equality of fixed-length authentication strings held by multiple wireless devices within constant time. The Chorus achieves a key authentication property, which prevents an adversary from tricking each device to believe that all strings are equal when they are not, which is enabled by exploiting the infeasibility of signal cancellation and unidirectional error detection codes. Chorus can be employed as a foundation to provide in-band group message authentication (GMA) and group authenticated key agreement (GAKA), that does not require any prior shared secret. Specifically, we design two GAKA protocols based on Chorus and formally prove their security. The most appealing features of our proposed protocols include: minimal hardware requirement (a common radio interface and a button), minimal user effort (pressing a button on each device on average), nearly constant running time, thus they are scalable to a large group of constrained wireless devices. Through extensive analysis and experimental evaluation, we show the security and robustness of Chorus under a realistic attack model, and demonstrate the high scalability of our GAKA protocols.

Categories and Subject Descriptors

C.2.0 [**General**]: Security and Protection; C.2.1 [**Network Architecture and Design**]: Wireless Communication

Keywords

Wireless Network, Trust Establishment, Message Authentication, Key Agreement, Security Protocols, Physical-layer

1. INTRODUCTION

Wireless networks are increasingly adopted by the emerging cyber-physical systems or "Internet-of-Things" (IoT) [1]. These networks typically consist of a large number of interoperable smart wireless devices that are constrained in resources (power, hardware, and user interfaces), such like wireless sensors. Their applications range from e-healthcare systems, smart home/building, to boarder monitoring in homeland security. Data transmitted by such wireless networks usually contain privacy-sensitive or safety-critical information, which are subjected to eavesdropping and malicious manipulations. Thus a fundamental problem is to securely initialize multiple wireless devices by establishing secret keys to protect the communication among them from the scratch.

Previously, a number of key pre-distribution based mechanisms have been proposed for establishing initial trust in Wireless Sensor Networks (WSNs) [13, 8, 11]. However, they all assume that nodes are loaded with some form of shared key materials before initial use. This may be a reasonable assumption in some scenarios but certainly not for all, especially for *ad hoc* formed wireless networks in user-centric applications. For example, a patient who purchases tens of wearable medical sensors and wants to deploy a body area network on her body, or a property manager who wants to setup a building monitoring system with hundreds of sensor nodes. The main reasons are three-fold: 1) Constrained wireless devices usually lack necessary user interfaces (e.g, USB ports) to configure keys manually. Even if they do, manual key deployment is not scalable to a large group of devices. 2) Commodity sensor devices are not sold with pre-loaded secrets, while the manufacturers are not always trusted by the users. 3) A global public key infrastructure (PKI) is not likely to exist as wireless devices can be produced by various manufacturers. Therefore, an important research task is to design *secure ad hoc trust initialization solutions that do not presume shared secrets*, and satisfy the following three properties: highly usable, scalable, and compatible with constrained resources.

In order to achieve secure ad hoc trust initialization, the main technical challenge is message authentication over the (insecure) wireless channel. It is well-known that the simple Diffie-Hellman key exchange over the wireless channel suffers from the Man-in-the-Middle (MitM) attack, as the unprotected wireless signal is subjected to malicious modifications (such like bit flipping and message overshadowing [7]). Thus, in the past decade, various researchers have proposed secure channel based approaches to work around this problem, which is usually called "secure device pairing". It relies on the security (authentication) properties of some auxiliary out-of-band (OOB) channel in one way or another. For example, well-known OOB channels include USB connection [38], infrared [2], visual [5, 32, 33, 27, 29, 9, 23, 22], audio [15], faraday cage [18], etc. However, all these schemes require non-trivial human support, and the devices to be paired should possess common additional hardware such like USB ports, screen, keypads, LEDs, accelerometers, etc. This assumption is often strong and impractical, because all these schemes are often obtrusive to use and not scalable, and are against the global trend for device miniaturization. Moreover, it is commonly believed that human implemented OOB channels can only tolerate up to 10 devices [9, 23, 30]. The human-implemented OOB channel and requirement for advanced hardware have been major obstacles against the practical adoption of those protocols.

Thus, it is very desirable to find alternative solutions that avoid the use of OOB channels, and merely operate over the wireless (in-band) while do not rely on any additional hardware. Ideally, it should work compatibly with any constrained device with a common wireless radio interface and require minimum user participation. Next we review recent advances in wireless physical-layer based secure communication initialization (including authentication and secrecy).

1.1 Related Works

Physical-layer trust establishment. The idea of this category of approaches is to derive trust using some physical layer characteristics unique to each link that cannot be easily eavesdropped/forged by others. Existing schemes have been mostly tackling the two issues of *key generation* and *device authentication* separately. On the former, Mathur et. al. [26] and Jana et. al. [17] first proposed to utilize the randomness in received signal strength (RSS) to extract a secret key between two devices. On the latter, related methods include ensuring close device proximity [6, 25, 35, 37], location distinction [36] and device identification, etc. Unfortunately, almost all of these techniques require costly advanced hardware, such like multiple-antennas [6] and wideband transceivers [35, 25]. This limits their applicability on constrained devices. In addition, the security notions of device proximity and location distinction are quite different from "entity and message authentication". They cannot uniquely bind a message to its originating entity. Furthermore, it is non-trivial to combine key generation with device authentication techniques.

Message Authentication and Integrity Protection. The closest works to ours are integrity code (I-code) proposed by Čapkun et. al. [7], and Tamper-Evident Pairing (TEP) proposed by Gollakota et. al [14]. The I-code primitive protects the integrity of every message sent over the insecure wireless channel. It assumes the infeasibility of sig-

nal cancellation, and exploits unidirectional error detection codes to provide message tamper-evidence. It can be applied to key establishment, satellite signal authentication, etc. On the other hand, TEP is an in-band device pairing protocol for 802.11 devices, which uses a tamper-evident announcement (TEA) that protects the message integrity by embedding cryptographic authentication information (e.g., a hash) into the physical signals, such that any tampering with it will be caught by the receiver.

Though the concept of the above is appealing, there are two limitations. First, their security are both based on the infeasibility of energy cancellation. But they only achieve a weak security guarantee, since recently Pöpper et. al [34] proposed a stronger yet practical correlated signal cancellation attack using a pair of directional antennas. Second, it is difficult to apply them to securely initialize multiple constrained devices such like medical sensors due to the scalability issue. I-code and TEA are both one-to-one message authentication primitives suitable for pairwise communication. If implemented on a sensor platform with 250kbps transmission rate, an I-coded message requires 0.5s to transmit 50 bits on a ZigBee sensor platform, given a slot length of 5ms [7]. While in TEA, each synchronization packet must be at least 19ms long [14]. In addition, the number of "ON_OFF" slots is large (roughly equals a hash length). This yields a total of more than 750ms for each TEA. Thus, direct usage or simple extension of I-code or TEP is not scalable to a large group of constrained devices, whereas the *delay* can be critical in many real-world applications [24].

1.2 Our Contributions

In this paper, we aim at making ad hoc trust initialization work strictly in-band and scalable to a group of devices, by firstly introducing a novel physical-layer primitive called "Chorus" which achieves authenticated message comparison over the insecure wireless channel, and use it to construct secure group authenticated key agreement (GAKA) protocols. The Chorus is partially inspired from I-code and TEP in that we also exploit the infeasibility of signal cancellation and unidirectional error detection codes; however, we combine a similar idea to I-code with the concept of empirical OOB channels used in message authentication protocols, to achieve key authentication and confirmation. We observe that in most of the group message authentication protocols (MAPs), the role of OOB channel is to achieve secure comparison: an authentication string (AS) s_i is typically derived by each device from the protocol transcript (messages to be authenticated); when all nodes' ASes are equal to each other all devices should output accept, and whenever any nodes' ASes are not equal all devices should output reject.

Thus, the key idea of Chorus is to let N devices compare the equality of their fixed-length strings by simultaneously emitting specially encoded signals, such that any differences among the strings will be detected by all the devices. It only outputs 1 bit of information (accept - all strings are equal, or reject - some strings are different). Due to the unidirectional property of the wireless channel (attacker can only flip a "0" to "1" but not vice versa), changing the comparison result from reject to accept is impossible except negligible probability. This makes Chorus an ideal replacement for traditional OOB channels. Based on Chorus, we design secure in-band GAKA protocols, where all the messages to be authenticated are exchanged using the normal high-bandwidth wireless transmission, with only one run of Chorus in the end of the protocols. Therefore our protocols achieve greater scalability than previous solutions and are suitable for constrained devices.

Specifically we make the following contributions:

(1) We introduce "Chorus", a primitive for authenticated equality comparison of strings from multiple devices over the wireless channel in constant time. We make Chorus resilient to the strong correlated signal cancellation attack using uncoordinated frequency hopping. Through extensive analysis, we show that the proposed design satisfies authenticated comparison with high probabilistic guarantees for real-world constrained devices, under a relatively strong attacker model. Our defense against the correlated signal cancellation attack is also of independent interest.

(2) Using Chorus, we construct two group message authentication protocols (MAPs) based on AS comparisons, which naturally yields two GAKA protocols. Because Chorus neither require human interaction nor is limited in the length of AS to be compared, we show that the Chorus greatly simplifies the trust initialization protocol design, by achieving an optimal number of rounds (two) and minimal amount of user interaction. Our GAKA protocols both run in nearly constant time, regardless of the number of devices.

(3) We provide thorough security proofs for our proposed group MAPs (and GAKAs), implement and evaluate the proposed protocols on 24 real-world wireless sensor devices. Experimental results demonstrate that our protocols are scalable and usable.

2. PROBLEM STATEMENT

We consider an ad hoc group $\mathcal{G}$ of N constrained wireless devices/nodes that share a common radio interface (e.g., ZigBee or WiFi), which is chosen by a user and will be deployed to form a wireless network. The devices do not share any secret key materials a priori. We assume the user either knows or can count the group size N correctly. The goal for trust initialization is to establish authenticated shared secret keys among them in the setup phase to support secure communication afterwards, which may include group or pairwise keys.

2.1 Design Requirements

Here we first give informal definitions for security requirements in ad hoc trust initialization. (1) Key authentication: the derived secret key is authentic and the same among all the devices in the intended group $\mathcal{G}$. This essentially requires both entity and message authentication, which means each message sent by a legitimate device in $\mathcal{G}$ should be identical to what is received by its intended recipient(s), and an attacker should not be able to impersonate any legitimate device. (2) Key secrecy: the derived secret key is not known by an attacker. (3) Key confirmation: every device in $\mathcal{G}$ should confirm the successful derivation of the same key if the above two properties are satisfied.

For practicality, the scheme should satisfy the following: (1) High scalability and efficiency. It should support a large number of devices up to the order of hundreds or even thousands. Ideally, the running time of the protocol shall be nearly constant regardless of the group size. In addition, the per-device communication, computation and storage overhead must be small. (2) High usability. The solution should involve as little human effort as possible, and be intuitive to use by non-expert users. (3) Low hardware requirement.

The solution shall be compatible with commercial-off-the-shelf (COTS) constrained devices with few interfaces (e.g., wireless sensors), and no advanced hardware such like multi-antennas or wide-band transceivers.

2.2 Attack Model and Assumptions

Our attack model is similar to the *Dolev-Yao model* [10], in that the adversary can take full control of the *normal wireless channel*, for example, it can eavesdrop, modify, remove, replay or inject messages (packets) transmitted over the wireless channel, and it can forge its identity (e.g., MAC address). However, the attacker cannot trivially disable the channel and block the transmission (e.g., using a Faraday cage). The signal cancellation attack is indeed possible for normal wireless channel as indicated in [34]. However, we will discuss ways to prevent this using specific mechanisms in more details in Sec. 3. The attacker can also jam the transmission so as to prevent the correct transmission of the information contained in a message. Further, we assume that the attacker is *computationally bounded*. We do not specifically address malicious denial-of-service/jamming attacks, which is an orthogonal problem; yet we do consider non-malicious interference from other nearby wireless devices operating within the same spectrum.

The attacker may possess powerful hardware such like software-defined radios and directional antennas. In addition, the attacker may have precise knowledge about the targeting environment and devices. For example, the exact location of each device, the channel status between each pair of wireless devices, and those between itself and the devices.

We assume that all legitimate devices are within direct communication range of each other. Furthermore, for key agreement protocols, we assume all the devices in $\mathcal{G}$ are benign (i.e., the manufacturer will not sneak spying devices when selling them). Otherwise if any device is compromised, no solution can achieve secrecy as it can send the key to an attacker. But if the protocol is merely for message authentication, this assumption is not necessary. Note that, our adversary model is relatively strong. Similar models have also been adopted by [7] and [14].

3. AUTHENTICATED COMPARISON OVER THE INSECURE WIRELESS CHANNEL

In this section, we first present the basic idea of authenticated equality comparison (AEC) over wireless channel (Chorus). Then we describe and analyze an enhancement which defends against known energy cancellation attacks.

3.1 The Basic Idea of Chorus

The Chorus does not directly authenticate a message that is sent and received over the wireless interface. Instead, it authenticates the equality comparison results for N bit strings over the wireless (derived from every node's messages as we will see in Sec. 4), as the truthful comparison of the equality of strings is the key to achieve authentication in group MAPs. We first define AEC.

DEFINITION 3.1. (*Authenticated Equality Comparison*) *Let there be $N \geq 2$ nodes that are within the communication range of each other, each holding a binary string $s_i, i \in (1, ..., N)$. AEC requires the following: 1) Non-spoofing: Whenever $\exists i, j, s_i \neq s_j$, then $\forall i \in (1, ..., N)$ outputs reject with high probability. 2) Correctness: If $\forall i, j \in (1, ..., N), s_i =*

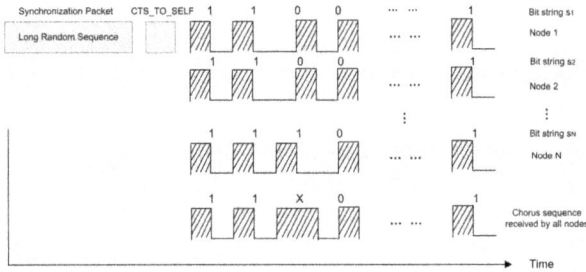

Figure 1: An example execution of basic Chorus using Manchester coding. Node N's string is different from others', which can be detected by all the N nodes. Shaded slots are random packets.

s_j, every node outputs accept. 3) Non-blocking: the AEC can neither be blocked from happening nor delayed, and the existence of it cannot be hidden.

That is, whenever a node outputs accept, it is assured (w.h.p.) that $\forall i, j, s_i = s_j$. Since only "accept" leads to successful message authentication in MAP protocols, the non-spoofing property is essential for security. AEC's properties are different from tamper-evidence in I-code [7] or TEA [14].

Examples of traditional OOB channels that satisfy AEC include the simultaneous LED blinking [33, 22]. However, we want to realize AEC over wireless. A straightforward idea is to let each of the N nodes broadcast its own s_i one-by-one using I-code; but this wastes bandwidth. Instead, we allow every node to broadcast their s_i simultaneously (Chorus), by encoding their strings using unidirectional error detection code and converting the encoded bits into ON-OFF keying. Detailed steps of the basic Chorus are as follows (instantiated using Manchester coding):

(1) It starts with a synchronization packet sent by one node (called coordinator), which contains random content and is longer than an usual packet. All other nodes detect the existence of this packet via threshold energy detection (i.e., the average RSSI is larger than a threshold T^2).

(2) After a short period when the sync packet ends, the coordinator broadcasts a short CTS_TO_SELF packet of length T_{cts}, which reserves the channel for the time period until Chorus concludes, by suppressing unwanted interference from other co-existing devices.

(3) Comparison phase: Each node i encodes its bit string s_i (of length l) using Manchester coding [40] to obtain an $2l$ bit string ($0 \rightarrow 01$ and $1 \rightarrow 10$), and map each encoded bit ($1/0$) into an ON/OFF slot respectively (of the same duration T_s). During each time slot $1 \leq j \leq 2l$, if it is an ON slot for a node, a short packet with random content is transmitted, simultaneously with everyone else (**"chorus"**); but if j is an OFF slot for a node, it remains silent and listens the channel. If $\forall 1 \leq j \leq 2l$, a node i does not detect energy in any of its own OFF slots, it outputs accept, otherwise outputs reject.

A sample timing diagram of a Chorus run is depicted in Fig. 1, where node N's string is "$1110 \cdots$" which differs from others' strings ("$1100 \cdots$") by one bit. The encoded strings are "$10101001 \cdots$" and "$10100101 \cdots$", respectively. This can be detected by all nodes (including N itself), because N will detect the aggregated signal of all other nodes during its 6th (OFF) slot, while all other nodes detect energy during their 5th (OFF) slot.

3.2 Security of the Basic Chorus

Different from I-code, in Chorus when each node sends its own signal, it cannot receive others' signals (we do NOT assume full-duplex transceivers). It seems that half of the information is lost. Thus the question is whether non-spoofing property can still be achieved. Next, we show that it is indeed the case as long as an adversary can only flip "0" to a "1" bit but not vice versa.

Claim 1: If signal cancellation is infeasible, the basic Chorus satisfies authenticated equality comparison.

First, for any two nodes' strings s_i, s_j that differ only in one bit, their respective Manchester encodings of that bit are either "01,10" or "10,01". Then, both nodes will detect a "1" during its OFF slot and output reject. Interestingly, each node can also decode its own ON slot as "1", and will obtain "11" which is not a correct codeword in Manchester code. Second, in general $\mathcal{G}$ can be divided into several subgroups $\mathcal{G}_1, ..., \mathcal{G}_k$ where strings in the same subgroup are equal but are pairwise different between different subgroups. For any node $i \in \mathcal{G}_{k'}$, its string s_i will differ from every one other subgroup's string by at least one bit. Thus it can be reduced to the two-node case.

Next, we consider an attacker that can only inject a signal generated by itself (type-I signal cancellation attacker).

LEMMA 3.1. *The realization of basic Chorus is secure against the type-I signal cancellation attacker.*

PROOF. The correctness is obvious.

According to Proposition 7.1 in [14], if the transmitted signal is unpredictable and the sender and receiver are within communication range, a type-I attacker cannot cancel the signal energy at the receiver even if she knows the channel function $h(t)$ between the sender and receiver and is perfectly synchronized with the sender. This is because the attacker needs to generate a signal with exactly the same content but the inverse phase in advance, which is infeasible.

Similarly, in our Chorus realization, the aggregated energy of packets sent during a "ON" slot cannot be canceled at any receiver by the adversary even if she knows the exact channel status. Because, during the jth slot, each node i in the chorus set $\mathcal{C}_j$ (an ON slot for them) sends a random packet denoted as signal $s_i(t)$, and the aggregated signal received by another node whose encoded bit $s'_j = 0$ is: $\sum_{i \in \mathcal{C}_j}(s_i(t) \star h_i(t)) + n(t)$ (where $n(t)$ is Gaussian noise), which is still a random signal. Thus, the soundness follows.

In addition, an adversary *cannot block or hide the existence* of a Chorus because it cannot cancel the energy of the sync packet by generating the same signal with an inverted phase by itself. Thus the non-blocking property follows. Note that, we do not consider denial-of-service attack as it does not affect authentication, i.e., flipping "0" slot to "1" only causes all the nodes to abort. □

Remarks. Note that, the above reasoning assumes that the aggregated chorus signal of ON slots of the same subgroup does not cancel out itself at receiving node i. But one may wonder whether this is the case in reality. Next, we show that the self-cancellation only happens with very small probability.

Intuitively, the more nodes transmit simultaneously, the higher the average total received power. This is similar to the phenomenon that more people speaking simultaneously

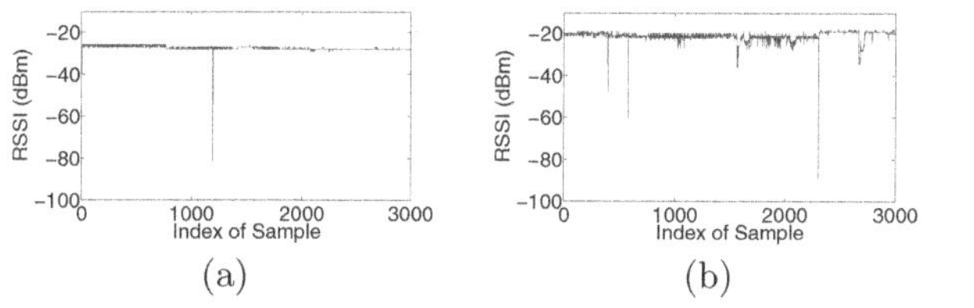

Figure 2: The self-cancellation effect. (a): 2 senders scenario. (b): 8 senders scenario.

in a room will more likely induce a louder sound. Specifically, we can model the signal received from each node as $\cos(\omega + \theta_i)$, where θ_i is a random phase delay. The superposition signal is $x_c(t) = \sum_{i=1}^{N} \cos(\omega t + \theta_i) = B_n \cos(\omega t + \tau)$, where B_N is the amplitude:

$$B_N^2 = N + \sum_{i=1}^{N} \sum_{j=1}^{N} \cos(\theta_i - \theta_j) \quad (1)$$

From Eq. (1), we can easily derive the expectation $\mathbb{E}[B_N^2)] = N$, which verifies our intuition of average power. To obtain the actual probability of self-cancellation (P_{sc}), we carry out two sets of experiments, in which $N(2$ or $8)$ sensor nodes transmit their Chorus signals simultaneously, and the receiver samples 3 RSSIs during each slot.

From Fig. 2, it can be seen that: (1). The average power of the aggregated received signal increases as N increases. (2). The signal self-cancellation phenomenon does exist; however, the occurrence of severe attenuation (-80dB) is very rare. We can derive an empirical value of P_{sc} from the experiment results (1 out of 1000 slots). In evaluation section, we will show that the self-cancellation does NOT affect the security of Chorus as P_{sc} is small.

3.3 Defending Against Powerful Signal Cancellation Attacks

A correlated signal cancellation attack is recently shown by Pöpper et. al. [34] to be practical, where the attacker does not generate its own signal. It is based on signal relaying, i.e., the attacker (Lucifer) is located at a distance away from both the sender (Alice) and receiver (Bob), and utilizes a pair of directional antennas to relay the sender's signal to the receiver. If he creates a phase delay for the carrier signal on the relay channel that is multiple of π and with the same signal amplitude, the received signal strength can be completely attenuated (see Fig. 4(a)). This attack doesn't depend on the packet content and modulation, while it mainly works under stable and predictable channel environments (e.g., static indoor scenarios). So it is important to consider this type of attack (we refer as Type-II) in the Chorus's design.

To defeat this type of powerful attack, we observe that the key factor for Lucifer to succeed is to create a phase difference of $\Delta\phi = (2k-1)\pi, k = 1, 2, ...$ Assuming the processing delay at Lucifer is negligible, we have:

$$\Delta\phi = \frac{2\pi f \Delta d}{c}, \quad (2)$$

where $\Delta d = d(A, L) + d(B, L) - d(A, B)$ is the distance difference between the relay channel and the direct channel of Alice and Bob, f is carrier frequency, and c is speed of light. Making $\Delta\phi \neq (2k-1)\pi$ will prevent the signal from

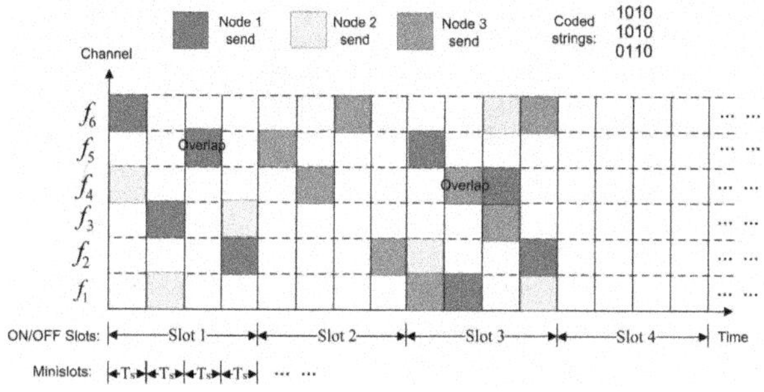

Figure 3: An example execution of the comparison phase of FH-Chorus.

being completely cancelled; but since one cannot predict the attacker's location, the only parameter Alice and Bob can control is f.

3.3.1 The Enhanced Chorus Scheme with Frequency Hopping (FH-Chorus)

We propose to make novel use of uncoordinated frequency hopping (UFH) [39] to protect Chorus from the Type-II attack. The basic idea is to make the probability of cancellation arbitrarily small by hopping over multiple frequencies.

Suppose the available spectrum for the radios of all the devices consists of n consecutive channels $f_1, ..., f_n$ with the range being Δf. In FH-Chorus, each ON/OFF slot is extended to m minislots of the same duration T_s. For each local ON slot of node i, i randomly hops among the set of available channels for m minislots, and sends a random packet during each minislot. For each local OFF slot of node i, it also randomly hops a channel for each minislot (at the same hopping rate), and listens on each channel. For each node, in each OFF slot, as long as it detects energy during at least one of the minislots, it will output reject. Otherwise, if it does not detect energy in any OFF slot, it outputs accept. Note that, in the comparison phase the packets do not contain meaningful content.

Synchronization is a little more complicated due to the need of frequency rendezvous. But again we can use UFH. Suppose k is the coordinator which randomly hops the channel with slot length T_s for a total of m slots, in each it sends a sync packet containing a *counter* which increases from one to m, along with some random padding bits. All other nodes randomly hop the channel with a longer slot length T_s' (not synchronized initially). If a node receives a sync packet on any channel, it decodes the counter and starts chorus after $(m - counter) * T_s + T_{cts}$ (seconds). In this way, if m is large enough such that every node receives at least one packet with high probability considering the energy cancellation attack, all nodes will be synchronized. The UFH not only prevents the sync packet being cancelled, but also naturally provides some degree of resistance to jamming/interference.

The CTS_TO_SELF packet does not need to be protected as it does not affect security.

A toy example of the comparison phase of FH-Chorus is depicted in Fig. 3, where there are three nodes 1, 2, 3 with $s_1 = s_2 = 11, s_3 = 01$, $m = 4$, and number of channels is 6. As long as node 3 hops to one of node 1 or 2's channels in slot 1, and nodes 1 and 2 hop to one of node 3's channels in slot 2, the bit difference will be detected by all nodes.

3.3.2 Analyzing the Attack Resilience of FH-Chorus

We analyze the successful signal detection probability P_d

P_d	successful signal detection probability at Bob
P_{nc}	probability of signal not being cancelled within one minislot by attacker across $\triangle f$
$\triangle f$	total hopping frequency range
$\triangle d_0$	minimum distance difference of attacker
h	total number of FH channels
b	number of channels cancelled by the attacker
m	number of FH minislots in one slot
B	amplitude of the signal from A received by B
T	receiver's signal detection threshold (amplitude)
η	cancellation margin: B^2/T^2 (in dB)

Table 1: Main notations.

at a node i if its string is different from some other nodes' strings. We look at the worst case where only one bit is different; in general, when multiple bits are different, P_d only becomes larger which benefits the receiver. So we constrain our analysis to a j-th bit. The nodes in $\mathcal{G}$ can be classified into two subgroups: those with $s_j = 0$ (denoted by $\mathcal{G}_0$) or $s_j = 1$ ($\mathcal{G}_1$). P_d is affected by the size of the group (e.g., $\mathcal{G}_0$) that the node is not in. Again, we consider the worst case where there is only one other node with a different string. This is because, due to random FH, i is more probable to detect energy in at least one minislot when there are many senders than only one sender.

Thus, we first focus on two nodes (Alice and Bob), given that their strings differ by one bit. Then we show that its result can be regarded as a lower bound to the detection probability when there are multiple nodes. The attacker's successful cancellation probability is $P_a = 1 - P_d$. Note that, to spoof all nodes in $\mathcal{G}_1$, the attacker needs to cancel out the energy of all $|\mathcal{G}_0| \cdot |\mathcal{G}_1|$ transmission links from nodes in $\mathcal{G}_0$ to $\mathcal{G}_1$, which requires at least $N-1$ pairs of directional antennas that increases with group size. Thus, it is reasonable to assume only one attacker for each link.

Detailed Model and Assumptions. We assume that the attacker can always choose an optimal location and antenna gain to achieve the maximum cancellation probability (minimize the RSSI of the signal at Bob), given the UFH strategy adopted by Alice and Bob. Lucifer also knows the channel status between (A, L) and (L, B). However, there are several practical restrictions for the attacker: (1) Lucifer cannot be located very close to the device group[1]. Its distance difference to Alice and Bob is: $\Delta d = d(A, L) + d(B, L) - d(A, B) \geq \Delta d_0$, which is the outer space of an ellipse (an *unsafe region*). (2) Lucifer cannot change his location in a short FH minislot (e.g., 5ms). (3) Lucifer is not capable of doing any real-time computation. In a word, he will choose a location to stay, and relay the signal sent by the legitimate transmitter.

Main Analytical Results. Our main result is, P_d can be made arbitrarily close to 1, with an increasing number of minislots m. To satisfy a given P_d, the required m can be derived as a function of $\triangle f$ and $\triangle d_0$. In reality, the feasible hopping range is often fixed, so we focus on the relationship of m and $\triangle d_0$.

(1). The Two-Node Case.

Let the signal received by Bob directly from Alice be $B \cos(2\pi f)$. Lucifer (at a fixed location) relays the signal such that the received relay signal by Bob is $B \cos(2\pi f_1 - \Delta\phi)$. The amplitude of the superposition signal is

$$x(f, \Delta d) = \sqrt{2B^2 + 2B^2 \cos(\Delta\phi)} \quad (3)$$

---

[1]This is reasonable, as in practice an attacker with directional antennas can be easily spotted by the user.

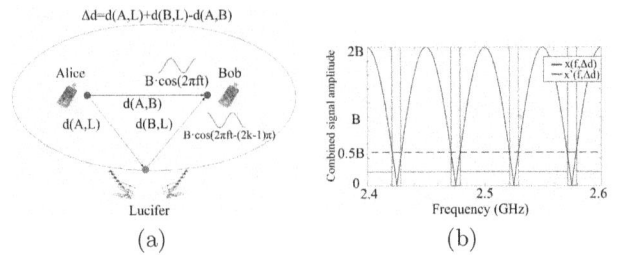

(a)

(b)

Figure 4: (a): illustration of correlated signal cancellation attack. (b): illustration of the superposition signal amplitude $x(f, \Delta d)$, and the corresponding wave function $x'(f, \Delta d)$ when choosing $T = 0.5B$.

in which $\Delta\phi = 2\pi f \Delta d/c$. The relation of $x(f, \Delta d)$ with f is illustrated in Fig.4(b). We can see that the attacker cannot cancel the signal at every frequency. By setting the detection threshold as T, the non-cancellation probability P_{nc} can be derived as $P_{nc} = \int_{f_a}^{f_b} x'(f, \Delta d) \, df/(f_b - f_a)$, in which $x'(f, \Delta d) = 1$ if $x(f, \Delta d) > T$, and $x'(f, \Delta d) = 0$ otherwise. For a fixed Δf, the P_{nc} is the ratio of the total length of non-cancelled frequency segments (L_1) to Δf. Combined with Eq. (3), we can also see the period of $x(f, \Delta d)$ monotonously decreases as Δd increases (intuitively, the larger the distance difference Δd, the more sensitive the phase difference $\Delta\phi$). Thus, as $\Delta d \to \infty$, this ratio will converge to some non-zero value.

As is shown by the simulation results in Fig. 5, P_{nc} first fluctuates and then converges as Δd increases. This implies that given any Δd_0, we can find a lower bound of P_{nc}: $P_{nc,min}$ for all $\Delta d \geq \Delta d_0$. Next we prove the existence of this lower bound in Theorem. 3.1.

THEOREM 3.1. (Lower Bound of P_{nc}) *Given a hopping range Δf, for any Δd_0, there is a lower bound $P_{nc,min} = (\lfloor x \rfloor - 1)L_1/(\lfloor x \rfloor L_0 + (\lfloor x \rfloor - 1)L_1)$, such that $\forall \Delta d \geq \Delta d_0$, we can guarantee $P_{nc} \geq P_{nc,min}$, where $L_1 = (\arccos((T^2 - 2B^2)/2B^2)c)/\pi\Delta d$, $L_0 = (-\arccos((T^2 - 2B^2)/2B^2)c + \pi c)/\pi\Delta d$, and $\lfloor x \rfloor$ is the maximum integer s.t. $\lfloor x \rfloor L_0 + (\lfloor x \rfloor - 1)L_1 \leq \Delta f$.*

The proof is in our technical report [16]. The above is a loose bound. In fact we can obtain a better actual lower bound $P_{nc,min}$ using numerical simulation. From Fig. 5, given any Δd_0, we can search all Δd within $[\Delta d_0, \Delta d_0 + W]$, where W is a large enough range. The minimum P_{nc} in this range will be taken as $P_{nc,min}$ for all $\Delta d \geq \Delta d_0$. We show both the theoretical and actual lower bounds of P_{nc} in Fig. 6. Because $\Delta d \geq \Delta d_0$ is an ellipse, for any Δd_0, we can guarantee a minimum P_{nc} by making sure that the attacker is out of this ellipse.

Next, we will derive the minimum number of minislots required to guarantee any P_d for FH-Chorus, based on the $P_{nc,min}$ derived above. Given a P_{nc}, we can obtain the cancellation probability on each FH channel. If the FH range includes h channels which span from f_a to f_b, then the maximum number of channels that can be cancelled by an attacker is $b = \lceil h(1 - P_{nc,min}) \rceil$. After deriving b, we can obtain the minimum number of minislots m to guarantee a given P_d:

THEOREM 3.2. (Minimum Number of FH Minislots) *Given a hopping range $\triangle f$ and Δd_0, an arbitrary detection probability threshold $P_d \to 1$ can be achieved by using a mini-*

172

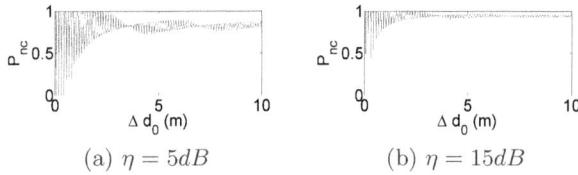

(a) $\eta = 5dB$ (b) $\eta = 15dB$

Figure 5: Non-cancellation probability w.r.t. $\triangle d_0$

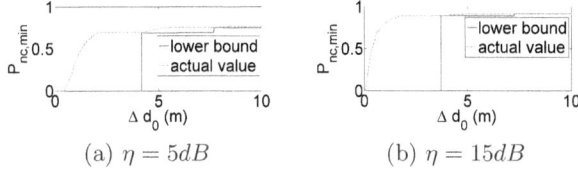

(a) $\eta = 5dB$ (b) $\eta = 15dB$

Figure 6: Lower bound of P_{nc}, given a $\triangle d_0$; $\Delta f = 80MHz$

mum number of FH minislots $m \geq \log_{(h^2-h+b)/h^2}(1 - P_d)$, in which h is the number of FH channels, and $b = \lceil h(1 - P_{nc,min}) \rceil$.

The proof is in [16]. From this we also found we can choose $h^* = \min\{2b, h\}$ to minimize m, therefore enhancing efficiency. Fig. 7 shows an example of the minimum number of minislots needed given a P_d and hopping range $\Delta f = 80MHz$. It can be seen that, if the cancellation margin η is larger, the required m is smaller to satisfy the same P_d.

(2). The General-Group Case. In the group case, the superposition signal sent simultaneously by multiple nodes is actually easier to be detected, even with self-cancellation. This is because, with random frequency hopping, a larger group of senders increases the probability of any of their signals being detected by a receiver. We will prove this intuition in the following.

THEOREM 3.3. (Minimum Number of Hopping Slots in the Group Case) *Assume in the two-node case, given a threshold P_d, Δd_0 and Δf, the required minimum number of minislots is m^*. Then in the group case, using the same m^*, and Δf, the probability of successful signal detection is lower bounded by P_d, if the attacker is located outside of the ellipses defined by every pair of nodes ($\forall i, j, \Delta d_{ij} \geq \Delta d_0$).*

The proof is in [16]. To sum up, given a feasible FH range Δf, we can satisfy any signal detection probability threshold P_d close to 1, by guaranteeing $\Delta d > \triangle d_0$ and its corresponding minimum number of minislots $m \geq \log_{(h^2-h+b)/h^2}(1 - P_d)$. As an example, when $\Delta d_0 = 2m$, $\Delta f = 80MHz$, $\eta = 15dB$, we obtain $m = 105$.

3.3.3 Robustness of Chorus

Chorus can be easily made robust against non-malicious interference. For example, the interference from 2.4G WiFi AP/laptop can generate a signal large enough to cause a false alarm in Chorus. Our first approach is to set a large enough RSSI detection threshold (e.g., −65dBm). This can filter out most of the ambient RF noise and cross-technology interference. In extreme cases (when the WiFi station is quite close) the interference could still induce a higher RSSI value than the threshold. However, through experiment (done in a campus building) we actually find that this type of strong interference is rare. We can only detect a small number

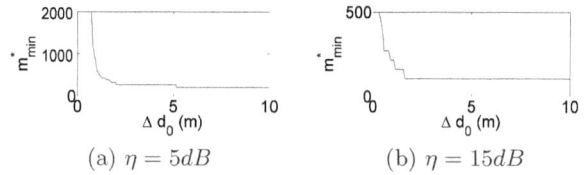

(a) $\eta = 5dB$ (b) $\eta = 15dB$

Figure 7: Minimum number of FH minislots required assuming different $\triangle d_0$; $\Delta f = 80MHz$

of false "ON" minislots in most cases (smaller than 3 when $m = 105$). Meanwhile, we observe that the expected number of true "ON" minislots is much larger (e.g., around 25 when $m = 105$) when any two nodes' comparison bit strings are different.

Therefore, our second approach to filter interference in FH-Chorus, is to impose a "ON" slot number threshold T_n (e.g 3) on the number of minislots where energy is detected, and ignore the "ON" alarm if that number is below T_n. In the following we analyze the robustness of Chorus given a chosen T_n. The tradeoff is that, the larger T_n, the higher the robustness, but the lower the detection probability P_d.

THEOREM 3.4. (Chorus's Robustness) *Given a "ON" slot number threshold T_n, the new successful signal detection probability P'_d is: $P'_d = P_d - \sum_{i=1}^{T_n} \binom{m}{i} \cdot ((1 - P_r)^{m-i}) \cdot ((P_r)^i)$, in which $P_r = (1/h) \cdot (1 - b/h)$.*

The proof is in [16]. For $T_n = 3$, the probability P'_d reduces to 0.9999 from 0.999999, when $\Delta d_0 = 2m$, $\Delta f = 80MHz$, $\eta = 15dB$.

3.3.4 Practical Issues

Two main practical issues in Chorus are: synchronization, and the message expansion (encoding overhead). For the former, the analysis of successful sync packet reception probability is similar to Sec. 3.3.2, and the same probability as P_d can be achieved using parameters in the previous example. For the latter, an optimal unidirectional code – Berger code can be used instead of Manchester code, where the check value size is $\lceil log_2(l + 1) \rceil$ bits for a message of l bits. We omit the details here due to lack of space.

3.4 Comparison with Empirical Channels

We model the properties of Chorus and compare with two major state-of-the-art types of authenticated empirical channels in Table. 2. The Dolev-Yao channel is taken as a reference. Examples of weak empirical channel are non-face-to-face human communication such like voice mail. Examples of weak empirical channel include face-to-face conversation. However, existing unspoofable empirical channels require non-trivial human support [29].

Chorus achieves comparable security properties as strong empirical channel, but uses in-band wireless communication without user involvement. We already showed that Chorus achieves unspoofability and non-blocking (with high probability), since energy cancellation can be made infeasible. Delaying also won't work as the nodes are assumed to be in each other's communication range. The attacker's ability to "create" and "replay" is more subtle. In a wireless channel, nothing prevents an attacker from injecting/replaying a sync signal and initiating a Chorus process among the legitimate nodes (nodes cannot distinguish where the signal comes from). However, when we integrate Chorus into GAKA protocols, such a problem can be easily avoided since

Attacker capabilities	Create	Modify	Delay	Block	Replay	Overhear	In-band
Dolev-Yao [10]	$\checkmark$	$\checkmark$	$\checkmark$	$\checkmark$	$\checkmark$	$\checkmark$	$\checkmark$
Weak empirical [41, 31, 29]	$\times$	$\times$	$\checkmark$	$\checkmark$	$\checkmark$	$\checkmark$	$\times$
Strong empirical [41, 31, 29]	$\times$	$\times$	$\times$	$\times$	$\times$	$\checkmark$	$\times$
Chorus	$\times^*$	$\times$	$\times$	$\times$	$\times^*$	$\checkmark$	$\checkmark$

Table 2: Comparison of Chorus with existing authenticated channels. *: explained in text.

a legitimate coordinator node will always send out its sync signal during each protocol run, so that all nodes will abort if they hear it more than once (similar to the synchronization mechanism in [14]).

4. IN-BAND TRUST INITIALIZATION PROTOCOLS FOR GROUPS OF DEVICES

Chorus enables the design of truly scalable in-band trust initialization protocols. Next we present two example protocols representing two extremes in the design space that use short and long authentication strings, respectively. As any GAKA protocol can be reduced to a message authentication protocol (MAP) [29], we will illustrate how to design efficient MAPs based on Chorus and focus on the group setting.

4.1 In-band MAP with Short Authentication String (SAS)

When the AS to be compared is a short string (e.g., 16 bits), there exists several traditional SAS-comparison based protocols both in the two-party setting [41, 19, 5] and the group setting [29, 20, 33]. The basic protocol structure follows the principle of "joint commitment before knowledge" (JCBK) [29], which consists of three rounds: 1) Commitment; 2) Decommitment; 3) Computing SAS and compare it over an OOB channel. We show that in-band group MAPs with short SAS (GMS) can be designed by using Chorus as a primitive to replace OOB channel-based SAS comparison.

Our GMS protocol is based on the SHCBK protocol proposed by Nguyen and Roscoe [28, 29]. The reason to choose their protocol is, unlike most of the group protocols [29, 20, 33], it does not need a heavy-weight non-malleable commitment scheme thus is much more computationally efficient. It uses a cryptographic hash ($H(x)$), and a *Digest* function:

DEFINITION 4.1 (DIGEST FUNCTION). $Digest(r, m): \{0, 1\}^L \times \{0, 1\}^n \to \{0, 1\}^\ell$ *is a mapping where m is the message to be digested and r is a key. It shall have two properties:*

(1) *(ϵ_u key-based uniformity) for any fixed m and y,*
$Pr_{r \in_R \{0,1\}^L}[Digest(r, m) = y] = \epsilon_u$.

(2) *(ϵ_r no uniform compensation) for any fixed θ and $m \neq m'$, $Pr_{r \in_R \{0,1\}^L}[Digest(r, m) = Digest(r \oplus \theta, m')] = \epsilon_r$.*

A concrete construction is given in [28] based on matrix product, where the ideal properties are achieved: $\epsilon_u = \epsilon_r = \frac{1}{2^\ell}$. Usually the output of a digest function is a short string, e.g., $\ell = 16$ bits. It is similar to a universal hash function, but the latter does not concern collision resistance under different keys.

Our GMS protocol is outlined in Fig. 8. Since the group is formed in an ad hoc way, the devices do not know the group $\mathcal{G}$ in advance. So in step 0, the user should count and enter the group size into a designated coordinator node. Steps 1 and 2 are the same with the SHCBK protocol. In round 3, each node computes an SAS and compare it via Chorus. $\mathcal{G}_i$ is

---

Input: message $INFO_i = \{i, m_i\}$, $i \in (1, ..., N)$
0. User picks coordinator k and enters group size N.
1. $\forall i \to_N \forall i'$: $INFO_i, H(i, r_i)$
   r_i is a random number;
2. $\forall i \to_N \forall i'$: r_i
3. k initiates Chorus by sending sync packet;
   $\forall i \Leftrightarrow_{Chorus} \forall i'$: $Digest(\oplus_{j \in \mathcal{G}_i} r_j, \{INFO_j\}_{j \in \mathcal{G}_i})$
   For k, if all SASes match and $|\mathcal{G}_k| = N$, accept.
   Otherwise output fail, send sync signal again.
   $\forall i \neq k$, if detected sync more than once, abort.
   Otherwise output accept.

Note: "$\to_N$": normal wireless channel;
"$\Leftrightarrow_{Chorus}$": Chorus channel.

Figure 8: In-band Group MAP with SAS (GMS)

the set of group IDs received by node i. Output confirmation is done by the coordinator sending another sync signal to all nodes (which cannot be removed by the attacker). This is because only the coordinator knows the correct group size. There is no need for the user to press buttons again.

Protocol Synchronization. Clearly, the GMS strictly follows the JCBK principle, where after round 1 all nodes are committed to their final SAS values. Synchronization is important to ensure JCBK. This can be done via several ways, for example, a strict message order can be imposed ([33]) where nodes with smaller ID send first, and the coordinator is the last one. In addition, all the nodes can set a timer to ensure the reception of sync packet in Chorus.

Security. We stress that the attacker takes full control of the normal wireless channel but not Chorus. The original SHCBK protocol was proven secure in [28], which uses a strong empirical OOB channel. We prove the security of GMS in Sec. 5. The intuition is that, the adversary cannot make all nodes' SASes equal by modifying the messages sent by legitimate nodes except with negligible probability. On the other hand, if SASes are not equal, Chorus ensures that all the nodes will reject with high probability.

4.2 In-band MAP with Long Authentication String (LAS)

Alternatively, if we let the input authentication string in the Chorus be longer, the protocol structure can be simplified into two rounds, eliminating the need for commitment/decommitment (see Fig. 9). Using a collision-resistant hash function $H(x)$ (for example, SHA-1 with 160 bits output), the devices in $\in \mathcal{G}$ can compute and compare a long authentication string (using Chorus) whose inputs include all the received messages. Our GML protocol can be viewed as an extension of Vaudenay's non-interactive two-party message authentication protocol (Vau05) [41] to the group case. Its security can also be reduced to the collision-resistance of the hash function. However, the Vau05 protocol requires the use of a secure OOB channel, and the human work to send/compare a 160-bit LAS is quite heavy.

Remarks. Correct device counting is required for secu-

```
Input: INFO_i = {ID_i, m_i} , i ∈ (1, ..., N)
0. User picks coordinator k and enters group size N.
1. ∀i →_N ∀i': INFO_i
2. k initiates Chorus by sending sync packet;
   ∀i ⇔_Chorus ∀i': H({INFO_j}_{j∈G_i})
The rest is the same with the GMS.
```

Figure 9: In-band Group MAP with LAS (GML)

rity purposes in our protocols, because the authenticated group member IDs are not known in advance. An incorrect count may facilitate an attacker to join the group. In fact, this is common underlying any ad hoc group MAP protocol ($N > 2$) [29, 20, 33]. However, we note that this adds only *minimum user effort*, since counting can be done while deploying the devices. If the network size is known in advance, or there is a machine counter, we can easily scale up to hundreds of nodes (unlike previous OOB-based methods). The count input can be easily implemented in sensors with buttons; otherwise it only requires one coordinator device which has richer interfaces such like a mobile phone.

From MAPs to GAKA Protocols. In the above protocols, if we change the message to be authenticated (m_i) of each node to a public number $X_i = g^{x_i}$ where x_i is a secret random number and g is a generator in $\mathbb{Z}_p$, then both of our protocols can establish $(N^2 - N)/2$ pairwise keys securely using Diffie-Hellman key exchange. Or, if a contributory group key agreement scheme is adopted (e.g.,[12]), the GMS and GML become group authenticated key agreement protocols (GAKAs) that establish a group key.

5. PROVING SECURITY PROPERTIES

Next we formally prove the security of the GMS and GML protocols. We define secure message authentication of a MAP based on the well-known Bellare-Rogaway model [4], which introduced the notion of "matching conversations" [4]. Essentially, if all the parties have matching conversations, all messages transmitted by them will be received unaltered, i.e., authentically.

DEFINITION 5.1 (SECURE MESSAGE AUTHENTICATION). *We say that Π is a (ϵ, T)-secure message authentication protocol with a group of participants $\mathcal{G}$ ($|\mathcal{G}| \geq 2$), if for any T-time adversary $\mathcal{A}$,*

(1) (Matching conversations ⇒ acceptance) If all pairs of parties in $\mathcal{G}$ have matching conversations, then all parties accept.

(2) (Acceptance ⇒ matching conversations) Letting $Adv_\Pi(\mathcal{A}) = Pr[\text{All-accept} \wedge \text{No-Matching}]$, where No-Matching refers to the event that the conversations are not matching, we have $Adv_\Pi(\mathcal{A}) \leq \epsilon$.

5.1 Security of the GMS Protocol

We have the following result, stated in concrete security guarantee. We use δ to denote the probability that all devices output accept when their SASes are not equal in Chorus.

THEOREM 5.1. *Assume that all devices in $\mathcal{G}$ are within range, group count is correct, and the coordinator is uncompromised. If the digest function satisfies $2^{-\ell}$-key-based uniformity and $2^{-\ell}$-no uniform compensation, and the hash function $H()$ is (ϵ_h, T_h)-preimage resistant and (ϵ_b, T_b)-second*

preimage resistant, the GMS is ($2^{-\ell} + \epsilon_h + 2\epsilon_b + \delta, T_b + T_h$)-secure.

The proof is in our technical report [16]. Essentially, this result bounds the adversary's (one-shot) deception probability to that of random guessing, as ϵ_h and ϵ_b are far smaller than $2^{-\ell}$ and can be neglected. In addition, δ is upper bounded by the attack success probability under single-bit difference: $P_a = 1 - P_d$ (derived in Sec. 3.3.2), which is around 10^{-6}. Thus, when SAS length $\ell = 16$ (32 slots in Chorus) this is about 10^{-5} (this can be freely tuned by the designer).

5.2 Security of the GML Protocol

THEOREM 5.2. *Assume that all devices in $\mathcal{G}$ are within range, group count is correct, and the coordinator is uncompromised. If the hash function $H()$ is (ϵ_c, T_c)-collision resistant, then the GML is ($\epsilon_c + \delta, T_c$)-secure.*

The proof is in [16]. In practice, $\epsilon_c \approx 2^{-80}$ if we use SHA-1 with output of 160-bits. This is much smaller than δ, so δ dominates the adversary's success probability in GML.

Reducing the number of minislots. Note that in GMS, P_a was computed by considering the worst case that only one LAS bit differs. However, we show that in GML, if the output of the cryptographic hash can be regarded as an ideal random mapping, δ is actually much lower than P_a; or equivalently, to maintain $\delta \approx 10^{-6}$, the actual required P_a can be much higher, which dramatically reduces the number of FH minislots (m) to represent each ON/OFF slot in Chorus. Intuitively, this is because with high probability around half of the LAS bits of two nodes will differ.

First, the number of different bits (hamming distance, D) between any two hash outputs follows a binomial distribution $B(l, 0.5)$ where l is the hash length. Second, δ^D is the probability that attacker can make two nodes output success in Chorus when their LASes differ in D bits. Thus, the probability

$$\delta = \sum_{d=1}^{l} Pr_{B(l,0.5)}[D = d]P_a^d. \quad (4)$$

We then find the maximum P_a: P_a^* s.t. $\delta \leq 10^{-6}$. Suppose $l = 160$, we obtain that $P_a^* \approx 0.83$. Considering the two node (worst) case, Let $P_c = 1 - P'_{nc}/h$ be the probability that one node fails to detect energy in one minislot when the other node is "ON", where $P'_{nc} = 1 - b/h$. And we have $P_a = P_c^m$. Thus, suppose $P'_{nc} = 0.5$ and $h = 4$, then $P_c = 0.875$, $m = 2$ (two FH minislots) should suffice.

5.3 Security of the GAKA Protocols

To show the security of GAKA protocol based on our group MAPs, the modular approach proposed by Bellare et. al. [3] can be applied. Specifically, it has two adversary models - the authenticated link model (AM) and unauthenticated link model (UM). If a protocol is proven to be secure under AM, then it can be shown to be secure in the UM, as long as each message transferred between the parties is authenticated by a protocol called message transfer (MT) authenticator. In our case, the GMS and GML protocols can be regarded as MT authenticators that authenticate all the nodes' messages. The group key agreement protocol in [12] is proven secure under the AM. Thus, our GAKA protocols are secure under the UM.

175

	Comm. time	Human Effort (bits)	Comp. cost/node	In-band
SPATE [23]	$O(N \cdot T_m)$	$O(N\ell)$	$O(N \cdot \text{hash})$	No
SAS-GMA [20, 21]	$O(N \cdot T_m)$	$O(\ell)$	$O(N \cdot \text{mod\_exp})$	No
I-Code (group) [7]	$N \cdot (2lT_S)$	0	$O(N \cdot \text{hash})$	Yes
TEP (group) [14]	$N \cdot (T_m + T_{sync} + lT_S)$	0	$O(N \cdot \text{hash})$	Yes
GMS (ours)	$2N \cdot T_m + T_{sync} + 2lT_S$	0	$O(N \cdot \text{hash})$	Yes
GML (ours)	$\leq N \cdot T_m + T_{sync} + 2lT_S$	0	$O(\text{hash})$	Yes

$l = 160$: hash length; $\ell = 16$; T_m: normal packet's duration; T_S: slot duration; T_{sync}: sync packet duration.

Table 3: Comparison of our protocols with representative existing ones.

6. EVALUATION

In this section, we analyze the complexity of our scheme first. Then we introduce our implementation and experiment results. Here we show the effectiveness of Chorus and the performance of GAKA protocols.

6.1 Complexity Analysis

We analyze and compare the complexity of our group message authentication protocols with previous representative ones based on OOB channels [23, 20, 21], and also with I-code [7] and TEP [14]. The latter two are extended to a group setting, by directly using I-code or TEA to authenticate each $INFO_i$. However, the same setup (device counting and push button on each device) is needed because the group is unknown in advance. To be fair in the security level, we assume that our UFH-based defense against the cancellation attack is also applied to both I-code and TEA (a slot will be extended to multiple minislots in the same way as ours). Note that I-code does not prevent an attacker from hiding the fact that a message was transmitted altogether using collisions or a capture effect [14].

We evaluate the costs only for the corresponding message authentication protocol (without key agreement computation and excluding human delay) in Table. 3. Our protocols are more scalable and efficient compared with them. For example, assume the example parameters we used before ($m = 105$, $\Delta d_0 = 2m$). On a 2.4GHz sensor platform, when $T_S = 5 * 105ms$, $T_m = 5ms$, $T_{sync} = T_S$, $N = 30$, our GMS protocol only needs 18.7s, while I-Code needs 5040s, and TEP-group requires nearly 2552s. The previous protocols [23, 20, 21] involve cumbersome user comparison of short digests, which incurs a large human delay linear with the SAS length.

6.2 Experimental Evaluation

Implementation. We implemented the FH-Chorus and the two GAKA protocols on a TinyOS 2.0 wireless sensor platform, with 24 Crossbow TelosB sensors. We choose node 1 as the controller. We turn off the CSMA to make our Chorus feasible. We set the length of each minislot as 5 ms. Each node will repeatedly check 5 RSSIs at each minislot of its OFF slots. The RSSI threshold is set to -65 dBm. Each of the 24 sensors will report its data to a gateway to help us evaluating the protocol's performance. We place 24 nodes on a table in an indoor environment. The experiments mainly consist of two parts. In the first part we evaluate the effectiveness of Chorus with SAS. In the second part we analyze the overhead of GAKA protocols.

6.2.1 Chorus Effectiveness Analysis

We use SAS to illustrate the effectiveness of Chorus. Based on our previous analysis, each slot contains 105 minislots. When the Chorus starts, each node will hop randomly within 4 channels out of all 16, send out random packets during the

Decomposition	Initialization	Crypto	Chorus	Total
GMS N=8 Time(s)	10	21	17	48
GMS N=24 Time(s)	30	28	17	75
GML N=8 Time(s)	10	20	8	38
GML N=24 Time(s)	30	24	8	62

Table 4: Time overhead of GAKA protocols

ON slots, and detect signal during OFF slots. We run Chorus 10 times, and show the results from a typical run.

First we verify the correctness of Chorus by setting the same SAS for all nodes before Chorus starts. In Fig. 10(a), we show the number of high-level RSSI minislots (N_{hd}) detected by each node during its OFF slot at each bit position. We can see if using the same SAS, the number of detected high-level RSSI minislots is almost zero for all nodes. Though some nodes will detect high-level RSSI brought by interference (such as Wi-Fi signal), the numbers are all below our interference threshold. In other words, the Chorus will output ACCEPT message at the end.

Next we verify the robustness of Chorus. In the worst case, only 1 bit is different for different SASes. In order to verify the robustness of Chorus even in the worst case, we intentionally pick a random node (node 3) to generate a totally different SAS from other nodes (all bits are reversed) before the Chorus starts. Then we check whether the nodes could detect the differences of all the 16 bits, i.e. whether the nodes are able to detect high-level RSSI ("ON") minislots during all the 16 OFF slots of these bits. We record the number N_{hd} detected within all OFF slots of each node.

This result is shown in 10(b). As we can see, all the nodes can detect multiple "ON" minislots during every OFF slot, which is also above the threshold T_n. Besides, most of the N_{hd} values detected by all nodes are consistent with their expected values through analysis. This consistency proves the correctness of our previous analysis in section 3.3. Meanwhile, the number N_{hd} detected by node 3 is much higher than that of other nodes. This also verifies our analysis in section 3.2: more simultaneous transmitting nodes will increase the possibility of a bit difference being detected, even considering the self-cancellation effect.

The above also indicates that all nodes can be synchronized, and maintain synchronization throughout the comparison phase in Chorus (at least for 32×105=3360 minislots). We note that some nodes' average N_{hd}s are slightly different from the expected value. For example, the average N_{hd} of node 2 (which is closest to node 3) is higher than the expected value, which may be caused by adjacent-channel interference. We observe that in rare cases some nodes experience small N_{hd} values, which can be caused by imperfect synchronization among devices or channel fading. However, this can be solved by increasing the length of each minislot.

Now we consider the signal cancellation attack. The effectiveness of Chorus under this attack can be directly inferred from the above results. Denote the number of high-level RSSI under the presence of relay signal cancellation attack as N_{hd_a}. Using the attacker's cancellation proba-

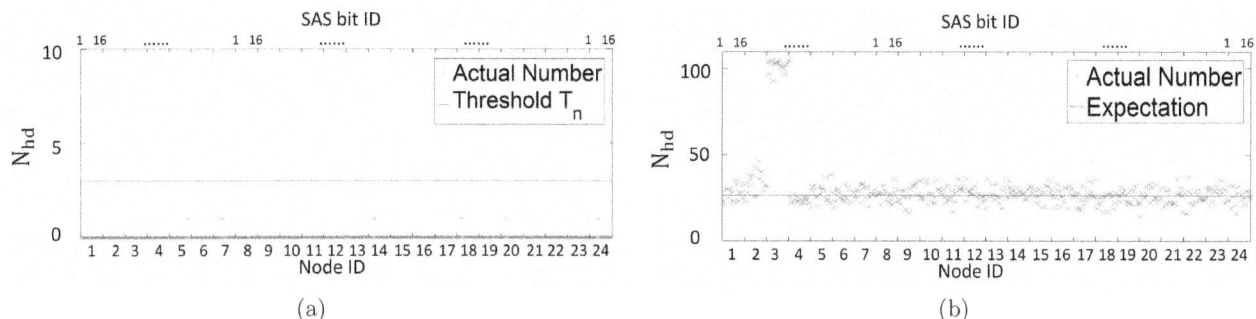

Figure 10: Verification of correctness and robustness of GAKA. All nodes are in the range 0.8m × 0.8 m. a) Number of high-level RSSI minislots N_{hd} detected using the same SAS. (b) Number of high-level RSSI minislots detected using SAS with one node's string inverted.

bility in one minislot $b/h < 1$, we can derive the relationship between N_{hd_a} and N_{hd} as $N_{hd_a} = (1 - b/h) \cdot N_{hd}$, and $\mathbb{E}[N_{hd_a}] = (1 - b/h) \cdot \mathbb{E}[N_{hd}]$, which is larger than T_n.

6.2.2 GAKA Efficiency Analysis

In this part we will analyze the time overhead of our GAKA protocols. We run the two GAKA protocols and recorded their running time in Table. 4. We can see that as the number of nodes increases, the time consumed by the cryptographic part (for key computation etc.) increases slowly. Besides, the time overhead of Chorus remains constant. Most of the overhead actually comes from system initialization, which involves one button press on each device (to start the device) and counting at the beginning. The average initialization time is about one second per device. Note that the GML uses fewer number of minislots ($m = 2$), so its Chorus time is lower than GMS.

6.2.3 Discussion

Our security model only considers two types of existing energy cancellation attacks. We believe that these types are complete, as the attacker either generates its own signal or does not. In theory, there could be a stronger correlated energy cancellation attack which requires more advanced hardware. That is, in addition to a pair of directional antennas, the attacker may use a wideband software defined radio (SDR) device which has a bandpass filter, and in real-time it computes and injects a phase delay corresponding to the legitimate signal's frequency. Ideally all channels' signals could get completely cancelled. But this attack can be difficult to carry out in practice as the phase delay must be very precisely generated.

We note that there is another similar type of attack – real-time reactive and selective jamming, where the the attacker can demodulate, interpret, and timely generate an interfering signal to a legitimate transmission that is already "on-the-air" [42]. When the interfering signal is the inverse of the legitimate one at the receiver, the latter can get cancelled. However, the attacker's response time needs to be extremely short. So far, it is reported that a reaction delay of $16\mu s$ can be achieved for sensor signals [42], which is nevertheless too large for complete signal cancellation.

7. CONCLUSION

In this paper we focused on the challenging problem of ad hoc trust initialization for a group of wireless devices without relying on an out-of-band channel. Our main contribution is Chorus, a novel primitive for authenticated equality com-

parison over the insecure wireless channel in constant time. Chorus achieves non-spoofable string equality comparison, which is based on the infeasibility of energy cancellation and unidirectional error detection codes. Through analysis, we show that Chorus is secure against all known signal cancellation attacks. Chorus can be readily applied to design group message authentication, and group authenticated key agreement protocols, which greatly enhances their scalability and simplifies the protocol structure. Future work will include extending Chorus to be robust under malicious jamming.

Acknowledgments This work is partially supported by NSF grants CNS-1116557 and CNS-1218085.

8. REFERENCES

[1] Top 50 internet of things applications - ranking. http://www.libelium.com/top_50_iot_sensor_applications_ranking/.

[2] D. Balfanz, D. K. Smetters, P. Stewart, and H. C. Wong. Talking to strangers: authentication in ad-hoc wireless networks. In *NDSS '02*, 2002.

[3] M. Bellare, R. Canetti, and H. Krawczyk. A modular approach to the design and analysis of authentication and key exchange protocols (extended abstract). In *ACM STOC'98*, pages 419–428. ACM, 1998.

[4] M. Bellare and P. Rogaway. Entity authentication and key distribution. In *Advances in Cryptology - CRYPTO'93*, pages 232–249. Springer, 1994.

[5] M. Cagalj, S. Capkun, and J.-P. Hubaux. Key agreement in peer-to-peer wireless networks. *Proceedings of the IEEE*, 94(2):467–478, Feb. 2006.

[6] L. Cai, K. Zeng, H. Chen, and P. Mohapatra. Good neighbor: Ad hoc pairing of nearby wireless devices by multiple antennas. In *NDSS 2011, San Diego, California, USA*. The Internet Society, 2011.

[7] S. Capkun, M. Cagalj, R. Rengaswamy, I. Tsigkogiannis, J.-P. Hubaux, and M. Srivastava. Integrity codes: Message integrity protection and authentication over insecure channels. *IEEE Transactions on Dependable and Secure Computing*, 5(4):208 –223, oct.-dec. 2008.

[8] H. Chan, A. Perrig, and D. Song. Random key predistribution schemes for sensor networks. In *IEEE S & P '03*, page 197, 2003.

[9] C.-H. O. Chen, C.-W. Chen, C. Kuo, Y.-H. Lai, J. M. McCune, A. Studer, A. Perrig, B.-Y. Yang, and T.-C. Wu. Gangs: gather, authenticate 'n group securely. In *MobiCom '08*, pages 92–103, 2008.

[10] D. Dolev and A. Yao. On the security of public key

protocols. *Information Theory, IEEE Transactions on*, 29(2):198 – 208, mar 1983.

[11] W. Du, J. Deng, Y. Han, P. Varshney, J. Katz, and A. Khalili. A pairwise key predistribution scheme for wireless sensor networks. *ACM Transactions on Information and System Security (TISSEC)*, 8(2):228–258, 2005.

[12] R. Dutta and R. Barua. Provably secure constant round contributory group key agreement in dynamic setting. *IEEE Trans. on Inf. Theory*, 54(5):2007–2025, May 2008.

[13] L. Eschenauer and V. D. Gligor. A key-management scheme for distributed sensor networks. In *CCS '02*, pages 41–47, 2002.

[14] S. Gollakota, N. Ahmed, N. Zeldovich, and D. Katabi. Secure in-band wireless pairing. In *USENIX*, SEC'11, pages 16–16, Berkeley, CA, USA, 2011. USENIX Association.

[15] M. T. Goodrich, M. Sirivianos, J. Solis, G. Tsudik, and E. Uzun. Loud and clear: Human-verifiable authentication based on audio. In *In IEEE ICDCS 2006*, page 10, 2006.

[16] Y. Hou, M. Li, and J. D. Guttman. Chorus: Scalable in-band trust establishment for multiple constrained devices over the insecure wireless channel. In *Technical Report*, Feb. 2013.

[17] S. Jana, S. Premnath, M. Clark, S. Kasera, N. Patwari, and S. Krishnamurthy. On the effectiveness of secret key extraction from wireless signal strength in real environments. In *MobiCom '09*, pages 321–332. ACM, 2009.

[18] C. Kuo, M. Luk, R. Negi, and A. Perrig. Message-in-a-bottle: user-friendly and secure key deployment for sensor nodes. In *SenSys '07*, pages 233–246, 2007.

[19] S. Laur and K. Nyberg. Efficient mutual data authentication using manually authenticated strings. *Cryptology and Network Security*, pages 90–107, 2006.

[20] S. Laur and S. Pasini. SAS-Based Group Authentication and Key Agreement Protocols. In *Public Key Cryptography - PKC '08*, LNCS, pages 197–213, 2008.

[21] S. Laur and S. Pasini. User-aided data authentication. *International Journal of Security and Networks*, 4(1):69–86, 2009.

[22] M. Li, S. Yu, W. Lou, and K. Ren. Group device pairing based secure sensor association and key management for body area networks. In *INFOCOM, 2010 Proceedings IEEE*, pages 1–9. IEEE, 2010.

[23] Y.-H. Lin, A. Studer, H.-C. Hsiao, J. M. McCune, K.-H. Wang, M. Krohn, P.-L. Lin, A. Perrig, H.-M. Sun, and B.-Y. Yang. Spate: small-group pki-less authenticated trust establishment. In *Mobisys '09*, pages 1–14, 2009.

[24] K. Lorincz, D. Malan, T. Fulford-Jones, A. Nawoj, A. Clavel, V. Shnayder, G. Mainland, M. Welsh, and S. Moulton. Sensor networks for emergency response: challenges and opportunities. *IEEE Pervasive Computing*, 3(4):16–23, Oct.-Dec. 2004.

[25] S. Mathur, R. Miller, A. Varshavsky, W. Trappe, and N. Mandayam. Proximate: proximity-based secure pairing using ambient wireless signals. MobiSys '11, pages 211–224, New York, NY, USA, 2011. ACM.

[26] S. Mathur, W. Trappe, N. Mandayam, C. Ye, and A. Reznik. Radio-telepathy: extracting a secret key from an unauthenticated wireless channel. In *MobiCom'08*, pages 128–139. ACM, 2008.

[27] J. M. McCune, A. Perrig, and M. K. Reiter. Seeing-is-believing: Using camera phones for human-verifiable authentication. In *IEEE S & P*, pages 110–124, 2005.

[28] L. Nguyen and A. Roscoe. Authenticating ad hoc networks by comparison of short digests. *Information and Computation*, 206(2-4):250–271, 2008.

[29] L. Nguyen and A. Roscoe. Authentication protocols based on low-bandwidth unspoofable channels: a comparative survey. *Journal of Computer Security*, 19(1):139–201, 2011.

[30] R. Nithyanand, N. Saxena, G. Tsudik, and E. Uzun. Groupthink: Usability of secure group association for wireless devices. In *Proceedings of the 12th ACM international conference on Ubiquitous computing*, pages 331–340. ACM, 2010.

[31] S. Pasini and S. Vaudenay. An optimal non-interactive message authentication protocol. CT-RSA'06, pages 280–294, 2006.

[32] S. Pasini and S. Vaudenay. SAS-based Authenticated Key Agreement. In *Public Key Cryptography - PKC '06*, volume 3958 of *LNCS*, pages 395 – 409, 2006.

[33] T. Perković, M. Čagalj, T. Mastelić, N. Saxena, and D. Begušić. Secure Initialization of Multiple Constrained Wireless Devices for an Unaided User. *IEEE transactions on mobile computing*, 2011.

[34] C. Pöpper, N. O. Tippenhauer, B. Danev, and S. Capkun. Investigation of signal and message manipulations on the wireless channel. ESORICS'11, pages 40–59, 2011.

[35] K. Rasmussen and S. Capkun. Realization of rf distance bounding. In *Proceedings of the USENIX Security Symposium*, 2010.

[36] K. Rasmussen, C. Castelluccia, T. Heydt-Benjamin, and S. Capkun. Proximity-based access control for implantable medical devices. In *ACM CCS*, pages 410–419. ACM, 2009.

[37] L. Shi, M. Li, S. Yu, and J. Yuan. Bana: body area network authentication exploiting channel characteristics. ACM WISEC '12, pages 27–38, 2012.

[38] F. Stajano and R. J. Anderson. The resurrecting duckling: Security issues for ad-hoc wireless networks. In *IWSP '00*, pages 172–194, 2000.

[39] M. Strasser, S. Capkun, C. Popper, and M. Cagalj. Jamming-resistant key establishment using uncoordinated frequency hopping. In *IEEE S & P*, pages 64–78. IEEE, 2008.

[40] A. S. Tanenbaum. *Computer networks (4. ed.)*. Prentice Hall, 2002.

[41] S. Vaudenay. Secure communications over insecure channels based on short authenticated strings. CRYPTO'05, pages 309–326, 2005.

[42] M. Wilhelm, I. Martinovic, J. Schmitt, and V. Lenders. Reactive jamming in wireless networks: how realistic is the threat. *Proc. of ACM WiSec*, 11:47–52, 2011.

Countermeasures against Sybil Attacks in WSN based on Proofs-of-Work

[*]Marek Klonowski
Institute of Mathematics and Computer Science,
Wrocław University of Technology
marek.klonowski@pwr.wroc.pl

[†]MichałKoza
Institute of Mathematics and Computer Science,
Wrocław University of Technology
michal.koza@pwr.wroc.pl

ABSTRACT

It has been shown that Sybil attack can be easily applied in Wireless Sensor Networks (WSN). An adversary capturing a few stations can corrupt most of classic protocols (e.g. leader election procedures) in most of considered models. Moreover, such attack is in practice undetectable in many realistic scenarios. In this paper we present an efficient countermeasure against Sybil attack. It is based on Proofs-of-Work technique – one of the methods of preventing sending spam. In contrast to previous solutions it is based only on limited computational power of the adversarial devices. Our approach does not require any restrictions on communication between adversarial stations.

Categories and Subject Descriptors

C.2.2 [**Computer-Communication Networks**]: Network Protocols

Keywords

leader election, adversary, proof of work

1. INTRODUCTION

In Sybil attack [9] an adversary having control over some network devices emulates a large number of virtual identities in order to gain disproportionately large influence on the network. Preventing Sybil attacks is especially difficult in Wireless Sensor Networks (WSN). There is no direct control over the devices. Some devices can be easily compro-

[*]The contribution is the results of realization of the project: "Detectors and sensors for measuring factors hazardous to environment - modeling and monitoring of threat". The project financed by the European Union via the European Regional Development Fund, within the framework of the Operational Programme Innovative Economy 2007-013. The contract for refinancing No. POIG.01.03.01-02-002/08-00.

[†]Supported by Foundation for Polish Science, MISTRZ project.

mised. Moreover, devices of such a system are often very constrained thus it is not feasible to implement solutions based on advanced cryptography. For this type of settings it has been shown that Sybil attack applied against standard protocols cannot even be detected [13]. The main result of our paper are protocols providing a *fair leader election* protocol. Proposed solutions are based on proofs-of-work [10]. They do not demand any restrictions on communication (as assumed in some previous works) and guarantees **optimal** level of security – the probability that an adversarial device is chosen as the leader is proportional to the fraction of total computational power of the adversarial devices to the power of all devices in the system.

Our solutions can be combined with some other techniques for protecting the network from Sybil-attacks e.g. based on communicational restrictions.

Notation. For a bit string x let $|x|$ be its length. For $k \leq |x|$, the first k bits of x are denoted $x_{[k]}$ and i-th bit of x by $x[i]$. A string x with the last bit truncated is denoted by x_-. Concatenation of bit-strings x and y is denoted by $x\|y$. By $x \oplus y$ we denote bitwise xor operation of bit stings x, y.

Network and communication. We consider a single-hop wireless network. Time is divided into rounds/slots. There are M devices controlled by the adversary and N honest devices. We assume that $N \in [N_{min}, N_{max}]$, but M is unknown. If more than one station transmits during a slot a *collision* occurs and no message can be reconstructed. Devices are able to distinguish between silence and collision in a slot. This description is equivalent to some popular MAC (*Multiple Access Channel*) models [6]. **Stations do not share any secret**. They are capable of computing some a priori agreed one way hash function and have stochastically independent pseudo random number generators (PRNG).

The aim of the stations is to **fairly** choose the leader. More precisely, after execution of the protocol: each station has a status - *leader* or *non-leader* known to it; exactly one station has the *leader* status; each honest station is chosen as a leader with probability at least $1/(N + M)$.

Adversary. Adversary controls M nodes. These nodes **can cooperate** in an arbitrary way. In contrast to some other protocols protecting against Sybil attacks, secret communication between adversarial nodes is allowed. The adversary wins if one of its stations becomes a leader. We assume that the adversary is not interested in blocking the protocol but gaining control over its execution. It is important to keep in mind that presence of adversarial devices cannot be detected as long as they act according to protocol. Thus by assumption - reducing adversary's winning chance to fraction of its devices in the system is considered an optimal solution.

We assumed that each adversarial devices has the same computational power as legitimate devices. Indeed – in our approach an identity is "authenticated" by its computational abilities. However, in most practical situations the adversary is able to attack the network using a stronger device (e.g. a laptop). One can easy check that from point of view of our analysis bringing in a k times stronger device is equivalent to bringing in k regular devices. Consequently the assumption can be relaxed and only the total computational power of the adversary is important.

Proofs-of-work. In 1992, Dwork and Naor [10] proposed a protocol that would discourage spammers from sending abundant mail. They suggested that e-mail messages can be accompanied by proofs of computational effort that is easy to check. Many different protocols realizing this idea have been proposed (e.g. Hashcash [1], Hokkaido [4], [17]). The idea was to force the sender to prove that it has made some significant effort devoted to sending a particular message.

Let us note that this idea can be realized using one-way hash function only, without necessity of applying asymmetric cryptography. For example, let m be a message. Let h, h' be two independent secure, one-way hash functions. The PoW for challenge M is such a value y that $h(M)_{[k]} = h'(y)_{[k]}$, where k is an appropriate constant.

Treating hash functions as random oracles, one can easily see that finding PoW for a given argument has expected running time 2^{k-1} and is strongly concentrated around its mean. On the other hand, a given solution can be verified very fast just by computing hash function for two arguments.

Hash function is just one possible way of constructing a PoW system (see [18]). In this paper the particular implementation is not specified. It is only assumed that some PoW function with desired properties exists.

DEFINITION 1. *Let PoW be a function such that in order to calculate its value ($y = \mathrm{PoW}(x)$) for a given challenge x one needs to spend a significant amount of time. But having y and x it is easy to check whether $y = PoW(x)$.*

2. PREVIOS RELATED WORK

Sybil attacks. Various countermeasures against Sybil attack have been proposed; including certification, auditing and resource testing. Many of these solutions are surveyed in [3]. Most of proposed techniques cannot be applied to protect systems considered in our paper due to their peculiarities – their ad hoc nature (i.e., no possibility of auditing, no secrets established a priori) and limited resources disqualifying public key cryptography. Paper [25] suggests a method of protecting sensor network from the Sybil attack based on the assumption that a single device is not able to broadcast on two different frequencies at the same time (slot). Presented approach is very innovative, however it cannot be applied our model In particular, it requires several channels and was designed for a different adversary model.

In [16, 13] algorithms for repelling Sybil attacks in WSN performed by advanced adversary are proposed. In these papers it is assumed that the adversary is able to take part in the protocol and transmit valid messages, it can also control multiple devices. Only assumptions are that devices cannot listen to the channel while they transmit and that adversar-

ial devices cannot communicate outside the shared channel in the runtime of the protocol. The proposed countermeasures are strongly related to physical limitations of working stations (e.g., inability to broadcast and listen at the same time) which are not valid for all types of devices and communication channel properties which may be difficult to secure. In contrast, the solution proposed in our paper is based only on computational limitations of adversarial devices.

Other important papers considering Sybil attack in the context of ad hoc networks are [5, 26]. However, none of proposed solutions seems to be applicable to the model considered in our paper. For example, the first paper is focused on mobile devices with multi-hop graph of connections. Also methods from the second paper are based on pre-distributing secrets and thus require some pre-deployment phase. In our paper we do not assume that stations share any secret.

Leader election protocols. There are many papers about leader election algorithms for wireless sensor networks [2, 14, 22, 23, 24, 27]. Bulk of these algorithms presented in the literature allow dishonest stations to significantly improve their chances to become the leader. Some of these attacks do not even require creating virtual stations emulated by the malicious station. Moreover in [13] it has been shown that the adversary attacking a leader election algorithm is in practice statistically undetectable after a single leader election round in most of protocols discussed in literature.

Adversary immune algorithms. There are several important papers wherein authors consider similar model of the network with a different model of the adversary. These solutions are focused on an *external adversary* who can merely disrupt communication by creating collisions [8]. These solutions do not help against an *internal adversary*, who may participate in the protocol, controlling a subset of stations.

A number of papers assume presence of an internal adversary. A randomized leader election algorithm running in time $O(\log^3 n)$ with energy cost (understood as a number of slots, when station is broadcasting) $O(\log n)$ is presented in [19]. Paper [20] presents a randomized initialization algorithm with runtime $O(n)$ and energy cost $O(\log n)$. In these papers the aim of the adversary is to make the execution as time/energy consuming as possible. In [21] authors consider aloha-type protocol with selfish users (the aim of each node is to broadcast its message as quickly as possible) from a game–theoretic perspective. Considered problems/models are to some extent similar to our analysis, however it seems hard to apply this methodology to our model. In [12] authors introduce efficient protocol for reliable, robust exchange of information in a single-hop, multi-channel radio network.

In their model, the adversary is allowed to simultaneously disrupt some of the available channels. The core of the proposed algorithm lies in a so-called multi-selector.

3. SIMPLE PROTOCOL

In this section we present a PoW-based leader election protocol. Conceptually it is much simpler than the extended protocol presented Section 4, however, it is efficient only for small-size networks. In contrast to the extended protocol the number of devices N or even its approximation does not need to be known in advance. The protocol consists of several phases. At the beginning, stations declare their presence. Each declared station is represented by its *identity*, or ID for short. Clearly, each honest station declares exactly one identity. In further phases, for each identity some actions are performed. Each honest station performs all actions for its corresponding identity. Adversarial stations can handle actions to be performed by declared IDs in an arbitrary way. In particular many adversarial stations can emulate one ID.

Phase: Registering identities. The aim of this phase is to list IDs of all stations. This can be realized for example by any initialization protocol restarted if after its completion there are still any stations willing to register their IDs (see eg. [16]). The only condition that has to be met is that the adversary cannot prevent honest stations from registration of its IDs. This can be easily realized in the assumed model with collision-detection. Indeed - the adversary is not able to cover up transmissions of honest stations. However the adversary can declare an arbitrary number of IDs.

Phase: Commitment. Each identity ID_i chooses two random values c_i and r_i. In the i-th time slot of this phase, ID_i announces the cryptographic commitments to c_i and r_i.

Phase: Generating the 1st common random value. In the i-th time slot of this phase, ID_i broadcasts c_i. The values c_i are verified against the commitments. If there are any inconsistencies, the corresponding value and identity are removed from the list. Finally, a common random value (crv) C is computed as $C = c_1 \oplus \ldots \oplus c_n$.

Phase: PoW- creation. Each ID_i is supposed to compute $x_i = \mathrm{PoW}(C, i)$.

Phase: Generating the 2nd common random value. In the i-th time slot of this phase, ID_i broadcasts r_i. These values are verified in the same way as the values c_i. The second common random value is $R = r_1 \oplus \cdots \oplus r_n \mod n$ (obviously omitting values of removed identities).

Phase: Choosing the leader candidate The identity on position R on the list, i.e. ID_R becomes a leader candidate. The leader candidate is required to present x_R. If it is correct, (i.e. $x_R = \mathrm{PoW}(C, R)$) ID_R becomes a leader. Otherwise the leader candidate election phase is repeated with $R := H(R) \mod n$. If there are more than T repetitions the leader is not chosen and the procedure is stopped.

3.1 Remarks

For simplicity the size of the random strings acting as commitments, random values and the length of the output of the hash functions are not specified. It is important that computing PoW is so time consuming that it is not possible for a device to compute PoW for two different challenges during the runtime of the protocol, even if the leader candidate has to be reelected many times. Since execution time can vary depending on number of registered identities it is reasonable to make PoW difficulty dependent on number of registered identities. Indeed, this will prevent the adversary from generating huge number of fake identities (without intention to authenticate them) in order to get more time for multiple PoW calculation. In practice, it is also necessary to limit the number of repetitions of the last phase by some threshold T. If the threshold is exceeded, then the system may assume DoS-type attack and act accordingly. Note that the procedure cannot be simply restarted in the case of negative verification of PoW. Indeed, in such a case the adversary could try to restart the procedure until its identity is chosen, improving significantly its chances to win.

3.2 Analysis of the protocol

For the analysis, it is assumed that each station can compute PoW for only one challenge during the runtime of the protocol. Let n be the total number of declared identities. If the adversary is cheating, then $n > M + N$. The following claim can be proven:

CLAIM 1. *Assume that it is impossible to present a valid response to given challenge without calculating the PoW. The probability that the adversarial station is chosen as a leader is at most $M/(M + N)$, independently of the strategy of the adversary.*

PROOF. Using standard reasoning the common value C is purely random and cannot be determined to any extent as long as there is at least one honest station. Thus, all challenges for the PoW that contain the value C have to be established during runtime of the protocol and the adversary cannot compute them in advance. Moreover each identity has to compute PoW for different challenge. As a consequence, having M devices, the adversary can compute PoW for at most M identities. On the other hand, the legitimate devices have N PoWs computed. Again the adversary cannot determine the value R. Thus each device has the same chance of having its index equal to R. Since only identities for which a PoW has been computed are taken into account when appointing a leader candidate, the probability that the adversary wins is $M/(M + N)$. $\square$

Note that in the above claim it is assumed that breaking the PoW is impossible whereas in fact it is possible, but with negligible probability.

4. GENERAL CASE PROTOCOL

The scheme presented in the previous section provides almost optimal security. However, its running time is $\Theta(n) = \Omega(N)$, which is impractical for big networks. In this section a randomized leader election protocol is presented. It is much faster at the cost of slightly reduced security level. We assume that number of stations is approximately known i.e., $N \in [N_{min}, N_{max}]$ (the assumption that N_{max} is known can be relaxed). Performance and security of the protocol depends on a parameter Γ which meaning will be explained later. The general idea of the protocol is the following: j-th identity chooses a pseudo random value ξ_j, and commits to it calculating PoW for this value; next, pseudo randomly and uniformly one of ξ values is selected. The identity that calculated PoW for this ξ becomes the leader. The protocol consists of the following phases:

Representatives listing phase. This phase is similar to listing phase in the previous protocol. However, each honest station decides to participate in this phase independently with some fixed probability p. The identities that decided to participate will be called *representatives*. Note that the adversary may put on the list an arbitrary number of representatives. Let $\tilde{n}$ be the number of declared representatives. The i-th representative on the list will be denoted by L_i.

Committing values phase. Each L_i chooses three random values: c_i, r_i and s_i. It broadcasts cryptographic commitments to these values in the i-th round of this phase.

Generating the 1st common random value phase. In the i-th slot of this phase, L_i broadcasts its value c_i. The revealed values are verified against commitments as in previous protocol. A crv is computed as $C = c_1 \oplus \ldots \oplus c_{\tilde{n}}$.

PoW- creating phase. Each **identity** ID_j in the network (not only the representatives) chooses randomly (independently of others) a value ξ_{ID_j} from the set $\{0, \ldots, \Gamma - 1\}$ where $\Gamma = 2^k$ for some k and computes $\text{PoW}(C, ID_j, \xi_{ID_j})$.

Generating the 2nd and 3rd crv phase. In the i-th slot of this phase, L_i broadcasts r_i and s_i. Finally, two crv are computed: $R = r_1 \oplus \cdots \oplus r_n$ and $S = s_1 \oplus \cdots \oplus s_n$.

Positions randomization. Using R and a hash function

$$h' : \{0,1\}^{|R| + |ID| + |\xi_{ID}|} \mapsto \{0, \ldots, \Gamma - 1\}$$

a translation table of identifiers is created. Namely, each identity ID_j calculates $\pi(ID_j) = h'(R, ID_j, \xi_{ID_j})$. The value $\pi(ID_j)$ is treated as the position of identity ID_j in the set $\{0, \ldots, \Gamma - 1\}$. We assume that values of h' are random with uniform distribution since R is a uniformly distributed random variable. In particular for fixed ID and ξ_{ID} the value $\pi(ID_j)$ is indistinguishable from uniformly distributed random variable from the set $\{0, \ldots, \Gamma - 1\}$. Obviously the translation table is created only virtually and is never transmitted. Each device knows only positions of its own identity (identities in case of an adversarial devices) in the table.

Binary random choice phase. The goal of this phase is to efficiently select one identity (by selecting $\pi(ID)$ value associated with it) from the set of all identities. On this stage each identity has three values associated with it: ID-its unique identifier; ξ_{ID} - the value for which PoW should be calculated; $\pi(ID)$ - the pseudo random position of ID in the set $\{0, \ldots, \Gamma - 1\}$. Note that the set of all identities and its associated values is not known to other stations.

A kind of binary search is performed among $\pi(ID)$ values. Intuitively, this can be presented as a walk from the root to a leaf in an incomplete binary tree (see Figure 1). The tree is built as a full binary tree upon elements from the set $\{0, \ldots, \Gamma - 1\}$ treated as leafs. Then, only occupied leafs (i.e. having values equal to $\pi(ID)$ for some ID) remain. All non-occupied leafs are removed.

The source of pseudo random bits for this phase is the value S. It is assumed that S is a bit string and consecutive bits of S are treated as random choices in respective rounds. That is, the walk in the tree is determined by the bits of S. Consecutive bits decide whether to go left or right on a given tree level. If the branch pointed by a bit is empty, then the bit is ignored and the walk goes to a non-empty branch. Let the following set be defined:

DEFINITION 2. *Let A_l be a subset of all identities such that an identity ID is in A_l if and only if $\pi(ID)$ has prefix l. That is $A_l = \{ID : \pi(ID)_{[|l|]} = l\}$.*

Set A_l can be interpreted as a set of all occupied leafs in the branch with the root value equal l. At the beginning l is the empty string ($l = \varepsilon$) and it grows (number of bits of l increases) at each level. The BRC used the subprocedure **TEST**: This procedure is invoked in order to determine the number of devices in a given branch. The result is **none**, **one**, or **noise**. In the second case an identifier is returned. For each l the procedure $\text{TEST}(l)$ is in fact a single time slot in which all devices in A_l are supposed to transmit in order to mark their presence. The adversary can cheat here by pretending to have an identifier in a given subbranch.

However, the important thing is that the adversary cannot pretend that some branch is empty, if it is not.

The BRC procedure is called recursively for some parameter l keeping track of the path travelled. On each level (unless the leafs level has been reached) the desired walk direction $b \in \{0, 1\}$ is determined. Next, it is checked whether the chosen branch (the set $A_{l||b}$) is nonempty by calling the TEST procedure. There are three possible options: If there are many IDs in the chosen branch, then the walk goes on recursively into this branch; If there is only one ID in the branch, it means that there is only one leaf and the walk can be considered finished. The ID is considered a leader candidate and the verification procedure is started; If the chosen branch is empty the other branch is checked analogously.

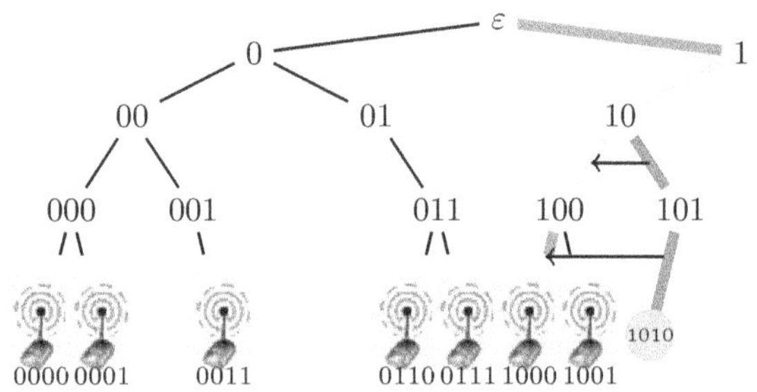

Figure 1: Execution of BRC for $S = 1110\ldots$ **In a full binary tree the walk would be right, right, right, left. However because in BRC(1) (called on depth 1) the TEST(11) procedure returns none the walk goes left ignoring $S[2] = 1$. A leaf with label 1010 indicates an adversary that tries to influence the execution - it generated the noise signal during TEST(101) called in BRC(10) because of which the walk goes right, and then during BRC(101) because of which the walk goes left. In the end the adversary is unable to present the valid PoW, so backtracking to BRC(101) and then to BRC(10) occurs. Finally the walk terminates at node 1000.**

Leader candidate verification phase. In this phase it is checked if the leader candidate ID correctly computed its PoW. The tested identity broadcasts the PoW response y and ξ_{ID}. The other stations check whether $(\pi(ID))_{|l|} = l$ and $y = \text{PoW}(C, ID, \xi_{ID})$. If the test is passed, ID becomes a leader. Otherwise, the ID is removed and the Binary Random Choice procedure is restarted one level up with parameter l_-. The walk will go back to the last level where there were two branches to choose and will go the other way.

4.1 Remarks on the construction

The idea of the general purpose protocol is the same as in the basic one. However, in order to avoid listing of all identities only representatives are involved in choosing the random values C, S and R. This is acceptable, because the proposed random values generation method is fair if there is even a single representative of honest station. Later, the leader is chosen from a greater set, but the set itself remains unknown thus a listing does not occur explicitly at all.

One of the most important phases is the leader selection phase – the BRC procedure. Note that the BRC procedure if fair only if leafs are assigned to identities at random. Indeed, it is easy to see that if we let the adversary choose the leafs freely it can easily increase its chances by choosing equally spread leafs. Intuitively, choosing leafs close to each other in a sparse tree will result in unnecessary competition between them. Random assignment of leafs is provided by calculation of $\pi(ID)$ value for each ID. It is important that calculation of $\pi(ID)$ value occurs after PoW creation phase. Otherwise the adversary could try many ξ values for each ID and calculate PoW for those giving the best distribution.

It is important that consecutive bits of S are responsible for decisions on consecutive levels of the tree and are reused if the walk returns to given level. For example, let us assume that $S[i]$ has been used to make decision on level i but later it turned out that adversary cheated and directed the walk to an empty branch on some earlier level $j < i$. When the cheat is detected and walk goes back to level j and this time goes to the correct branch. Again the bit $S[i]$ will be used to decide on level i but this time in the correct branch. This is very important, since otherwise adversary could try to use up some bits for decisions in fake branches and alter the actual path of the walk.

It should be stressed that the idea of binary search with collisions has been used (in a different context) in some *singulation* protocols for RFID-systems [15].

It should never happen that on a given level both subbranches are empty. It would mean that the entire branch is empty. The BRC procedure should not enter an empty branch. However, this may happen, if there are some dishonest devices concealing their presence or pretending their presence in empty branches. If the procedure ends in a leaf that is taken by more than one identity all the identities are verified. An exemplary execution of the Binary Random Choice procedure is presented at Figure 1. With some probability there will be no honest station on the representatives list. Thus the adversary will be able to actually choose some values that are supposed to be pseudo random.

4.2 Security Analysis

Let M be the number of adversarial stations and N the number of honest stations. Assume that $N_{min} \leq N \leq N_{max}$ and that $M \leq \gamma N$, in particular $M \leq \gamma N_{max}$. For the analysis let the parameters be fixed as $\Gamma = (N_{max})^{\beta}$ and $p = \alpha \log(N_{min})/N_{min}$. In the analysis below we use the following definition.

DEFINITION 3 (AUTHORIZED IDENTITY). *An authorized identity is such ID that has chosen a value ξ_{ID} and some device has computed* $\mathrm{PoW}(C, ID, \xi_{ID})$.

CLAIM 2. *With probability at least $1 - 1/(N_{min})^{\alpha}$ there is at least one honest station that has its ID on the list of representatives.*

PROOF. Note that the probability that none of the legitimate devices is a representative does not exceed

$$(1-p)^{N_{min}} = (1 - \alpha \log(N_{min})/N_{min})^{N_{min}}$$
$$\leq \exp(-\alpha \log(N_{min})) = 1/(N_{min})^{\alpha}.$$

$\square$

CLAIM 3. *If there is at least one legitimate representative, then for any subset of dishonest stations the values C, R and S are indistinguishable from a uniformly distributed random variables.*

CLAIM 4. *If there is at least one representative of honest stations, then for any assignment of authorized identities to leafs, the result of BRC procedure cannot be altered.*

PROOF. The result of the BRC procedure is determined by the value S and results of consecutive TEST procedure executions. By Claim 3 from the point of view of each identity the value S is chosen uniformly at random and thus cannot be altered. However, the result of the TEST procedure can be manipulated by the adversary. Namely, direction of walk can occur only if the TEST result will be changed from **none** to **one**, or **noise**. This can be achieved only by emulating some fake occupied leafs in an empty branch. However, in such a case the cheating attempt is inevitably detected in the Leader Candidate Verification phase. Consequently, the algorithm will roll back to the cheated TEST procedure state and will continue the other (correct) direction. $\square$

CLAIM 5. *If there is at least one legitimate representative, then each identity is assigned to a leaf chosen uniformly at random.*

PROOF. Each authorized identity's assignment depends on R. Thus ξ, can be treated as a random variable having uniform distribution. At the time of choosing ξ position of each authorized identity can also be treated as a random variable having uniform distribution. $\square$

CLAIM 6. *If there is at least one legitimate representative, then before R and S values are revealed all authorized identities have the same probability of being chosen during the binary random walk phase.*

PROOF. The above claim is a simple consequence of symmetry, Claim 4 and Claim 5. Indeed, since authorized identities assignment is random and uniform and the BRC procedure result cannot be altered, then each authorized identity has the same chance of being chosen. $\square$

From the above discussion the following theorem emerges:

THEOREM 1. *Assuming that each device can calculate at most one valid PoW during runtime of the algorithm and that devices decide to become representatives with probability $p = \alpha \log(N_{min})/N_{min}$, the adversary is chosen as a leader with probability that does not exceed*

$$(1 - 1/(N_{min})^{\alpha}) \cdot M/(M + N) + 1/(N_{min})^{\alpha} .$$

PROOF. If there is no legitimate representative (by Claim 2 this occurs with probability $1/(N_{min})^{\alpha}$), the adversary can become the leader. Otherwise, the adversary can authorize at most M identities while there will be exactly N identities representing honest stations. Consequently, by Claim 6 if there is a legitimate representative the adversary's winning chances are equal $M/(M + N)$. $\square$

From construction of the algorithm one can easy prove that:

CLAIM 7. *Let us assume that the adversary is not trying to delay the execution of the protocol. Then, the expected execution time of the general case protocol is $O(\log(N + M) + (N + M) \cdot p \cdot sel)$ where sel is a single identity selection time. Assuming $p = \frac{\alpha \log(N_{min})}{N_{min}}$, and that the algorithm from [27] is used to elect each representative, then finally the expected execution time of the general case protocol is $O(\log(M + N) \log(\log(N + M)))$.*

It can be easily shown that the expected execution time of the simple protocol is $\Omega(N + M)$. Thus from above claim one can see that the time complexity of the general case protocol is dramatically better than the time complexity of the simple protocol (Section 3).

5. CONCLUSIONS

We presented the first Sybil attack immune leader election algorithm for wireless ad hoc networks based only on computational limitations of devices.Described approach can be combined with other methods as an additional layer of security. There are two most important ideas. The first is using the Proof of Work concept in order to make it resource consuming to maintain an identity. The second is the binary random search concept that saves the effort of listing all devices (BRC technique combined with randomization of commitments). Most of identities never participate actively in this protocol (they do not transmit any message), but they have to be prepared just for the case of being chosen. That is, they just perform some local computations. We believe that the method proposed in this protocol can be a powerful when combined with other approaches - especially based on location [7] or restrictions of communication [13, 16].

6. REFERENCES

[1] A. Back. Hashcash - a denial of service counter-measure. Technical report, 2002.

[2] J. L. Bordim, Y. Ito, and K. Nakano. Randomized leader election protocols in noisy radio networks with a single transceiver. In M. Guo, L. T. Yang, B. D. Martino, H. P. Zima, J. Dongarra, and F. Tang, editors, *ISPA*, volume 4330 of *Lecture Notes in Computer Science*, pages 246–256. Springer, 2006.

[3] C. S. Brian Neil Levine and N. B. Margolin. A survey of solutions to the sybil attack, 2006.

[4] F. Coelho. Exponential memory-bound functions for proof of work protocols, 2005.

[5] B. L. C.Piro, C.Shields. Detecting the sybil attack in ad hoc networks. In *Proc. IEEE/ACM SecureComm*, pages 1–11, 2006.

[6] J. Czyzowicz, L. Gasieniec, D. R. Kowalski, and A. Pelc. Consensus and mutual exclusion in a multiple access channel. *IEEE Trans. Parallel Distrib. Syst.*, 22(7):1092–1104, 2011.

[7] S. Delaët, P. S. Mandal, M. A. Rokicki, and S. Tixeuil. Deterministic secure positioning in wireless sensor networks. *Theor. Comput. Sci.*, 412(35):4471–4481, 2011.

[8] S. Dolev, S. Gilbert, R. Guerraoui, and C. C. Newport. Secure communication over radio channels. In R. A. Bazzi and B. Patt-Shamir, editors, *PODC*, pages 105–114. ACM, 2008.

[9] J. R. Douceur. The sybil attack. In *IPTPS*, pages 251–260, 2002.

[10] C. Dwork and M. Naor. Pricing via processing or combatting junk mail. In E. F. Brickell, editor, *CRYPTO*, volume 740 of *Lecture Notes in Computer Science*, pages 139–147. Springer, 1992.

[11] V. Geffert, J. Karhumäki, A. Bertoni, B. Preneel, P. Návrat, and M. Bieliková, editors. *SOFSEM 2008: Theory and Practice of Computer Science, 34th Conference on Current Trends in Theory and Practice of Computer Science, Nový Smokovec, Slovakia, January 19-25, 2008, Proceedings*, volume 4910 of *Lecture Notes in Computer Science*. Springer, 2008.

[12] S. Gilbert, R. Guerraoui, D. Kowalski, and C. Newport. Interference-Resilient Information Exchange. In *IEEE InfoCom 2009*, 2009.

[13] Z. Golebiewski, M. Klonowski, M. Koza, and M. Kutylowski. Towards fair leader election in wireless networks. In P. M. Ruiz and J. J. Garcia-Luna-Aceves, editors, *ADHOC-NOW*, volume 5793 of *Lecture Notes in Computer Science*, pages 166–179. Springer, 2009.

[14] T. Hayashi, K. Nakano, and S. Olariu. Randomized initialization protocols for packet radio networks. In *IPPS/SPDP*, pages 544–. IEEE Computer Society, 1999.

[15] A. Juels, R. L. Rivest, and M. Szydlo. The blocker tag: selective blocking of rfid tags for consumer privacy. In S. Jajodia, V. Atluri, and T. Jaeger, editors, *ACM Conference on Computer and Communications Security*, pages 103–111. ACM, 2003.

[16] M. Klonowski, M. Koza, and M. Kutylowski. Repelling sybil-type attacks in wireless ad hoc systems. In *ACISP*, 2010.

[17] M. Klonowski and T. Struminski. Proofs of communication and its application for fighting spam. In Geffert et al. [11], pages 720–730.

[18] M. Klonowski and T. Struminski. Proofs of communication and its application for fighting spam. In Geffert et al. [11], pages 720–730.

[19] M. Kutyłowski and W. Rutkowski. Adversary immune leader election in ad hoc radio networks. In G. D. Battista and U. Zwick, editors, *ESA*, volume 2832 of *Lecture Notes in Computer Science*, pages 397–408. Springer, 2003.

[20] M. Kutyłowski and W. Rutkowski. Secure initialization in single-hop radio networks. In C. Castelluccia, H. Hartenstein, C. Paar, and D. Westhoff, editors, *ESAS*, volume 3313 of *Lecture Notes in Computer Science*, pages 31–41. Springer, 2004.

[21] A. MacKenzie and S. B. Wicker. Stability of multipacket slotted aloha with selfish users and perfect information. In *INFOCOM*, 2003.

[22] R. M. Metcalfe and D. R. Boggs. Ethernet: distributed packet switching for local computer networks. *Commun. ACM*, 19(7):395–404, 1976.

[23] K. Nakano and S. Olariu. Randomized o (log log n)-round leader election protocols in packet radio networks. In K.-Y. Chwa and O. H. Ibarra, editors, *ISAAC*, volume 1533 of *Lecture Notes in Computer Science*, pages 209–218. Springer, 1998.

[24] K. Nakano and S. Olariu. Randomized leader election protocols in radio networks with no collision detection. In D. T. Lee and S.-H. Teng, editors, *ISAAC*, volume 1969 of *Lecture Notes in Computer Science*, pages 362–373. Springer, 2000.

[25] J. Newsome, E. Shi, D. X. Song, and A. Perrig. The sybil attack in sensor networks: analysis & defenses. In K. Ramchandran, J. Sztipanovits, J. C. Hou, and T. N. Pappas, editors, *IPSN*, pages 259–268. ACM, 2004.

[26] A. Perrig, J. A. Stankovic, and D. Wagner. Security in wireless sensor networks. *Commun. ACM*, 47(6):53–57, 2004.

[27] D. E. Willard. Log-logarithmic selection resolution protocols in a multiple access channel. *SIAM J. Comput.*, 15(2):468–477, 1986.

Energy Attacks and Defense Techniques
for Wireless Systems

Sheng Wei   Jong Hoon Ahnn   Miodrag Potkonjak
Computer Science Department
University of California, Los Angeles (UCLA)
Los Angeles, CA 90095
{shengwei, jhahnn, miodrag}@cs.ucla.edu

ABSTRACT

This paper addresses the energy attacks towards wireless systems, where energy is the most critical constraint to lifetime and reliability. We for the first time propose a hardware-based energy attack, namely energy hardware Trojans (HTs), which can be well hidden in the wireless systems and trigger ultra-high energy increases at runtime. Then, we develop a non-destructive HT detection approach to identify the energy attack by remotely sampling the power profiles of the system and characterizing the gate-level temperatures. Our evaluation results on ISCAS benchmarks indicate the effectiveness of the proposed energy attacks and defense techniques.

Categories and Subject Descriptors

K.6.5 [**Management of Computing and Information Systems**]: Security and Protection—*Physical Security*

General Terms

Security

Keywords

Wireless security, hardware Trojan, leakage energy

1. INTRODUCTION

Wireless communication systems, especially mobile devices, have been widely used in a variety of personal and commercial applications, including the traditional wireless phone services and the emerging Internet mobile applications, such as social networking, mobile banking, and multimedia applications. With the ever increasing popularity of these applications, security and integrity of the devices have become a critical concern. Due to the mobile and wireless nature, security primitives for wireless systems are much more challenging and vulnerable than for traditional computer systems.

Both research and practical efforts have been directed towards wireless security at various levels, including software applications, communication channels, as well as hardware systems. Among them, many conventional and new security attacks [5][9][18][1][24][35][16] have been well identified, analyzed, and resolved. However, we note that the security towards one of the most crucial and fundamental components in wireless systems, namely energy consumption, has rarely been discussed until now.

It is well known that all wireless devices are energy constrained, especially with the more and more popular uses of mobile devices in computation-intensive applications, such as high definition video streaming and mobile computing. Although huge efforts have been made on energy reduction and optimizations, the systems are vulnerable to energy attacks that intend to cause high energy increase and reduce the lifetime of the system. For example, an adversary may implant a malware that runs in the background of a smart phone and consumes a large amount of battery power. Besides software attacks, energy attacks at the hardware level pose a more severe threat to the wireless systems, because they are much more challenging to be detected and disabled. For example, an untrusted foundry of the cell phone chip or untrusted third party manufacturer of the phone may have hidden hardware Trojans in the hardware that leak additional energy from the phone.

We target on the energy attacks on wireless systems caused by hardware Trojans (HTs), which are malicious modifications to the hardware systems conducted by an untrusted foundry or manufacturer during the manufacturing process [25][29][32][33]. We show that such an energy attack can be implemented by embedding an ultra-small hardware Trojan trigger to the integrated circuit (IC) of the system, which hides in the circuit and results in huge energy increase on the device. In particular, we for the first time develop and demonstrate two types of hardware energy attacks by manipulating the input vectors and biasing the power supply, respectively. We argue that such energy attacks are extremely difficult to detect. On one hand, the attacker tends to hide the HT trigger in the target circuit by either sizing it ultra-small or by activating it only when an extremely rare event occurs. On the other hand, the leakage energy by itself has a high dependence on the environment temperatures and, therefore, the attacker could attribute any malicious energy attack to the variations of environment temperatures. It is significantly difficult for the detection process to distinguish between the two cases.

Based on the investigation of the powerful energy attacks, we develop defense approaches for energy attacks by leveraging power profiling and temperature characterizations. The idea is to sample the total leakage power consumption of the wireless system and characterize the gate-level temperature profile by assuming that there is no energy attack. In the case where there is indeed malicious energy attack, the characterized temperature profile will not meet the normal spatial and temporal thermal distributions on an IC. We identify the possible discrepancy in temperatures quantitatively by defining a hardware Trojan indicator concerning the spatial inconsistency of the gate-level temperatures. Note that our energy HT detection approach requires no instrumentation to the wireless system, which can be conducted remotely without having direct physical access to the target device.

Figure 1 shows the overall flow of our approaches for energy attack and defense, which spreads over the wireless system manufacturing, post-silicon testing, and real time (in-field) operation. Firstly, we investigate on the energy HT attack that can be conducted by an adversary during the manufacturing process, where the design objective is to maximize the potential damage (i.e., the leakage energy consumption) and minimize the probability of being detected (i.e., placing the trigger of the HT at a rarely switching or low leakage location, as well as enabling the sequential-based activation event). Note that we develop two attack techniques, namely input vector control and forward body biasing, which significantly increase the leakage energy in the sleep mode and operation mode of the wireless system, respectively. Secondly, we employ a gate-level temperature characterization method to monitor and identify abnormal leakage energy variations at the runtime. Our approach characterizes the gate-level temperature profiles via energy profiling and thus distinguishes the abnormal energy hike due to security attacks and the normal energy variations due to environmental factors.

To the best of our knowledge, this is the first complete analysis and implementation of hardware-based energy attacks and defense techniques in wireless systems. In summary, our technical contributions include:

- **Attack**: a powerful and well hidden hardware energy attack to wireless systems by embedding and hiding hardware Trojan components in the target circuit;

- **Defense**: a temperature-aware power profiling approach to detect energy hardware Trojans in the wireless systems without embedding any additional hardware to the system.

The remainder of this paper is organized as follows. In Section 2, we summarize the existing research work regarding hardware Trojan detection, thermal-aware IC design, and adaptive body bias. Section 3 introduces the leakage energy model and the gate-level characterization approach that we employ in this work. In Section 4, we introduce the design of the high leakage HT in both the sleep mode and the system operation mode; Section 5 discusses our detection techniques against the high energy attack, which identifies the presence of attacks via gate-level temperature characterization. We show our experimental results in Section 6 and conclude the paper in Section 7.

2. RELATED WORK

In this section, we summarize the existing research efforts in the areas of hardware Trojan detection, thermal-aware IC design, and adaptive body bias techniques, with the emphasis of our new contributions compared to the previous approaches and techniques.

2.1 Hardware Trojan Detection

Hardware Trojans (HTs) [25][29][32][33][13][26] are malicious modifications to integrated circuits (ICs) that are possibly conducted by untrusted CAD tools at design time, or untrusted foundries during manufacture. With the ever increasing trend of IC outsourcing, the security and reliability concerns caused by hardware Trojans have drawn a great deal of attention in the IC design and security community. Furthermore, since ICs are fundamental building blocks of embedded systems, where the security primitives are much more challenging and vulnerable than traditional hardware systems, it is crucial to evaluate and detect the HTs.

Tehranipoor et al. [25] provided a comprehensive survey of HT detection. Recently, more and more efforts have been made in developing side-channel (e.g., power and delay) based detection techniques [2][19][12][15][29][31][28].

The existing HT detection efforts targeted only on HTs that cause direct security attacks, such as implanting a backdoor in the circuit and extracting confidential information from the system at runtime. However, we note that the security toward one of the most crucial and fundamental components in embedded systems, namely energy consumption, has rarely been discussed until now. Our energy HT detection approach is new compared to the existing HT detection techniques. We for the first time analyze and address hardware Trojans of energy attacks towards wireless systems, which has not been discussed in the current literature.

2.2 Thermal-aware IC Design and Analysis

Thermal effect has become a crucial factor being considered in IC design and manufacturing because of the interdependency between IC properties (i.e. delay and leakage power) and temperature. Recently, many research efforts have been made on thermal-aware leakage model and leakage reduction techniques. Li et al. [20] developed an architectural model for subthreshold and gate leakage that explicitly defines the relationship between leakage power and temperature. It shows that the subthreshold leakage currents are exponentially dependent on temperature and voltage. Besides leakage power, Liao et al. [21] showed that the performance of IC also depends on temperature.

Temperature monitoring of IC systems has drawn a great deal of attention in the IC design and manufacturing community. Finite element analysis (FEA) [38] and resistor networks [7] approaches employ heat transfer model for semiconduct materials and calculate the IC temperature profiles at design time. Power blurring technique [14] reduces the computation time of FEA by using matrix convolution. However, these techniques are designed for temperature calculation and prediction at the design time, which is not resilient to the process variation and cannot be used for runtime temperature monitoring. At runtime, thermal sensor-based approaches [27] have been proposed to measure the temperatures in the real time. However, these approaches introduce a high overhead and instrumentation to the target circuit.

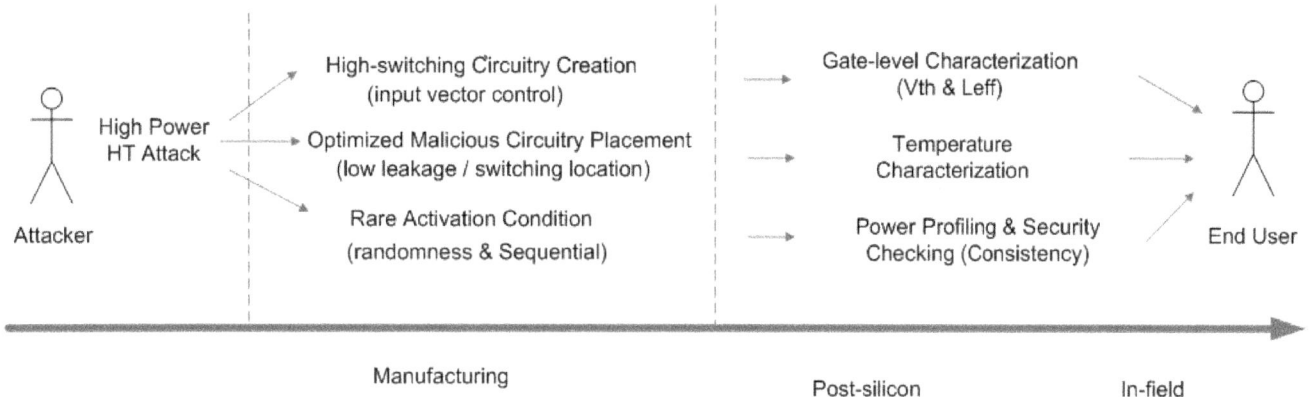

Figure 1: Overall flow of high power attack and defense.

Compared to the existing approach, our gate-level temperature profiling technique is new in the following aspects. First, our approach is non-destructive, as we do not require any additional hardware to be added in the target circuit, which minimizes the overhead; Second, we take into consideration of the process variation and measurement error, which enables us to monitor the real time temperatures for remote wireless systems. For example, our technique can also be used to facilitate remote hardware identification or watermarking [11][17][39].

2.3 Adaptive Body Bias

Adaptive body bias (ABB) has been widely adopted as an efficient post-silicon approach in the research efforts of leakage energy reduction and performance optimization [23][10][6]. For example, Nabaa et al. [23] proposed the use of ABB through the use of the new FPGA architecture that includes an additional characterizer circuit to reduce the leakage energy by 3 times. Gregg et al. [10] proposed using ABB to compensate for the process variation and improve delay and leakage. Chen et al. [6] compare the effectiveness of adaptive supply voltage (ASV) and ABB. Furthermore, similar with the pre-silicon dual V_{th} approach, researchers have proposed multiple body bias values in the target circuit, each drives a subset of the gates. For example, Xu et al. [36] cluster the gates at a finer-grained level and apply multiple ABB values to control the leakage energy consumption.

Compared to the existing work, We investigate on an unconventional use of the ABB techniques, possibly by an adversary, to increase the leakage energy consumption of wireless systems. Instead of compensating for the process variation and reducing leakage energy consumption, an attacker may apply forward bias voltages to increase the leakage energy exponentially.

3. PRELIMINARIES

In this section, we summarize the preliminaries and key observations that serve as the foundation of our proposed energy attack and defense techniques, including the energy model we employ for evaluating the leakage energy, as well as the gate-level characterization scheme that recovers the gate-level physical properties from global leakage energy measurements.

3.1 Energy Model

We consider leakage energy and switching energy that are major sources of energy consumption during IC operations. The leakage energy is dependent on the IC physical properties such as effective channel length L_{eff} and threshold voltage V_{th}. Equation (1) is the gate-level leakage energy model [22], where W is gate width, L is gate length, V_{th} is threshold voltage, and T is the temperature. The rest of the parameters are considered as constants and are discussed in details in [22].

$$P_{leakage} = 2 \cdot n \cdot \mu \cdot C_{ox} \cdot \frac{W}{L} \cdot (kT/q)^2 \cdot D \cdot V_{dd} \cdot e^{\frac{\sigma \cdot V_{dd} - V_{th}}{n \cdot (kT/q)}} \quad (1)$$

Equation (1) indicates that the leakage energy of logic gate depends on the temperature in a non-linear manner. Therefore, any temperature changes in the environment, or due to the switching of the gates in the circuit, will have impact on the energy consumption of the IC system.

The gate-level switching energy model [22] is described by Equation (2), where the switching energy is dependent on gate width W, gate length L, and supply voltage V_{dd}. where α is the switching probability.

$$P_{switching} = \alpha \cdot C_{ox} \cdot W \cdot L \cdot V_{dd}^2 \quad (2)$$

3.2 Gate-level Characterization

In gate-level characterization (GLC), we recover the gate-level IC properties from global side-channel measurements under the application of various input vectors [29][30][34]. For example, when J input vectors have been applied on a target circuit with K gates, the gate-level leakage energy values can be solved using the following linear program (LP):

$$Objective: \quad \min_{1 \le j \le J} \mathcal{F}(err_j) \quad (3)$$

$$Constraints: \quad \Sigma_{k=1}^{K} E_{jk} = \tilde{E}_j + err_j$$
$$j = 1, \dots, J$$

where E_{jk} is the leakage energy of gate k ($k = 1, \dots, K$) when input vector j ($j = 1, \dots, J$) is applied; $\tilde{E}_j$ is the measured total leakage energy when the input vector j is applied; err_j is the measurement error; $\mathcal{F}$ is a metric for quantifying the measurement errors, such as l_1 or l_2 norm. In this LP formulation, E_{jk} can be expressed as a product of its constant nominal value $E_{nom,jk}$ and a scaling factor (due to PV) δ_k, i.e., $E_{jk} = \delta_k E_{nom,jk}$. By solving the LP with δ_k as the variables, we can obtain the value of E_{jk} for each gate k ($k = 1, \ldots, K$). Furthermore, by following the energy models (i.e., Equations (1) and (2)), we can formulate a system of nonlinear equations and solve for the physical-level properties (i.e., threshold voltage and effective channel length) of each individual gate.

4. HIGH LEAKAGE ENERGY ATTACK

In this section, we design and analyze energy attacks on wireless systems via malicious hardware modification during the manufacturing process (i.e., energy hardware Trojans). We analyze two major components of the hardware Trojans, namely (1) *action*, which indicates what attacks the HT could impose and how the attacks are implemented; and (2) *trigger*, which indicates the activation condition that would trigger the energy attack. By analyzing these two components, we aim to quantify the energy increase caused by the attacks as well as the probability of activation, which are quantitative indicators of the effectiveness of attacks and the difficult level for detection, respectively. An attacker would tend to maximize the action while reducing the trigger probability to impose damaging and well hidden energy attacks.

4.1 Sleep-mode Energy Attack: Input Vector Manipulation

For most of the wireless systems and applications, the circuit of the system would stay in the sleep mode for a large portion of the time. For example, in the case of a cell phone, the major components are only exercised once it is in a voice call. Similarly, in a wireless sensor network, the communication circuitry only operates during the data collection process. Consequently, the input vectors that are being applied during the IC sleep mode becomes crucial with regard to the leakage energy consumption, because of the fact that the leakage energy of a logic gate highly depends on the input vectors [37]. For example, as shown in Figure 2, the leakage energy of a NAND gate can vary up to 12 times with different input vectors. Although this phenomenon can provide us with an opportunity for leakage energy reduction, it is more easily leveraged by an attacker for energy attack.

Input	Leakage Current
0	100.3 nA
1	227.2 nA

Input Vector	Leakage Current
00	37.84 nA
01	95.17nA
10	100.3 nA
11	454.5 nA

Figure 2: Leakage current of an inverter and a NAND gate under different input vectors [37].

Figure 3 shows a motivational example of energy attack using input vector manipulation. In this case, the HT component applies an input vector that sets the maximum number of gates in the high energy state, which results in the

highest energy consumption (i.e., 2.96 times of the minimum energy consumption). If this situation continues over time without being identified by the user or tester of the system, it will cause several times more energy consumption, which is considered significant in a power hungry system.

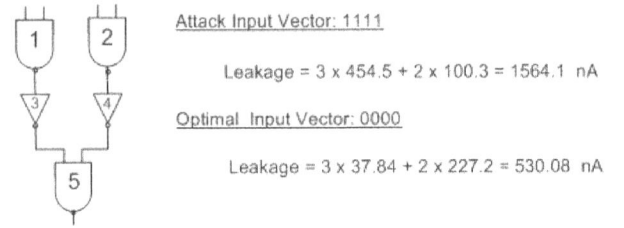

Attack Input Vector: 1111

Leakage = 3 x 454.5 + 2 x 100.3 = 1564.1 nA

Optimal Input Vector: 0000

Leakage = 3 x 37.84 + 2 x 227.2 = 530.08 nA

Figure 3: Example of energy attack via input vector manipulation. The energy consumption caused by the attack input vector is 2.96 times compared to the optimal input vector.

4.2 Operation-mode Energy Attack: Forward Adaptive Body Bias

During IC operation, the supply voltage plays an important role to the total leakage power consumption. According to Equation (1), the leakage energy of a transistor increases exponentially with the increase of supply voltage. We argue that this phenomenon can be leveraged for powerful energy attacks, since an exponential energy increase could cause huge impact to the wireless system and thus becomes the best interest of an attacker. In particular, we argue that an attacker could possibly apply a forward adaptive body bias (FBB, or forward ABB) voltage that, instead of compensating for the process variation as in the normal use of body biasing techniques [6][10][36], would increase the energy exponentially in the circuit under attack.

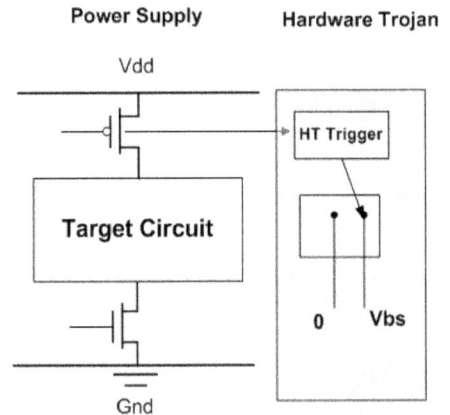

Figure 4: Example of energy attack using forward body biasing.

With this consideration, we implement a sample malicious circuitry that triggers the FBB-based energy attack, as shown in Figure 4. The shaded part of the circuit is the malicious component, or HT, embedded by an adversary. During the operation mode, once triggered, the HT can select and apply a forward body bias voltage, which increases the supply voltage and maximizes the leakage energy without compromising the functionality of the system.

4.3 HT Triggers: Rare Activation of Attacks

In the previous two subsections, we have shown that HT-based energy attack could cause huge energy increase, either linearly in the sleep mode or exponentially in the operation mode. However, the attacks would not take effective unless the triggers are well hidden from the common detection approaches. In this subsection, we discuss in details how an attacker could design the trigger such that the resulting HT attack has a low probability to be detected by the HT detection attempts. The intuitions of hiding the HT trigger, from the attacker's perspective, include the following: (1) hide the HT trigger in the circuit both physically and in terms of their observable properties, such as delay and power; and (2) minimize the activation probability of the HT.

4.3.1 Hiding the HT trigger

In order to bypass the most commonly used side channel-based detection approaches, an attacker aims to place the HT trigger in such a way that it is non-observable via the commonly considered side channels, including delay, leakage power, and switching power. In order to achieve this goal, we add only one single gate in the target circuit that serves as the trigger. With this single gate HT trigger, we ensure that any resulting delay or power variation is minimum to increase the difficulty level for detection. Furthermore, to further complicate the side channel-based detection approaches, we place the single gate HT trigger at a circuit location where the delay and power are non-observable or difficult to measure. For example, in Figure 5(a), the delay of the HT trigger is non-observable due to the parallel reconvergent paths. One can measure the delay between the two endpoints x and y. However, it is not possible to determine whether the measured delay is for path 1 or path 2 and, therefore, the HT can be hidden under the delay measurements.

4.3.2 Minimizing the activation probability

Based on the HT trigger placement that is difficult to detect, we further reduce its activation probability to bypass the security checks that are based on generated test vectors, such as automatic test pattern generation (ATPG) [2]. The idea is to set the activation condition in such a way that it is only known to the attacker and very rarely triggered during a normal IC operation or test. We achieve this goal by using two approaches. Firstly, we select the fan-in gates from the target circuit in such a way that the HT trigger is rarely switched, as shown in an example in Figure 5(b). The activation probability of the NAND HT trigger is $1/2^n$, where n is the number of inputs that can be customized by the attacker to balance the trade-off between the size of the HT trigger and the activation probability. Secondly, we leverage sequential elements (i.e., flip-flops) that create temporal-based activation conditions in a finite state machine (FSM). In this way, the activation probability of the HT trigger can be further reduced exponentially based on the results from the first approach. For example, as shown in Figure 5(c), the 5-state FSM serves as the activation condition, which triggers the HT only when all 5 states are satisfied in 5 consecutive clock cycles. As a result, the activation probability is $\prod_{i=1}^{m} P_i$, where P_i is the activation probability of the vector in state i, and m is the number of states.

5. TEMPERATURE-AWARE CONSISTENCY-BASED HT DETECTION

In this section, we discuss our detection approach to identify the HT-based energy attacks. In order to exclude the possible impact of temperature in energy increase, which is likely to be claimed by the attacker, we conduct gate-level temperature characterization to recover the temperature profile of the circuit. Then, we calculate the value of our defined HT indicator, which is the spatial inconsistency of temperatures among physically adjacent gates, to determine the presence of energy attack.

5.1 Energy Paradox

The most straightforward detection approach towards the energy attack is by sampling the energy profiles of the operating wireless systems on a regular basis and observe the abnormal energy increase. Although remote sampling and data collection is a common practice for wireless system performance or status monitoring, the collected power profile is not an effective indicator for energy attacks, due to the following energy paradox, which can be leveraged by energy attackers:

Energy Paradox. Due to the exponential dependence of leakage energy on temperatures, as indicated in Equation (1), it is not certain to the system user whether the energy increase is caused by normal temperature variations or malicious energy attacks. As a matter of fact, it is common that the target wireless system, such as a wireless sensor network, is deployed in a hazardous environment where the temperature varies in an unknown pattern. In this case, the power profiling approach by itself is not sufficient to reach a conclusive judgment of whether any energy attack exists or not.

5.2 Gate-level Temperature Characterization

In order to address the energy paradox and obtain accurate energy HT detection results, in the case of a high energy profile, we must measure or characterize the temperature of the target IC to either exclude its impact or report that the energy increase is due to temperature. Several approaches have been proposed in monitoring the temperatures of IC systems, such as FEA [38], power blurring [14], and sensors [27]. However, the FEA and power blurring approaches work at the IC design stage without the taking account of the impact of process variation and are not resilient for post-silicon attacks. The sensors-based approaches provide real-time measurements of temperatures, but they require additional sensor circuitry in the target IC, which greatly increases the complexity and cost of the system.

We develop a non-destructive gate-level temperature characterization approach using power profiling, which does not require additional hardware circuitry being added to the target IC. The approach is based on the physical-level GLC concerning threshold voltage (V_{th}) and effective channel length (L_{eff}). We show the flow of temperature characterization in Pseudocode 1. Firstly, before the release and deployment of the wireless system, we characterize the gate-level V_{th} and

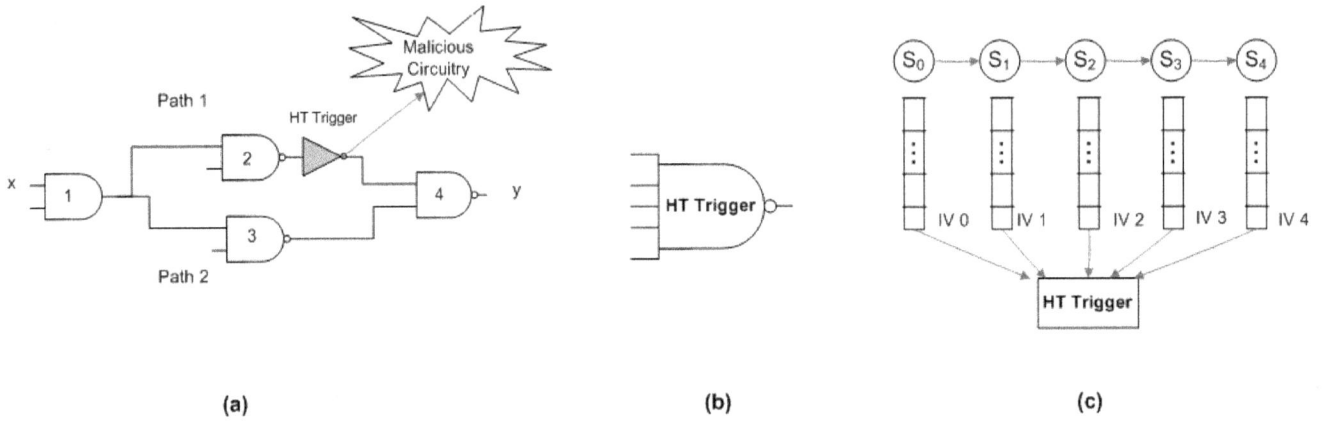

Figure 5: Example of hiding the HT triggers: (a) a HT trigger embedded in the reconvergent paths where delay is non-observable; (b) a HT trigger activated by a 5-state finite state machine, which reduces the activation probability exponentially.

L_{eff} at room temperature, where we assume the temperature T in Equation (1) as a constant value. Then, after the system has been deployed and in operation, we take M power measurements and characterize the temperature (T_i) of each gate using Equation (1) based on the V_{th} and L_{eff} that are already known. Finally we conduct online security checking to determine the presence of energy attack using the HT indicator, as defined and discussed in the next subsection.

Pseudocode 1 Gate-level temperature characterization via power profiling.

1: *Post-silicon*:
2: Gate-level characterization to solve V_{th}, L_{eff} of each gate at room temperature following Equations (1) and (2);
3: *Runtime*:
4: Take M power measurements via sampling;
5: **for all** Gates g_i in the circuit **do**
6:     Solve for temperature T_i following Equations (1);
7: **end for**
8: *Security Check*:
9: Conduct security check on temperature T_i over all gates;

5.3 HT Indicator

The problem we face in inspecting the characterized temperature profile for HT detection is that the normal temperature profile, or the "golden model", is not available in the case of wireless systems. It is because the system is often deployed in unknown environments (e.g., wireless sensor network) or has a mobile nature and a high probability of environmental changes (e.g., smart phones). Therefore, the online temperature security inspection cannot be done via simple comparisons.

We solve the problem by defining a HT indicator that represents the temperature inconsistency over gates that are adjacent to each other in the target IC. Our intuition is that the heat transfer process would create spatial correlations in the temperatures of gates that are physically close to each other. Therefore, if we ever observe that there is an abnormally large deviation between the temperatures of two or

more adjacent gates, it is an indicator that the energy increase is not likely caused by temperature changes but by malicious energy attacks. This is based on the assumption that it is computationally impossible for an attacker to emulate the heat transfer model and impose energy attacks following exactly the same pattern.

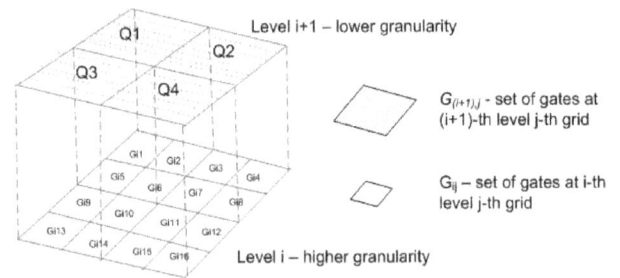

Figure 6: Principal component analysis (PCA) model to define the hardware Trojan indicator.

We define the HT indicator using the principal component analysis (PCA) models [8] that were originally used for modeling spatial correlations in IC process variations. As shown in Figure 6, we group the gates into multiple grids at different levels in order to capture the inconsistency of the temperatures between various boundaries of adjacent gates. At each specific level, we define the HT indicator as the average standard deviation, over all grids, of temperatures among all gates within each grid. In particular, at the i-th level, the HT indicator H_i can be calculated as the following:

$$H_i = (\sum_{j=1}^{N_i} stddev(G_{ij}))/N_i \qquad (4)$$

where N_i is the number of grids at level i, and G_{ij} is the set of gates in the j-th grid at level i. We use H_i to evaluate the temperature deviation over adjacent gates at different granularities. Depending on the sizes and physical properties of the circuit under test, different levels of H_i plays different roles in the final evaluation of the temperature deviations. Therefore, we define the following weighted function for the overall HT indicator:

$$H = (\sum_{i=1}^{L} w_i H_i)/L \qquad (5)$$

where L is the number of levels that we divide using the PCA model, and w_i is the weight factor at level i concerning the physical properties of the circuit.

6. EXPERIMENTAL RESULTS

We evaluate our temperature-aware energy HT detection approach on a set of ISCAS benchmarks that are widely used in the IC design and hardware security community [4][3]. In this section, we discuss in details our evaluation results in the energy HT attack and defense.

6.1 Effectiveness of Energy HT Attack

We evaluate the effectiveness of energy HT attack from two aspects that are essential for a hardware Trojan: (1) HT action, i.e., how much energy increase the energy HT causes to the circuit under attack; and (2) HT trigger, i.e., the probability of activation of the HTs, which indicates how well they can be hidden from common detection approaches.

6.1.1 Energy Increase

We evaluate the energy increase caused by forward ABB-based energy attack by inserting a HT trigger that selects ABB voltages up to 1.0V. Figure 7 shows the energy increase due to the attack on a set of ISCAS benchmarks. We observe that the energy consumption grows exponentially with the linear increase of the ABB voltage, creating huge impacts on the circuit under attack.

6.1.2 Activation Probabilities of the Energy HTs

We further evaluate the activation probabilities of the HT triggers in order to quantify the difficulty levels for detecting these energy attacks. Table 1 shows our simulation results on a set of ISCAS benchmarks. We evaluate two cases with the number of fan-in gates for the HT trigger selected from the target circuit varying between 5 and 10. In each case, we employ a 5-state finite state machine that randomly generates 5 sequential activation input vectors that serve as the activation condition of the HT trigger. Our results show that the activation probabilities are extremely low and decrease exponentially with the increase of fain-in signals and the number of states in the finite state machine. The probabilities are low enough to create challenging attacks that are computationally infeasible to be covered by any existing functional test detection schemes.

6.2 Effectiveness of Temperature-aware HT Detection

We evaluate the effectiveness of our temperature-aware HT detection approach from two aspects. Firstly, we evaluate the accuracy of the gate-level temperature characterization, which is an indicator of how accurate we can capture the abnormal temperature variations. Then, we evaluate the HT detection approach by comparing the HT indicator values in two cases where HTs are present and where there are no HTs, in order to determine the false positives and false negatives in HT detection.

6.2.1 Accuracy of Gate-level Temperature Characterization

We evaluate the accuracy of the gate-level temperature characterization by comparing the characterized temperatures and the actual temperatures and quantifying the average characterization errors. Figure 8 shows the distribution of the characterization errors for each gate in a set of ISCAS benchmarks. The results indicate that characterization errors of all gates are controlled within the 2% mark except for very few outliers gates. Also, the accuracy does not decrease with the increase of the circuit sizing, indicating the scalability of our detection approach.

6.2.2 Effectiveness of Detection Using HT Indicator

In order to evaluate the effectiveness of HT detection, we characterize the gate-level temperature profiles in two cases where there are no energy attacks (i.e., HT-free) and where there are forward ABB-based energy HTs embedded and triggered in the target circuit (i.e., HT-present). For each case, we calculate the value of the HT indicator as defined in Section 5.3 to observe the difference between the two cases. In our simulation, we use 3 level of grids in the PCA model (L=3) and use evenly assigned weight factors for the w_i (i.e., $w_i = 1/3, i = 1, 2, 3$) in calculating H. Figure 9 shows our evaluation results of HT indicator H in both the HT-present and HT-free cases. There is an obvious and large difference between the HT indicators in the two cases. In the HT-present case, the HT indicator is significantly larger than that of the HT-free case. Therefore, the HT indicator is a good metric for differentiating the two cases, which provides us with zero false positives and zero false negatives in the detection of energy HTs.

7. CONCLUSION

We developed the first hardware Trojan-based energy attack towards wireless systems using input vector manipulation and forward body biasing. To complicate the detection process, we hide the trigger of the energy hardware Trojan by embedding it in unobservable paths and minimizing its activation probability. Then, as defense, we developed a temperature-aware gate level characterization approach that distinguishes between the energy increase caused by normal temperature variations and due to malicious attacks. Our simulation results on ISCAS benchmarks verified the effectiveness of the energy attack and defense approaches. To the best of our knowledge, this is the first attempt to analyze and address hardware Trojan-based energy attacks in wireless systems.

8. ACKNOWLEDGEMENTS

This work was supported in part by the NSF under Award CNS-0958369, Award CNS-1059435, and Award CCF-0926127, and in part by the Air Force Award FA8750-12-2-0014.

9. REFERENCES

[1] B. Awerbuch, D. Holmer, C. Nita-Rotaru, and H. Rubens. An on-demand secure routing protocol resilient to byzantine failures. In *ACM workshop on Wireless security (WiSe)*, pages 21–30, 2002.

[2] M. Banga and M. Hsiao. A region based approach for the identification of hardware Trojans. In *IEEE*

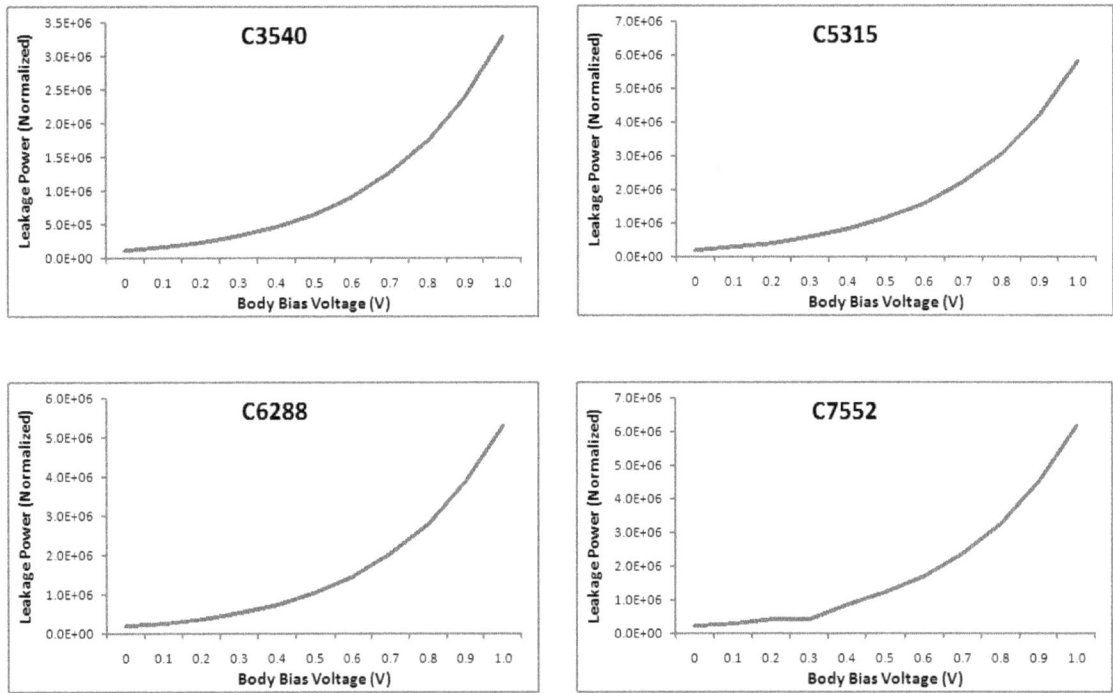

Figure 7: Leakage power increase due to forward ABB attack.

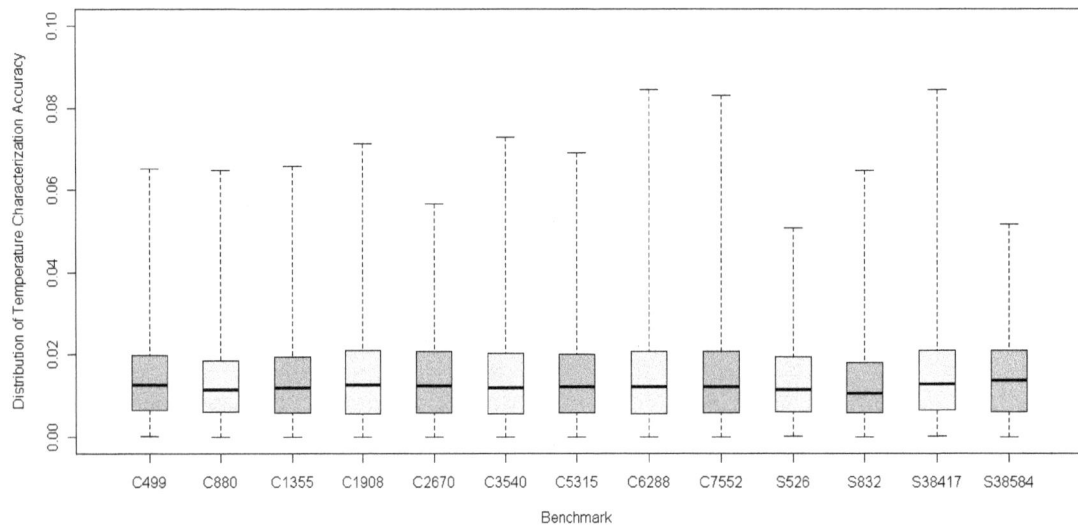

Figure 8: Accuracy of temperature characterization.

Table 1: Activation probability of hardware Trojans using low-switching and sequential triggers.

Benchmark	# Gates	# inputs	# outputs	Probability of Activation (5-trigger)	Probability of Activation (10-trigger)
C432	160	36	7	4.8E-17	2.4E-43
C499	202	41	32	8.9E-24	2.0E-39
C880	383	60	26	2.9E-16	1.1E-35
C1355	546	41	32	2.1E-37	5.9E-44
C1908	880	33	25	3.8E-26	4.8E-43
C3540	1669	50	22	2.3E-23	2.8E-46
C5315	2307	178	123	9.5E-27	7.0E-49
C7552	3512	207	108	7.6E-27	2.3E-47

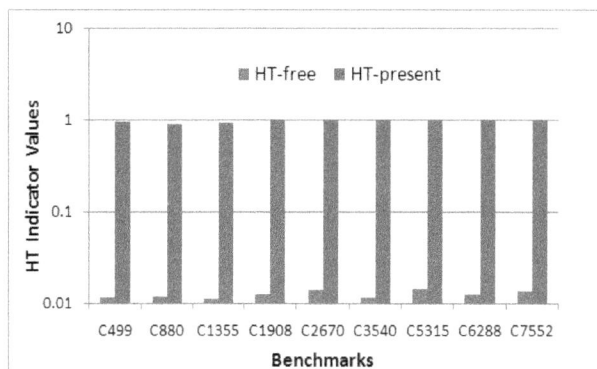

Figure 9: Energy HT detection results: HT indicator values in HT-free and HT-present cases.

International Workshop on Hardware-Oriented Security and Trust (HOST), pages 40–47, 2008.

[3] F. Brglez, D. Bryan, and K. Kozminski. Combinational profiles of sequential benchmark circuits. In *IEEE International Symposium on Circuits and Systems (ISCAS)*, pages 1929–1934, 1989.

[4] F. Brglez and H. Fujiwara. A neutral netlist of 10 combinational benchmark circuits and a target translator in FORTRAN. In *IEEE International Symposium on Circuits and Systems (ISCAS)*, pages 677–692, 1985.

[5] S. Capkun, L. Buttyan, and J. Hubaux. Self-organized public-key management for mobile ad hoc networks. *IEEE Transactions on Mobile Computing*, 2(1):52–64, 2003.

[6] T. Chen and S. Naffziger. Comparison of adaptive body bias (ABB) and adaptive supply voltage (ASV) for improving delay and leakage under the presence of process variation. *IEEE Transactions on Very Large Scale Integration (VLSI) Systems,*, 11(5):888–899, 2003.

[7] Y. Cheng, C. Tsai, C. Teng, and S. Kang. Temperature-driven cell placement. In *Electrothermal Analysis of VLSI Systems*, pages 157–179, 2000.

[8] B. Cline, K. Chopra, D. Blaauw, and Y. Cao. Analysis and modeling of CD variation for statistical static timing. In *IEEE/ACM International Conference on Computer-Aided Design (ICCAD)*, pages 60–66, 2006.

[9] L. Eschenauer and V. Gligor. A key-management scheme for distributed sensor networks. In *ACM Conference on Computer and Communications Security (CCS)*, pages 41–47, 2002.

[10] J. Gregg and T. Chen. Post silicon power/performance optimization in the presence of process variations using individual well-adaptive body biasing. *IEEE Transactions on Very Large Scale Integration (VLSI) Systems*, 15(3):366–376, 2007.

[11] I. Hong and M. Potkonjak. Behavioral synthesis techniques for intellectual property protection. In *Design Automation Conference (DAC)*, pages 849–854, 1999.

[12] Y. Jin and Y. Makris. Hardware Trojan detection using path delay fingerprint. In *IEEE International Workshop on Hardware-Oriented Security and Trust (HOST)*, pages 51–57, 2008.

[13] R. Karri, J. Rajendran, K. Rosenfeld, and M. Tehranipoor. Trustworthy hardware: Identifying and classifying hardware Trojans. *IEEE Computer Magazine*, 43(10):39–46, 2010.

[14] T. Kemper, Y. Zhang, Z. Bian, and A. Shakouri. Ultrafast temperature profile calculation in IC chips. In *International Workshop on Thermal investigations of ICs*, pages 1–5, 2006.

[15] F. Koushanfar and A. Mirhoseini. A unified framework for multimodal submodular integrated circuits Trojan detection. *IEEE Transactions on Information Forensics and Security*, 6(1):162–174, 2011.

[16] F. Koushanfar and M. Potkonjak. CAD-based security, cryptography, and digital rights management. In *Design Automation Conference (DAC)*, pages 268–269, 2007.

[17] J. Lach, W. Mangione-Smith, and M. Potkonjak.

FPGA fingerprinting techniques for protecting intellectual property. In *Custom Integrated Circuits Conference*, pages 299–302, 1998.

[18] L. Lazos and R. Poovendran. SeRLoc: secure range-independent localization for wireless sensor networks. In *ACM workshop on Wireless security (WiSe)*, pages 21–30, 2004.

[19] J. Li and J. Lach. At-speed delay characterization for IC authentication and Trojan horse detection. In *IEEE International Workshop on Hardware-Oriented Security and Trust (HOST)*, pages 8–14, 2008.

[20] Y. Li, D. Parikh, Y. Zhang, K. Sankaranarayanan, M. Stan, and K. Skadron. State-preserving vs. non-state-preserving leakage control in caches. In *Design Automation and Test in Europe (DATE)*, pages 22–29, 2004.

[21] W. Liao, L. He, and K. Lepak. Temperature and supply voltage aware performance and power modeling at microarchitecture level. *IEEE Transactions on Computer-Aided Design of Integrated Circuits and Systems*, 24(7):1042–1053, 2005.

[22] D. Markovic, C. Wang, L. Alarcon, T. Liu, and J. Rabaey. Ultralow-power design in near-threshold region. *Proceedings of the IEEE*, 98(2):237–252, 2010.

[23] G. Nabaa, N. Azizi, and F. Najm. An adaptive FPGA architecture with process variation compensation and reduced leakage. In *Design Automation Conference (DAC)*, pages 624–629, 2006.

[24] A. Sadeghi, I. Visconti, and C. Wachsmann. Anonymizer-enabled security and privacy for RFID. In *International Conference on Cryptology and Network Security (CANS)*, pages 134–153, 2009.

[25] M. Tehranipoor and F. Koushanfar. A survey of hardware Trojan taxonomy and detection. *IEEE Design Test of Computers*, 27(1):10–25, 2010.

[26] M. Tehranipoor, H. Salmani, X. Zhang, X. Wang, R. Karri, J. Rajendran, and K. Rosenfeld. Trustworthy hardware: Trojan detection and design-for-trust challenges. *IEEE Computer Magazine*, 44(7):66–74, 2011.

[27] A. Vahdatpour, S. Meguerdichian, and M. Potkonjak. A gate level sensor network for integrated circuits temperature monitoring. In *IEEE Sensors*, pages 652–655, 2010.

[28] S. Wei, K. Li, F. Koushanfar, and M. Potkonjak. Provably complete hardware trojan detection using test point insertion. In *IEEE/ACM International Conference on Computer-Aided Design (ICCAD)*, pages 569–576, 2012.

[29] S. Wei, S. Meguerdichian, and M. Potkonjak. Gate-level characterization: Foundations and hardware security applications. In *Design Automation Conference (DAC)*, pages 222–227, 2010.

[30] S. Wei, S. Meguerdichian, and M. Potkonjak. Malicious circuitry detection using thermal conditioning. *IEEE Transactions on Information Forensics and Security*, 6(3):1136–1145, 2011.

[31] S. Wei and M. Potkonjak. Scalable segmentation-based malicious circuitry detection and diagnosis. In *IEEE/ACM International Conference on Computer-Aided Design (ICCAD)*, pages 483–486, 2010.

[32] S. Wei and M. Potkonjak. Scalable consistency-based hardware Trojan detection and diagnosis. In *International Conference on Network and System Security (NSS)*, pages 176–183, 2011.

[33] S. Wei and M. Potkonjak. Hardware trojan horse benchmark via optimal creation and placement of malicious circuitry. In *Design Automation Conference (DAC)*, pages 90–95, 2012.

[34] S. Wei and M. Potkonjak. Scalable hardware Trojan diagnosis. *IEEE Transactions on Very Large Scale Integration (VLSI) Systems*, 20(6):1049–1057, 2012.

[35] S. Wei and M. Potkonjak. Wireless security techniques for coordinated manufacturing and on-line hardware trojan detection. In *ACM conference on Security and Privacy in Wireless and Mobile Networks (WiSec)*, pages 161–172, 2012.

[36] H. Xu, R. Vemuri, and W. Jone. Temporal and spatial idleness exploitation for optimal-grained leakage control. In *IEEE/ACM International Conference on Computer-Aided Design (ICCAD)*, pages 468–473, 2009.

[37] L. Yuan and G. Qu. A combined gate replacement and input vector control approach for leakage current reduction. *IEEE Transactions on Very Large Scale Integration (VLSI) Systems*, 14(2):173–182, 2006.

[38] L. Zhang, N. Howard, V. Gumaste, A. Poddar, and L. Nguyen. Thermal characterization of stacked-die packages. In *Semiconductor Thermal Measurement and Management Symposium*, pages 55–63, 2004.

[39] D. Ziener and J. Teich. Power signature watermarking of IP cores for FPGAs. *Journal of Signal Processing Systems*, 51(1):123–136, 2008.

Subtle Kinks in Distance-Bounding:
An Analysis of Prominent Protocols

Marc Fischlin
TU Darmstadt
Mornewegstrasse 30
64293 Darmstadt, Germany
marc.fischlin@gmail.com

Cristina Onete
CASED and TU Darmstadt
Mornewegstrasse 30
64293 Darmstadt, Germany
cristina.onete@gmail.com

Abstract. Distance-bounding protocols prevent man-in-the-middle attacks by measuring response times. The four attacks such protocols typically address, recently formalized in [10], are: (1) mafia fraud, where the adversary must impersonate to a verifier in the presence of an honest prover; (2) terrorist fraud, where the adversary gets some offline prover support to impersonate; (3) distance fraud, where provers claim to be closer to verifiers than they really are; and (4) impersonations, where adversaries impersonate provers during lazy phases. Dürholz et al. [10] also formally analyzed the security of (an enhancement of) the Kim-Avoine protocol [14].

In this paper we quantify the security of the following well-known distance-bounding protocols: Hancke and Kuhn [13], Reid et al. [16], the Swiss-Knife protocol [15], and the very recent proposal of Yang et al. [17]. Concretely, our main results show that (1) the usual terrorist-fraud countermeasure of relating responses to a long-term secret key may enable so-called *key-learning mafia fraud* attacks, where the adversary flips a single time-critical response to learn a key bit-by-bit; (2) though relating responses may allow mafia fraud, it sometimes enforces distance-fraud resistance by thwarting the attack of Boureanu et al. [5]; (3) none of the three allegedly terrorist-fraud resistant protocols, i.e. [15, 16, 17], *is* in fact terrorist fraud resistant; for the former two schemes this is a matter of syntax, attacks exploiting the strong formalization of [10]; the attack against the latter protocol of [17], however, is almost trivial; (4) unless key-update is done *regardless* of protocol completion, the protocol of Yang et al. is vulnerable to Denial-of-Service attacks. In light of our results, we also review definitions of terrorist fraud, arguing that, while the strong model in [10] may be at the moment more appropriate than mere intuition, it could be too strong to capture terrorist attacks.

Categories and Subject Descriptors

K.6.5 [**Management of Computing and Information Systems**]: Security and Protection—*Authentication*

Keywords

distance-bounding, provable security, cryptographic protocol

1. INTRODUCTION

Designed in 1993 [6], distance-bounding protocols address man-in-the-middle (MITM) relay attacks in authentication, also called mafia fraud by Desmedt [9]. Essentially, distance bounding enhances authentication such that a verifier accepts if the prover authenticates *and* it can prove it is within a pre-set distance of the verifier (here associated with a clock). Classical distance-bounding protocols consist of *phases*, either *lazy* (not using a clock), or *time-critical* (where the verifier measures the time elapsed during a challenge-response phase). As relaying causes processing delays, pure MITM relay is detected by the clock. When the clock is used, we say it is *online*; by contrast, when it is not used, we call it *offline*; thus, when we say offline communication, it does not mean it takes place when the prover is powered off.

Most existing distance-bounding protocols come with their own security models, in which they are proved secure; however, an easier comparison and understanding of security requires a unified framework. Only two such frameworks exist: an earlier model by Avoine et al. [2] and a later one by Dürholz et al. [10]. Both models formalized four attacks against distance-bounding protocols; the latter paper focuses on RFID distance bounding where provers are RFID *tags*, and verifiers are RFID *readers*. The two frameworks differ slightly in their definitions; in this paper, we use the notions of the more recent framework of [10]:

MAFIA FRAUD. The adversary impersonates to the reader while communicating with the genuine tag. Here the clock prevents pure relaying.

TERRORIST FRAUD. The tag helps the adversary authenticate by disclosing useful information in offline phases. However, the tag should not reveal trivial information like the secret key.

DISTANCE FRAUD. The (malicious) tag claims to be closer to the reader than it actually is.

OFFLINE IMPERSONATION RESISTANCE. The adversary impersonates the honest tag during lazy phases only.

195

Note that the security notions in distance bounding, though first formalized by [2], were first coined by Desmedt [9]. Also note that offline impersonation security was introduced by Dürholz et al. particularly for resource-constrained RFID tags, which cannot support many time-critical phases. If a protocol is *not* offline impersonation secure, it only has an impersonation security roughly equivalent to its mafia-fraud resistance. In other words, if such a protocol is deployed in resource-constrained devices, it is highly susceptible to regular impersonation in authentication. Note that the model of Avoine et al. [2] considers impersonation resistance to be the equivalent of offline impersonation resistance combined with time-critical impersonation resistance (i.e. mafia-fraud resistance).

Recently, Cremers et al. [8] also introduced a new attack, called *distance hijacking*. The framework in [10] models security for a single reader and a single tag; by contrast, distance hijacking requires two provers, and has not yet been formalized in the model of [10]. We leave the investigation of this attack for further work.

1.1 RFID Distance-Bounding Protocols

One reason to choose the framework of Dürholz et al. [10] is that it accounts for generic settings including resource-constrained RFID distance bounding; this framework was already used to assess the security of an enhancement of the Kim-Avoine distance-bounding protocol [14]. On the one hand, formal analysis can detect protocol flaws, such as those exploited by e.g. [1]. On the other hand, analyzing security properties in a common, formal framework enables easier comparison between protocols. In this paper we continue the assessment work of [10] by quantifying the security of several known distance-bounding protocols, with quite surprising results. We in fact show a few subtle kinks in various protocols, which compromise security.

Most distance-bounding protocols consist of a number of slow (lazy) phases, followed by a number N_c of time-critical phases, where a clock is used by the reader to time bit-exchanges between itself and the tag. Whereas lazy phases can be computationally expensive, time-critical phases add to the communication complexity. In resource-constrained scenarios it is unclear how many time-critical phases the prover can even support.

Since time-critical phases are typically just bit exchanges, the best known mafia-fraud resistance is about $\frac{1}{2}$ per time-critical phase, totalling $\left(\frac{1}{2}\right)^{N_c}$ over N_c phases. This bound was first achieved by Brands and Chaum [6]. However, [6] use computationally-expensive signature schemes, unsuitable for RFID tags. The same resistance can be achieved by the Swiss-Knife protocol [15] at the cost of a second lazy phase and a minor modification see Section 3.4. The very efficient, well-known Hancke-Kuhn protocol [13] only has a $\left(\frac{3}{4}\right)^{N_c}$ resistance, being vulnerable to the so-called *Go-Early strategy*, where a mafia adversary queries the tag in advance and learns about half of the correct time-critical responses and can guess the others (see also Section 3.1). The same bound holds for the very recent protocol of Yang et al. [17], which also thwarts so-called *key-learning mafia-fraud* attacks, where the adversary tries to learn the secret key from related time-critical responses. By contrast, the protocol of Reid et al. [16] *is* susceptible to key-learning attacks, and so is *not* mafia-fraud resistant. Interestingly, the more efficient symmetric distance-bounding protocols do not, in view of

the recent attack of Boureanu et al. [5], achieve *distance fraud resistance*. We describe this attack in detail in the following sections.

Some existing protocols also address terrorist fraud. An early scheme due to Reid et al. [16] uses related time-critical responses to attain terrorist-fraud resistance: one output by a PRF, the other, an encryption of the secret key under the first response. If an adversary knows both responses, he can recover the secret key. Also addressing terrorist fraud is the Swiss-knife protocol of Kim et al. [15] and the very recent proposal due to Yang et al. [17]. We show that, whereas the terrorist-fraud proof for the latter protocol is simply *flawed*, the former two schemes are also susceptible to a terrorist-fraud attack where the adversary learns *some* information about the long-term secret, but cannot use this information directly at full efficiency. This last attack is more a question of *syntax*, exploiting the formalism used by the (simulation-based) model of [10]. Both protocols intuitively achieve a measure of terrorist-fraud resistance; however, they are not *provably* secure. This might indicate the model of [10] is too strong. We discuss the relation between model strength and intuitive security at length in Appendix A. In our syntactic attack the prover forwards one response, the PRF output, and lets the adversary guess the responses for the other rounds. A successful adversary learns some bits of the secret key, but when no longer able to query the prover, his success probability drops.

Notably, though all these schemes use pseudorandom functions, the PRF attack of Boureanu et al. [5] does not always apply. The protocol in [16] computes only one time-critical response by means of the PRF; the other is an encryption of the first response under a long-term secret key. Reid et al. use a generic symmetric encryption scheme; this does *not* automatically grant distance-fraud resistance, but particular instantiations of such schemes e.g. one-time-pad encryption *do* grant distance-fraud resistance. In our proof, we require the further assumption that the secret key is chosen honestly by a trusted party uniformly from a distribution computationally indistinguishable from the uniform random distribution (of appropriate length). The Swiss-Knife protocol [15] is also distance-fraud resistant under the same assumption (see below). In both cases, the idea is that the time-critical responses computed by the prover and verifier are *not* both output by a PRF; the second response is the XOR of the PRF output and a (nearly) random value. Thus the Hamming distance between the responses is most likely high, regardless of any tampering with the PRF output.

1.2 Our Contributions

In this paper we quantify the security of the following schemes: Hancke-Kuhn [13], Reid et al. [16], Kim et al. [15], and Yang et al. [17]. For each protocol, we (1) show concrete bounds for any attained properties, thus (2) disproving claims of distance-fraud resistance for the schemes of Hancke and Kuhn and Yang et al. (the attack is a direct consequence of the generic attack of Boureanu et al. [5]), (3) disproving claims of terrorist-fraud resistance for all the three allegedly terrorist-fraud resistant schemes (though the attack against [16] and [15] is contrived, exploiting the simulation-based model of [10]), (4) disproving claims of mafia-fraud resistance for the protocol due to Reid et al. [16] (we show a key-learning mafia-fraud attack against this protocol), (5) noting that a small modification must be made to the Swiss-

196

Knife and Yang et al. protocols in order to prove any mafia and impersonation resistance properties.

Let us take a closer look at our result (3). Aside from the flaw in the proof of Yang et al. [17], we show that the intuitively terrorist-fraud resistant of [15] and [16] are, in fact, *insecure* in the model of [10]. Our attack (against both schemes) exploits the syntax of the framework in [10], using the fact that, by yielding relatively little information about a long-term secret key, the malicious tag can (1) help the adversary authenticate; and also (2) diminish its probability to later authenticate on its own. The question of whether these attacks are a weakness of the protocols or of the model of Dürholz et al. is discussed in the Appendix.

In fact we compare the intuition behind terrorist fraud with the strong model of [10]. Our analysis indicates that the problem lies somewhere between model strength and protocol security. Our attack against the scheme of Reid et al. assumes that the malicious tag agrees to yield *partial* information about the secret key so as to aid the adversary, but not if the adversary can directly use the data for future authentication. However, the attack is in itself somewhat impractical, or flawed, since (a) when the tag helps, the adversary's success probability is not overwhelming (just significantly larger than guessing probability); (b) when the tag stops helping, the adversary has a *lower* probability to succeed than before, but not an insignificant one. Thus, it seems the model of Dürholz et al. is too strong. However, at the moment, this is the most thorough security framework for distance bounding in the literature. By contrast, Avoine et al. [2], resp. [3] require information-theoretical hiding properties for the tag's secret key. This, however, seems too strong a restriction: intuitively, an adversary may learn some information about the key without gaining any advantage in future attacks. We leave two open questions regarding this issue: (1) Should we aim to design protocols which are provably secure in the framework of Dürholz et al.?; and (2) Is it possible to capture terrorist-fraud resistance better in a different model?

The results of our protocol analysis are summarized in Table 1, where we show (only the loose) security bounds for the attacks described above, ignoring the small terms. A precise quantification can be found in Section 3.

	Mafia	Terror	Distance	Impersonation
[13]	$\left(\frac{3}{4}\right)^{N_c}$	×	×	×
[16] [1]	×	×	$\left(\frac{3}{4}\right)^{N_c-T}$	×
[15] [2]	$\left(\frac{1}{2}\right)^{N_c-T}$	×	$\left(\frac{3}{4}\right)^{N_c-T}$	$\left(\frac{1}{2}\right)^{|V|}$
[17] [2]	$\left(\frac{3}{4}\right)^{N_c}$	×	×	×

Figure 1: Distance Bounding at a glance. [1] **This protocol is only distance-fraud resistant for some instantiations.** [2] **The results concern a minor modification of these protocols. Here, N_c is the number of time-critical rounds, T is a fault-tolerance level, and $|V|$ is the bit-length of an authentication string V.**

2. PRELIMINARIES

2.1 Model Overview

The security model in [10] considers single-verifier-single-prover distance bounding, particularly for RFID settings;

here, a single prover (tag) $\mathcal{T}$ and a single verifier (reader) $\mathcal{R}$ share a secret key sk generated by an algorithm Kg. We explicitly require that Kg is run at random from the space of all keys, which we denote K. We thus require that the tag does not choose its own key (this ensures that the protocols are distance-fraud resistant). The reader uses a clock to measure the time elapsed between sending a challenge and receiving the response. Dürholz et al. consider round-based distance-bounding protocols, where rounds are *time-critical* if the clock is used, and *lazy* otherwise. We briefly review the main attacks here and refer to [10] for more details.

Mafia Fraud. Mafia-fraud resistance is a man-in-the-middle (MITM) attack, where pure relay is prevented by the reader's clock. Informally, the MITM consists of two adversaries: a *leech*, which impersonates the reader to an honest tag, and a *ghost*, which impersonates the tag to an honest reader. The ghost must authenticate to the reader; however, the clock detects processing delays in the MITM resulting from pure relay. Both the reader and the tag are unaware of the MITM attack.

Dürholz et al. [10] formalize mafia fraud by introducing an abstract clock denoted clock, which keeps track of the messages sent in several protocol executions, called *sessions*. The sessions can be: reader-tag (the adversary is a passive eavesdropper), reader-adversary (the adversary impersonates the tag to the reader), and adversary-tag (the adversary impersonates the reader to the tag). Relaying is considered round-wise (actually, phase-wise, since [10] considers time-measurements over possibly many rounds). A phase is *tainted* if the adversary *purely* relays communication between a reader-adversary and an adversary-tag session. An adversary *purely* relays messages if, having received a message in a session sid, it relays the exact, same message in a session sid'; then, upon receiving a response, he relays it back again, for all the rounds in the tainted phase. If the adversary changes any messages sent between sessions, this is not pure relay. Also, if the adversary queries one session with some message m *before* receiving the same m in the other session, this is not relaying. See [10] for more details.

Terrorist Fraud. In terrorist attacks the adversary is helped to authenticate by a dishonest tag, whose information shouldn't allow the adversary to win on its own. In [10], the dishonest tag is only queried in lazy phases[1]. Querying the tag during a time-critical phase of a reader-adversary sessions taints this phase.

A crucial part of the terrorist-fraud definition is formalizing how much a tag can help the adversary. In [10], the attack is defined in terms of a simulator: a scheme is terrorist-fraud resistant if a adversary aided by the tag is as likely to win as a simulator using the adversary's internal information. To be fair, the simulator has as many impersonation attempts as the adversary, but it only begins its attack once the adversary succeeds to authenticate. For instance, session-specific data is useless to the simulator; however, if the tag forwards the secret key, then the simulator can recover it and also succeed. Formally, for every successful terrorist-fraud adversary $\mathcal{A}$, there must exist a simulator $\mathcal{S}$ whose success probability is at least as large as the adversary's. As we discuss in the Appendix, this definition is very strong; in fact in light of our results, it seems *too* strong. One

---

[1]This approach is in agreement with the previous, more informal model of Avoine et al. [2] and with the intuition behind terrorist fraud.

advantage of it is that it allows formal proofs; on the other hand, other existing definitions, e.g. [3], seem to require that *no* information at all about the secret key is leaked, which restricts the attack very much, i.e. the model appears to be too weak.

Distance Fraud. Distance fraud adversaries are the malicious tags themselves, who are outside the reader's proximity, but aim to prove the contrary. Since the reader's clock measures time accurately, the adversary must anticipate the reader's challenges and respond in advance. In [10], the adversary must commit in advance to each time-critical response; else, the phase is tainted. For distance fraud, reader-tag and adversary-tag sessions are no longer relevant as the adversary *is* the tag.

(Offline) Impersonation Resistance. Finally, offline impersonation resistance refers to lazy-phase tag authentication. The idea, introduced by [4], is that even without the time-critical phases the prover is still authenticated (though replays are still allowed). We use the formal definition in [10], noting that this notion refers *only* to lazy phases, whereas Avoine et al. [2] define impersonation security for the entire protocol. We call the notion of [10] *offline* impersonation resistance so as to emphasize the difference between the two notions. Note that in mafia-fraud resistant protocols the Avoine-impersonation security equals the Dürholz offline impersonation security combined with the mafia-fraud resistance during time-critical phases.

Notation. Following the notation [10], we denote the advantage of an adversary $\mathcal{A}$ against property pty $\in$ $\{\mathrm{mafia, dist, terror, imp}\}$ as $\mathbf{Adv}_{\mathsf{ID}}^{\mathrm{pty}}(\mathcal{A})$. To relate the distance bounding security levels to the security of the underlying cryptographic primitives, we often transform a successful distance-bounding adversary $\mathcal{A}$ into one or more adversaries $\mathcal{A}'$ against the primitive(s). We use standard notions for these primitives, letting $\mathbf{Adv}_{\mathcal{S}}^{\mathrm{Exp}}(\mathcal{A}')$ denote the maximal probability that an adversary $\mathcal{A}'$ breaks a cryptographic scheme $\mathcal{S}$ in some experiment Exp. For example, $\mathbf{Adv}_{Sign}^{\mathrm{Unf}}(\mathcal{A}')$ denotes the probability of forging signatures, $\mathbf{Adv}_{\mathsf{PRF}}^{\mathrm{d}}(\mathcal{A}')$ denotes the advantage of distinguishing a pseudorandom function PRF from random, and $\mathbf{Adv}_{\mathcal{E}}^{\mathrm{IND\text{-}CPA}}(\mathcal{A}')$ is the advantage of breaking the IND-CPA security in encryption. The parameters of $\mathcal{A}'$ in these experiments will be specified in terms of the parameters of $\mathcal{A}$.

Parameters. We note that most existing protocols do not provide for flaws in communication; by contrast, apart from the upper-bound $t_{\max}$ of the roundtrip transmission time and the number of time-critical rounds N_c, [10] also allows for max. $T_{\max}$ phases with delayed responses and max. $E_{\max}$ phases with wrong responses. However, note that fault tolerance is essential in resource-constrained environments, e.g. RFID.

We quantify adversaries in terms of their runtime t (including honest party processing time, as the adversary waits for responses) and the numbers $q_{\mathcal{R}}$, $q_{\mathcal{T}}$, resp. q_{obs} of reader-adversary, adversary-tag, and resp. reader-tag sessions.

3. PROTOCOL ASSESSMENT

In this section, we analyze the security of the following distance-bounding protocols: Hancke and Kuhn [13], Reid et al. [16], the Swiss-Knife protocol [15], and the protocol of Yang et al. [17]. Note that many protocol are susceptible to the distance-fraud attack of Boureanu et al. [5]; our analysis

shows that provable terrorist-fraud resistance is also harder to attain than intuition indicates.

Most distance-bounding protocols do not feature fault tolerance, assuming thus that the prover's behavior, transmissions, and transmission times are all reliable and constant. It is thus assumed that: the verifier's challenges always arrive at the prover (in constant time), the prover's responses also always arrive at the verifier (in constant time), and the prover always has the same processing delay. This is not always the case, especially for resource-constrained devices like RFID, where transmissions are not always reliable, and transmission times and processing delays may vary. This is why Dürholz et al. introduce fault tolerance parameters as outlined in Section 2.

3.1 Hancke and Kuhn

The protocol due to Hancke and Kuhn addresses mafia and distance fraud, and uses a PRF implemented as HMAC. However, the attack of Boureanu et al. [5] breaks its distance-fraud resistance; its mafia-fraud resistance is suboptimal. The protocol consists of: (i) a lazy phase, where the parties exchange nonces and pre-compute an HMAC value, divided in a left and a right half, and (ii) N_c time-critical phases where the reader sends a random bit and the tag responds with a bit either from the left or the right half of response. This is depicted in Figure 2.

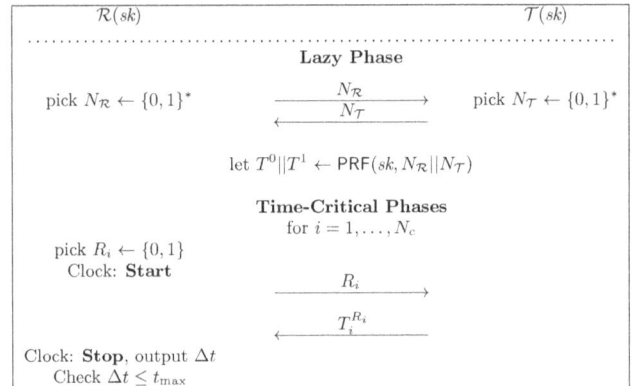

Figure 2: The Hancke and Kuhn protocol

The PRF output size is $2N_c$ bits, divided equally into a left and right half. Note that the value $\mathsf{PRF}(sk, N_{\mathcal{R}}||N_{\mathcal{T}})$ cannot be computed by a mafia-fraud adversary without knowledge of sk; however, the lack of reader authentication decreases its mafia-fraud resistance. There is no offline impersonation resistance, as we also state more formally below.

THEOREM 3.1 (HANCKE-KUHN PROPERTIES). *Let* ID *be the distance-bounding authentication scheme in Figure 2 with parameters* $(t_{\max}, N_c)$*. This scheme has the following properties:*

- *It is not offline impersonation resistant, distance-fraud resistant, nor resistant to terrorist fraud.*

- *For any* $(t, q_{\mathcal{R}}, q_{\mathcal{T}}, q_{\mathrm{obs}})$*-mafia-fraud adversary* $\mathcal{A}$ *against the scheme there exists a* (t', q')*-distinguisher* $\mathcal{A}'$ *against* PRF *(where* $\mathcal{A}'$ *runs in time* $t' = t + O(n)$ *and makes at most* $q' = q_{\mathcal{R}} + q_{\mathcal{T}} + q_{\mathrm{obs}}$ *queries) such* *that*

$$\boldsymbol{Adv}_{\mathsf{ID}}^{mafia}(\mathcal{A}) \leq q_{\mathcal{R}} \cdot \left(\tfrac{3}{4}\right)^{N_c} + \binom{q_{\mathcal{R}} + q_{\mathrm{OBS}}}{2} \cdot 2^{-|N_{\mathcal{R}}|}$$
$$+ \boldsymbol{Adv}_{\mathsf{PRF}}^{d}(\mathcal{A}') + \binom{q_{\mathcal{T}}}{2} \cdot 2^{-|N_{\mathcal{T}}|}.$$

PROOF. The first statement follows easily: an offline impersonation adversary sending a random nonce during the lazy phase wins with probability 1 as there is no lazy-phase authentication. In terrorist fraud, a malicious tag forwards the adversary during the lazy phase of session sid the string $T^0||T^1$, which allow the adversary to win with probability 1, responding to time-critical challenges like a legitimate tag. However, when a simulator attempts to authenticate, its session is fresh, and so is the reader's nonce, thus either the PRF output is different, or we can find a PRF collision. Finally, the attack of Boureanu et al. [5] makes the protocol not distance-fraud resistant: indeed, the distance-fraud adversary knows the secret key and can choose a weak nonce $N_{\mathcal{T}}$, such that $T^0 = T^1$ with high probability (note that the pseudorandomness of PRF only requires that it is indistinguishable from random on the average, not for every input). Once the adversary forwards this $N_{\mathcal{T}}$ to the reader, it can just commit to the value $T_i^0 = T_i^1$ for every round and win with high probability.

The proof of the last statement consists of the following high-level steps: (1) replace the PRF output of honest parties by independent random strings $T^0||T^1$ for each new nonce pair $(N_{\mathcal{R}}, N_{\mathcal{T}})$; (2) show that nonce pairs are (almost) unique, except for possibly one adversary-tag session sid* sharing the same nonce pair as a reader-adversary session sid (these sessions are created by MITM relays); and (3) bound the probability that the adversary passes the time-critical phases for at most one adversary-tag interaction.

We first claim that replacing the PRF-values by random (but consistent) values can at most decrease the adversary's success probability by the distinguishing advantage for PRF. Indeed, we can construct adversary $\mathcal{A}'$ against PRF via black-box simulation of $\mathcal{A}$, each time applying the random or pseudorandom oracle to nonce pairs on behalf of honest parties. Finally, $\mathcal{A}'$ checks if $\mathcal{A}$ succeeds in some reader-adversary session and outputs 1 if this happens. The distinguishing advantage of $\mathcal{A}'$ corresponds to the decrease of $\mathcal{A}$'s success probability when switching to random $T^0||T^1$.

For (2) we consider all the nonce-pairs in a mafia-fraud attack. Assume that there exist two sessions (between adversary and tag or reader, or between both honest parties) with the same pair $(N_{\mathcal{R}}, N_{\mathcal{T}})$. We claim that this can only be a reader-adversary session and an adversary-tag session (created by relaying nonces), except with probability

$$\binom{q_{\mathcal{R}} + q_{\mathrm{OBS}}}{2} \cdot 2^{-|N_{\mathcal{R}}|} + \binom{q_{\mathcal{T}}}{2} \cdot 2^{-|N_{\mathcal{T}}|}.$$

Indeed, fresh sessions (not created by relay) have at least one honest party, generating its nonce honestly. If three identical nonce pairs appear, then two of them are either in the at most $q_{\mathcal{R}} + q_{\mathrm{OBS}}$ executions with the reader, or in the $q_{\mathcal{T}}$ executions with the tag. Such collisions occur with the above probability. We now let $\mathcal{A}$ lose if a collision appears, decreasing its success probability by this negligible term, and we consider collision-free executions. Thus, except for

the matching session, all values $T^0||T^1$ in the attack are independent.

Now consider a reader-adversary session sid in which $\mathcal{A}$ successfully authenticates to $\mathcal{R}$. By assumption the same nonce pair appears in at most one other adversary-tag session which we denote sid* (if no such session exists, we have the case below, where $\mathcal{A}$ does not make use of any matching session). We claim that session sid* taints at least one phase of sid with high probability; as for this protocol $T_{\max} = 0$ (i.e. at most one phase *may* be tainted), sid is invalidated.

Suppose, to the contrary, that the matching session sid* taints no time-critical phase in sid. Consider an untainted time-critical phase of sid where $\mathcal{R}$ sends $R_i = b$ and expects T_i^b, after the adversary has successfully passed the first $i-1$ time-critical phases. Now $\mathcal{A}$ can do one of the following in phase i:

THE GO-EARLY STRATEGY. $\mathcal{A}$ sends some bit R_i^* to $\mathcal{T}$ in session sid* before having received R_i; i.e., $\mathsf{clock}(\mathsf{sid}^*, i+2) > \mathsf{clock}(\mathsf{sid}, i+2)$ in the notation of [10]. Since R_i is random and independent of all other data, $R_i^* \neq R_i$ w.p. $\tfrac{1}{2}$; then, $\mathcal{A}$ does *not* receive T_i^* in sid* and can only guess T_i in sid. If $b = R_i = R_i^*$, however, $\mathcal{A}$ sends the correct reply T_i^b with probability 1.

THE GO-LATE STRATEGY. In session sid the adversary sends some T_i^* to $\mathcal{R}$ before receiving $(T_i^b)^*$ in session sid* (i.e., $\mathsf{clock}(\mathsf{sid}, i+3) < \mathsf{clock}(\mathsf{sid}^*, i+3)$). Now $\mathcal{A}$ succeeds only with probability $\tfrac{1}{2}$.

THE MODIFY-IT STRATEGY. The adversary chooses its strategy such that it receives R_i in sid, sends some $R_i^* = b$ in sid*, receives $(T_i^b)^*$ in sid*, and forwards some T_i^* in sid. This scheduling is that of pure relay, but $R_i \neq R_i^*$ or $T_i^* \neq (T_i^b)^*$. If $(b = R_i^*) \neq R_i$, then $(T_i^b)^*$ is never sent by $\mathcal{T}$ in sid*; thus $\mathcal{A}$ only guesses T_i^* (with probability $\tfrac{1}{2}$); if $b = R_i^* = R_i$ then $T_i^* \neq (T_i^b)^*$ makes the reader reject.

THE TAINT-IT STRATEGY. The adversary taints this phase of sid through sid*. Then $\mathcal{A}$ loses sid.

The Taint-it strategy may be ignored, as it disables sid. The Go-Late and Modify-it strategies both succeed with probability at most $\tfrac{1}{2}$. The Go-Early strategy succeeds w.p. $\tfrac{3}{4}$. This accounts for an adversary making use of a session sid*. If $\mathcal{A}$ makes no use of such a session, then the adversary can do no better than guessing for each phase, succeeding w.p. $\tfrac{1}{2}$. We take into account the best strategy (Go-Early) for N_c rounds, and account for the $q_{\mathcal{R}}$ attempts, obtaining the claimed bound. □

3.2 Reid et al.

The protocol of Reid et al. [16] adds a symmetric encryption scheme to the Hancke-Kuhn construction (the authors suggest a one-time-pad xor operation[2]). This protocol inherits the offline impersonation vulnerability of [13]; however, distance-fraud resistance *can* be achieved for some implementations of the encryption scheme (but not for all), as the attack of Boureanu et al. [5] cannot be extended to arbitrary schemes.

We first describe this protocol. In their paper [16] use a *key derivation function* denoted KDF, which can be viewed

---

[2]Reid et al. also show protocol versions where the symmetric encryption is done differently; our analysis here considers generic instantiations of the encryption scheme.

as a PRF. We denote it PRF as for the Hancke-Kuhn protocol. Reid et al. also use a symmetric IND-CPA encryption
scheme denoted $\mathcal{E}$. To this scheme they associate a symmetric ephemeral secret key eph and a long term secret key sk.
The notation $\mathcal{E}_{\mathsf{eph}}(sk)$ denotes the encryption under eph of
the plaintext sk. We denote the corresponding decryption
process by $\mathcal{D}$. Furthermore, Reid et al. [16] associate public
identities $\mathcal{ID}_A, \mathcal{ID}_B$ to the tag and reader. The main idea
is that both reader and tag compute a symmetric ephemeral
key eph as the output of PRF; then eph is used to encrypt
the long-term secret key sk with $\mathcal{E}$. For each time-critical
round, $\mathcal{R}$ challenges $\mathcal{T}$ with a random bit, and the tag responds with a bit either from the encrypted value or from
eph. Terrorist fraud resistance should result from the fact
that $\mathcal{A}$ requires either the value sk (which the simulator can
later use to authenticate) or both eph and the encryption
of sk (which the simulator can use to find $\mathcal{D}_{\mathsf{eph}}(sk)$, thus
learning sk). The scheme is depicted in Figure 3.

Before stating the full security properties of this scheme,
note that a terrorist attack is intuitively successful if an adversary authenticates aided by a malicious tag, but it cannot authenticate later *without* this aid. However, this informal definition is deceptive: many protocols claiming to
be terrorist-fraud resistant are actually resistant to a weak
attack, if the prover does not forward *any* sensitive information to the adversary; in other words, $\mathcal{A}$ learns nothing
about the secret key [3]. By tying any knowledge (or accurate guessing) of both responses to the secret key, protocols
can attain this form of terrorist-fraud resistance. An attack
where the prover forwards even a single bit of one response
is ruled out, as it gives $\mathcal{A}$ a better success probability in later
sessions; however, in practice this definition restricts an adversary very much, as some information about the secret key
will not significantly aid $\mathcal{A}$ in future authentication.

However, the definition of [10] allows partial attacks as
valid, concretely requiring that the prover's help does not
give $\mathcal{A}$ an *equal* advantage for future sessions, i.e. it only
excludes attacks where the prover gives the adversary a fragment of the *secret key itself*, exclusively. Whereas the previous definition demands that *no* information about the secret key leaks to the adversary, the model of Dürholz et
al. restricts the adversary only insofar $\mathcal{A}$ has a (possibly
marginally) lower probability to win after the prover stops
helping. We discuss the merits of both definitions in Appendix A, and in the following we show that the protocol in
Figure 3 is *not* terrorist-fraud resistant in the sense of [10].

THEOREM 3.2 (REID ET AL. PROPERTIES). *Let* ID *be
the distance-bounding authentication scheme in Figure 3
with parameters* $(t_{\max}, N_c)$. *This scheme has the following
properties:*

- *It is neither offline impersonation resistant, nor
  terrorist-fraud resistant (assuming the pseudorandomness of* PRF*).*

- *If the symmetric encryption scheme in this protocol is
  instantiated as bitwise XOR, this scheme is* not *mafia-
  fraud resistant.*

- *If the symmetric encryption scheme used in this protocol is instantiated as bitwise XOR, and furthermore if
  the secret key* sk *is generated honestly at random from a
  distribution* $\mathcal{D}$ *computationally indistinguishable from
  the uniform distribution* U *with length* $|sk| = N_c$, *for*

any $(t, , q_{\mathcal{R}}, q_{\mathcal{T}}, q_{\mathrm{OBS}})$-*distance-fraud adversary* $\mathcal{A}$ *there
exists a distinguisher* $\mathcal{A}^*$ *between* $\mathcal{D}$ *and* U *it holds
that:*

$$\boldsymbol{Adv}_{\mathsf{ID}}^{dist}(\mathcal{A}) \leq q_{\mathcal{R}} \cdot \left(\frac{3}{4}\right)^{N_c} + \boldsymbol{Adv}_{\mathsf{Kg}}^{\mathrm{d}(\mathcal{D},U)}(\mathcal{A}^*).$$

PROOF. The impersonation resistance proof is the same
as before, and we do not reiterate it.

We first look at the distance-fraud resistance of this protocol (i.e. the second statement). Contrarily to the Hancke
and Kuhn protocol, the responses computed here by the
tag are eph and c, where only eph is output by a PRF. If
the symmetric encryption function $\mathcal{E}$ is instantiated as bitwise XOR, the Hamming distance between responses eph
and c is exactly sk at each execution. The proof goes as
follows: (1) replace sk by a uniform random value (and lose
a term equalling the distinguishing advantage between the
uniform random distribution U and the distribution $\mathcal{D}$ that
sk is chosen from; this term is $\boldsymbol{Adv}_{\mathsf{Kg}}^{\mathrm{d}(\mathcal{D},U)}(\mathcal{A}^*)$); (2) for each
time-critical phase i in the attack, it holds with probability
$\frac{1}{2}$ that c_i is different from $\mathsf{eph}_i = c_i \oplus sk$; when $c_i = \mathsf{eph}_i$
(w.p. $\frac{1}{2}$), the adversary sends c_i to win the round, and else,
when $c_i \neq \mathsf{eph}_i$, the adversary must choose one value to send
(essentially guessing the challenge), winning w.p. $\frac{1}{2}$.

However, if $\mathcal{E}$ *is* instantiated as a one-time-pad XOR operation, i.e. $c \leftarrow \mathsf{eph} \oplus sk$, we cannot prove the protocol
mafia-fraud secure. This statement is discussed to an extent
in [15]. We show an attack, also shown in [5], which we call
key-learning mafia fraud. Here the adversary tries to reconstruct the key sk in a string sk'; to this purpose it starts a
reader-adversary session sid_1 and an adversary-tag session
sid_1^*, then relays all the lazy and time-critical rounds up to
the $(N_c - 1)$-th round (this is possible since the no-tainting
restriction applies only for the session where $\mathcal{A}$ authenticates
successfully; indeed, the attacker can run this attack when
$\mathcal{T}$ is in proximity of $\mathcal{R}$). Finally, in the last round $\mathcal{A}$ receives
the challenge $b = R_{N_c} \in \{0, 1\}$ in session sid_1 and forwards
$\bar{b} = b \oplus 1$ in session sid_1^*, receiving $T^{\bar{b}}$. Now $\mathcal{A}$ forwards $T^{\bar{b}}$
to $\mathcal{R}$ in session sid and waits to see if it is authenticated by
the reader; if so, it sets $sk'_{N_c} = 0$, else $sk'_{N_c} = 1$.

There are two cases:

$b = 0$. Then, in sid_1, the reader expects c_{N_c} as a response.
The adversary has learned $\mathsf{eph}_{N_c} = c_{N_c} \oplus sk_{N_c}$ from
session sid_1^*, where it sent $R^{\bar{b}} = R^1$; if eph_{N_c} is accepted, $sk_{N_c} = 0$ and the adversary's guessed bit is cor-
rect, i.e. $sk'_{N_c} = sk_{N_c}$. Else, if $\mathcal{R}$ rejects, then $sk_{N_c} = 1$,
and $\mathcal{A}$ has again guessed correctly.

$b = 1$. In this case, the reader expects eph_{N_c}; the value forwarded by the adversary is $c_{N_c} = \mathsf{eph}_{N_c} \oplus sk_{N_c}$. The
same reasoning applies as above.

The adversary continues the attack for the other bits, thus
recovering the key. Once $\mathcal{A}$ learns the full $sk' = sk$, it opens
a "challenge" session sid with the verifier, and a parallel session sid^* with the prover. The adversary relays the lazy-
phase communication, then queries $\mathcal{T}$ in advance to learn
eph (i.e. it sends $R_i = 1$ for every round $i \in 1, \ldots, N_c$ in
sid^*). This is the Go-Early strategy, and $\mathcal{A}$ taints none of
the rounds. For every phase in sid, if $\mathcal{R}$ sends $R_i = 1$ then
$\mathcal{A}$ sends eph_i; else, if $\mathcal{R}$ sends $R_i = 0$, then it forwards
$\mathsf{eph}_i \oplus sk'_i$, in both cases responding correctly. Now $\mathcal{A}$ wins
with probability 1.

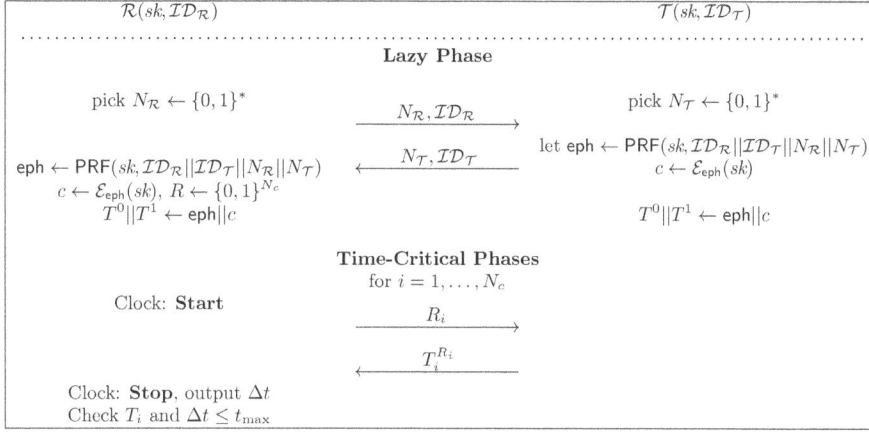

$$\mathcal{R}(sk, \mathcal{ID}_\mathcal{R}) \qquad\qquad\qquad\qquad \mathcal{T}(sk, \mathcal{ID}_\mathcal{T})$$

Figure 3: The Reid et al. protocol

However, this attack is not extendable to arbitrary symmetric encryption schemes. In order to achieve a mafia-fraud resistance equal to that of the Hancke-Kuhn protocol, the encryption scheme must not leak *any* information about the key even if it learns both time-critical responses for a given round. Finally we show a terrorist adversary $\mathcal{A}$ for which there exists no simulator such that $\mathbf{Adv}_{\mathsf{ID}}^{\text{terror}}(\mathcal{A}, \mathcal{S}, \mathcal{T}) \leq 0$. This would prove that the scheme is insecure against terrorist fraud. The idea is for the malicious tag $\mathcal{T}'$ to facilitate the adversary's attack without revealing information that can be used directly later in authentication. If $\mathcal{A}$ receives the value eph from $\mathcal{T}'$ in each impersonation attempt, it correctly answers any challenges $R_i = 1$, and can guess the other responses. Thus, its success probability is $\frac{3}{4}^{N_c}$. However, the simulator has no access to $\mathcal{T}$; it may run $\mathcal{A}$ internally, but assuming PRF is pseudorandom, there is only a negligible probability that $\mathcal{A}$ knows eph for any of the $q_\mathcal{R}$ sessions where $\mathcal{S}$ attempts to authenticate. Also, the simulator has at most negligible probability to learn sk from the value eph forwarded to $\mathcal{A}$ by $\mathcal{T}'$. Thus, the simulator wins with at most $\frac{1}{2}^{N_c} + \mathbf{Adv}_{\mathsf{PRF}}^{\mathrm{d}}(\mathcal{A}') + \mathbf{Adv}_{\mathcal{E}}^{\text{IND-CPA}}(\mathcal{A}'')$ probability for a distinguisher $\mathcal{A}'$ against the pseudorandomness of PRF and the most successful adversary $\mathcal{A}''$ against the IND-CPA of $\mathcal{E}$ (however, $\mathbf{Adv}_{\mathsf{PRF}}^{\mathrm{d}}(\mathcal{A}')$ and $\mathbf{Adv}_{\mathcal{E}}^{\text{IND-CPA}}(\mathcal{A}'')$ are negligible). Thus, $\mathcal{A}$ has an advantage over any simulator $\mathcal{S}$. □

3.3 The Swiss-Knife Protocol

The Swiss-Knife protocol [15] aims to achieve privacy as well as mafia, terrorist, and distance-fraud resistance, and employs two lazy phases: the first precedes the time-critical phase, the second follows it. In the first part, the reader and tag exchange nonces and runs a PRF on input a system constant const and a tag-chosen nonce $N_\mathcal{T}$; the output value a is XORed with the long-term secret key sk, thus obtaining response vectors T^0 and T^1. Bits of these responses are sent in the time-critical phases depending on the reader's challenges. After the time-critical rounds, the tag authenticates in the second lazy phase by running the PRF on all the received challenges, its identity, and the session's nonces. The reader may then authenticate by running the PRF on the tag's nonce.

We note that the responses T^0, T^1 have a circular dependency on sk. Thus, the value $T^0 = a$ is output by the PRF run under sk, and the second response T^1 masks a under sk. This dependency makes any mafia-fraud and impersonation proof very tricky: we are unable to prove (or disprove) any such result for the original version of the Swiss-Knife protocol. However, a very minor modification *does* enable us to prove some strong properties for the protocol: we break the circular dependency on sk by changing the key generation algorithm Kg, such that it now outputs *two* keys, sk and sk^*. Then sk is used to generate a as before, while T^1 is computed as $T^0 \oplus sk^*$, i.e. we use sk^* for this. Note that generating and storing sk^* on the prover only causes a reasonable overhead. In the original Swiss-Knife protocol sk is XORed with a; thus, $|sk| = |a| = N_c$, i.e. the number of time-critical protocol phases. Since resource-constrained provers, e.g. RFID tags, can support only few time-critical phases, the keys must be short; thus the two short keys would not take up much storage space.

We can prove a high mafia-fraud resistance for the modified Swiss-Knife protocol: the second lazy phase is essential in preventing key-learning attacks on sk^*. This second PRF computation also grants a mafia-fraud resistance of about $(\frac{1}{2})$ per round (i.e. optimal security). Each tag is associated with an identity ID, stored by $\mathcal{R}$ together with sk and sk^*, and never sent in clear in order to achieve anonymity; thus the reader needs to search the database exhaustively to find it. The protocol has some fault tolerance, i.e. the reader counts a total number of errors consisting of: (1) the number of faulty challenges R_i that the tag receives; (2) the number of faulty responses T_i that the reader receives; and (3) the number of rounds in which the tag's response exceeds the time threshold $t_{\max}$. The modified protocol is depicted in Figure 4. We note that Kim et al. also another version of the protocol, where $\mathcal{T}$ only sends selected bits out of the authentication string, allowing for longer (thus safer) keys; the security analysis for that version follows the same steps as below, with the key-guessing advantages growing exponential in the number of added (hidden, unused) bits of sk.

THEOREM 3.3 (SWISS-KNIFE PROPERTIES). *Let* ID *be the distance-bounding authentication scheme in Figure 4 with parameters* $(t_{\max}, N_c)$. *Assume furthermore that the*

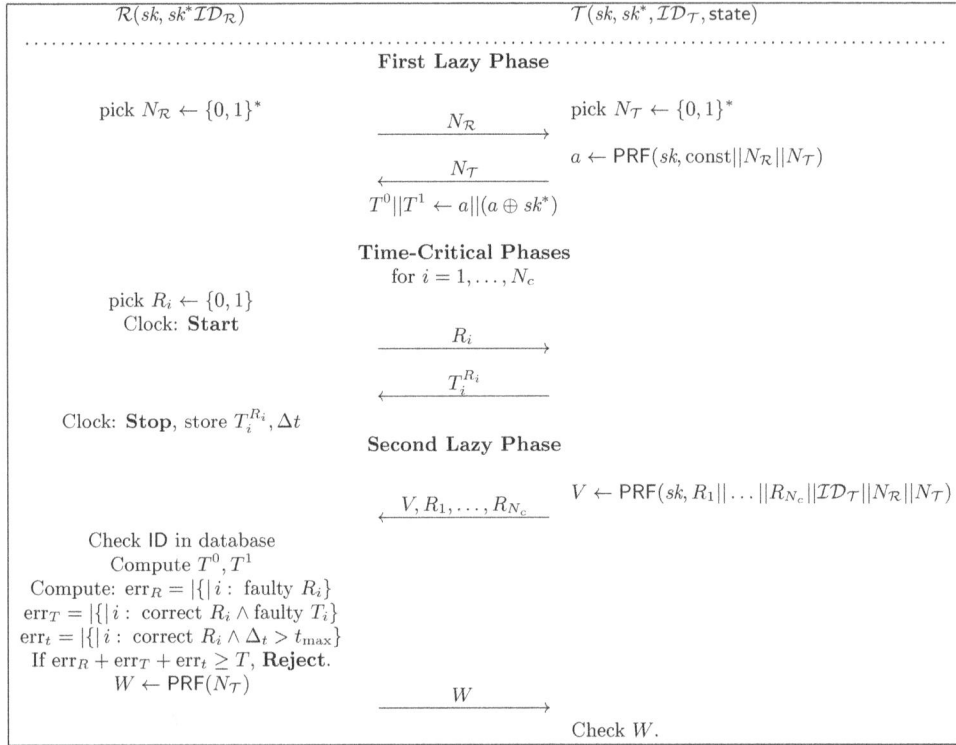

| $\mathcal{R}(sk, sk^*\mathcal{ID}_{\mathcal{R}})$ | $\mathcal{T}(sk, sk^*, \mathcal{ID}_{\mathcal{T}}, \text{state})$ |

First Lazy Phase

pick $N_{\mathcal{R}} \leftarrow \{0,1\}^*$ pick $N_{\mathcal{T}} \leftarrow \{0,1\}^*$

$\xrightarrow{\quad N_{\mathcal{R}} \quad}$

$a \leftarrow \mathsf{PRF}(sk, \text{const}||N_{\mathcal{R}}||N_{\mathcal{T}})$

$\xleftarrow{\quad N_{\mathcal{T}} \quad}$

$T^0||T^1 \leftarrow a||(a \oplus sk^*)$

Time-Critical Phases
for $i = 1, \ldots, N_c$

pick $R_i \leftarrow \{0,1\}$
Clock: **Start**

$\xrightarrow{\quad R_i \quad}$

$\xleftarrow{\quad T_i^{R_i} \quad}$

Clock: **Stop**, store $T_i^{R_i}, \Delta t$

Second Lazy Phase

$\xleftarrow{\quad V, R_1, \ldots, R_{N_c} \quad}$ $V \leftarrow \mathsf{PRF}(sk, R_1||\ldots||R_{N_c}||\mathcal{ID}_{\mathcal{T}}||N_{\mathcal{R}}||N_{\mathcal{T}})$

Check ID in database
Compute T^0, T^1
Compute: $\text{err}_R = |\{|\, i : \text{faulty } R_i\}$
$\text{err}_T = |\{|\, i : \text{ correct } R_i \wedge \text{faulty } T_i\}|$
$\text{err}_t = |\{|\, i : \text{ correct } R_i \wedge \Delta_t > t_{\max}\}|$
If $\text{err}_R + \text{err}_T + \text{err}_t \geq T$, **Reject**.
$W \leftarrow \mathsf{PRF}(N_{\mathcal{T}})$

$\xrightarrow{\quad W \quad}$

Check W.

Figure 4: The Swiss-Knife protocol

secret keys sk and sk^* output by Kg are chosen uniformly at random from a distribution that is computationally indistinguishable from the random distribution on bitstrings of length $|sk|$. This scheme has the following properties:

- It is not terrorist-fraud resistant (assuming the pseudorandomness of PRF).

- For any $(t, q_{\mathcal{R}}, q_{\mathcal{T}}, q_{\text{OBS}})$-distance-fraud adversary $\mathcal{A}$ against the scheme, there exists an adversary $\mathcal{A}'$ distinguishing sk^* from a truly random value such that:

$$\boldsymbol{Adv}_{\mathsf{ID}}^{dist}(\mathcal{A}) \leq q_{\mathcal{R}} \cdot \left(\tfrac{3}{4}\right)^{N_c - T} + \boldsymbol{Adv}_{\mathsf{Kg}}^{d(\mathcal{D},U)}(\mathcal{A}').$$

- For any $(t, q_{\mathcal{R}}, q_{\mathcal{T}}, q_{\text{OBS}})$-mafia-fraud adversary $\mathcal{A}$ against the scheme there exists a (t', q')-distinguisher $\mathcal{A}'$ against the computational indistinguishability of the distribution $\mathcal{D}$ from Uniform, and a (t'', q'')-distinguisher $\mathcal{A}''$ against PRF (where $t', t'' = t + O(n)$ and $q'.q'' = q_{\mathcal{R}} + q_{\mathcal{T}} + q_{\text{OBS}}$) such that:

$$\begin{aligned}
\boldsymbol{Adv}_{\mathsf{ID}}^{mafia}(\mathcal{A}) &\leq \left(\tfrac{1}{2}\right)^{N_c - T} + 2\boldsymbol{Adv}_{\mathsf{Kg}}^{d(\mathcal{D},U)}(\mathcal{A}') \\
&+ \binom{q_{\mathcal{R}} + q_{\text{OBS}}}{2} \cdot 2^{-(|N_{\mathcal{R}}| + \lceil \frac{N_c}{2} \rceil - T)} \\
&+ \binom{q_{\text{OBS}} + q_{\mathcal{T}}}{2} \cdot 2^{-(|N_{\mathcal{T}}| + \lceil \frac{N_c}{2} \rceil - T)} \\
&+ \boldsymbol{Adv}_{\mathsf{PRF}}^{d}(\mathcal{A}'') + 2^{-|V|}.
\end{aligned}$$

- For any $(t, q_{\mathcal{R}}, q_{\mathcal{T}}, q_{\text{OBS}})$-offline impersonation adversary $\mathcal{A}$ against the scheme there exists a (t', q')-distinguisher $\mathcal{A}'$ against PRF (where $t' = t + O(n)$ and

$q' = q_{\mathcal{R}} + q_{\mathcal{T}} + q_{\text{OBS}}$) such that

$$\begin{aligned}
\boldsymbol{Adv}_{\mathsf{ID}}^{imp}(\mathcal{A}) &\leq q_{\mathcal{R}} \cdot 2^{-|V|} + \boldsymbol{Adv}_{\mathsf{PRF}}^{d}(\mathcal{A}') \\
&+ \binom{q_{\mathcal{R}} + q_{\text{OBS}}}{2} \cdot 2^{-(|N_{\mathcal{R}}| + N_c - T)} \\
&+ \binom{q_{\mathcal{T}} + q_{\text{OBS}}}{2} \cdot 2^{-(|N_{\mathcal{T}}| + N_c - T)}.
\end{aligned}$$

PROOF. The proof of statement 1 follows the same lines as for the scheme of Reid et al. in the previous section. In this case, the adversary is given a by the dishonest prover and must guess the responses for the rounds where the challenge is 1. Finally, the tag forwards the correct authentication string V. The adversary's winning probability is again $\left(\tfrac{3}{4}\right)^{N_c}$. By contrast, the simulator cannot reconstruct the correct authentication string V unless it guesses it or unless it learns sk, which happens with probability $\left(\tfrac{1}{2}\right)^{N_c} + \left(\tfrac{1}{2}\right)^{sk}$ (note that in the modified version of the protocol, the value sk cannot be learned by any MITM interaction; in particular, it is the second key sk^* that is susceptible to key-learning attacks).

For the second statement, note that the tag does *not* choose the keys sk and sk^*; in particular, if sk^* is chosen at random from a distribution computationally indistinguishable from the uniform random distribution, the attack of Boureanu et al. [5] is thwarted. Indeed, with great probability, whatever the value a output by the PRF for this session, the value $a \oplus sk^*$ is with high probability at a large Hamming distance from a. The formal proof goes as follows. First we replace sk^* by a uniform random value (and lose a term $\boldsymbol{Adv}_{\mathsf{Kg}}^{d(\mathcal{D},U)}(\mathcal{A}')$). Now we consider each round

i in a distance-fraud attack. For this phase it holds with probability $\frac{1}{2}$ that the bit $T_i^0 = a_i$ is different from the corresponding bit of $T_i^1 = a_i \oplus sk^*$. If $T_i^0 = T_i^1$ (this happens with probability $\frac{1}{2}$), the adversary forwards T_i^0 and wins the round; else, if the two values are unequal, then the adversary has to guess which value to send (essentially predicting the challenge) and is successful with probability $\frac{1}{2}$. After we account for the fault tolerance level T we attain the above-stated bound.

For the third statement, the proof goes slightly differently than for previous protocols, e.g. for the scheme due to Hancke and Kuhn in Section 3.1. In particular, the response strings T^0 and T^1 are now related by means of the secret key sk^*. Note that both keys are used for authentication: sk is always used to key the PRF and compute T^0 and V (since it is only used for the PRF computation, one can argue that this key is in some sense "safe"), and sk^* is used to compute T^1 given T^0. If an adversary can somehow learn sk^*, it can successfully commit mafia fraud as follows: after learning this key, it opens a new session with the reader and a parallel session with the tag, then uses the Go-Early strategy to query the tag about critical-phase responses. By XOR-ing each response with the corresponding bit of the learned sk^*, this mafia adversary has both time critical responses for every phase and can thus authenticate. Thus, we must argue that the adversary cannot learn (apart from guessing probability) the value of sk^*.

Our proof goes as follows. In a first step, we replace the values of sk and sk^* by truly random values, thus losing a term $2\mathbf{Adv}_{\mathsf{Kg}}^{\mathsf{d}(\mathcal{D},\mathcal{U})}(\mathcal{A}')$. In a second step, we account for nonces shared between the sessions. In particular, note that if two sessions involving the reader share both nonces (and thus use the same T^0, T^1), the adversary may learn up to a half of the bits of sk^* (accounting for about $\frac{1}{2}N_c$ time-critical phases where the challenges will differ in the two sessions, and for the responses which the adversary either allows an honest prover to forward, or gets from the prover —with e.g. the Go-Early strategy). We can assume that nonces are unique for every session involving the reader (thus being shared with at most a single parallel adversary-tag session created by relays), except with probability $\binom{q_{\mathcal{R}} + q_{\text{OBS}}}{2} \cdot 2^{-(|N_{\mathcal{R}}| + \lceil \frac{N_c}{2} \rceil - T)} + \binom{q_{\text{OBS}} + q_{\mathcal{T}}}{2} \cdot 2^{-(|N_{\mathcal{T}}| + \lceil \frac{N_c}{2} \rceil - T)}$.

Now we can replace the response string T^0 and the verification string V by random values, losing a term $\mathbf{Adv}_{\mathsf{PRF}}^{\mathsf{d}}(\mathcal{A}'')$ (per reader-adversary session). This is the step where we take advantage of the fact that we have two independent keys, sk and sk^*, because we can make this transition by the pseudorandomness of the PRF, even when knowing sk^*. Next, we argue that if the adversary opens parallel reader-adversary and adversary-tag sessions such that they do not share the exact time-critical transcripts (challenges and responses), they cannot learn bits of the secret key sk^* (except with negligible probability). In particular, if at least one challenge or response is changed, the verification string V computed by the tag will not be correct, thus authentication always fails (regardless of whether the time-critical responses are correct or not). The adversary could still, for such sessions, try to guess the value of V, which it succeeds with probability $2^{-|V|}$. Note that the adversary cannot have seen the string V in any other sessions, since we assumed that the nonces are random for each session involving the reader.

At this point, thus, we have argued that, if the adversary opens reader-adversary and adversary-tag sessions where the transcripts are not exactly identical, it fails to learn any information about sk^*. We also note that no such parallel sessions yield bits of sk^* (since the adversary only learns a single time-critical response per phase). At this point, we can replace the second value, T^1, by a random value, since sk^*, which is considered to be random, does not leak. Now we look at how parallel sessions with identical transcripts can be created. There are two ways: (a) the adversary taints at least T phases (and somehow gets the other challenges and responses correctly); in this case, it cannot win in this particular reader-adversary session; (b) the adversary taints at most T phases, and in the other phases it either guesses the challenge in a Go-Early strategy (if it sent an incorrect challenge, then the transcript will be changed in the adversary-tag session compared to the reader-adversary session); this is equivalent to guessing a total of $N_c - T$ bits, and in this case the adversary wins the game in this reader-adversary session. Thus, accounting for $q_{\mathcal{R}}$ sessions, the overall probability is: $2^{N_c - T}$. Summing up and accounting for all the reader-adversary sessions we obtain the claimed bound.

The proof for offline impersonation security runs more or less as in the previous proof, with the following main steps: (1) account for repeating nonces; (2) replace V by a truly random value; and (3) accounting for the probability that an adversary can generate V without relaying it (this is a guessing probability only). $\square$

3.4 The Protocol of Yang et al.

Similarly to the Swiss-Knife protocol, the recent scheme of Yang et al. [17] aims to achieve mafia, distance, terrorist, and offline impersonation security, as well as privacy and mutual authentication. This scheme only has one, rather than two lazy authentication phases (as in the Swiss-Knife protocol). Privacy is achieved by means of tag-id updates, where the reader can resynchronize with the tag at every authentication attempt. In the distance-bounding part, the reader authenticates first in the lazy phase (though the tag does not). Instead, tag authentication is limited to the time-critical phases.

Both the reader and the tag run a PRF on input two nonces (interchanged by the parties) and the tag's identifier; the output is split into three parts: a string v used for reader authentication, and two response strings T^1 and T^2, used to compute a third string T^3, which is the bitwise XOR of T^1, T^2, and sk, i.e. the secret key shared by the reader and the tag. As in the Swiss-Knife protocol, we add a second key sk^* to break the circular key dependency and achieve provable security. If the tag checks (and accepts) v, then the time-critical phases begin: in each phase $i = 1, \ldots, N_c$, the reader sends a random challenge R_i. The tag responds depending on the value of $R_i || v_i$, where v_i is the i-th bit of the reader-authentication string v. If $R_i || v_i = 0 || 0$, then the response bit is T_i^1; if $R_i || v_i = 1 || 1$, then the tag responds T_i^2; else, it sends T_i^3. Denote by PRF the pseudorandom function used to generate T^0 and T^1, and by H a different PRF used for key updates. The scheme is depicted in Figure 5.

Using an implicit second challenge bit (from v_i) thwarts key-learning mafia fraud, but the absence of a second authentication phase may enable Denial-of-Service (DoS) attacks, which break privacy; this depends on when the tag updates its key. If the tag updates state only after run-

$\mathcal{R}(sk, sk^*, \mathcal{TID}, \mathcal{TID}^*)$ $\mathcal{T}(sk, sk^*, \mathcal{ID})$

....

First Lazy Phase

pick $N_{\mathcal{R}} \leftarrow \{0,1\}^*$ $\xleftarrow{\quad N_{\mathcal{T}}, \mathcal{ID} \quad}$ pick $N_{\mathcal{T}} \leftarrow \{0,1\}^*$

If $\mathcal{ID} \notin \{\mathcal{TID}, \mathcal{TID}^*\}$, abort
Else, if $\mathcal{ID} = \mathcal{TID}^*$, then
Set $\mathcal{TID} \leftarrow \mathcal{TID}^*$ and $\mathcal{TID}^* \leftarrow H(\mathcal{ID}, sk)$
$T^1 || T^2 || v \leftarrow \mathsf{PRF}(sk, N_{\mathcal{T}} || N_{\mathcal{R}} || \mathcal{ID})$

$\xrightarrow{\quad N_{\mathcal{R}}, v \quad}$ Compute T^1, T^2, v
               If v does not verify, abort

$T^3 \leftarrow T^1 \oplus T^2 \oplus sk^*$

Time-Critical Phases
for $i = 1, \ldots, N_c$

pick $R_i \leftarrow \{0,1\}$
Clock: **Start**

$\xrightarrow{\qquad R_i \qquad}$

     If $R_i || v_i = 0 || 0$ then set $T_i \leftarrow T_i^1$
     Else if $R_i || v_i = 1 || 1$ set $T_i \leftarrow T_i^2$
     Else set $T_i \leftarrow T_i^3$

$\xleftarrow{\qquad T_i \qquad}$

Clock: **Stop**, store $T_i, \Delta t$

If $\forall i$, T_i is correct and $\Delta t \geq t_{\max}$ Update $\mathcal{ID} \leftarrow H(\mathcal{ID}, sk)$
then set $\mathcal{TID} \leftarrow H(\mathcal{ID}, sk)$ and $\mathcal{TID}^* \leftarrow H(\mathcal{TID}, sk)$
Else reject prover.

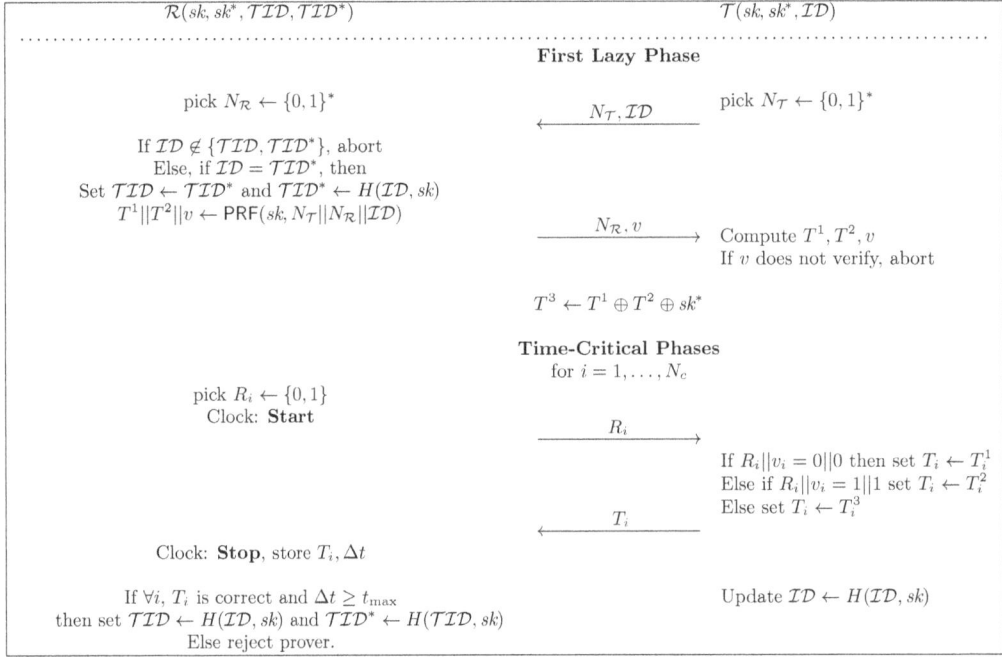

Figure 5: The Yang et al. protocol

ning the time-critical phases, an adversary can simply intercept the last challenge and guess the response, sending it to the reader. As the tag does not receive the last challenge, it aborts; the reader will update state (and be permanently desynchronized from the tag) with probability $\frac{1}{2}$ (the probability that $\mathcal{A}$ guessed the correct response). However, personal communication with Vincent Zhuang, a co-author in [17], assure us that the tag should update the id at every session where the string v verifies, regardless of whether the session is complete or not.

The terrorist-fraud resistance proof in [17] is flawed; a malicious tag can forward bits from different strings, depending on the string v (which is a known value), without revealing the whole strings or the secret key. In personal communication, Vincent Zhuang assured us that the authors of [17] also later found this attack, a fact that will be acknowledged in future publications.

THEOREM 3.4 (YANG ET AL. PROPERTIES). *If, for the distance-bounding authentication scheme* ID *in Figure 5 with parameters* $(t_{\max}, N_c)$, *the secret keys* sk, sk^* *output by* Kg *are chosen uniformly at random from a distribution* $\mathcal{D}$ *computationally indistinguishable from the uniform random distribution on bitstrings of length* $|sk|$, *the following holds for* ID:

- *It is not terrorist-fraud resistant, nor distance-fraud resistant, nor (tag) offline-impersonation secure.*

- *For any* $(t, q_{\mathcal{R}}, q_{\mathcal{T}}, q_{\mathrm{OBS}})$-*mafia-fraud adversary* $\mathcal{A}$ *against the scheme there exist: a* (t', q')-*distinguisher* $\mathcal{A}'$ *against the computational indistinguishability of* $\mathcal{D}$, *and a* (t'', q'')-*distinguisher against* PRF *(*$t', t'' =$*

$t + O(n)$ *and* $q', q'' = q_{\mathcal{R}} + q_{\mathcal{T}} + q_{\mathrm{OBS}}$ *queries) such that:*

$$\boldsymbol{Adv}_{\mathsf{ID}}^{mafia}(\mathcal{A}) \leq \left(\frac{3}{4}\right)^{N_c} + 2\boldsymbol{Adv}_{\mathsf{Kg}}^{d(\mathcal{D},U)}(\mathcal{A}')$$
$$+ \binom{q_{\mathcal{R}} + q_{\mathrm{OBS}}}{2} \cdot 2^{-(|N_{\mathcal{R}}| + \lceil \frac{N_c}{2} \rceil)}$$
$$+ \binom{q_{\mathrm{OBS}} + q_{\mathcal{T}}}{2} \cdot 2^{-(|N_{\mathcal{T}}| + \lceil \frac{N_c}{2} \rceil)}$$
$$+ 2\boldsymbol{Adv}_{\mathsf{PRF}}^{d}(\mathcal{A}'').$$

PROOF. We begin with the second statement, showing how to change the proof of the Swiss-Knife protocol (taking into account the three responses). We begin as follows: (1) we replace sk and sk^* by uniform random strings; (2) we account for shared nonces between sessions; (3) replace the PRF output T^1, T^2, and v by truly random values; (4) argue that, except with negligible probability, the adversary cannot tamper with the values $N_{\mathcal{R}}, v$ sent by the reader, this being equivalent to an adversary breaking the pseudorandomness of PRF; (5) we argue that simple observation does not yield any information about sk^*. Now comes a modification to our proof of the Swiss-Knife protocol: we argue that, if $\mathcal{A}$ does perform an active attack, it learns at most two of the three inter-related response bits, and thus it does *not* learn a bit of the secret sk^*. Indeed, the tag will only provide a single response bit, depending on the challenge forwarded by the MITM adversary, and the reader will only receive a single response bit forwarded by the MITM adversary. The most an adversary can learn is whether the bits from two responses are equal to each other or not; this, however, says nothing about sk^*, since the third possible response bit masks it. Thus we have again argued that no bits of sk^* are leaked, so we can replace T^3 by a truly random string. Now we account for the success probability of an adversary in a mafia-fraud attack where T^1, T^2, T^3, and v

are all truly random strings, for unique nonces. Note that the string v cannot be tampered with (see step (4) above), and that from the adversary's point of view, the protocol is now similar to the Hancke-Kuhn protocol: indeed, given the corresponding bit of v (which is already set for this session), there are two possible responses, which are inter-dependent and pseudo-random. If $v_i = 0$, the possible responses are T_i^1, for $R_i = 0$, and T_i^3, for $R_i = 1$. If $v_i = 1$, then the responses can be either T_i^3 if $R_i = 0$, or T_i^2 otherwise. This yields the same time-critical success probability as in the Hancke-Kuhn protocol.

We now focus on the first statement. First, the protocol is not distance-fraud resistant. In an attack similar to those shown by Boureanu et al. [5], a malicious prover could choose a weak nonce $N_\mathcal{T}$ such that, for every reader-chosen nonce $N_\mathcal{R}$, it would hold that $T^1 = T^2 = sk^*$. In this case, it holds that $T^3 = sk^* \oplus sk^* \oplus sk^*$, and thus the malicious prover can always respond to a randomly generated challenge, regardless of the value of v.

The protocol is also trivially not (tag) offline-impersonation secure in the definition of [10], since it involves no lazy-phase authentication for the tag.

Thirdly, the protocol is not terrorist-fraud resistant. After the malicious tag computes v, it forwards the following values to the adversary: for $i = 1, \ldots, N_c$, if $v_i = 0$, then it forwards $0\|T_i^1$ and $1\|T_i^3$, but *not* T_i^2; and else, if $V_i = 1$, then it forwards $1\|T_i^2$ and $0\|T_i^3$, but *not* T_i^1. Here, the prepended bits indicate to $\mathcal{A}$ which response to send in each time-critical phase. Now $\mathcal{A}$ authenticates with probability 1; however, it learns nothing about the key, which is still bitwise hidden by the one unknown bit. Thus, a simulator cannot hope to compete with the adversary in this setting. □

Acknowledgements

We thank the anonymous reviewers and Vincent Zhuang for their valuable comments.

4. REFERENCES

[1] M. R. S. Abyneh. Security analysis of two distance-bounding protocols. In *Proceedings of RFIDSec 2011*, volume 7055 of *Lecture Notes in Computer Science*, pages 94–107. Springer, 2011.

[2] G. Avoine, M. A. Bingol, S. Karda, C. Lauradoux, and B. Martin. A formal framework for cryptanalyzing RFID distance bounding protocols. http://eprint.iacr.org/2009/543.pdf, 2009.

[3] G. Avoine, C. Lauradoux, and B. Martin. How secret-sharing can defeat terrorist fraud. In *Proceedings of the Fourth ACM Conference on Wireless Network Security WISEC 2011*, pages 145–156. ACM Press, 2011.

[4] G. Avoine and A. Tchamkerten. An efficient distance bounding rfid authentication protocol: Balancing false-acceptance rate and memory requirement. In *Information Security*, volume 5735 of *Lecture Notes in Computer Science*, pages 250–261. Springer-Verlag, 2009.

[5] I. Boureanu, A. Mitrokotsa, and S. Vaudenay. On the pseudorandom function assumption in (secure) distance-bounding protocols, 2012.

[6] S. Brands and D. Chaum. Distance-bounding protocols. In *Advances in Cryptology — Eurocrypt'93*, Lecture Notes in Computer Science, pages 344–359. Springer-Verlag, 1993.

[7] P. H. Cole and D. C. Ranasinghe. *Networked RFID Systems and Lightweight Cryptography*. Springer-Verlag, 2008.

[8] C. Cremers, K. B. Rasmussen, and S. Čapkun. Distance hijacking attacks on distance bounding protocols. pages 113–127. IEEE Computer Society Press, 2012.

[9] Y. Desmedt. Major security problems with the 'unforgeable' (Feige)-Fiat-Shamir proofs of identity and how to overcome them. In *SecuriCom*, pages 15–17. SEDEP Paris, France, 1988.

[10] U. Dürholz, M. Fischlin, M. Kasper, and C. Onete. A formal approach to distance bounding RFID protocols. In *Proceedings of the 14th Information Security Conference ISC 2011*, Lecture Notes in Computer Science, pages 47–62. Springer-Verlag, 2011.

[11] Editors, Y. Zhang, and P. Kitsos. *Security in RFID and Sensor Networks*. CRC Press, 2009.

[12] T. H-Security. Chip-based ID cards pose security risk at airports. http://www.h-online.com/security/news/item/Chip-based-ID-cards-pose-security-risk-at-airports-905662.html, 2010.

[13] G. P. Hancke and M. G. Kuhn. An rfid distance bounding protocol. In *SECURECOMM*, pages 67–73. ACM Press, 2005.

[14] C. H. Kim and G. Avoine. Rfid distance bounding protocol with mixed challenges to prevent relay attacks. In *Proceedings of the 8th International Conference on Cryptology and Networks Security (CANS 2009)*, volume 5888 of *Lecture Notes in Computer Science*, pages 119–131. Springer-Verlag, 2009.

[15] C. H. Kim, G. Avoine, F. Koeune, F.-X. Standaert, and O. Pereira. The swiss-knife RFID distance bounding protocol. In *Proceedings of the 14th Information Security Conference ISC 2011*, Lecture Notes in Computer Science, pages 98–115. Springer-Verlag, 2009.

[16] J. Reid, J. M. G. Nieto, T. Tang, and B. Senadji. Detecting relay attacks with timing-based protocols. In *ASIACCS*, pages 204–213. ACM Press, 2007.

[17] A. Yang, Y. Zhuang, and D. S. Wong. An Efficient Single-Slow-Phase Mutually Authenticated RFID Distance-Bounding Protocol with Tag Privacy. In *Information and Communications Security*, volume 7618 of *Lecture Notes in Computer Science*, pages 285–292. Springer-Verlag, 2012.

APPENDIX

A. THE CASE FOR TERRORIST FRAUD RESISTANCE

In this paper, we prove that two schemes claiming to be terrorist-fraud resistant, and intuitively achieving some degree thereof, are insecure in the framework of Dürholz et al. [10]. In particular, our attack against schemes in [16] and [15] uses the fact that *partial* key-related information gives an adversary some advantage over the simulator (representing an unaided adversary).

Our results may be viewed from two separate points of view. It can be argued, on the one hand, that the model in [10] is too strong, and does not accurately capture the notion of terrorist-fraud resistance. On the other hand, our results may be viewed as proof that terrorist-fraud resistance is in fact a very powerful attack, which is difficult to counteract in practice. We present and assess both points of view below.

Model strength. As noted in Section 3.2, the notion achieved by [16] is very weak since it excludes even prover information that significantly aids adversary authentication while disclosing a relatively insignificant *part* of the secret key. Note that previous definitions, e.g. that of Avoine et al. [2], are ambiguous on this point, requiring, literally, that the prover's help gives the adversary no advantage in future attempts. It is unclear, however, what "further" means in this context: does it refer to the adversary's success probability *after* the prover helped it, compared to the adversary's success *before* the prover helped it, or rather to the notion captured by [10], i.e. the success probability of the adversary *after* the prover helped it compared to *while* the prover helped it?

The recent work of Avoine et al. [3] indicates that a construction is terrorist-fraud resistant if it leaks no information about the secret key (the proof of [3] looks at the conditional entropy of the secret). However, we can argue that this restriction on the adversarial capacity may be too strong: the adversary *may* learn information about the secret *as long as this does not increase its success probability*.

If we take the intuitive notion of Avoine et al. [2], the protocol due to Reid et al. [16] is intuitively terrorist fraud resistant. However, note that no existing formal definition covers the weaker definition of terrorist-fraud resistance presented informally above. Thus, it is hard to say how secure such protocols are, or how they compare in terms of adversarial advantage.

A further question is which definition best captures the intuition behind terrorist-fraud attacks. A strong degree of terrorist-fraud resistance is always more desirable; in this sense, the definition of [10] sets the standard for protocol design. However, this definition seems hard to achieve, as it allows attacks where indirect information about the key is given to the adversary (see Sections 3.2 and 3.4).

The intuition of terrorist-fraud resistance is that the malicious prover will help the adversary authenticate, but wants to control his access. Thus, the adversary must not be able to authenticate without the prover's help. Note, however, that the adversary always has some (usually negligible in the number of time-critical rounds) probability to authenticate by itself: this is equivalent to the probability that he guesses the responses or, equivalently, the secret key.

How far does the model in [10] cover this intuitive notion? Dürholz et al. quantify the adversary's success probability in the presence of the malicious prover, and then the simulator's probability (where the simulator cannot query the prover, but can access the adversary's view, or state when it queried the tag). The scheme is terrorist-fraud resistant if the simulator's probability of success equals (or is greater than) the adversary's probability of success. In other words, an attack is successful if the prover's help enables the adversary to succeed in one session with some probability, but this probability diminishes in future sessions if the prover is no longer available. In this case the prover has the guarantee that the adversary can later authenticate only with less probability. This definition seems too strong, as Dürholz et al. accept an attack where the prover authenticates with probability 75% (3 out of 4 times), but the simulator can only authentication with probability 50% (1 out of 2 times). This contradicts the spirit of terrorist-fraud resistance as it is understood in the literature.

A middle way would be to define a so-called tolerance level for the simulator, i.e. accept attacks as long as the simulator's success probability does not exceed this tolerance level. Note, however, that the attack presented in Section 3.2 can be tweaked so that the adversary still has an advantage over the simulator, whereas the simulator succeeds with a probability within the tolerance level (instead of giving half the response, the prover would forward only a number of bits of this response, thus easing the adversary's job).

It is our opinion that the notion described in [10], though strong, does capture the intuition of terrorist-fraud resistance better than the weaker definition which these protocol seem to attain. A common approach in security is to be conservative and to ask for strong(er) security, rather than to label insecure protocols as secure.

Constructive aspects. A second perspective in which to view our result is a constructive one, i.e. if we consider that the model by Dürholz et al. captures the correct notion of terrorist fraud resistance, then clearly achieving this definition requires a stronger construction. One might argue that the strong requirement posed by the model in [10] would lead to inefficient constructions. We argue, however, that the notion of terrorist fraud resistance, is, in its own right, a very strong notion: here, the (dishonest) prover *helps* the adversary authenticate. The challenge is thus to ensure that *any* information leaked to the adversary automatically will carry over to the simulator.

Also note that there is a clear separation between distance-bounding realizations for RFID and for more powerful devices. Indeed, terrorist fraud resistance might be more easily achieved if one can use, say, public key cryptography. In this sense, we could wonder how realistic a threat terrorist fraud attacks are on RFID systems and whether it is worth addressing them directly in protocol design. With RFID tags used in the pharmaceutical industry, in general logistics, and in public transport [7, 11], it seems that terrorist fraud attacks are quite likely in practice in these settings. In fact, RFID systems are also used in airport security in many German airports: impersonation MITM attacks have already been mounted on these systems by the Chaos Computer Club (CCC) [12]. Though these attacks were not real-time relay attacks, the incentive to mount mafia and terrorist fraud attacks on RFID authentication protocols is rather high. It remains an open question whether RFID systems can be efficiently protected against terrorist fraud in practice, however. The results in this paper show that terrorist fraud resistance is not trivial to achieve, and that achieving it may be inefficient for RFID devices. As terrorist fraud resistance is, however, both a very strong, and a very desirable goal, the authors of this paper interpret their results as an incentive to construct protocols that *are*, in fact, terrorist fraud resistant in the notion of Dürholz et al.

Efficient, Secure, Private Distance Bounding without Key Updates

Jens Hermans   Roel Peeters
KU Leuven, ESAT/COSIC & iMinds
Kasteelpark Arenberg 10
3001 Leuven, Belgium
firstname.lastname@esat.kuleuven.be

Cristina Onete
CASED & TU Darmstadt
Mornewegstrasse 30
64293 Darmstadt, Germany
cristina.onete@gmail.com

ABSTRACT

We propose a new distance bounding protocol, which builds upon the private RFID authentication protocol by Peeters and Hermans [25]. In contrast to most distance-bounding protocols in literature, our construction is based on public-key cryptography. Public-key cryptography (specifically Elliptic Curve Cryptography) can, contrary to popular belief, be realized on resource constrained devices such as RFID tags. Our protocol is wide-forward-insider private, achieves distance-fraud resistance and near-optimal mafia-fraud resistance. Furthermore, it provides strong impersonation security even when the number of time-critical rounds supported by the tag is very small. The computational effort for the protocol is only four scalar-EC point multiplications. Hence the required circuit area is minimal because only an ECC coprocessor is needed: no additional cryptographic primitives need to be implemented.

Categories and Subject Descriptors

K.6.5 [**Management of Computing and Information Systems**]: Security and Protection—*Authentication*

Keywords

RFID, Distance bounding, Privacy, Cryptographic protocol

1. INTRODUCTION

Authentication protocols are used for a wide range of applications, such as tracing goods for logistics, payment for public transport, Passive Keyless Entry and Start (PKES) systems used in cars, and personal identification for access control. While authentication protocols provide protection against impersonation, relay attacks are not considered. In relay attacks, an adversary just forwards data between the prover and the verifier. Francillon *et al.* [16] showed that for PKES, this vulnerability can be exploited in practice: one can simply drive off in another person's car by forwarding

messages between the car and the owner's passive authentication device.

In order to prevent relay attacks (also called mafia fraud by Desmedt [13]), Brands and Chaum [7] proposed the first distance-bounding protocol, using the fact that pure relaying over a large distance introduces a processing delay for the adversary, which the reader can detect if equipped with a clock. Most distance-bounding protocols are round based and may be grouped into phases, which are called either lazy (no clock is used) or time-critical. In time-critical phases, the roundtrip time between sending a challenge and receiving a response is measured; the measured timings provide an upper bound on the distance between the communicating parties.

There are four main security threats for distance bounding protocols:

1. **Impersonation Security.** The adversary attempts to impersonate the prover during the lazy phases, but without pure relay.

2. **Distance Fraud.** The adversary is a malicious prover that tries to prove that it is closer to the honest verifier than it really is.

3. **Mafia Fraud.** The adversary impersonates an honest prover in the presence of this prover and the honest verifier.

4. **Terrorist Fraud.** The adversary impersonates a prover with that prover's aid to the honest verifier. However, the adversary is unable to impersonate the prover when unaided.

Lazy phase impersonation security was only recently classified as a desirable property. In general, impersonation security used to be achieved only during time-critical rounds. However, as noted by Avoine and Tchamkerten [4], resource constrained devices such as RFID tags cannot support many time-critical rounds and thus lazy-phase impersonation resistance is required.

Distance fraud is quite easy to achieve when considered independently of other properties. Optimal distance-fraud resistance can be achieved by having the verifier send random challenge bits to the prover and waiting for the prover to echo them back. However, any party within the legitimate distance (whether or not in possession of legitimate credentials) can echo challenge bits in time. This breaks in particular mafia-fraud resistance. In order to attain mafia-fraud resistance, distance-bounding protocols in the symmetric setting usually employ a PseudoRandom Function

(PRF), returning bits from the PRF's output depending on the reader's challenges. However, Boureanu *et al.* [6] recently showed that the PRF assumption is not sufficient to prove distance-fraud resistance in such cases.

Distance-bounding protocols in the literature address two or more of the above threats [4, 7, 10, 18, 20, 21, 28, 30]. Note that such protocols are also hard to design, since they should be as lightweight as possible so as to be implementable on resource-constrained devices. Furthermore, one needs to take measures to keep the processing at the device as small as possible. Rasmussen *et al.* [27] and Ranganathan *et al.* [26] proposed the first practical implementations of such protocols by using analog components, which allows for the necessary small processing delay.

That most of the proposed lightweight authentication and distance-bounding protocols use symmetric cryptography, is a result of the myth that public-key cryptography cannot be implemented on resource constrained devices such as RFID tags. However, Lee *et al.* [22, 23] and Wenger and Hutter [33] showed that public-key cryptography *can* be implemented. More specifically, these papers proposed efficient dedicated coprocessors that can do elliptic curve arithmetic on curves suitable for Elliptic Curve Cryptography (ECC).

Most distance bounding protocols do not address privacy. An exception is the Swiss-Knife protocol of Kim *et al.* [21], where tags have a shared secret with the reader which is used for authentication. However, from the moment the prover's secret is compromised, the tag is traceable, since the shared secret is never updated. In recent work Onete [24] shows how to achieve private distance bounding, but only with key updates; the same approach is taken by Yang *et al.* [34], whose protocol also fails to attain the claimed distance-fraud and terrorist-fraud resistance [15].

Vaudenay [32] showed that stronger privacy requires key agreement, and thus cannot be achieved in the symmetric setting (as is the case in [34]). Intermediate privacy levels can be achieved in the symmetric setting with key updates. However, key updates are unsuitable for resource-constrained devices as the write operation typically requires high energy. Moreover, protocols relying on key updates can be vulnerable to desynchronization attacks.

The notion of 'wide-forward-insider' [25] privacy covers the case where an adversary uses the internal state from a corrupted tag to attack the privacy of other tags. These insider attacks where described by van Deursen *et al.* [31], clearly showing that wide-forward privacy protocols are not sufficient. For two wide-forward private protocols it was shown that the adversary can link uncorrupted tags if he can to learn the outcome of the protocol and the state of a legitimate 'insider' tag. Note that an adversary can easily get a legitimate e.g. a legitimate public transportation ticket.

Our Contribution. We propose the first distance-bounding protocol that attains wide-forward-insider privacy in the sense of Hermans *et al.* [19]. Our protocol relies on the recently-proposed public-key (ECC) secure, wide-forward-insider RFID authentication protocol due to Peeters and Hermans [25], such that the resulting protocol resists distance- and mafia-fraud attacks while remaining secure and wide-forward-insider private. The proposed scheme has nearly optimal mafia-fraud resistance and a very high impersonation security, resulting from the soundness of the underlying protocol.

2. PRELIMINARIES

Our work builds on the work of Peeters and Hermans [25]. For this reason we briefly recall the privacy model and definitions used in [25] in the following sections. We also give an overview of the definitions related to distance bounding, as defined by Dürholz et al. [14].

2.1 Notation

To relate the security level of a protocol to the security of the underlying cryptographic primitives or number theoretic assumptions, we often transform a successful adversary $\mathcal{A}$ into one or more adversaries $\mathcal{A}'$ against the primitive(s) or assumption(s). Let $\mathbf{Adv}_{\mathcal{S}}^{\mathrm{Exp}}(\mathcal{A}')$ denote the advantage of an adversary $\mathcal{A}'$ in breaking a cryptographic scheme or assumption $\mathcal{S}$ in some experiment Exp. An overview of the relevant number theoretic assumptions is given in Sect. 2.2.

In this paper we consider elliptic curves $\mathbb{E}$ and subgroups $\mathbb{G}_\ell$ of points on $\mathbb{E}$ of prime order ℓ over $\mathbb{F}_p$, usually generated by a point P. Points on the elliptic curve are denoted by uppercase characters. In general, we denote scalars by lowercase letters. We denote by aP the scalar multiplication of the point P by the scalar $a \in \mathbb{Z}_\ell^*$. For a any scalar $x \in \mathbb{Z}_\ell^*$, the corresponding uppercase letter X is defined as xP. The key-generation algorithm of our scheme outputs a pair $(priv, Pub)$ such that $priv \in_R \mathbb{Z}_\ell^*$ and $Pub = priv \cdot P \in \mathbb{G}_\ell$. We denote by O the point at infinity of the elliptic curve.

Our construction relies on the $\texttt{xcoord}(\cdot)$ function, which is the DSA conversion function [8]. This function, returns the x-coordinate of a point. For a point $Q = \{q_x, q_y\} \in \mathbb{G}_\ell$, with $q_x, q_y \in [0 \ldots p-1]$, $\texttt{xcoord}(Q)$ maps Q to $q_x \mod \ell$. Additionally, we define $\texttt{xcoord}(O) = 0$.

For a bitstring x, we denote by $[x]_k$ the least significant k bits of the string.

2.2 Number Theoretic Assumptions

Discrete Logarithm (DL). Let A be a given, arbitrarily chosen element of $\mathbb{G}_\ell$. The discrete logarithm (DL) problem is to find the unique integer $a \in \mathbb{Z}_\ell^*$ such that $A = aP$. The DL assumption states that it is computationally hard to solve the DL problem.

One More Discrete Logarithm (OMDL). The one more discrete logarithm (OMDL) problem was introduced by Bellare *et al.* [5]. Let $\mathcal{O}_1()$ be an oracle that returns random elements $A_i = a_i P$ of $\mathbb{G}_\ell$, and let $\mathcal{O}_2(\cdot)$ be an oracle that returns the discrete logarithm of a given input base P. The OMDL problem is to return the discrete logarithms for each of the elements obtained from the m queries to $\mathcal{O}_1()$, while making strictly less than m queries to $\mathcal{O}_2(\cdot)$ (with $m > 0$). The OMDL assumption is that it is computationally hard to solve the OMDL problem.

x-Logarithm (XL). Brown and Gjøsteen [9] introduced the x-Logarithm (XL) problem: given an elliptic curve point, determine whether its discrete logarithm is congruent to the x-coordinate of an elliptic curve point. The XL assumption states that it is computationally hard to solve the XL problem. Brown and Gjøsteen also provided some evidence that the XL problem is almost as hard as the DDH problem (see below).

Diffie Hellman (DH). Let aP, bP be any two given arbitrary elements of $\mathbb{G}_\ell$, with $a, b \in \mathbb{Z}_\ell^*$. The computational Diffie Hellman (CDH) problem is, given P, aP and bP, to find abP. The 4-tuple $\langle P, aP, bP, abP \rangle$ is called a Diffie Hellman (DH) tuple. Given a fourth element $cP \in \mathbb{G}_\ell$, the decisional Diffie Hellman (DDH) problem is to determine if $\langle P, aP, bP, cP \rangle$ is a valid Diffie-Hellman tuple or not. The DDH assumption states that it is computationally hard to solve the DDH problem.

Oracle Diffie Hellman (ODH). Abdalla *et al.* [1] introduced the ODH assumption:

Definition 1 *Oracle Diffie Hellman (ODH) Given $A = aP$, $B = bP$, a function H and an adversary $\mathcal{A}$, consider the following experiments:*
Experiment $\mathbf{Exp}_{H,\mathcal{A}}^{odh}$:

- $\mathcal{O}(Z) := H(bZ)$ *for* $Z \neq \pm A$
- $g = \mathcal{A}^{\mathcal{O}(\cdot)}(A, B, H(C))$
- *Return* g

The value C is equal to abP for the $\mathbf{Exp}_{H,\mathcal{A}}^{odh-real}$ experiment, chosen at random in $\mathbb{G}_\ell$ for the $\mathbf{Exp}_{H,\mathcal{A}}^{odh-random}$ experiment.
We define the advantage of $\mathcal{A}$ violating the ODH assumption as:

$$\left| \Pr\left[\mathbf{Exp}_{H,\mathcal{A}}^{odh-real} = 1 \right] - \Pr\left[\mathbf{Exp}_{H,\mathcal{A}}^{odh-random} = 1 \right] \right|.$$

The ODH assumption consists of the plain DDH assumption combined with an additional assumption on the function $H(\cdot)$. The idea is to give the adversary access to an oracle $\mathcal{O}$ that computes bZ, without giving the adversary the ability to compute bA, which can then be compared with C. To achieve this one restricts the oracle to $Z \neq \pm A$, and moreover, only $H(bZ)$ instead of bZ is released, to prevent the adversary from exploiting the self reducibility of the DL problem.[1] The crucial property that is required for $H(\cdot)$ is one wayness. In the following part we use a one way function based on the DL assumption. We define the function $H(Z) := \texttt{xcoord}(Z)P$.

Theorem 1 *The function $H(\cdot)$ is a one-way function under the DL assumption.*

2.3 Privacy Model

Hermans *et al.* [19] provided a general game-based privacy model for RFID, which is robust and easy to apply. For more details on the different existing RFID privacy models and a comparison between these, the reader is referred to [19].

The intuition behind the RFID privacy model of Hermans *et al.* is that of tag indistinguishability, i.e. privacy is guaranteed if an adversary cannot distinguish with which one of two RFID tags (of its choosing) it is interacting by means of a set of oracles. The main ideas of this model resemble previous frameworks: the adversary interacts with the tags by means of handles, called virtual tags (vtags), and privacy is defined as an indistinguishability game (or experiment $\mathbf{Exp}$) between a challenger and the adversary.

This game is defined as follows. First the challenger picks a random challenge bit b and then sets up the system $\mathcal{S}$ with a security parameter k. Next, the adversary $\mathcal{A}$ can use a subset (depending on the privacy notion) of the following oracles to interact with the system:

- $\texttt{CreateTag}(ID) \rightarrow T_i$: on input a tag identifier ID, this oracle creates a tag with the given identifier and corresponding secrets, and registers the new tag with the reader. A reference T_i to the new tag is returned.
- $\texttt{Launch}() \rightarrow \pi$: this oracle launches a new protocol run on the reader R_j, according to the protocol specification. It returns a session identifier π, generated by the reader.
- $\texttt{DrawTag}(T_i, T_j) \rightarrow vtag$: on input a pair of tag references, this oracle generates a virtual tag reference, as a monotonic counter, $vtag$ and stores the triple $(vtag, T_i, T_j)$ in a table $\mathcal{D}$. Depending on the value of b, $vtag$ either refers to T_i or T_j. If one of the two tags T_i or T_j is in the list of insider tags $\mathcal{I}$, $\perp$ is returned and no entry is added to $\mathcal{D}$. If T_i is already references as the left-side tag in $\mathcal{D}$ or T_j as the right-side tag, then this oracle also returns $\perp$ and adds no entry to $\mathcal{D}$. Otherwise, it returns $vtag$.
- $\texttt{Free}(vtag)_b$: on input $vtag$, this oracle retrieves the triple $(vtag, T_i, T_j)$ from the table $\mathcal{D}$. If $b = 0$, it resets the tag T_i. Otherwise, it resets the tag T_j. Then it removes the entry $(vtag, T_i, T_j)$ from $\mathcal{D}$. When a tag is reset, its volatile memory is erased. The non-volatile memory, which contains the state S, is preserved.
- $\texttt{SendTag}(vtag, m)_b \rightarrow m'$: on input $vtag$, this oracle retrieves the triple $(vtag, T_i, T_j)$ from the table $\mathcal{D}$ and sends the message m to either T_i (if $b = 0$) or T_j (if $b = 1$). It returns the reply from the tag (m'). If the above triple is not found in $\mathcal{D}$, it returns $\perp$.
- $\texttt{SendReader}(\pi, m) \rightarrow m'$: on input π, m this oracle sends the message m to the reader in session π and returns the reply m' from the reader (if any) is returned by the oracle.
- $\texttt{Result}(\pi)$: on input π, this oracle returns a bit indicating whether or not the reader accepted session π as a protocol run that resulted in successful authentication of a tag. If the session with identifier π is not finished yet, or there exists no session with identifier π, $\perp$ is returned.
- $\texttt{Corrupt}(T_i)$: on input a tag reference T_i, this oracle returns the complete internal state of T_i. Note that the adversary is not given control over T_i.
- $\texttt{CreateInsider(ID)} \rightarrow T_i, S$: create an insider tag T_i. This runs $\texttt{CreateTag}$ to create a new tag T_i and $\texttt{Corrupt}$ on the newly created tag. The tag T_i is added to the list $\mathcal{I}$ of insider tags.

By using the $\texttt{DrawTag}$ oracle the adversary $\mathcal{A}$ can arbitrarily select which two tags to interact with. Based upon the challenge bit b chosen initially, a virtual tag is then associated to either the 'left' tags T_i or the 'right' tags T_j. At the end of the adversary's interaction, $\mathcal{A}$ outputs a guess bit g. The outcome of the game will be $g \overset{?}{=} b$, *i.e.*, 0 for an incorrect and 1 for a correct guess. Thus, the adversary wins the game if it can distinguish whether it has interacted with the 'left' or the 'right' world.

The advantage of the adversary $\mathbf{Adv}_{\mathcal{S},\mathcal{A}}(k)$ is defined as:

$$\left| Pr\left[\mathbf{Exp}_{\mathcal{S},\mathcal{A}}^0(k) = 1 \right] + Pr\left[\mathbf{Exp}_{\mathcal{S},\mathcal{A}}^1(k) = 1 \right] - 1 \right|.$$

---

[1] The adversary can set $Z = rA$ for a known r and compute $r^{-1}(bZ) = bA$.

The following privacy notions were introduced by Vaudenay [32] and are also present in Hermans *et al.*'s framework. *Strong* attackers are allowed to use all the oracles available. *Forward* attackers are only allowed to do other corruptions after the first corruption, protocol interactions are no longer allowed. *Weak* attackers cannot corrupt tags. Independently of these classes, there is the notion of *wide* and *narrow* attackers. A *wide* attacker is allowed to get the result from the reader, *i.e.* whether the identification was successful or not; while a *narrow* attacker does not.

If an adversary is allowed to call `CreateInsider` the privacy notion is called 'insider', so we can speak of forward-insider and weak-insider adversaries. For strong and destructive the `CreateInsider` can be simulated using the normal `CreateTag` and `Corrupt` oracles, i.e. strong-insider and destructive-insider are equivalent to strong and destructive respectively. The privacy notions are related as follows:

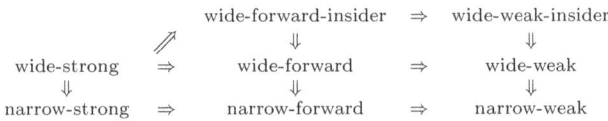

$$
\begin{array}{ccccc}
& & \text{wide-forward-insider} & \Rightarrow & \text{wide-weak-insider} \\
& \nearrow\!\!\!\!/ & \Downarrow & & \Downarrow \\
\text{wide-strong} & \Rightarrow & \text{wide-forward} & \Rightarrow & \text{wide-weak} \\
\Downarrow & & \Downarrow & & \Downarrow \\
\text{narrow-strong} & \Rightarrow & \text{narrow-forward} & \Rightarrow & \text{narrow-weak}
\end{array}
$$

We use arrows between two notions to denote that any protocol that is private in the sense of the first notion is also private in the sense of the second notion.

For most practical applications, wide-forward-insider privacy is sufficient. By contrast, the weaker notion of wide-forward privacy is *not* sufficient; indeed, in e.g. transportation systems an adversary has easy access to an insider tag and can thus abuse any privacy guarantees of the system. Furthermore, it seems that the wide-strong notion captures a scenario exceeding the practical requirements for privacy, where an adversary may first corrupt a tag and then release it again for future tracking. However, in practice this can be done more easily, without physically tampering with the tag itself (i.e. corrupting it). For instance the attacker could, when having physical access to the tag, attach his own tracking device to it.

Note that we further restrict the `Corrupt` oracle, such that it only returns the non-volatile state of the tag. This restriction allows to exclude trivial privacy attacks on multi-pass protocols, that require the tag to store some information in volatile memory during the protocol run.

2.4 Private Authentication Protocol

The definition of a private authentication protocol is due to Peeters and Hermans [25]. This definition is specific for the RFID setting in the sense that it assumes that concurrent attacks are impossible, since tags can only participate in one session at the time. Furthermore their security definition does not model physical distance, as a result relay attacks are not considered.

Private authentication protocols have the following three properties: *correctness*, *soundness*, and *privacy*. Correctness and soundness are necessary to establish the security of the authentication protocol. Correctness ensures that the reader (verifier) does not reject legitimate tags (provers). Soundness ensures that an illegitimate tag (*i.e.* an adversary not in possession of legitimate credentials) is always rejected by the reader. Privacy will ensure that all parties cannot infer any information on the tag's identity from the protocol messages, except the reader to which the tag is authenticating.

Only the content of the exchanged messages is taken into account, not the physical characteristics of the radio links as studied by Danev *et al.* [12], which should be dealt with at the hardware level.

Definition 2 *Correctness. A scheme is correct if the authentication of a legitimate tag only fails with negligible probability.*

Definition 3 *Soundness. A scheme is resistant against impersonation attacks if no polynomially bounded strong adversary succeeds, with non-negligible probability, in being authenticated by the reader as the tag it impersonates. Adversaries may interact with the tag they want to impersonate prior to, and with all other tags prior to and during the protocol run. All tags, except the impersonated tag, can be corrupted by the adversary.*

Definition 4 *Privacy. A privacy protecting protocol, modeled by the system $\mathcal{S}$, is said to computationally provide privacy notion X, if and only if for all polynomially bounded adversaries $\mathcal{A}$, it holds that $\boldsymbol{Adv}_{\mathcal{S},\mathcal{A}}^{X}(k) \leq \epsilon$, for negligible ϵ.*

2.5 Distance Bounding

The security model of Dürholz *et al.* [14], which formalizes security notions for distance-bounding protocols (in particular taking into consideration relay attacks) considers a single verifier and a single prover, in particular for the RFID setting. Here, the single prover $\mathcal{P}$ is an RFID tag and the verifier $\mathcal{V}$ is the reader. The reader uses a clock to measure the time elapsed between sending a challenge and receiving the response. Dürholz et al. consider round-based distance-bounding protocols, where rounds are called *time-critical* if a clock is used to measure the roundtrip time, and *lazy* otherwise.

In the following, we provide intuitive descriptions of impersonation security, mafia fraud, distance fraud, and terrorist fraud. For the formal definitions and for further insight, we refer the reader to the original paper [14].

Impersonation Resistance. Impersonation resistance refers to lazy-phase tag authentication. The idea, introduced by Avoine and Tchamkerten [4], is that even without the time-critical phases (relay attacks are not considered), the prover is still authenticated. By contrast, Avoine *et al.* [2] define impersonation security for the entire protocol. Note that impersonation resistance as defined by Avoine *et.al.* is also achieved by protocols that are lazy-phase impersonation secure and mafia-fraud resistant.

Distance Fraud. Distance-fraud adversaries control the tags themselves. The adversary is further away than allowed from the reader, but aims to convince the reader of the contrary. Since the reader's clock measures time accurately, the adversary must anticipate the reader's challenges and respond in advance.

Mafia Fraud. Mafia-fraud resistance considers a Man-In-The-Middle (MITM) attack, where pure relay is prevented by the reader's clock. Informally, the attacker consists of two parts: a *leech*, which impersonates the reader to an honest tag, and a *ghost*, which impersonates the tag to an honest reader. Both the reader and the honest tag are unaware of the MITM attack.

Terrorist Fraud. In terrorist-fraud attacks, the dishonest prover cooperates with an adversary in order to enable this adversary to authenticate. The informal restriction is that the prover does not forward trivial data, like the secret key. This attack is rather controversial, as we discuss it at length in Section 3.2.

Attacks in [14]. All the attacks above are formalized by Dürholz et al. [14] by introducing an abstract clock, which keeps track of the messages sent in several protocol executions called sessions. These sessions can be: reader-tag (the adversary is a passive eavesdropper), reader-adversary (the adversary impersonates the tag to the reader), and adversary-tag (the adversary impersonates the reader to the tag). Relaying is considered round-wise. In mafia-fraud attacks, a phase is called *tainted* if the adversary purely relays communication between a reader-adversary and an adversary-tag session. Here, *pure* relay refers to an adversary receiving a message in a session sid and *then* relaying the exact, same message in a session sid'. Having received a response, the adversary relays it back again between sid' and sid, for all subsequent rounds in the tainted phase. If the adversary changes any of the messages in one session before it forwards them in the other session, this is not pure relay. Also, if the adversary queries one session with some message m before receiving the same m in the other session, this is not relaying. In distance fraud, phases are *tainted* if the adversary does not commit in advance to the responses of time-critical phases before the phase has started. In terrorist fraud, the adversary taints a time-critical phase by querying the adversary during that phase.

Attack Parameters. Apart from the upper bound $t_{\max}$ of the roundtrip transmission time and the number of time-critical rounds n, we also allow for at most $T_{\max}$ phases with delayed responses and at most $E_{\max}$ phases with wrong responses. Though most existing protocols do not provide for erroneous/delayed communication, fault tolerance is essential in resource-constrained environments, e.g. RFID.

When specifying the adversary's characteristics one considers its runtime t and the number $q_{\mathcal{V}}$, $q_{\mathcal{P}}$, q_{OBS} of respectively reader-adversary, adversary-tag, and reader-tag sessions.

3. THE PROTOCOL

In several distance-bounding protocols (e.g., Hancke-Kuhn [18]), the tag and reader use a long-term shared secret to compute an ephemeral, session-specific shared secret. Afterwards, during each of the n time-critical rounds, the reader sends a challenge bit and expects a single response bit, either from the left or from the right half of the computed ephemeral value, depending on the challenge. This ephemeral secret is the result of a PseudoRandom Function (PRF), often instantiated with an H-MAC. Our proposed protocol follows this structure; very importantly, however, our protocol is in the asymmetric setting. Recall from the introduction that the need for the asymmetric setting arises from our desire to design a protocol without key updates[2] that

---

[2]For tags updating their keys, it is important that the reader and tag stay synced, meaning that measures should be taken to prevent desynchronization attacks. Moreover, updating keys requires high energy.

guarantees strong privacy and also protects against distance fraud.

Our proposed protocol is depicted in Fig. 1. All tags are initialized with a private/public key pair $(x, X = xP)$ and the tags' public keys are registered in the reader's database. The reader's private/public key pair is $(y, Y = yP)$ of which the public key is known to all tags.

To generate the ephemeral shared secret, an anonymous Diffie-Hellman key agreement, with fresh random values from both sides ($R_1 = r_1 P$ and $R_3 = r_3 P$), takes place, resulting in a shared point $r_1 r_3 P$ on the elliptic curve. To map this point to a uniformly distributed element in $\mathbb{Z}_\ell^*$, a cryptographic hash function can be used. Unfortunately, current hash functions [29] require at least 50% of the circuit area of the most compact ECC coprocessor implementation. Instead, we propose to use the ECDSA conversion function [8], which comes almost for free when using elliptic curves. This function simply returns the x-coordinate of a point on the elliptic curve. Note that the set of x-coordinates does not span $\mathbb{Z}_\ell^*$ entirely, as such the x-coordinates are not uniformly-randomly distributed in $\mathbb{Z}_\ell^*$. However, we only need $2n$ bits. Chevalier *et al.* [11] showed that binary truncation of the x-coordinate of the last element of an instance of the DDH problem is statistically indistinguishable from the uniform distribution :

$$\langle aP, bP, U_k \rangle \approx_S \langle aP, bP, \mathrm{lsb}_k(\mathrm{xcoord}(cP)) \rangle .$$

If there are no transmission errors and no Man-in-the-Middle (MITM) interference, then $t^0 \| t^1 = u^0 \| u^1$; these values will be used by the tag to answer the reader's subsequent challenges. Now the reader chooses a random challenge $e \in \mathbb{Z}_\ell^*$ and sends the first n bits, one per round, as its time critical challenges, expecting a bit from the corresponding response vector, i.e. for a challenge bit b sent in the ith round, the tag should respond with t_i^b. The round trip time is measured and compared with a maximal round trip time t_{max}.

Finally, the protocol ends in a second lazy phase, in which the last messages of the underlying private authentication protocol are broadcast. This underlying private authentication protocol is due to Peeters and Hermans [25] and has the following structure: commit, exam, response. The tag's commitment is now the point R_2 sent in the first lazy phase. We cannot reuse the point R_1; indeed, if R_1 is used, an attacker could impersonate the verifier and send the prover Y instead of a random $r_3 P$, thus having a better probability to distinguish the tag. The reader's full exam value e is sent to the tag, which in turn compares this to the received bit challenges in the time-critical phase. As such, we can enforce a higher level of mafia-fraud resistance. The tag must also verify that $e \neq 0$, to prevent trivial attacks. The response is similar to the Schnorr authentication protocol, providing a very high level of impersonation resistance. To achieve privacy, the response contains an additional blinding factor d. This blinding factor is computed using a static DH key exchange, with the randomness r_2 of the tag it already committed to (by sending R_2) and the public key Y of the reader. To map this point $r_2 Y$ to a scalar while breaking the homomorphisms that exist between the input and the output, again the $\texttt{xcoord}(\cdot)$ function is used. Due to the non-uniformity subgroup of x-coordinates in $\mathbb{Z}_\ell^*$, a privacy adversary could build a distinguisher. However, this adver-

State: x_j, Y
Tag T_j

Secrets: $y, \mathtt{DB} = \{X_j\}$
Reader

$r_1, r_2 \mathbb{Z}_\ell^*$

$R_1 = r_1 P, R_2 = r_2 P$ →

$R_1, R_2 \neq O?$
$r_3 \mathbb{Z}_\ell^*$

← $R_3 = r_3 P$

$R_3 \neq O?$
$t^0 || t^1 = [\mathtt{xcoord}(r_1 R_3)]_{2n}$

$u^0 || u^1 = [\mathtt{xcoord}(r_3 R_1)]_{2n}$
$c = [e]_n \quad \text{for} \quad e \mathbb{Z}_\ell^*$

n fast rounds

timer δ_i

← c_i

$f_i = (c_i - 1)t_i^0 + c_i t_i^1$ →

$\forall i = 0 \ldots n - 1 : \delta_i < t_{max} \quad \wedge$
$f_i \overset{?}{=} (c_i - 1)u_i^0 + c_i u_i^1$

← e

$c_{n-1} || \ldots || c_1 || c_0 \overset{?}{=} [e]_n$
$d = \mathtt{xcoord}(r_2 Y)$

$s = x_j + er_1 + r_2 + d$ →

$\tilde{d} = \mathtt{xcoord}(y R_2)$
$\tilde{X} = (s - \tilde{d})P - eR_1 - R_2 \in \mathtt{DB}?$

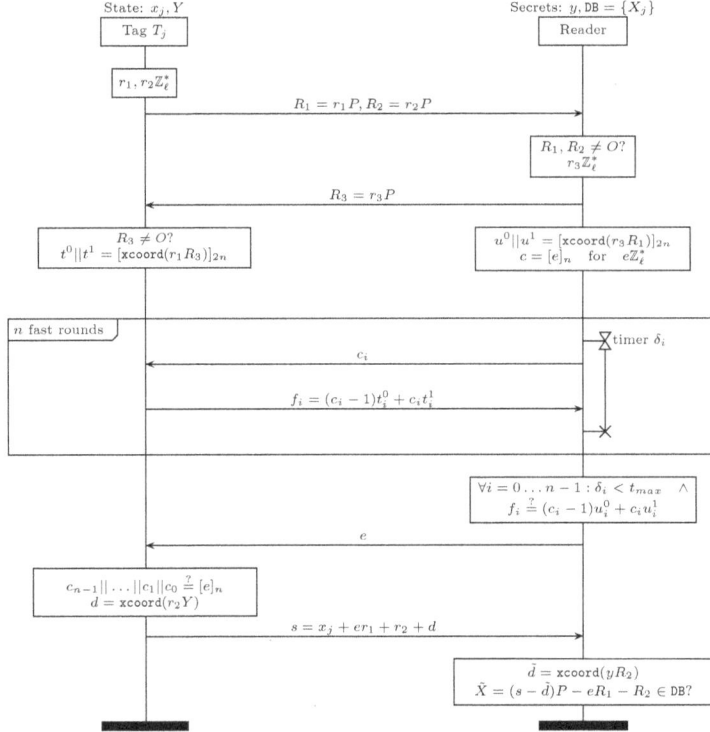

Figure 1: Efficient, secure, wide-forward-insider private distance bounding protocol.

sary has no information on d directly, only on $d + r_2$. Given the XL assumption, this poses no threat.

3.1 Properties

Let DB be the distance-bounding authentication scheme in Fig. 1 with parameters (t_{max}, n). We proceed to give the formal security statements for this protocol.

Impersonation Resistance. This property follows directly from the correctness and soundness of the underlying private authentication protocol, due to Peeters and Hermans [19]. As a result, our protocol has very high impersonation security that does not depend on the number of performed round in the time-critical phase.

Theorem 2 (Correctness) DB *is correct in the sense of Def. 2.*

The proof of correctness is trivial due to the fact that our protocol does not make use of key updates. As a result, desynchronization attacks need not to be taken into account. Therefore, this proof is omitted.

Theorem 3 (Soundness) DB *is sound according to Def. 3 under the OMDL assumption.*

PROOF. Assume an adversary $\mathcal{A}$ that can break the extended soundness with non-negligible probability, i.e. that can perform a fresh, valid authentication with the verifier. Without loss of generality we will assume the target tag is

known at the start of the game. [3] We construct an adversary $\mathcal{B}$ that wins the OMDL game as follows:

- Set $X = \mathcal{O}_1()$; this value will be used as the public key of the target tag.

- $\mathcal{B}$ executes $\mathcal{A}$. During the first phase of $\mathcal{A}$, $\mathcal{B}$ simulates the SendTag oracles for the target tag as follows (all other oracles are simulated as per protocol specification):

  - On the first SendTag($vtag$) query of the i'th protocol run:
    return $R_{1,i} = \mathcal{O}_1()$ and $R_{2,i} = r_{2,i} P$.

  - On the third SendTag($vtag, e_i$) query of the i'th protocol run:
    set $d_i = \mathtt{xcoord}(y R_{2,i})$ and return $s_i = \mathcal{O}_2(X + d_i P + e_i R_{1,i} + R_{2,i})$

- During the second phase of $\mathcal{A}$, $\mathcal{B}$ proceeds as follows:

  - On the first call of $\mathcal{A}$ to Result(π), compute $d = \mathtt{xcoord}(y R_2)$ and store (s, d). Next, rewind $\mathcal{A}$ until right before the call to SendReader(π, R).

---

[3]Otherwise, the proof can be adapted by choosing the public keys of the tags as $X_i = \mathcal{O}_1()$. All tag queries are simulated as for the target tag, until the tag is corrupted. When corrupting a tag, call $\mathcal{O}_2(X_i)$ for that tag and use the result as private key for simulating all following queries to that tag. At the end of the game, use the $\mathcal{O}_2(\cdot)$ oracle to extract all remaining discrete logarithms, except for the target tag.

On the next call to `SendReader`(π, R), return a new random e'.

- On the next call of $\mathcal{A}$ to `Result`(π): compute $r_1 = {(s-s')}/{(e-e')}$ and $x = s - d - er_1 - r_2$ return $(x, e_1^{-1}(s_1 - x - d_1 - r_{2,1}), \ldots, e_k^{-1}(s_k - x - d_k - r_{2,k}))$.

The simulation by $\mathcal{B}$ is perfect during both phases. At the end of the game $\mathcal{B}$ will successfully win the OMDL with non-negligible probability, unless $s = s'$, which happens with negligible probability since both e and e' are randomly chosen after $R \neq O$ is fixed. $\square$

Distance-Fraud Resistance. Intuitively, distance-fraud resistance requires both the unpredictability of challenges and that the response has sufficient entropy, even with respect to a party having the secret key, i.e. a dishonest prover. The flaws in the proofs for distance-fraud resistance as identified by Boureanu *et al.* [6], have not yet been resolved in the symmetric setting. By contrast, in our case, we use a public-key setting, where the ephemeral secret is the truncation of the x-coordinate of a point on the elliptic curve. The prover first selects an integer nonce r_1 and sends the value $R_1 = r_1 P$. Then the verifier (honestly) selects another nonce r_3 and truncates the x-coordinate of $r_3 R_1$; the output is then a bitstring which is distributed according to the uniform random distribution.

Theorem 4 (Distance-Fraud Resistance) *For any* $(t, q_\mathcal{V}, q_\mathcal{P}, q_{\text{OBS}})$ *distance fraud adversary $\mathcal{A}$, it holds that*

$$\boldsymbol{Adv}_{\text{DB}}^{dist}(\mathcal{A}) \leq q_\mathcal{V} \cdot \left(\tfrac{3}{4}\right)^n.$$

PROOF. First we argue that $r_1 \cdot r_3$ can be replaced by a random integer r. Second, we argue that the bits in the binary truncation $t^0 \| t^1$ are distributed according to the random distribution. Finally, we argue that the given bound holds. The first part holds because, on the one hand, the reader is honest in this attack and thus r_3 is chosen an random *after* the tag has committed to r_1, and on the other hand because the reader checks that $R_1 \neq O$, thus that $r_1 \neq 0$. The second part holds in view of the results of Chevalier et al. [11]. Thus, for every i, it holds that $Pr[t_i^0 = t_i^1] = \frac{1}{2}$. Finally, it holds that the adversary wins every round where $t_i^0 = t_i^1$ with probability 1. However, if $t_i^0 \neq t_i^1$, the adversary has only guessing probability to win, i.e. $\frac{1}{2}$. This adds up to a success probability of $\frac{3}{4}$ per round. Accounting for $q_\mathcal{V}$ attempts, we have the bound above. $\square$

Mafia-Fraud Resistance. The basic-most requirement for mafia-fraud resistance is that the ephemeral secret is hard to compute without knowing the long-term secret key. However, in order to increase mafia-fraud resistance we need to prevent the adversary from performing a MITM attack, which we call the Go-Early strategy following the notation of Dürholz et al. [14] and Fischlin and Onete [15]. Briefly, this attack works as follows: having first forwarded the lazy-phase messages between a prover and a verifier (i.e. the adversary opens a reader-adversary session sid and an adversary-prover session sid* that are "related" in the sense that the time-critical responses will be the same), the MITM adversary will then be queried by the reader with a challenge bit c and will expect a response bit r. However, in the Go-Early

strategy, the adversary first queries the prover in session sid* with a random bit c^*, receiving r^* in response, and will use this response to answer to the reader subsequently, in session sid. In other words, if $c = c^*$, then the adversary wins the round by forwarding $r = r^*$, else, if $c = c^*$, it guesses the correct response with probability $\frac{1}{2}$, totaling a success probability of $\frac{3}{4}$ per round. In our protocol we reduce this success probability by using a strategy similar to [7, 21], i.e. we add a lazy authentication phase depending on the challenges received by the prover. Thus, as soon as the adversary mis-guesses one challenge, it makes the prover compute a different response in this lazy phase, which cannot be used by the adversary in session sid. In particular, we merge authentication with distance bounding and use the reader's challenge bits in order to compute the authentication string.

Theorem 5 (Mafia-Fraud Resistance) *For any* $(t, q_\mathcal{V}, q_\mathcal{P}, q_{\text{OBS}})$-*mafia-fraud adversary $\mathcal{A}$ against the scheme there exist: a (t', q')-distinguisher $\mathcal{A}'$ that can distinguish the truncated output of the x-coordinate of the last element of a DDH element from random; an adversary $\mathcal{A}''$ that can solve the DL problem; and an adversary $\mathcal{B}$ against the soundness of the underlying protocol such that:*

$$\begin{aligned}
\boldsymbol{Adv}_{\text{DB}}^{mafia}(\mathcal{A}) \leq\ & q_\mathcal{V} \cdot \left(\tfrac{1}{2}\right)^n + \binom{q_\mathcal{V} + q_{\text{OBS}}}{2} \cdot 2^{-\ell} + \\
& \boldsymbol{Adv}_{dist}(\mathcal{A}') + 2q_\mathcal{V}\boldsymbol{Adv}_{DL}(\mathcal{A}'') + \\
& \boldsymbol{Adv}_{Sound}(\mathcal{B}) + \binom{q_\mathcal{P}}{2} \cdot 2^{-\ell},
\end{aligned}$$

with ℓ the order of the elliptic curve subgroup $\mathbb{G}_\ell$.

PROOF. The proof proceeds as follows:

1. We show that one can safely replace the output $T^0 \| T^1$ by truly random values, for each new nonce pair (r_1, r_3).

2. Show that nonce pairs are (almost) unique, except for possibly one adversary-tag session sid* having the same nonce pair as a reader-adversary session sid (here the adversary relays the nonces between sessions).

3. Bound the probability that the adversary passes the time-critical phases for at most one adversary-tag interaction.

The first step goes as follows. First, note that the adversary can learn the values r_1 and r_3 (and thus compute the ephemeral secret by using the public keys) with at most probability $2\boldsymbol{Adv}_{DL}(\mathcal{A}''')$ per authentication attempt. In this case, the adversary can bypass tainting the phase by querying the prover *after* it has successfully completed the time-critical phases, in order to learn the final authentication string. We assume now that the adversary cannot guess these values. By the results of Chevalier *et al.* [11], indicating that binary truncation of the x-coordinate of the last element of an instance of the DDH problem is statistically indistinguishable from the uniform distribution. Thus, replacing $t_0 \| t_1$ by random values decreases the adversary's success probability by at most $\boldsymbol{Adv}_{dist}(\mathcal{A}')$.

Next we consider all the nonces appearing in an attack of the adversary $\mathcal{A}$ mounting a mafia fraud attack. Assume that there exist two sessions (between adversary and tag or reader, or between both honest parties) with the same pair (r_1, r_3). This can only be a reader-adversary session and an adversary-tag session, except with probability (see [14]):

$$\binom{q_{\mathcal{V}} + q_{\text{OBS}}}{2} \cdot 2^{-\ell} + \binom{q_{\mathcal{P}}}{2} \cdot 2^{-\ell}.$$

Now declare $\mathcal{A}$ to lose if a collision appears, decreasing its success probability by this negligible term, but allowing us to consider collision-free executions. In particular, except for the matching session, all values $T^0 \| T^1$ in the attack are independent.

Now consider a reader-adversary session sid in which $\mathcal{A}$ successfully impersonates the tag $\mathcal{P}$ to $\mathcal{V}$, such that the same nonce pair appears (by assumption) in at most one other adversary-tag session. If such a (unique) matching adversary-tag session sid* exists, then this session (we claim) must taint sid with high probability (if sid* does not exist we have the case below, where the adversary does not use the additional session). If even a single phase of the protocol is tainted, this invalidates session sid. Thus, suppose to the contrary, that the matching session sid* taints no time-critical phase in sid.

Consider an untainted time-critical phase of sid where $\mathcal{V}$ sends $c_i = b$ and expects t_i^b. The adversary has thus successfully passed the first $i - 1$ time-critical phases and can choose to do one of the following in the i-th phase:

THE GO-EARLY STRATEGY. In session sid* the adversary has sent some bit c_i^* to $\mathcal{P}$ before having received $\{t_i^b\}^*$. The probability that $c_i^* \neq c_i$ is $\frac{1}{2}$, in which case $\mathcal{A}$ does not know the value t_i^b and must guess it or taint the round. However, note that if the adversary sends $c_i^* \neq c_i$ in sid*, this invalidates the lazy authentication step following the protocol, where the value s is computed based on the received challenges. Thus, this strategy invalidates the attack with probability $1/2$ per round.

THE GO-LATE STRATEGY. In session sid the adversary responds to c_i with some $\{t_i^{c_i}\}^*$ before receiving $\{t_i^{c_i}\}^*$ in session sid*. Now $\mathcal{A}$ succeeds only with probability $\frac{1}{2}$ for this phase.

THE MODIFY-IT STRATEGY. The adversary schedules the message such that it receives c_i in sid, sends some $c_i^* = b$ in sid*, receives t_i^b in sid*, and forwards some t_i^* in sid. Hence, the scheduling corresponds to a pure relay attack, but $c_i \neq c_i^*$ or $t_i^* \neq t_i^b$. If $b = c_i^*$ is wrong then t_i^b is never sent by $\mathcal{P}$ in sid* and the adversary can thus only guess t_i^* with probability $\frac{1}{2}$; if $b = c_i = c_i^*$ then $t_i^* \neq t_i^b$ makes the reader reject.

THE TAINT-IT STRATEGY. The adversary taints this phase of sid through sid*. This is equivalent here to losing in sid.

Thus, the most successful strategy is the Go-Early strategy, which, however, invalidates the attack with high probability. The overall success probability thus amounts to the value claimed in the theorem. □

Privacy. Since the underlying authentication protocol is wide-forward-insider private, we merely have to ensure that

the challenge and response strings reveal no information about the secret key of the tag x. The challenges are chosen at random; furthermore, we use a binary truncation of the x-coordinate output for the ephemeral secret, which ensures that the response is indistinguishable from random and reveals no information about the secret.

The privacy of the protocol can be shown under an extended ODH assumption where the adversary, in addition to $A = aP, B = bP, \texttt{xcoord}(C)P$ and the oracle $\mathcal{O}(Z)$, is also given $\texttt{xcoord}(C) + a$.

Before giving the privacy proof we first introduce a conjecture that is used as building block for obtaining wide-forward-insider privacy.

Conjecture 1 *Assume a set* $\mathcal{X} = \{x_1, \ldots, x_n\}$ *and* $\mathcal{I} = \{\iota_1, \ldots, \iota_m\}$ *with* $x_i, \iota_j \in_R \mathbb{Z}_\ell^*$. *The game proceeds as follows:*

1. $b \in_R \{0, 1\}$.

2. *The adversary* $\mathcal{A}$ *is given* $\mathcal{I}$ *and can interact with the system through the following oracles:*

   (a) $\mathcal{O}_1(\alpha, \beta) := \begin{cases} (i, \tilde{r}_i + x_\alpha) & \text{if} \quad b = 0 \\ (i, \tilde{r}_i + x_\beta) & \text{if} \quad b = 1 \end{cases}$
   with $\tilde{r}_i \in_R \mathbb{Z}_\ell^*$ *and let* i *be a counter that is incremented at every call*

   (b) $\mathcal{O}_2(s, i) := s - \tilde{r}_i \in \mathcal{X} \cup \mathcal{I}$

   (c) $\mathcal{O}_3(s) := s \in \mathcal{X}$ [4]

3. *The adversary* $\mathcal{A}$ *is given* $\mathcal{X}$ *and outputs a bit* g.

The adversary wins the game if $b \stackrel{?}{=} g$.

We conjecture that the adversary has negligible probability in winning the above game.

The intuition behind the experiment described above is that the adversary has a set of insider tags for which it knows the secret keys ($\mathcal{I}$) and that there is a set of tags for which the keys remains secret ($\mathcal{X}$). Through $\mathcal{O}_1$ the adversary can obtain output of the non-corrupted tags, which is a random value added to the tag secret. Just as in the privacy definition, a random bit determines which tag secret x_i is selected. Since a fresh random value $\tilde{r}_i$ is added to every tag output, it is obvious that the adversary has negligible advantage in winning the game when only given $\mathcal{O}_1$.

The oracles $\mathcal{O}_2$ and $\mathcal{O}_3$ let the adversary verify the tag output. Both oracles only return a binary value indicating whether validation succeeded. The random $\tilde{r}_i$'s are used in $\mathcal{O}_2$ to verify the input. Intuitively, the only way that the adversary can win the game is by either guessing some x_i and checking it through oracle $\mathcal{O}_3$ or by giving an input (s, i) to $\mathcal{O}_2$ that did not directly originate from a call to $\mathcal{O}_1$ (i.e. that maps to a different x_i than the call to $\mathcal{O}_1$ did). The probability of both these events happing however seems negligible.

Theorem 6 (Privacy) DB *is narrow-strong and wide-forward-insider private according to Def. 4 under an extended ODH, the XL assumption and Conjecture 1.*

---

[4]Due to a technicality in the privacy proof, we need to replace this oracle by $\mathcal{O}_3(S) := dlog(S) \in \mathcal{X}$. Note that it is the challenger, which is computationally unbounded, that computes the discrete logarithm in this oracle. This definition is equivalent to the one given here, since the adversary can always call $\mathcal{O}_3$ with sP instead of s.

PROOF. Assume an adversary $\mathcal{A}$ that wins the privacy game with non-negligible advantage. Using a standard hybrid argument [35, 17], we construct an adversary that breaks the ODH-assumption. We set $Y = B$. $\mathcal{B}_i$ plays the privacy game with $\mathcal{A}$. $\mathcal{B}_i$ selects a random bit $\tilde{b}$, which will indicate which world is simulated to $\mathcal{A}$. All oracles are simulated in the regular way, with the exception of the `SendTag` and `Result` oracle for the target tag:

- `SendTag`($vtag$):

  - $j \neq i$: Generate $r_1, r_2 \in_R \mathbb{Z}_\ell^*$. Take $R_1 = r_1 P$, $R_2 = r_2 P$. Return R_1, R_2.
  - $j = i$: Generate $r_1 \in_R \mathbb{Z}_\ell^*$ Take $R_1 = r_1 P$ and $R_2 = A$. Return R_1, R_2.

- `SendTag`($vtag, e$), j'th query: retrieve the tuple $(vtag, T_0, T_1)$ from the table $\mathcal{D}$. Take the key x for tag $T_{\tilde{b}}$.

  - $j < i$: Generate $r \in_R \mathbb{Z}_\ell^*$. Take $d = \texttt{xcoord}(rP)$. Return $s = x + er_1 + d + r_2$.
  - $j = i$: Return $s = x + er_1 + (\texttt{xcoord}(C) + a)$.
  - $j > i$: Take $d = \texttt{xcoord}(r_1 Y)$. Return $s = x + er_1 + d + r_2$.

- `Result`(π): If the received R_2 in session π matches A from the ODH problem take $\dot{d}P = \texttt{xcoord}(C)P$. If not, check if R_2 matches any of the R_2's generated during the first $i-1$ `SendTag` queries. If so, use the r generated in that query and compute $\dot{d}P = \texttt{xcoord}(rP)P$. Otherwise, take $\dot{d}P = \mathcal{O}(R_1)$. Finally, compute $\dot{X} = sP - (\dot{d}P) - eR_1 - R_2$. Check $\dot{X}$ with the database, return true if $\dot{X}$ is found, false otherwise.

At the end of the game $\mathcal{A}$ outputs its guess g for the privacy game. $\mathcal{B}_i$ outputs $(\tilde{b} \overset{?}{=} g)$.

The above simulation to $\mathcal{A}$ is perfect, since validation is done in the same way as the protocol specification. If $R_2 = A$, the oracle $\mathcal{O}(\cdot)$ cannot be used. However, in this case we know the corresponding value of d by directly using $\texttt{xcoord}(C)P$, which gives the same result.

We use $\mathcal{A}^i$ (with $i \in [1 \ldots k]$) to denote the case that $\mathcal{A}$ runs with the first i `SendTag` queries random instances, and the other queries real instances. This is the case when $\mathcal{B}_{i+1}$ runs with a real ODH instance, or $\mathcal{B}_i$ with a random ODH instance.

By the hybrid argument we get that

$$\| \Pr\left[\mathcal{A}^0 \text{ wins}\right] - \Pr\left[\mathcal{A}^k \text{ wins}\right] \| \leq \sum \mathbf{Adv}_{\mathcal{B}_i}.$$

Note that $\mathcal{A}^i$ wins if $\tilde{b} \overset{?}{=} g$.

In the case of $\mathcal{A}^0$, it is clear $\Pr\left[\mathcal{A}^0 \text{ wins}\right] = \Pr\left[\mathcal{A} \text{ wins}\right]$ since all oracles are simulated exactly as in the protocol definition.

In the case of $\mathcal{A}^k$, all `SendTag` queries are simulated with $r \in_R \mathbb{Z}_\ell^*$ and $d = \texttt{xcoord}(rP)$.

Narrow-strong privacy Since $s = x + er_1 + d + r_2$ and $R_1 = r_1 P, R_2 = r_2 P$, it follows under the XL assumption that $(x + er_1 + d + r_2, e, R_1 = r_1 P, R_2 = r_2 P)$, with d a random value from the x-coordinate distribution, is indistinguishable from $(\tilde{r}, e, R_1 = r_1 P, R_2 = r_2 P)$, with $\tilde{r}$ a uniformly random value. Hence it follows that s is indistinguishable from a uniformly random value independent of x, as long as $e, d \neq$

0. Note that this only holds in the absence of a `Result` oracle (which is able to distinguish $\tilde{r}$ from random).

So $\mathcal{A}^k$ has probability $1/2$ of winning the privacy game, since it obtains no information at all on x from a tag.

$$
\begin{aligned}
\| \Pr\left[\mathcal{A}^0 \text{ wins}\right] - \Pr\left[\mathcal{A}^k \text{ wins}\right] \| &= \| \Pr\left[\mathcal{A} \text{ wins}\right] - \tfrac{1}{2} \| \\
&= \tfrac{1}{2} \mathbf{Adv}_{\mathcal{A}}^{privacy} \\
&\leq \sum \mathbf{Adv}_{\mathcal{B}_i}
\end{aligned}
$$

It follows that at least one of the $\mathcal{B}_i$ has non-negligible probability to win the ODH game.

Wide-forward-insider privacy For proving wide-forward-insider privacy, we also have to simulate the `Result` oracle, which was ommitted in the case of narrow-strong privacy. After applying the XL assumption to show that $(d + r_2, R_2)$ is indistinguishable from $(\tilde{r}, R_2)$, we can now do a straightforward reduction to the game from Conjecture 1. All `SendTag`($vtag, e$) calls are simulated using $\mathcal{O}_1(i, j)$ for the tags T_i and T_j passed to `DrawTag`. Calls to `Result` are simulated using $\mathcal{O}_2(sP - eR_1 - R_2, i)$ if the R_2 received by the server matches an R_2 resulting from a `SendTag`(), otherwise $\dot{d}$ is computed as in the original protocol and $\mathcal{O}_3(sP - eR_1 - R_2 - \dot{d}P)$ is used to validate the resulting secret. □

3.2 Terrorist Fraud

General distance-bounding models mention four main security threats: impersonation security, distance fraud, mafia fraud and terrorist fraud. Our protocol is resistant against the former three attacks, but not to the latter one. Indeed, a dishonest prover can simply send the values $t_0 \| t_1$ to the adversary for a given session sid, thus helping it win with probability 1; however, these values cannot be reused to win a different session with independent nonces. We discuss here the notion of terrorist-fraud resistance, its applicability, and its attainability.

Terrorist-fraud attacks are in general very strong, as they consider a misbehaving, or malicious prover, willing to aid the adversary. Such attacks could be considered for instance when two entities wish to share the same identity without being caught. For instance, one entity, say Alice, might want to share her public transport privileges with another entity, called Bob, but only for a given amount of time. However, Alice does not wish to let Bob abuse her kindness; thus, she wants to make it hard for Bob to authenticate later, without her permission.

Formal models of terrorist-fraud resistance disagree about what constitutes a valid terrorist-fraud attack. Indeed, the model due to Avoine et al. [2] stipulates that the attack is only valid if the adversary has no further advantage from the information forwarded by the prover. In latter work, Avoine et al. [3] rely on the fact that adversary strategies need to information-theoretically hide the secret. However, this restriction is unnecessarily strong, since the prover could forward information about the secret which does not help the adversary in future authentication sessions. The model due to Dürholz et al. [14] is more lenient towards the adversary: the malicious prover can forward any information to the adversary, provided that this information does not help a simulator (given the adversary's view) authenticate with the same probability. In this model, the adversary may be willing to leak some information about the secret key, as long as these cannot be used directly by the adversary to authenticate.

In the symmetric setting, protocols aiming to attain terrorist-fraud resistance (e.g. [3, 10, 21, 28]) relate the two responses used during time-critical phases by means of a secret key. If the prover reveals both time-critical responses (corresponding to the response bits for a 0 and 1 challenge bits) for any given round, the adversary learns a bit of the secret key that relates the two responses. Since the prover only helps the adversary offline, it cannot know the challenges that the adversary will receive at every round. As a result, it cannot help the adversary authenticate by forwarding only one of the responses. While such protocols might attain some form of terrorist fraud resistance, they are not terrorist-fraud resistant in the definition of [14], as proved in the recent result of Fischlin and Onete [15].

In the context of public-key cryptography, one could use similar strategies in order to attain the same intuitive form of terrorist-fraud resistance. In particular, the tag would compute a binary truncation of the x-coordinate of $r_1 R_3$ as t^0, and then set $t^1 = t^0 \oplus priv$ for some the private key $priv$. Note that this notion might be too weak. Indeed, Fischlin and Onete [15] show a generic attack in which the adversary forwards one of the session responses, say t^0 (this does not reveal any information about the secret). Thus, the adversary is able to respond correctly to any round in which the challenge is 0. For the other rounds, the adversary can guess the response; thus the overall winning probability is roughly $\frac{3}{4}$ per time-critical round. However, once the prover withdraws its support, the adversary (more formally a simulator having access to the adversary's view) is unable to use the information learned during the prover-aided phase of the attack. Thus, if we used the same strategy for our protocol, this attack would still apply.

It is unclear whether such an attack captures the intuition of terrorist fraud. On the one hand, the model of Avoine et al. [2] seems too restrictive: indeed, it seems unreasonable to require the prover to forward no information at all about the secret key. Since terrorist fraud resistance is the strongest type of attack against distance bounding protocols, it may be quite feasible for a prover to accept leakage of a few bits of the secret key as the price of a successful attack. On the other hand, the stronger notion due to Dürholz et al. [14] also seems too strong, enabling a prover to forward quite a lot of information.

In order to achieve provable terrorist-fraud resistance it seems the protocol must include a weakness, i.e. a back door for the simulator to authenticate. This is why we do not aim to address terrorist-fraud resistance here. It remains an open question which model captures the intuition behind terrorist-fraud resistance best and how to attain this property.

3.3 Allowing for Errors

The protocol as shown in Fig. 1 does not allow for transmission errors or delays in communication. However, communication is not that reliable in the typical RFID environments. In particular, transmissions are susceptible to delays and they might also be incorrect, *e.g.*, in the case of collisions.

In order to account for such weaknesses, tolerance parameters are introduced for faulty and for delayed transmissions, respectively, as outlined in Sect. 2.5. Our protocol can also be modified to be robust with respect to transmission errors.

The tag will check, upon receiving the value e from the reader, whether the Hamming distance between the first n bits of this value and the concatenation of the received challenges c_i is greater than the tolerance level E_{max}. If so, the tag may choose to abort or simply forward a random value for s. The reader allows for a maximum of E_{max} erroneous time-critical responses (with respect to its computed values $u^0 || u^1$). Furthermore, the reader also allows for a maximum of T_{max} number of rounds where the roundtrip times exceeds t_{max}. If there are too many delayed or erroneous rounds, the reader rejects the tag.

3.4 Performance of the Protocol

General Infrastructure Our protocol assumes that the tag is able to know the public key Y of the reader to which it attempts to authenticate. In practice, this can be achieved by storing (a small number of) public keys on the tag itself. At the reader side, the public keys of the tags are either stored locally or kept at a central server that is connected to the readers.

Protocol Complexity and Parameters Our protocol requires the tag to generate randomness of bitlength $\log_2(l)$. During each protocol run, the tag must store, apart from its own secret key and the public key(s) of the reader(s), four registers of size $\log_2(l)$ to store the necessary information to complete the protocol. The tag performs four (costly) EC point multiplications and some scalar arithmetic. Note that the time-critical responses do not require arithmetic and is done by a simple if-else statement. The bottleneck in the implementation constitutes of the EC point multiplications. For a 80 bit security level, an elliptic curve over a field of about 160 bits is needed. As an indication, the ECC co-processor of Lee *et al.* [23], implementing such a curve, requires less than 15 kGEs (kilo-Gate Equivalents), consumes around $13.8 \mu W$ of power, and requires 85 ms for a single point-multiplication.

4. CONCLUSION

We proposed a new distance bounding protocol and provide rigorous proofs for all achieved properties. Our protocol achieves a very high impersonation resistance independent of the number of rounds in the time-critical phase. The protocol has distance fraud resistance of about $\frac{3}{4}$ per time-critical round. The proof bypasses the flaws identified by Boureanu *et al.* [6] which affect most distance bounding protocols in the literature. Our protocol achieves mafia fraud resistance at a near optimal rate of about $\frac{1}{2}$ per time-critical round. However, it does not achieve terrorist fraud resistance since we are not willing to introduce weaknesses as argued in Sect. 3.2. Finally, the protocol achieves one of the strongest possible degrees of privacy, namely wide-forward-insider privacy.

Acknowledgements

The authors would like to thank the anonymous reviewers for their helpful comments. We would like to thank Saartje Verheyen, without whom this paper would not have been possible.

This work was supported by the Flemish Government, IWT SBO MobCom, FWO G.0360.11N Location Privacy, and by the Research Council KU Leuven: GOA TENSE;

and by the European Commission through the FIDELITY project (contract number 284862) and the ICT programme under contract ICT-2007-216676 ECRYPT II. Jens Hermans is a research assistant, sponsored by the Fund for Scientific Research - Flanders (FWO).

5. REFERENCES

[1] M. Abdalla, M. Bellare, and P. Rogaway. The Oracle Diffie-Hellman assumptions and an analysis of DHIES. In D. Naccache, editor, *Cryptographer's Track at RSA Conference*, volume 2020 of *LNCS*, pages 143–158. Springer, 2001.

[2] G. Avoine, M. A. Bingol, S. Karda, C. Lauradoux, and B. Martin. A formal framework for analyzing RFID distance bounding protocols. In *Journal of Computer Security - Special Issue on RFID System Security, 2010*, 2010.

[3] G. Avoine, C. Lauradoux, and B. Martin. How secret-sharing can defeat terrorist fraud. In *Proceedings of the Fourth ACM Conference on Wireless Network Security WISEC 2011*, pages 145–156. ACM Press, 2011.

[4] G. Avoine and A. Tchamkerten. An efficient distance bounding RFID authentication protocol: Balancing false-acceptance rate and memory requirement. In *Conference on Information Security 2009*, volume 5735 of *LNCS*, pages 250–261. Springer, 2009.

[5] M. Bellare, C. Namprempre, D. Pointcheval, and M. Semanko. The One-More-RSA-Inversion problems and the security of Chaum's blind signature scheme. *Journal of Cryptology*, 16:185–215, 2003.

[6] I. Boureanu, A. Mitrokotsa, and S. Vaudenay. On the pseudorandom function assumption in (secure) distance-bounding protocols. In *Progress in Cryptology – LATINCRYPT 2012*, volume 7533 of *LNCS*, pages 100–120. Springer, 2012.

[7] S. Brands and D. Chaum. Distance-bounding protocols. In *Advances in Cryptology — EUROCRYPT'93*, volume 765 of *LNCS*, pages 344–359. Springer, 1993.

[8] D. R. Brown. Generic groups, collision resistance, and ECDSA. *Designs, Codes and Cryptography*, 35(1):119–152, 2005.

[9] D. R. L. Brown and K. Gjøsteen. A security analysis of the NIST SP 800-90 elliptic curve random number generator. In A. Menezes, editor, *Advances in Cryptology — CRYPTO*, volume 4622 of *LNCS*, pages 466–481. Springer, 2007.

[10] L. Bussard and W. Bagga. Distance-bounding proof of knowledge to avoid real-time attacks. In *Security and Privacy in the Age of Ubiquitous Computing*, volume 181 of *IFIP AICT*, pages 222–238. Springer, 2005.

[11] C. Chevalier, P.-A. Fouque, D. Pointcheval, and S. Zimmer. Optimal randomness extraction from a Diffie-Hellman element. In *Advances in Cryptology – EUROCRYPT '09*, number 5479 in LNCS, pages 572–589. Springer-Verlag, 2009.

[12] B. Danev, T. S. Heydt-Benjamin, and S. Čapkun. Physical-layer identification of RFID devices. In *USENIX*, pages 125–136. USENIX, 2009.

[13] Y. Desmedt. Major security problems with the 'unforgeable' (Feige)-Fiat-Shamir proofs of identity and how to overcome them. In *SecuriCom*, pages 15–17. SEDEP Paris, France, 1988.

[14] U. Dürholz, M. Fischlin, M. Kasper, and C. Onete. A formal approach to distance bounding RFID protocols. In *Proceedings of the 14th Information Security Conference ISC 2011*, volume 7001 of *LNCS*, pages 47–62. Springer, 2011.

[15] M. Fischlin and C. Onete. Provably secure distance-bounding: an analysis of prominent protocols. 6th Conference on Security and Privacy in Wireless and Mobile Networks ACM WISec 2013, 2013.

[16] A. Francillon, B. Danev, and S. Čapkun. Relay attacks on passive keyless entry and start systems in modern cars. Cryptology ePrint Archive, Report 2010/332, 2010. http://eprint.iacr.org/.

[17] O. Goldreich. *Foundations of Cryptography: Volume 1, Basic Tools*. Cambridge University Press, 2001.

[18] G. P. Hancke and M. G. Kuhn. An RFID distance bounding protocol. In *Conference on Security and Privacy for Emergency Areas in Communication Networks 2005*, pages 67–73. IEEE, 2005.

[19] J. Hermans, A. Pashalidis, F. Vercauteren, and B. Preneel. A new RFID privacy model. In V. Atluri and C. Diaz, editors, *ESORICS 2011*, volume 6879 of *LNCS*, pages 568–587. Springer, 2011.

[20] C. H. Kim and G. Avoine. RFID distance bounding protocol with mixed challenges to prevent relay attacks. In *Conference on Cryptology and Networks Security 2009*, volume 5888 of *LNCS*, pages 119–131. Springer, 2009.

[21] C. H. Kim, G. Avoine, F. Koeune, F. Standaert, and O. Pereira. The swiss-knife RFID distance bounding protocol. In *Information Security and Cryptology (ICISC) 2008*, LNCS, pages 98–115. Springer, 2008.

[22] Y. K. Lee, L. Batina, K. Sakiyama, and I. Verbauwhede. Elliptic curve based security processor for RFID. *IEEE Transactions on Computers*, 57(11):1514–1527, 2008.

[23] Y. K. Lee, L. Batina, D. Singelée, and I. Verbauwhede. Low-cost untraceable authentication protocols for RFID. pages 55–64. ACM, 2010.

[24] C. Onete. Key updates for RFID distance-bounding protocols: Achieving narrow-destructive privacy. Cryptology ePrint Archive, Report 2012/165, 2012. http://eprint.iacr.org/.

[25] R. Peeters and J. Hermans. Wide strong private RFID identification based on zero-knowledge. Cryptology ePrint Archive, Report 2012/389, 2012. http://eprint.iacr.org/.

[26] A. Ranganathan, N. O. Tippenhauer, D. Singelée, B. Skoric, and S. Capkun. Design and Implementation of a Terrorist Fraud Resilient Distance Bounding System. In S. Foresti, F. Martinelli, and M. Yung, editors, *ESORICS 2012*, volume 7459 of *LNCS*, pages 415–432. Springer-Verlag, 2012.

[27] K. B. Rasmussen and S. Čapkun. Realization of RF Distance Bounding. In *USENIX*, pages 389–402. USENIX, 2010.

[28] J. Reid, J. M. G. Nieto, T. Tang, and B. Senadji. Detecting relay attacks with timing-based protocols. In *ACM symposium on information, computer and*

communications security (ASIACCS) 2007, pages 204–213. ACM Press, 2007.

[29] SHA-3 Zoo. Overview of all Candidates for the Current SHA-3 Hash Competition Organized by NIST. `http://ehash.iaik.tugraz.at/wiki/The_SHA-3_Zoo`.

[30] R. Trujillo-Rasua, B. Martin, and G. Avoine. The Poulidor distance-bounding protocol. In *RFIDSec 2010*, volume 6370 of *LNCS*, pages 239–257. Springer, 2010.

[31] T. van Deursen and S. Radomirović. Insider attacks and privacy of RFID protocols. In S. Petkova-Nikova, A. Pashalidis, and G. Pernul, editors, *EUROPKI*, volume 7163 of *LNCS*, pages 65–80. Springer, 2011.

[32] S. Vaudenay. On privacy models for RFID. In *Advances in Cryptology — Asiacrypt 2007*, volume 4883 of *LNCS*, pages 68–87. Springer, 2007.

[33] E. Wenger and M. Hutter. A hardware processor supporting elliptic curve cryptography for less than 9 kGEs. In *CARDIS 2011*, volume 7079 of *LNCS*, pages 182–198. Springer, 2011.

[34] A. Yang, Y. Zhuang, and D. S. Wong. An efficient single-slow-phase mutually authenticated RFID distance-bounding protocol with tag privacy. In *Information and Communications Security*, volume 7618 of *LNCS*, pages 285–292. Springer, 2012.

[35] A. C.-C. Yao. Theory and applications of trapdoor functions (extended abstract). In *FOCS 1982*, pages 80–91. IEEE Computer Society, 1982.

Author Index

www.ingramcontent.com/pod-product-compliance
Lightning Source LLC
Chambersburg PA
CBHW061414210326
41598CB00035B/6205